Vibration Testing and Applications in System Identification of Civil Engineering Structures

This book covers vibration testing and identification of dynamic structural systems with the aim of preparing researchers in the growing field of structural health monitoring research. It starts with the fundamentals of structural dynamics, then covers the methods of modal analysis and system identification, vibration tests, and the related experimental setups. It concludes with an outline of the authors' software, demonstrating practical applications, and illustrated with real-world case studies of full-scale structures. Theory is presented and derived step-by-step, with a detailed measurement system developed for vibration tests.

- Emphasizes practical applications with a strong theoretical foundation.
- Provides detailed procedures and experimental setups for vibration tests.
- Provides online Support Material with MATLAB® codes for modal analysis and finite element model updating, based on measured data from vibration tests.

Vibration Testing and Applications in System Identification of Civil Engineering Structures will suit students and researchers aiming to contribute to the field of structural health monitoring of civil engineering structures.

Vibration Testing and Applications in System Identification of Civil Engineering Structures

Heung-Fai Lam and Jia-Hua Yang

CRC Press is an imprint of the
Taylor & Francis Group, an **informa** business

MATLAB® is a trademark of The MathWorks, Inc. and is used with permission. The MathWorks does not warrant the accuracy of the text or exercises in this book. This book's use or discussion of MATLAB® software or related products does not constitute endorsement or sponsorship by The MathWorks of a particular pedagogical approach or particular use of the MATLAB® software.

First edition published 2023
by CRC Press
6000 Broken Sound Parkway NW, Suite 300, Boca Raton, FL 33487-2742

and by CRC Press
4 Park Square, Milton Park, Abingdon, Oxon, OX14 4RN

CRC Press is an imprint of Taylor & Francis Group, LLC

© 2023 Taylor & Francis Group, LLC

Reasonable efforts have been made to publish reliable data and information, but the author and publisher cannot assume responsibility for the validity of all materials or the consequences of their use. The authors and publishers have attempted to trace the copyright holders of all material reproduced in this publication and apologize to copyright holders if permission to publish in this form has not been obtained. If any copyright material has not been acknowledged please write and let us know so we may rectify in any future reprint.

Except as permitted under U.S. Copyright Law, no part of this book may be reprinted, reproduced, transmitted, or utilized in any form by any electronic, mechanical, or other means, now known or hereafter invented, including photocopying, microfilming, and recording, or in any information storage or retrieval system, without written permission from the publishers.

For permission to photocopy or use material electronically from this work, access www.copyright.com or contact the Copyright Clearance Center, Inc. (CCC), 222 Rosewood Drive, Danvers, MA 01923, 978-750-8400. For works that are not available on CCC please contact mpkbookspermissions@tandf.co.uk

Trademark notice: Product or corporate names may be trademarks or registered trademarks and are used only for identification and explanation without intent to infringe.

ISBN: 978-1-138-33288-1 (hbk)
ISBN: 978-1-032-29839-9 (pbk)
ISBN: 978-0-429-44586-6 (ebk)

DOI: 10.1201/9780429445866

Typeset in Sabon
by Deanta Global Publishing Services, Chennai, India

Access the Support Material: www.routledge.com/9781138332881

Contents

Preface

The two authors of this book have contributed equally, and they are co-first authors. The aim of this book is to provide fundamental theories and basic skills for researchers who want to start their research in the field of structural model updating and structural health monitoring through measured vibration data of target structures, such as buildings and bridges. In the past 10 to 20 years, this research area has become more and more popular, attracting not only many publications but also lots of research funding. Under the umbrella of structural model updating and structural health monitoring, there are many essential topics including vibration measurement (experiments under both laboratory conditions and, more importantly, field test conditions), modal analysis (including conversion of measured time-domain data to the frequency domain and extraction of modal parameters, such as natural frequencies and mode shapes, from the measured data), modeling (including analytical modeling at the structural element level or finite element modeling of the entire structural system), analytical and numerical optimization (e.g., minimizing the discrepancy between the measured and model-predicted structural responses with the selected sets of model parameters as design variables), and probability theory (explicitly addressing the uncertainty problem mainly due to measurement noise and modeling errors). Those topics are distributed in the various chapters in this book starting with the fundamentals of structural dynamics in Chapter 2, which covers the necessary knowledge and skills for understanding other relatively more advanced topics (e.g., model updating using modal-domain information in Chapter 6). In particular, Chapter 2 is the easiest chapter about structural dynamics at undergraduate level (mainly based on the authors' lecture notes). Chapters 3 and 4 are about modal analysis at postgraduate level. Chapters 5 and 6 are more difficult chapters about system identification and model updating (and damage detection or health monitoring) at the research level (mainly based on the authors' publications following the deterministic approach). Chapter 7 is the most difficult, stepping into the probabilistic approach mainly for researchers (mainly based on the authors' publications following the Bayesian approach). Hopefully, this book will be helpful for students and researchers at different levels in the area of structural model updating and health monitoring of civil engineering structures. It must be pointed out that the data and analysis results in this book have not been published. Only the ideas and methods are from published papers.

One of the best ways to learn a subject is by practice. After introducing a theory or a method/algorithm, examples are given for the implementation of the newly introduced concepts or skills. One important feature of this book is that some MATLAB codes are available, not only for solving the problems but also for visualizing the analysis results. Note that the source codes are provided (not the p-code in binary format), so the readers can learn from the MATLAB codes (this is another language that may help students in understanding difficult theories). This certainly speeds up the learning process, and readers can quickly and accurately adopt the introduced techniques. Furthermore, many case studies are available in this book for both structural systems under laboratory conditions and existing civil engineering structures. The measured data together with the analysis results certainly help students to fully understand the theories and help researchers to implement the techniques in the correct manner.

About the development in structural model updating and health monitoring, many tall buildings and long-span bridges are equipped with different types of sensors for continuous monitoring of both the static (e.g., strain) and dynamic (e.g., acceleration) responses. Some of them even measure the system input (e.g., wind speeds by anemometers and/or ground accelerations by seismometers). The technology is well-developed and relatively mature from the hardware viewpoint. However, the development in software for extracting useful information for decision-making (e.g., structural damage detection) from the set of measured data is not as mature as that of the hardware. From the literature, many methods perform successfully under ideal situations but exhibit limited success (or complete failure) in real applications. The main reasons are the problems of measurement noise and modeling errors, which complexly introduce uncertainties into the results of system identification (including modal identification, model updating, and damage detection). Pinpointing a solution of the inverse problem of system identification is almost hopeless in real applications. An appropriate way to handle this difficulty is to qualify the uncertainty following the probability theory. Hopefully, Chapter 7 of this book can give readers some insight and future directions of research in this area.

Based on the experience of the authors, it is extremely difficult, or impossible, to develop a general model updating (or damage detection) method that is applicable to all types of structures (e.g., buildings *and* bridges) under all types of excitations (e.g., ambient excitation *and* forced vibration) and all types of purposes (e.g., detecting cracks on beams, fracture at connections, *and* corrosion at supports). For a successful methodology, the characteristics of the structure, excitation, and purpose must be taken into consideration in the formulation and implementation. Furthermore, the quality and quantity of measurement must also be considered (i.e., optimal sensor configuration techniques) to ensure that the obtained information is qualified for fulfilling the purpose. There is certainly much work to be done to cost-effectively apply structural model updating and health mentoring techniques towards solving industrial problems.

Heung-Fai Lam, Ph.D.
Associate Professor, Department of Architecture and Civil Engineering, City University of Hong Kong
Chair Professor (Pengcheng Scholar), Harbin Institute of Technology, Shenzhen
Associate Editor, Engineering Structures

Jia-Hua Yang, Ph.D.
Associate Professor, School of Civil Engineering and Architecture, Scientific Research Center of Engineering Mechanics, Guangxi University
December 2021

Acknowledgments

We would like to take this opportunity to thank the following teachers, collaborators, and friends. The professional technical support from Mr. Chi-Kin Lai, City University of Hong Kong is greatly appreciated. Whenever we approached Mr. Lai with problems related to field tests, he always came through with numerous original solutions and suggestions. Vibration tests on civil engineering structures require successful teamwork. We would like to express our sincere and heartfelt thanks to all the research team members, including Professor Feng-Liang Zhang, Harbin Institute of Technology, Dr. Yan-Chun Ni, Tongji University, Dr. Man-Tat Wong, Mass Transit Railway (MTR) Corporation Limited, Dr. Qin Hu, Huazhong University of Science and Technology, Dr. Hua-Yi Peng, Harbin Institute of Technology, Dr. Jun Hu, Wuhan University of Technology, Dr. Stephen Adeyemi Alabi, Federal University of Technology Akure, Dr. You-Hua Su, Xi'an Jiaotong University, and Dr. Mujib Olamide Adeagbo, City University of Hong Kong.

Some personal thoughts and acknowledgments are in the following.

HEUNG-FAI LAM

I would like to take this opportunity to thank my former MPhil supervisor, Professor Jan-Ming Ko, The Hong Kong Polytechnic University, and my former Ph.D. supervisor, Professor Lambros Katafygiotis, The Hong Kong University of Science and Technology. I began my research path on vibration tests, structural model updating, and structural health monitoring during my MPhil study, following the deterministic approach, and extending to the probabilistic approach during my Ph.D. study. Professor Ko equipped me with a series of vibration test skills and allowed me to participate in many interesting projects, including the vibration monitoring of the Tsing Ma bridge during its construction. My understanding of the Bayesian approach was totally gleaned from Professor Katafygiotis. Both Professor Ko and Professor Katafygiotis are perfect role models of not only ideal supervisors but also honorable individuals. I have been very lucky to meet so many outstanding teachers in my research life. I would like to thank Professor James L. Beck, California Institute of Technology, for his inspiring discussions and lectures in the use of sampling techniques in system identification. Furthermore, the help from Professor Costas Papadimitriou, University of Thessaly, in the development of various optimal sensor configuration methods is highly appreciated. I would also like to say thank you to my two old friends, Professor Siu-Kui Ivan Au, Nanyang Technological University, and Professor Ka-Veng Kelvin Yuen, University of Macau.

The crucial turning point of my career is joining City University of Hong Kong (CityU) in 2003. Thank you so much to Professor Sritawat Kitipronchai, University of Queensland, for recruiting me when he was the Head of the Department of Building and

Construction of CityU. He is unquestionably a great boss and an incredible friend to me. I would also like to thank Professor Wai Ming Lee, and Ms. Prudence Lau of CityU for their friendship. Part of this book was written during my sabbatical leave at Harbin Institute of Technology, Shenzhen, with the financial support from the Shenzhen government through the Pengcheng Scholar award (鹏城学者). Without the help from Professor Hong-Jun Lui and Dr. Hua-Yi Peng, of the Harbin Institute of Technology, Shenzhen, I would not have been awarded the Pengcheng Scholar.

For the experimental setups and the implementation of the series of laboratory and field tests, the financial support from Research Grants Council of the Hong Kong Special Administrative Region, China (GRF 9042509 (CityU 11210517), GRF 9041889 (CityU 115413), and GRF 9041568 (CityU 115510) are highly appreciated.

The support from my family is essential for me in writing this book and developing my career. I would like to express my appreciation to my wife, Mrs. Wan Yee Amy Li, for her approvals of my various conference leaves, despite missing many of her and my birthdays together. Since you are educating our children, I can focus on educating my students. My deep appreciation goes to my wonderful parents, Mrs. Pui Yin Chan, and Mr. Kee Kwong Lam, for their continuous support and protection. I am deeply thankful to my younger brother, Mr. Ho Man Lam, and his wife, Ms. Erica Lai, for supporting me during both the ups and downs. Finally, special thanks to my wife, my daughter, Miss Muk Wai Lam, and my son, Mr. Muk Kin Lam. Without your love and support, I am just a silly monkey on a tree. I am the luckiest man in the world.

JIA-HUA YANG

I am lucky to have had wonderful teachers in my career. I would like to express my special appreciation and thanks to my Ph.D. advisor, Professor Heung-Fai Lam. He has been a tremendous mentor for me. When I was like a baby fish wandering in the ocean, he guided me with his wisdom and patience to the scientific world. I still remember when I visited his office for the first time, and he patiently explained his computer code for dynamic analysis of framed structures. I greatly appreciate his continuous availability and unfailing support that allowed me to grow as a research scientist. Our regular meetings every Wednesday, during which I learned teaching and research methods and obtained valuable ideas from him, were most enjoyable. I will never forget his loud and characteristic laugh in the meetings. I have benefited a lot from his life philosophy: be serious as a scientist and be optimistic for life. During my doctoral study, I took the great course offered by Professor Siu-Kui Au (Nanyang Technological University), Structural Dynamics and Its Applications. Professor Au showed me the beauty of mathematics in this course. The difficult and time-consuming assignments in his course (usually, for one week I needed 3 days for theoretical derivation and 3 more days for writing codes) have become my useful research tools. I am grateful for his continuous encouragement of my research on Bayesian system identification and uncertainty quantification. It is a privilege to work with Professor James Beck (California Institute of Technology) on Bayesian system identification. I learned from him how to do serious research, which is influenced by his philosophy of Bayesian methods. I am deeply grateful for Professor Beck's word-by-word editing of my papers.

This book receives financial support from several grants, including National Natural Science Foundation of China (Grant No. 51808400), National Key Research and

Development Program of China (Grant No. 2020YFC1512504), Shanghai Sailing Program (Grant No. 18YF1424500), Science and Technology Base and Talent Program of Guangxi (Grant No. GuikeAD21220031), and High-level Talent Program of Guangxi University (Grant No. A3030051030). The generous support is greatly acknowledged.

Part of this book was written when I was with the Department of Disaster Mitigation for Structures at Tongji University. The generous administrative help from Professor Ying Zhou (Head of College of Civil Engineering) and Professor Wen-Sheng Lu (department leadership) is much appreciated. Thanks also go to Professor Zheng Chen (former Deputy Head of School of Civil Engineering and Architecture) and Professor Yong-Hui An (Head of School of Civil Engineering and Architecture) for their generous support after I moved to Guangxi University.

Writing a book is a challenging task. It means continuous dedicated and intense work. This book could not have been done without support from my family. I greatly appreciate their love. My father is always there when I need him. I am grateful for every dish he has cooked for me and all of the sacrifices that he has made on my behalf. My deepest gratitude goes to my wife, Xu Zeng (Eileen). I thank her for being always my great support in the moments when there was no one to answer my queries. She was pregnant during the writing of this book, and then our second child, Yi-Cheng, was born. I am in debt to my wife, our first child, Le-Xuan, and our second child, Yi-Cheng, for not being around much and not being able to take good care of them. I am incredibly fortunate to have their tolerance and understanding. I would also like to thank my parents-in-law. Their unconditional love is what sustains me thus far. I sincerely thank my uncles, aunties, and cousins for always giving me the best festival gathering in the world and keeping me warm all the time. Thanks also go to my best friends, Ling-Jun Wei and Rui-Qi Yang (also my cousin). I will always remember every beer we have together, especially during my ups and downs.

Authors

Heung-Fai Lam is currently Associate Professor at City University of Hong Kong and Associate Editor of Engineering Structures. He earned a B.Eng. and an M.Phil. from The Hong Kong Polytechnic University in 1992 and 1994, respectively, and a Ph.D. from The Hong Kong University of Science and Technology in 1999. Due to his outstanding research in structural health monitoring of civil engineering structures, Prof Lam was awarded the Pengcheng Scholar Chair Professor in Harbin Institution of Technology, Shenzhen, China.

Jia-Hua Yang is Associate Professor at Guangxi University. He earned a B.Eng. (2010) from Central South University and a Ph.D. (2015) in civil engineering from City University of Hong Kong. Before joining Guangxi University, he was Assistant Professor at Tongji University. He is a recipient of the EASEC Young Researcher/Engineer Award (2019), ACMSM Stan Shaw Best Paper Award (2018), and HKIE Commendation Merit Award of Structural Excellence Award (2019, 2021). He is a committee member of the Dynamics Committee of Engineering Mechanics Institute, ASCE.

Chapter 1

Introduction

1.1 FUNDAMENTALS OF STRUCTURAL VIBRATION TESTS

Structural vibration tests are useful means for extracting important mechanical properties from the tested sample. They are applicable in many different industries, such as aerospace, automotive, pump, and heating, ventilation, and air conditioning (HVAC). This book focuses on vibration tests for civil engineering structures.

Figure 1.1 shows the schematic of a typical experimental setup of structural vibration tests (impact hammer tests) of a scaled-down structural model under laboratory conditions. The setup can be easily modified for implementing free and forced vibration tests under laboratory conditions or ambient vibration field tests. In the figure, the target structure is a scaled-down two-story building model (or a shear building model) fixed onto the ground. It should be noted that support conditions are essential for civil engineering structures (a significant component of the structure). This is not the case for mechanical/aerospace engineering structures like cars and airplanes.

The journey starts by setting the structure in vibration using the impact hammer. A load cell is installed at the tip of the hammer (as shown in Figure 1.2), and it records the impact force applied onto the structure as a function of time. Several hammer tips are available with different stiffness, which controls the contact time duration. A stiffer hammer tip will result in a shorter contact time, and the resulting excitation covers a higher frequency range that is suitable when the targeted natural frequencies of the structure are high. If the target frequency is low (e.g., 1 to 2 Hz), a soft hammer tip that increases the contact time is more suitable.

The impact-induced vibration of the target structure will be picked up by accelerometers, as shown in Figure 1.1. Consider the two-story shear building model as an example; the lateral vibrations for both floors are important for identifying the dynamic characteristics (e.g., the natural frequencies and mode shapes) of the structure. Therefore, two sensors are used in this example, and each measures the lateral vibration of a floor. The working principle of a piezoelectric accelerometer is simplified and illustrated in Figure 1.3. It is important to fix the accelerometer onto the target structure through the base, so that it vibrates together with the structure. There is a mass inside the sensor case, and a layer of piezoelectric material is installed under the mass. When acceleration is applied to the sensor (through the structure), the inertia force of the mass will induce pressure onto the layer of piezoelectric material, which releases charges proportional to the applied pressure. By measuring the charge, the acceleration of the sensor and the target structure can be determined. It should be noted that the mass of accelerometers must be very small compared to the mass of the target structure. Otherwise, it will certainly affect the dynamics of the structure.

DOI: 10.1201/9780429445866-1

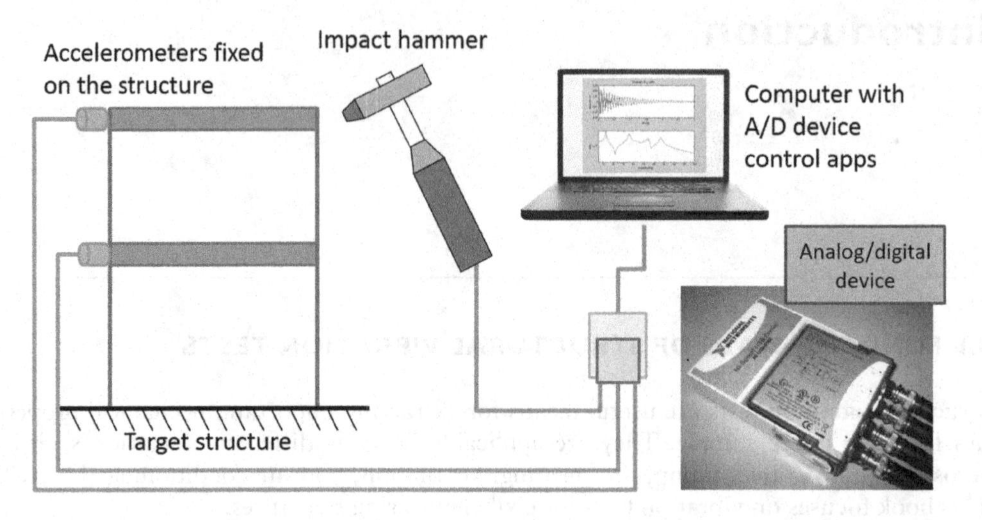

Figure 1.1 A typical experimental setup of a structural vibration test (impact hammer test).

Both the responses and impact force signals will be sent to the data acquisition system (i.e., the analog/digital device in Figure 1.1) through the coated cables. As the charge signals are very weak, they must be amplified and converted to voltage before transfer. Otherwise, the error will be very large. In modern vibration tests, the charge amplifier is usually integrated into the analog/digital (A/D) device (or the data acquisition system). The main purpose of an A/D device is to convert the continuous analog signal in voltage into data points that can be stored on a computer. This A/D process is essential as it is impossible for a computer to handle a continuous signal requiring an infinite capacity of hard disk space. An important parameter in the digitization process is the sampling frequency (Hz), f_s, which means the number of equally spaced data points to be recorded per second. For a given sampling frequency, the time step size, dt (i.e., the spacing between data points) can be calculated as:

$$dt = \frac{1}{f_s} \tag{1.1}$$

With a high sampling frequency, many data points are recorded, implying that a high level of information is retained (or the loss of information is small). However, lots of information (especially high-frequency information) will be lost if a low sampling frequency is adopted. Of course, the higher the sampling frequency, the larger the hard disk capacity required, and the longer the analysis time required. Therefore, deciding an appropriate sampling frequency sometimes controls the success of the vibration test (and the follow-up analysis).

Figure 1.4 is an example to illustrate the importance of the sampling frequency in the A/D process. The top sub-figure shows the continuous signal to be digitized. It is clear from the sub-figure that the continuous signal consists of high-frequency and low-frequency components, and both components are decaying. The middle sub-figure shows the digitized data with a sampling frequency of 2 Hz (considered as high in this example). The data points are represented by red crosses in the sub-figure, and the straight lines connecting consecutive data points can be considered as the linear approximation of the

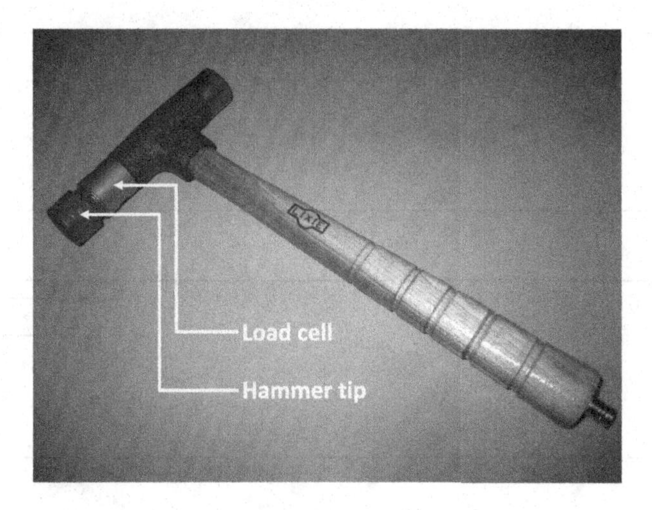

Figure 1.2 A typical impact hammer.

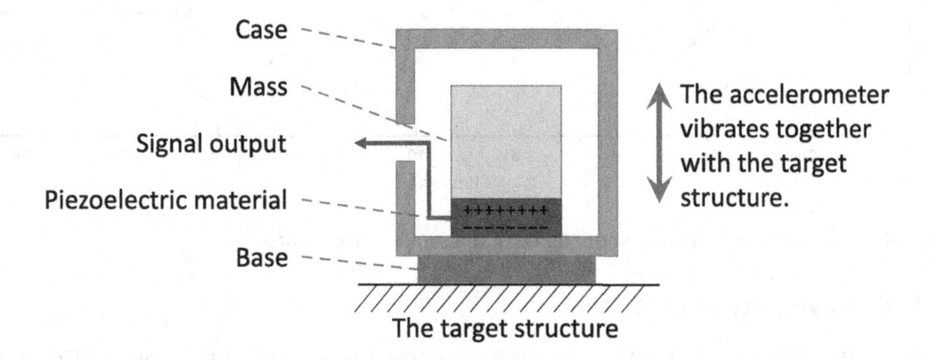

Figure 1.3 Working principle of a typical accelerometer.

signal in between measured data points. Although the digitized curve looks very different from the original continuous signal, it can, to a certain extent, retain the high-frequency information of the original signal. Next, the same signal is digitized with a sampling frequency of 1 Hz (considered as low in this example). In the lower sub-figure, it is very clear that the high-frequency information in the original signal is completely lost, and only the low-frequency information can be retrained. It must be pointed out that there is no way to restore the original signal from the digitized data points. Thus, the sampling frequency must be decided with great care.

After the A/D process, the set of digitized data is stored on the computer, which is also used as a controller through specific software applications, such as LabVIEW (National Instruments 2003) or MATLAB®. It must be pointed out that software drivers must be installed for the operating system to communicate with the hardware of the data acquisition system. Apart from the digitization process (determined by the sampling frequency), the controller also governs the start and end of the measurement, and the location in the hard disk for storing the data file. Since different accelerometers have different sensitivities (unit = mV/g), the sensitivities for all sensors must be input to the controller so that the digitized data points from different channels (i.e., sensors) can be directly compared.

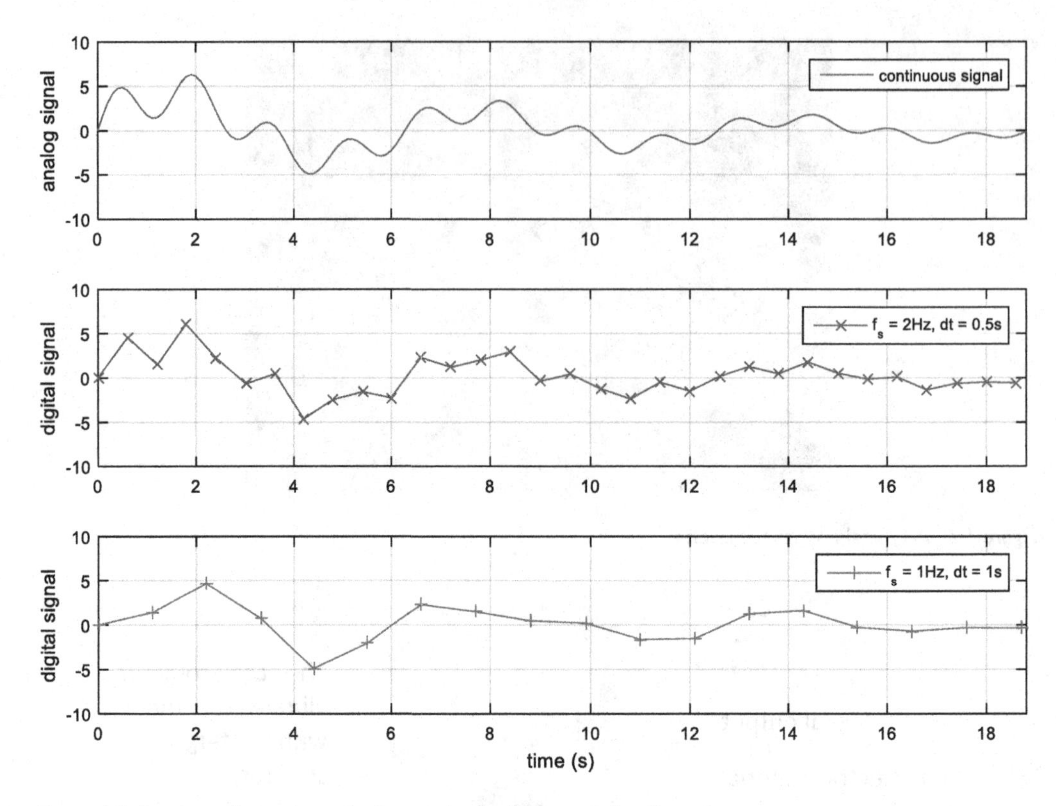

Figure 1.4 An example to show the importance of the sampling frequency.

1.1.1 Free vibration test

Free vibration test is to measure the vibration responses of the target structure under the action of initial conditions, i.e., initial displacement and/or initial velocity. The initial velocity may be induced by an impulse (using an impact hammer) that changes the momentum of the system. As there is no external disturbance during the measurement of structural responses, it is called a "free" vibration test. If an impact hammer is used to induce initial velocity to the target structure, the schematic in Figure 1.1 can be modified by removing the cable connecting the hammer to the data acquisition system (as no force information is needed in a free vibration test). The free vibration test is very attractive, as the measured vibration responses show the solo characteristics of the target structure (there is no external force).

Under laboratory conditions, the initial conditions are easy to apply and measure. The situation for field tests is very different. That is, one can easily induce a 1 cm initial lateral displacement on a 1 m building model under laboratory conditions, and the induced initial displacement can be easily measured. However, how can one create 1 m lateral displacement at the top of a 40-story building? The following are some methods to induce initial displacement and/or velocity for civil engineering structures (e.g., buildings, bridges, and dams).

- If the target structure is not too heavy (e.g., a footbridge), one may still use an impact hammer (much bigger than the one in Figure 1.2) to induce an initial velocity at a given location and direction of the target structure.

- The initial displacement can be induced by hanging a heavy weight at the target location using a cable. The vibration of the structure can be initialized by cutting the cable.
- For very heavy civil engineering structures, such as dams, one may generate the initial velocity by a carefully controlled explosion at a particular location on the structure.
- One may set up an exciter at the location of initial displacement to apply a sinusoidal excitation at the natural frequency of the target structure. With the damping effects of civil engineering structures, the large vibration amplitude will converge to a certain level. After converging, one can switch off the exciter, and the structure will undergo free vibration.

No matter which of the above methods is employed, it is very difficult to ensure the pre-defined initial displacement and/or velocity are accurately induced. Furthermore, it is very difficult, if not impossible, to directly measure the induced initial conditions. Fortunately, the initial displacement and velocity can be estimated from the measured time-domain responses through least-squares curve fitting techniques.

1.1.2 Forced vibration test

It is difficult to induce the initial conditions on civil engineering structures for free vibration tests. Furthermore, the initial displacement and velocity at a given location and direction can usually stimulate only one or a few vibration modes. If the damping of the structural system is high, the vibration responses induced by the initial conditions will be damped out very fast, and the number of data points may not be adequate for identifying the required structural and/or dynamic characteristics with acceptable accuracy. Aside from the free vibration test, another popular type of test is the forced vibration test. By continuously inputting energy to the target structure, the vibration responses can be continuously maintained at a detectable level for a pre-defined time duration. One of the advantages of this method is the feasibility of increasing the force magnitude to improve the signal-to-noise ratio of the measured structural responses. This test, however, must be treated with great care to avoid structural damage due to excessive vibration.

For a typical forced vibration test, one needs to modify the schematic in Figure 1.1 by replacing the impact hammer with an exciter (or vibration actuator). Forced vibration tests under laboratory conditions are relatively easy to control as electricity supply is not a problem. One may hang an exciter at an appropriate position related to the target structure (usually a scaled-down model in laboratory tests). Figure 1.5 shows a possible arrangement to hang an exciter for forced vibration tests under laboratory conditions. The vibration force can be applied at the pre-defined location and direction of the structure through a steel rod connected to the exciter. A load cell must be installed to measure the applied force as a function of time, as it is the essential information in a forced vibration test. The arrangement for field tests is much different and more difficult. For example, one usually cannot hang the exciter at the required position; and it is not easy to get a continuous electricity supply outdoors.

Figure 1.6 shows an example of a forced vibration test of a footbridge. The main purpose of this test is to identify the modal parameters, including the natural frequencies, mode shapes, and damping ratios of the bridge. Several accelerometers were installed at the pre-defined locations on the bridge, and the exciter (in the upper part of Figure 1.6)

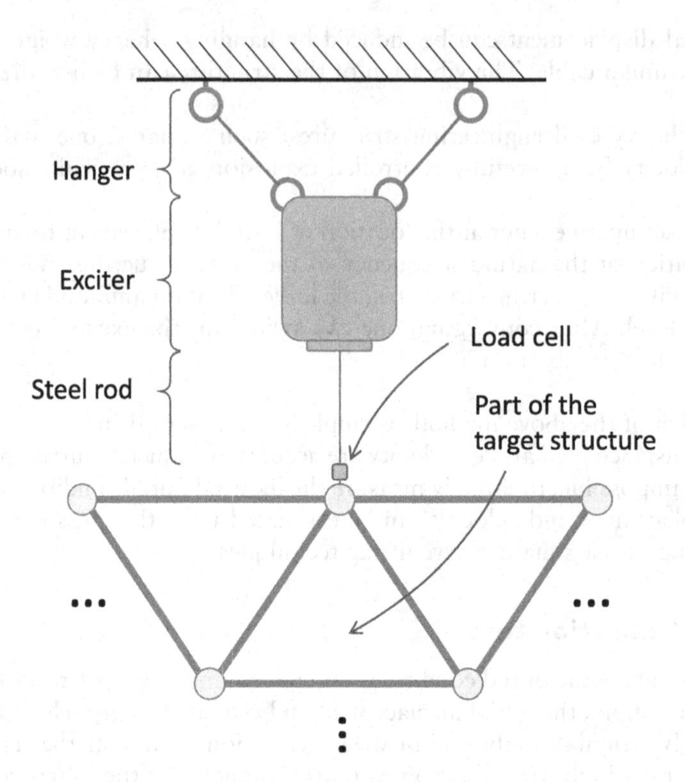

Figure 1.5 A schematic to show an example of forced vibration test under laboratory conditions.

was located at the mid-span of the bridge (favorable to stimulate the first bending mode of the footbridge). There are many different types of exciters; the adopted one in this example generates forces onto the target structure by using a reaction mass, as shown in Figure 1.7. As the self-weight of the exciter is very high, it can be simply placed on the target structure through the rigid base without installation through bolts and nuts. Due to the dynamics of the exciter and the target structure, the signal input to the exciter is not the same as the force applied to the structure. Thus, it is necessary to measure the acceleration of the reaction mass (as shown in Figure 1.7) for calculating the force input to the target structure as a function of time. All sensors and the exciter were connected to the self-developed data acquisition system (with built-in power supply), which was connected to the computer (at the lower part of Figure 1.6) with LabVIEW installed. The data acquisition system together with the computer (i.e., the controller) is called the "console" in a vibration test. A sine sweep test was carried out as part of the forced vibration test. The basic idea is to apply a sinusoidal force with varying frequency to the structure. A frequency range must be defined (say, from 0 to 50 Hz), and the excitation frequency will first be linearly increased from the lowest frequency to the highest one within a pre-defined time duration (say, 2 mins). Then, the excitation frequency will be linearly reduced from the highest frequency to the lowest one within the same duration. In Figure 1.6, the upper figure on the computer screen shows the acceleration of the reaction mass (i.e., the force input), and the lower figure shows the acceleration responses from different accelerometers. It is clear from the figures on the computer screen that the vibration responses at different locations are either in-phase or out-phase with respect to the excitation. The excitation frequency

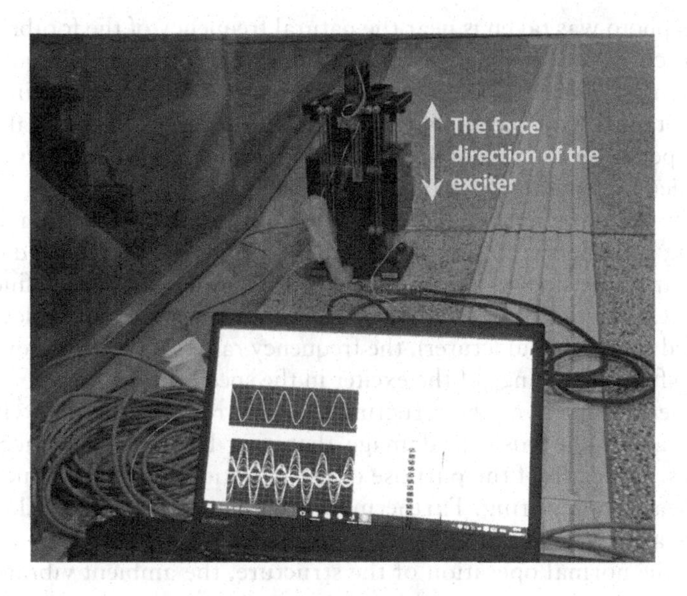

Figure 1.6 A example of a forced vibration test of a footbridge.

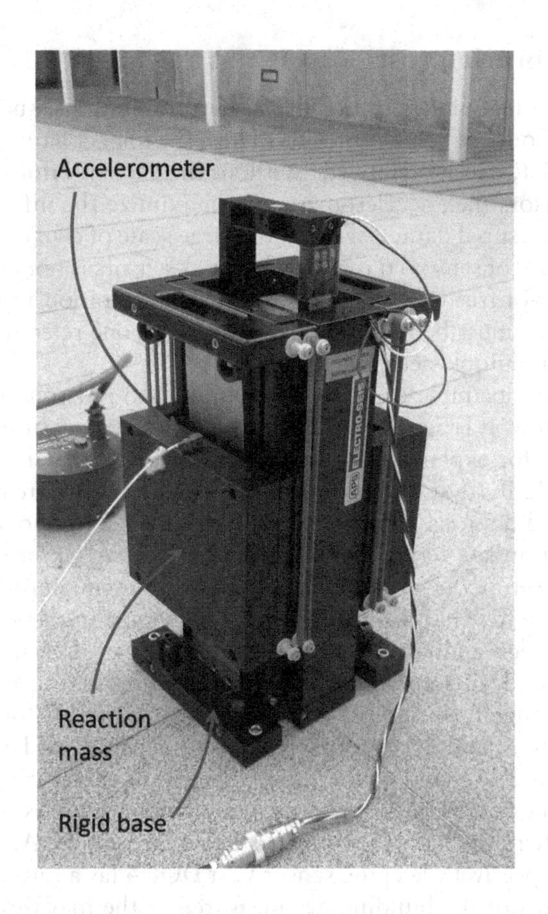

Figure 1.7 An exciter.

at the time this photo was taken is near the natural frequency of the footbridge. Since the sine sweep test covers all frequency components (as defined by the frequency range), it is believed that most of the vibration modes of the target structure within the frequency range can be obtained with acceptable accuracy. The accuracy of the analysis results, of course, also depends on the modal identification method (please refer to Chapters 3, 4, and 5 for detailed information on modal identification/analysis).

Another very popular method of forced vibration test is the random vibration test. Instead of using the sine sweep function, a random function is employed as the forcing function. Similar to the sine sweep test, the frequency range must be defined in the random vibration test. As the exciter can only function in a specific frequency range (information provided by the manufacturer), the frequency range of the vibration tests must be a sub-set of the frequency range of the exciter in the specification.

In fact, property owners of the structures usually reject the forced vibration tests, as they worry about the possible damage that may be induced or accumulated on their structures, especially if the purpose of the vibration test is for structural damage detection or health monitoring. Furthermore, the detectable force-induced vibration will very likely affect the normal operation of the target structure. To avoid influence of the tests on the normal operation of the structure, the ambient vibration test is the best solution.

1.1.3 Ambient vibration test

An ambient vibration test makes use of the natural excitations, such as wind and surrounding traffic, without artificial excitation. Thus, the impact hammer in the schematic, as shown in Figure 1.1, can be removed. Without the exciter, efforts can be focused on the placement of sensors (i.e., accelerometers) to maximize the information that can be extracted from the measured data. Owing to the large scale of civil engineering structures and the limited number of sensors (i.e., the number of sensors is not large enough to cover all measured degrees-of-freedom (DOFs)), an ambient vibration test is usually divided into multiple setups. It must be pointed out that at least one reference sensor/channel is needed per setup in a multiple-setup vibration test.

Figure 1.8 shows an example of a 10-story building to illustrate the idea of multiple setup in a vibration test. It is assumed that there is only one measured DOF per floor (to simplify the situation for explanation purposes), and only four wired accelerometers are available (labeled as A, B, C, and D in the figure). Vibration tests are usually started from the roof to the ground floor, as vibration amplitudes at the top floors are relatively large which is favorable for on-site verification of measured data. After measuring the vibration responses for a given setup, the measured data (or part of them) must be verified to ensure the data are applicable for analysis. If any error is observed in the verification, the error source must be identified and corrected. Then, a new set of data for the corresponding setup must be measured. In setup 1, sensors A to D were used to measure DOFs 10 to 7, respectively. To change from setup 1 to 2, sensors A, B, and C for DOFs 10, 9, and 8 are moved to DOFs 6, 5, and 4, respectively, keeping sensor D at DOF 7 (as a reference between setups 1 and 2). Owing to the limited length of cables connecting the sensors to the console (i.e., the data acquisition system and computer), it is better to put all four sensors near each other. To change from setup 2 to 3, sensors D, A, and B are moved to DOFs 3, 2, and 1, respectively, keeping sensor C at DOF 4 (as a reference between setups 2 and 3). All 10 DOFs of the building are measured in the first three setups. Based on

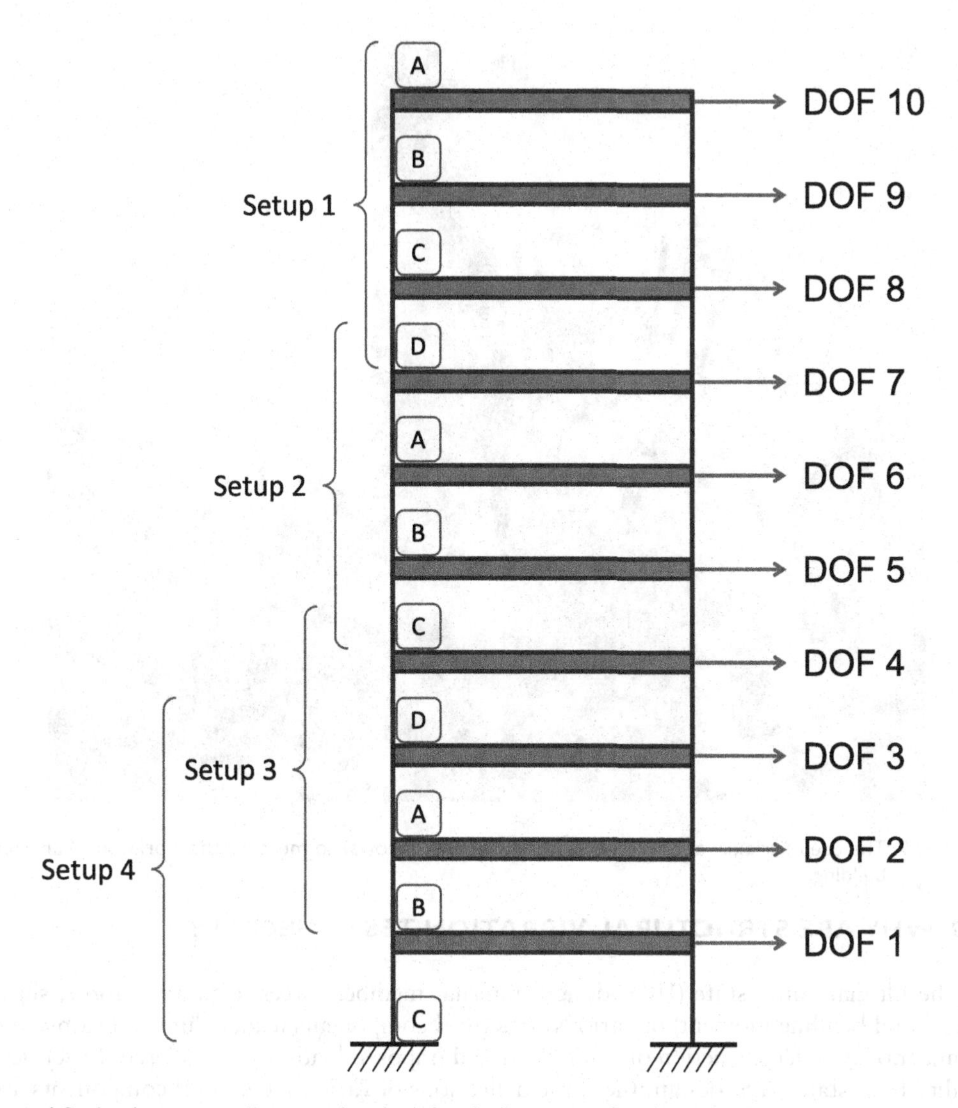

Figure 1.8 A schematic to illustrate the idea of multiple setups of a vibration test.

the experience of the authors, there is a vibration at the "ground" floor of most buildings (although the amplitude of vibration is usually small when compared to that of other floors). Hence, it is strongly recommended to also measure the vibration of the ground floor. To do this, sensor C is moved to the ground floor while sensors D, A, and B are kept as the references between setups 3 and 4. This measurement schedule has been adopted in many field tests (Hu & Yang 2019; Lam et al. 2019, 2019a; Hu et al. 2017; Lam et al. 2017, 2017a) and was proved to be efficient. Figure 1.9 shows the situation of a multiple-setup ambient vibration test of an aged building (Lam et al. 2017), which is very similar to the situation in this example.

If wireless sensors are employed, one may prefer to keep sensor A at DOF 10 as a reference for all setups, as the corresponding vibration amplitude is the largest (low signal-to-noise ratio), and it is believed to be the most accurate.

Figure 1.9 The use of triaxial accelerometers (indicated by arrows) to measure the vibration of an aged building.

1.2 WHY ARE STRUCTURAL VIBRATION TESTS NEEDED?

In the ultimate limit state (ULS) design, internal member forces (e.g., axial force, shear force, and bending moment) of various structural components (e.g., columns, beams, and connections) under the action of static dead and imposed loads are considered. In serviceability limit state (SLS) design, the static deflections of various structural components are considered. For regular buildings (this is usually the case in the classroom examples), the wind effects on the target structure are considered as the equivalent static wind load to simplify the design process. Under such a situation, engineers need to calculate the top drift and inter-story drift of a tall building for SLS (stiffness design) on top of the ULS requirements. The vibration of a structural system seems unconnected to the structural design viewpoint. This is, of course, not true! In general, subjects related to structural dynamics are considered electives in most degree programs. This gives students the impression that structural vibration is not related to their careers. This is, of course, also not true!

To show the truth, the importance of structural vibration is illustrated through a series of real projects (some details and analysis results of the projects are not covered in this book, as they are confidential) in the following sections.

1.2.1 Building vibration

Considering the SLS, engineers must ensure that the target structure functions under normal operations. One of the design criteria is to consider the lateral vibration of tall buildings under wind action. If the natural frequency of the fundamental lateral vibration mode of a tall building falls into the excitation frequency range of wind under normal weather conditions, the wind-induced vibration will be amplified. Residents will complain or even move out of the building if the vibration level is so high that it affects human comfort. Under such a situation, the design of the tall building is considered as a failure from the SLS viewpoint. Of course, the lateral vibration of buildings under the action of very strong winds is unavoidable. For example, the signal number 10 super typhoon Mangkhut (山竹) attacked Hong Kong on September 16, 2018. Many reports about large amplitude swaying of buildings were received on that day (e.g., Chiu 2018). Under normal operations, however, the lateral vibration of buildings must be at an acceptable level that will not influence human comfort.

For some aged buildings, the degradation of materials or accumulated structural damage may affect the lateral stiffness of the system resulting in unexpected vibration problems for the occupants. Under such a situation, a vibration test is a possible way to gather structural information for identifying the problem. Figure 1.9 shows an example of using triaxial accelerometers (i.e., each sensor consists of three output channels with two horizontal directions and one vertical direction) for measuring the lateral vibration of various floor slabs of an aged building. As the number of sensors is usually limited, and it is impossible to measure the vibration of all floors with a single setup, the vibration test must be carefully planned and divided into multiple setups to cover all floors of the building structure. Figure 1.9 shows only one of the setups. Each setup must have some reference sensors to ensure that the measurements from various setups can be assembled (as discussed in Section 1.1.3). As occupants usually use the lifts, carrying out the vibration test at the staircases can minimize the influence on the normal operation of the building.

Apart from the lateral vibration of a building, the vertical vibration of structural components, such as the floor slabs, is also important from the SLS viewpoint. Suppose an average person walks back and forth on a floor slab, the pacing frequency is about 2 to 3 Hz (i.e., stepping onto the floor about 2 to 3 times per second). If the natural frequency for vertical vibration of the floor slab is equal to 2 to 3 Hz, the vibration amplitude of the floor may become larger and larger (for a low damping system) due to resonance. From experience, the second harmonic effect may also cause vibration problems (i.e., the slab vibrates two times for every step) if the natural frequency of the floor slab is equal to 4 to 6 Hz. As the natural frequency for the first vertical vibration mode of a floor slab with a normal span (e.g., 6 to 8 m) is usually higher than 6 Hz, the vertical vibration problem is, in general, not an issue for normal floor slabs. Figure 1.10 shows an example of a setup for roof vibration monitoring. The motors for lift services are usually installed on the roof of a building. If the vibration frequency of the motors matches or is very close to the natural frequencies (the first two to three modes are important) of the roof slab, the vertical vibration of the slab may become serious and cause discomfort for occupants at the top story. This kind of problem is more common for footbridges.

1.2.2 Bridge vibration

One can easily recall the vibration problem of the London Millennium Bridge during its opening (Dallard et al. 2001). During the opening ceremony of this elegant footbridge on June 10, 2000, the bridge was extremely crowded (as expected), and people could only move slowly. The in-phase stepping from a large group of pedestrians induced a large amplitude lateral vibration of the footbridge. The bridge was then closed for investigation.

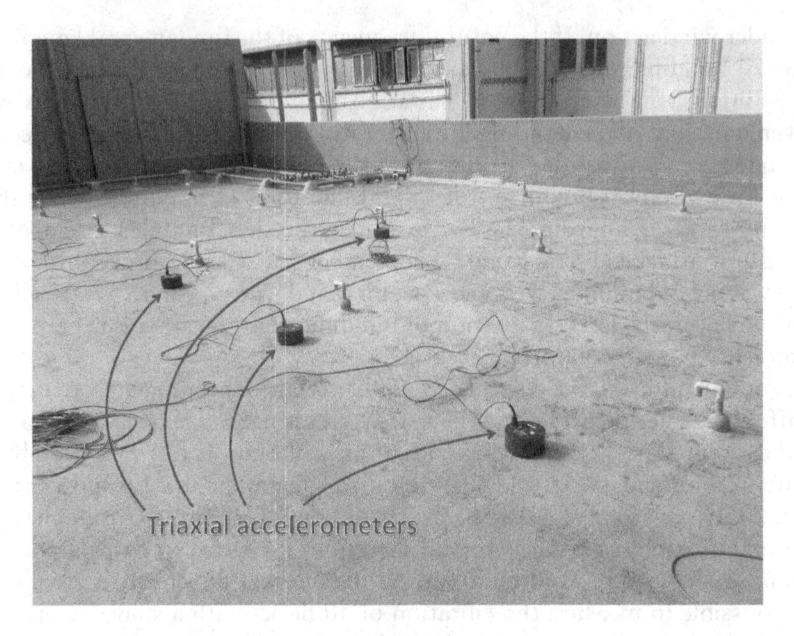

Figure 1.10 Monitoring the vertical vibration of the roof slab.

To identify and solve this kind of structural vibration problem, vibration measurement is necessary. The most popular way to implement a vibration test is to use accelerometers. In general, several accelerometers are preferred for monitoring the vibration, at different locations and directions (or different DOFs) on the target structure, to obtain not only the natural frequencies and damping ratios but also the mode shapes. The mode shape information is important to confirm if the first measured mode is the fundamental one. Furthermore, the mode shape information is essential for matching the measured modes to the finite element predicted ones. When the modal parameters and the maximum accelerations are obtained from the field tests, possible vibration control solutions can be proposed. For example, one may increase the stiffness (e.g., by modifying the support or boundary conditions) to increase the natural frequencies of the structure in avoiding the excitation frequency; or one may install "tuned" mass dampers on the structure to shift the natural frequency of the selected mode (usually the fundamental one). To overcome the vibration problem of the London Millennium Bridge, tuned mass dampers were installed under the bridge, as shown in Figure 1.11.

As an example, Figure 1.12 shows the experimental setup for the vibration test of a footbridge. In this test, six triaxial accelerometers (marked by arrows in the figure), which measure the vibration in three orthogonal directions, were placed at the pre-defined locations on the bridge. The vibration test was carried out after midnight to minimize inconvenience for pedestrians.

1.2.3 Determination of cable tension of long-span bridges

Cable-based long-span bridges are very popular worldwide (e.g., suspension bridges and cable-stayed bridges). For safety considerations, the tension force on all cables must be monitored (e.g., once per 5 to 6 years). An abnormal loss of tension on stay-cables (or suspenders), or unexpected changes in force distributions on different cables, is an

Figure 1.11 The tuned mass dampers under the London Millennium Bridge for controlling the vibration due to pedestrian load.

Figure 1.12 Experimental setup for vibration measurement of a footbridge (an example of vibration testing).

indication of a structural problem of the bridge (e.g., damage to cables that reduces their load-carrying capacities, damage to the bridge deck that alters the flexure stiffness along the deck and affect the tension force distributions on the supporting cables). The question is how to measure the tension force on existing cables. Direct measurement using a dynamometer is impossible, as one cannot cut the cable. The use of strain gauges is possible only if they are installed before the application of tension forces on the cables. The most popular method available for estimating the tension force of bridge cables (including vertical suspenders and inclined stay cables) is by calculating through the natural frequencies of the target cables. From the literature, the tension force on a vertical suspender can be calculated using the natural frequency of the cable (Zui et al. 1996; Ren et al. 2005; Kim & Park 2007; Huang et al. 2014). In general, the higher the tension force on a cable (e.g., a vertical suspender), the higher the natural frequencies of the cable will be. Consider a vertical suspender with both ends fixed; if a lateral displacement is induced at the mid-length of the suspender, the tension force in the suspender will generate restoring forces. Thus, an increase in tension force will increase the external force required to produce a unit lateral displacement on the suspender. In other words, an increase in suspender tension force will increase the stiffness of the system and the corresponding natural frequencies.

Figure 1.13 shows the installation of two accelerometers (for vibration at two orthogonal directions perpendicular to the cable) onto a stay cable for estimating its tension force. Consider the first vibration mode of a cable: the amplitude of vibration near the mid-length of the cable is usually large. Therefore, the location of sensor installation is usually far away from the support on the bridge deck. Figure 1.14 shows an example of the measured acceleration (in g) from one of the two accelerometers (i.e., channels). The vibration is mainly due to wind and is random in nature. It is difficult to obtain any information directly from the time-domain data. Next, the power spectral density (PSD) of the two sets of time-domain data was calculated (see Chapter 3) and plotted in Figure 1.15. It is clear that the PSD curves for both channels show "peaks" at particular frequencies. Physically, a peak in PSD shows the frequency at which the amplitude of vibration is large. In this case, those frequencies can be considered as the natural frequencies of the cable. For example, the first and second peaks are at about 0.9 Hz and 1.8 Hz, respectively. These two are the natural frequencies of the first and second lateral vibration modes of the cable. The corresponding cable tension can then be calculated based on the formulation given by Huang et al. (2014).

Figure 1.13 The installation of two accelerometers onto a bridge cable for estimating its tension force.

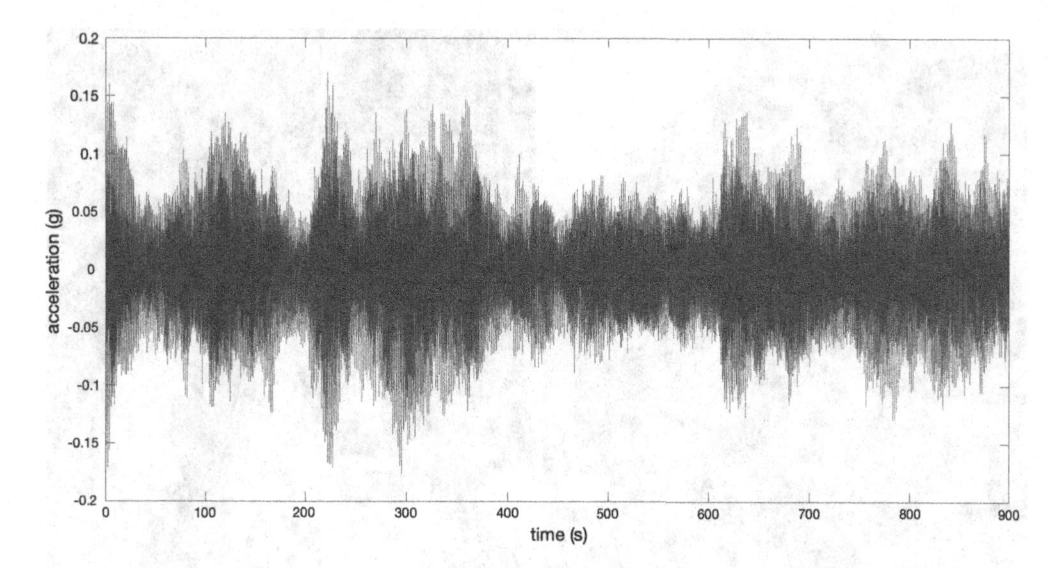

Figure 1.14 An example of measured time-domain acceleration from one of the two channels.

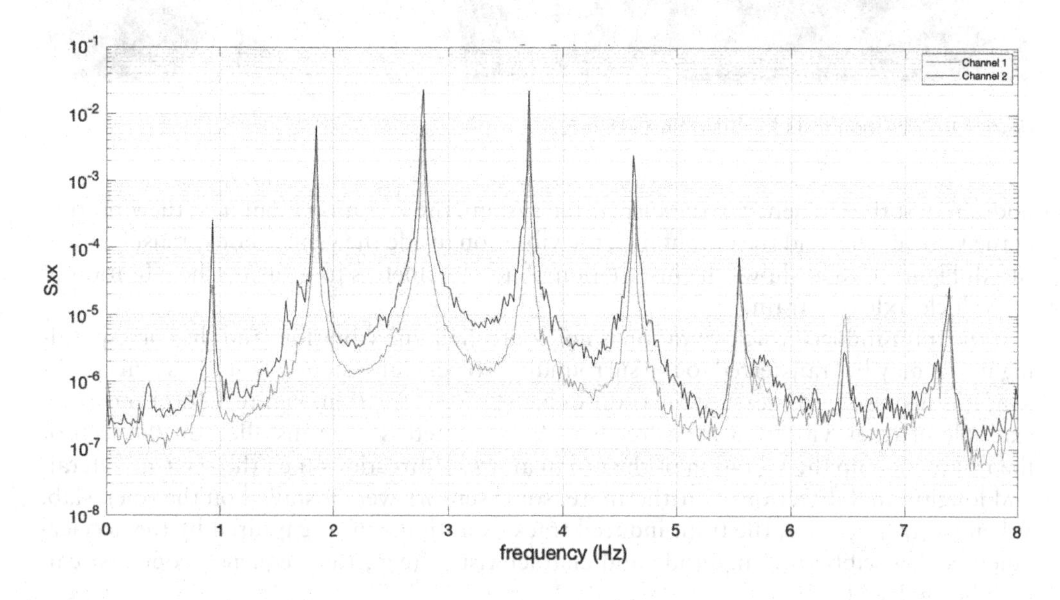

Figure 1.15 An example of power spectral density (PSD) curves for the two channels. (a) train vibration. (b) track vibration.

1.2.4 Train-induced vibration

Another example to show the importance of vibration tests is the measurement of train and track vibrations. Uncontrolled train-induced vibration will cause many problems. The continuous vibration of the train during operation will cause passenger discomfort and reduce ride quality. Sudden vibration in large amplitude at certain locations may cause unstable objects (e.g., luggage) to fall and even cause standing passengers to fall. Train vibration can be reduced by suitable designs of the suspension systems of the train

Figure 1.16 Vibration tests for the train and track.

body. To test the efficiency of the suspension system, the vibration input (i.e., the vibration at the vehicle axle) and the output (i.e., the vibration inside the vehicle body) must be measured. Figure 1.16(a) shows the installation of accelerometers to measure the vibration of the vehicle axle of a train.

If the train-induced track vibration is not controlled, the vibration (and the corresponding noise) may be transferred to the surrounding structural systems. To address this problem, it is necessary to measure the train-induced track vibration. Figure 1.16(b) shows an example of track vibration measurement. Accelerometers were installed on the web of the rail to pick up the vibration of the rail in all three directions (i.e., the vertical, lateral, and longitudinal directions). Furthermore, some sensors were installed on the track slab. When the train passes, the train-induced track vibration can be captured by the accelerometers. The vibration amplitude and characteristics (e.g., the frequency contents) can then be analyzed.

1.3 WHAT IS SYSTEM IDENTIFICATION?

System identification is a large research area that covers many techniques for identifying a system (a general one) that defines the input-output relationship based on the observed (or measured) input and output (denoted as input-output system identification) or just the observed output (denoted as output-only system identification). In general, the input and output signals are continuous in nature. With the advances in computers, calculations and analysis are usually carried out numerically. Thus, the discretization of the continuous input and output signals is required. In this book, it is assumed that the signals are digitized into discrete quantities.

Under the umbrella of system identification, this book focuses on two main techniques and their applications in structural health monitoring (SHM). They are the modal system identification (i.e., modal identification/analysis) and structural system identification (i.e., model updating).

In modal system identification, the modal parameters (system), such as the natural frequencies, mode shapes, and damping ratios, are identified based on the measured time-domain applied force (input) and responses (output). In real application on civil engineering structures, it is expensive to apply external forces or to introduce initial disturbance onto the structure to induce detectable vibration responses. It is also dangerous to do so, as the applied forces or disturbance may damage the structural and/or non-structural components of the structure. Thus, the structure's vibration under the action of ambient excitations is considered, and the force (input) cannot be measured in general. Therefore, the modal analysis method introduced in this book is applicable in both forced vibration and ambient vibration situations.

In structural system identification, the model parameters (system), such as the stiffness coefficients of beams and columns of the finite element model, are identified based on the measured time-domain applied force (input) and responses (output). Due to the same reason in the modal system identification, system input is usually unavailable. Under such a situation, the identified modal parameters are usually treated as the responses of the structural system (i.e., the system output).

1.3.1 Modal system identification—modal analysis

Modal analysis, in general, is the technique used to calculate the modal parameters of a structural system by utilizing the measured dynamic data from vibration tests. If forced vibration tests are conducted (see Section 1.1.2), both the measured excitation and responses are available. Modal testing can be used to identify natural frequencies, mode shapes, and damping ratios of the system. However, both free and forced vibration tests are usually difficult to arrange for civil engineering structures, as discussed earlier. As a result, modal analysis for civil engineering structures is usually done with ambient vibration tests. As it is not necessary to apply external force or introduce disturbance to the target structure, the vibration test and the corresponding modal identification can be carried out under the operational condition of the structure, and therefore, this technique is called "operational modal analysis." Operational modal analysis is convenient, as only structural responses (system output) are needed to be measured, and its implementation is cost-effective. As the vibration level under ambient excitation is usually very low, the identified modal parameters can only reflect the structural characteristic under low-level vibration. Furthermore, the accelerometer employed must be very sensitive due to the low vibration level. Without using the force information and the relatively low signal-to-noise ratio, the uncertainties associated with the operational modal analysis results are believed to be high.

Consider the aged building in Figure 1.9 as an example: the measurement plan of an ambient vibration test is discussed to give readers some ideas in ambient modal identification of an existing building. The target building is a 14-story aged building made of reinforced concrete. Figure 1.17 shows the orientation of the building. For modal analysis of this kind of building (i.e., tall rectangular buildings), one is expected to identify

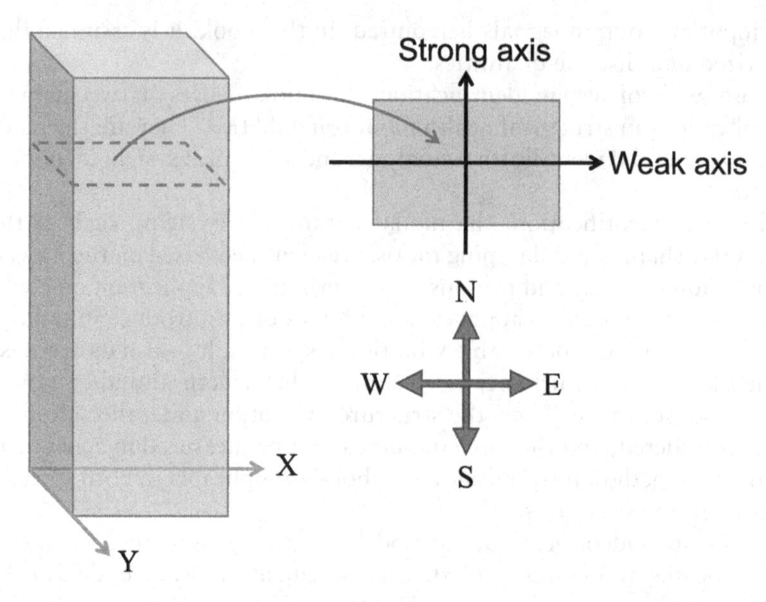

Figure 1.17 The orientation of a 14-story aged building and the definition of strong and weak axes.

(1) the translational vibration modes about the weak axis (as defined in Figure 1.17); (2) the translational vibration modes about the strong axis; (3) the torsional modes about the vertical axis; and, optionally, (4) the vertical vibration modes (i.e., the tension and compression vibration of columns). As rotational accelerometers are uncommon, the vibration of the torsional modes must be calculated from the measured translational vibrations. To do this, at least three measurement points per floor are needed. In general, tall buildings have at least three staircases, which are perfect locations for installing the accelerometers. In this example, the vibration test was carried out in the three staircases from the roof to the ground floor. The three sets of measured data (one for each staircase) were combined through the measurement on the roof (refer to Section 6.3.3.4 for the method to calculate the torsional vibration through measured translational vibrations at the three staircases). In this case study, the triaxial accelerometers, as shown in Figure 1.10, were employed.

By following the modal identification method introduced in Chapter 3, the natural frequencies, mode shapes, and damping ratios of the first six modes were identified and summarized in Figure 1.18. In the figure, the dashed lines and dots represent the undeformed building and the measurement points, respectively. The triangle at the top represents the roof (simply connecting the three staircases on the roof). The black solid lines represent the identified mode shapes. The red arrows on the roof show the vibration direction. Mode 1 of the building is the first translational mode for vibration along the North-South (NS) direction, as shown in Figure 1.18(a). The identified natural frequency (1.5 Hz) and damping ratio (0.6%) are shown in the title of the figure. Mode 2 is the first translational mode for vibration along the East-West (EW) direction, as shown in Figure 1.18(b). The first torsional mode is shown in Figure 1.18(c). The second NS translational, EW translational, and torsional modes are shown in Figure 1.18(d), (e), and (f), respectively.

The natural frequencies and mode shapes are important modal parameters that reflect the stiffness distribution of the structure. Since most structural damage can be represented by the reduction in stiffness of certain structural components, these modal

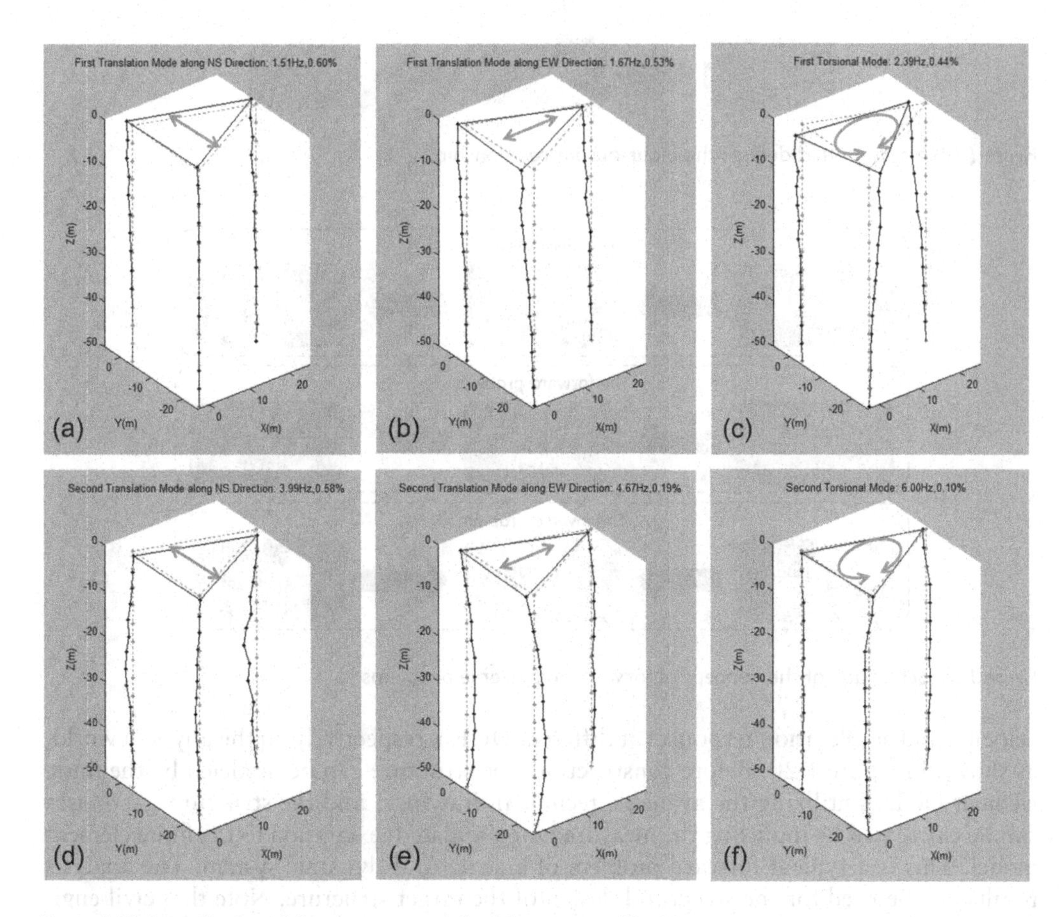

Figure 1.18 Identified modal parameters of the 14-story aged building.

parameters are commonly used as means for structural damage detection in the literature (Das et al. 2016; Avci et al. 2021). Damping ratios, as another modal parameter, show the ability of the structure to dissipate energy: that is, the conversion of the mechanical energy (=kinetic energy+potential energy) to other forms of energy (e.g., heat energy). It is believed that some types of structural damage, such as cracks, will provide additional channels for dissipating mechanical energy. Therefore, some researchers proposed monitoring the damping of a structural system for the purpose of structural damage detection. However, the measurement noise in the identified damping ratios is much higher than that in the identified natural frequencies and mode shapes. Due to the high uncertainties, damping ratios will not be considered as a means for structural damage detection or health monitoring in this book.

1.3.2 Structural system identification—model updating

In many fields of science and engineering, the input-output relationship is controlled by a system, as shown in Figure 1.19. Considering the structural dynamics of a civil engineering structure, the system, input, and output are represented by the structure, the excitation (e.g., wind and traffic loads), and the structural responses (e.g., the displacement,

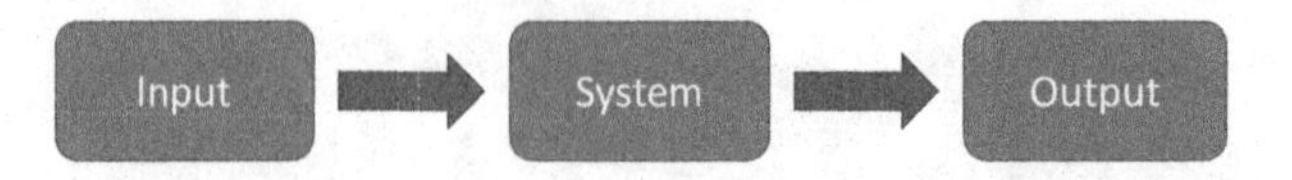

Figure 1.19 A system that defines the input-output relationship.

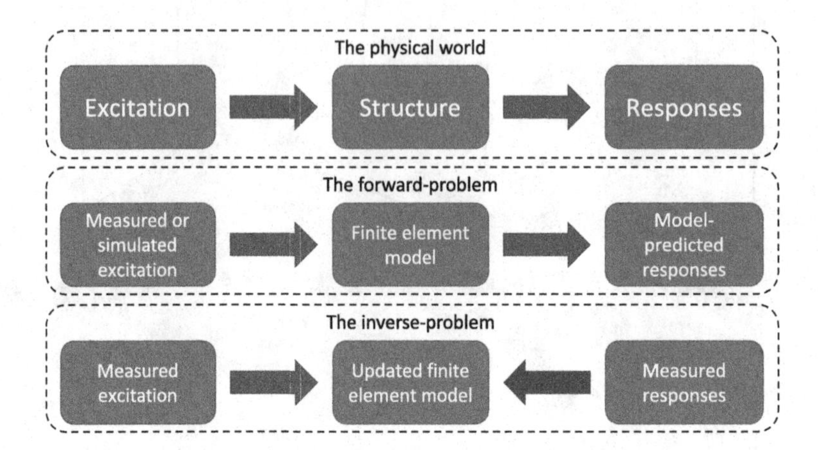

Figure 1.20 Schematic of the concept of forward and inverse problems.

velocity, and acceleration responses at different DOFs), respectively, in the physical world, as shown in Figure 1.20. Before construction, the structure can be modeled by the finite element method utilizing the available technical drawings, and the structural responses can be calculated by inputting the measured (or simulated) excitation to the finite element model. This is a typical forward problem of a structural dynamic system. The analysis results can be used for the structural design of the target structure. Note that civil engineering structures are, in general, too complicated to be modeled by differential equations. Thus, the finite element model is believed to be the most appropriate choice. Once the construction is completed, one may measure the responses of the structure under a given excitation. When the model-predicted structural responses are compared to the measured ones, the discrepancy is usually large due to the problems of measurement noise and modeling error. This discrepancy clearly implies that the finite element model cannot accurately represent the mechanical behavior of the structure. The most popular way to address this problem and to "correct" the model is based on the set of measured responses (and excitation), such that the discrepancy between the model-predicted and measured responses is minimized. This process is called "model updating," and it is a typical inverse-problem as the information flow is opposite to that in the forward problem (see Figure 1.20).

It is believed that the discrepancy between the model-predicted and measured responses is mainly due to measurement noise and modeling error. With the advances in vibration test equipment, the effect of measurement noise is very low, and it usually can be well modeled by using Gaussian distributions. Modeling error is the key problem in structural model updating. The modeling error is at least due to three different sources of errors. The first one is the error in idealizing the physical structure to formulate the simplified analytical or finite element model (e.g., assuming the connections are either pinned or rigid, while they are semi-rigid in the real situation). The error induced during the discretization of the structure to form the finite element model also belongs to this source. The second one is the error induced in the parameterization of the finite element model. Consider the finite element model of a reinforced concrete building frame as an example; it is usually assumed that the modulus of elasticity for all beams and columns is the same. However, if one carries out core tests for samples from various beams and columns, the measured modulus of elasticity for different structural components is different. It is clear that the value used in the finite element model is in the average sense. From the global viewpoint, the effect of this error is minimal, as some structural components have higher stiffness values while the stiffness values of some other components are lower. On average, the effects tend to cancel each other out. However, this error cannot be neglected when the local stiffness of individual structural components is of interest (e.g., in structural damage detection). The third error source is due to the error in the assigned values of model parameters (e.g., the stiffness of individual structural components). It must be pointed out that the first two error sources can be addressed by model class selection techniques (Beck & Yuen 2004; Lam et al. 2017), while the third one can be handled by model updating methods (see Chapter 6).

To illustrate the basic concept of model updating, a simple example of a two-story steel frame is employed, as shown in Figure 1.21. The steel frame on the left of the figure is the physical structure. Without loss in generality, the natural frequencies of the structure are employed as the structural responses. By modal analysis, the natural frequencies, $\hat{f}_n$, for $n = 1,2,\ldots,N$ were identified from the measured time-domain accelerations from the vibration test, where N is the number of modes to be considered in the model updating process. To maximize the information available for model updating, N is usually the highest measured mode with acceptable accuracy.

As the material properties of steel sections are very stable, the stiffness values of beams and columns are not considered as uncertain. To reduce the effect of modeling error, all

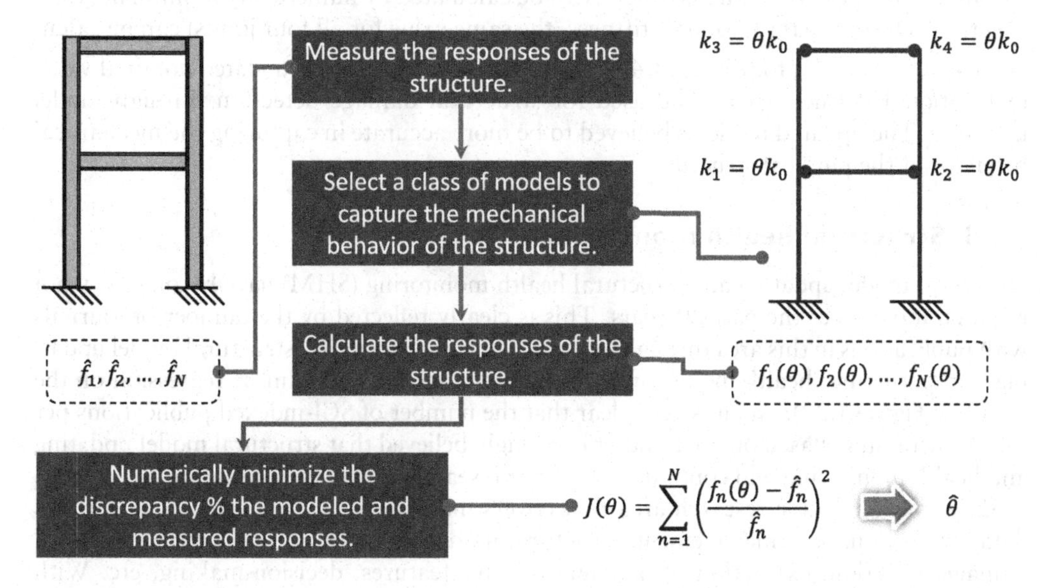

Figure 1.21 An example to illustrate the basic concept of model updating.

beam-column connections are considered as semi-rigid in the finite element model. Their rotational stiffness, k_i, for $i = 1,2,3,4$, of joints 1 to 4 are calculated by multiplying a non-dimensional scaling factor θ to the nominal rotational stiffness, k_0. For an intact steel frame, the rotational stiffness values for all beam-column connections are assumed to be the same. Thus, only one scaling factor is used for all four joints (this is the parameterization). Here, θ is the unknown to be calculated by model updating. The numerical value of the nominal rotational stiffness can be estimated by experience or obtained from the literature. The finite element model of the steel frame is shown on the right-hand side of the figure. The dots at the end of the beams represent the rotational spring.

For a given numerical value of θ, the model-predicted natural frequencies, $f_n(\theta)$, for $n = 1,2,\ldots,N$, can be calculated by solving the eigenvalue problem of the system stiffness (depends on θ) and mass matrices. As an initial trial, $\theta = 1$ is a reasonable guess. Note that the order of modes from experiment may not necessarily be the same as that of modes from computer-simulation. Thus, it is essential to match the mode orders from different sources (Moller & Friberg 1998).

As the estimated nominal value of rotational stiffness may not be very accurate, the identified natural frequencies (from modal analysis) are not the same as the model-predicted ones. For illustration purposes, the simplest model updating method is adopted here. That is, to formulate the measure-of-fit function, $J(\theta)$, as:

$$J(\theta) = \sum_{n=1}^{N} \left(\frac{f_n(\theta) - \hat{f}_n}{\hat{f}_n} \right)^2 \tag{1.2}$$

The J function is the sum of squares of the fractional errors in natural frequencies, and it measures the discrepancy between the measured and model-predicted responses. The J function is considered as the objective function, and the scaling factor, θ, is the design variable. The optimal scaling factor, $\hat{\theta}$, can be calculated by numerically minimizing the J function. The optimal rotational stiffness (the same value for all four joints) can be calculated as $\hat{k}_i = \hat{\theta} k_0 = \hat{k}_0$, for $i = 1,2,3,4$, where $\hat{k}_0$ is considered as the updated nominal value of rotational stiffness. It can be used for structural damage detection through model updating. The updated model is believed to be more accurate in capturing the mechanical behavior of the physical structure.

1.3.3 Structural health monitoring

Structural model updating and structural health monitoring (SHM) have become very hot research topics over the past 20 years. This is clearly reflected by the number of journals with publications in this area through a search using the keywords "structural model updating" or "structural health monitoring" in the title. One may do a linear regression on the data (see Figure 1.22), and it is very clear that the number of SCI-indexed publications per year is increasing. Based on the trend, it is strongly believed that structural model updating and health monitoring will continue to be a hot research area in the coming years.

SHM is a very wide research area that covers many technologies, such as sensor and data acquisition, data management, structural modeling and model updating, structural damage detection, extraction of damage-sensitive features, decision-making, etc. With the development in artificial intelligence (AI), many SHM methods are developed on deep learning (or deep neural networks). This book focuses only on the model-based structural

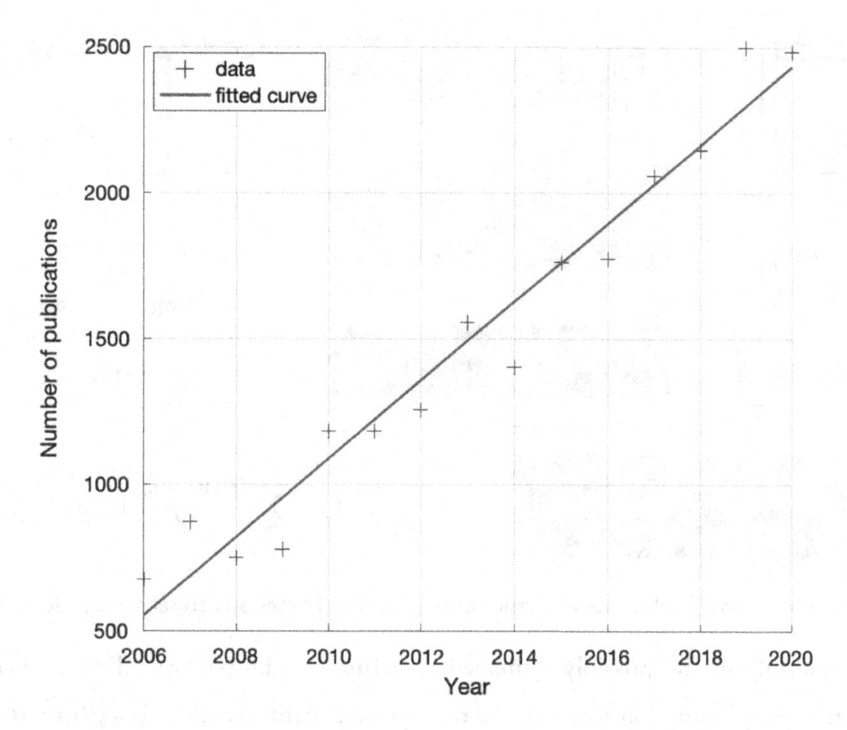

Figure 1.22 Number of publications per year with "structural model updating" or "structural health monitoring" in the title in each year (data extracted from Web of Science).

damage detection techniques under SHM. Thus, it is more accurate to use the term structural damage detection rather than SHM.

To illustrate the concept of model-based structural damage detection, the simple steel frame example in Figure 1.21 is adopted again and extended to the example as shown in Figure 1.23.

For structural damage detection, the vibration responses of the target structure are continuously monitored. By modal analysis, the set of natural frequencies, $\hat{f}_n^D$, for $n = 1,2,\ldots,N$, for the possibly damaged structure can be identified, where the superscript D represents "possibly damaged." The class of models for model updating (in Figure 1.21) cannot be used for the detection of joint damage, as it assumes that the rotational stiffness values for all four beam-column connections are the same. Figure 1.23 (right-hand side) shows the modified class of models for damage detection of the joints. The rotational stiffness values of all four joints are calculated as $k_i = \theta_i \hat{k}_0$, for $i = 1,2,3,4$, where $\hat{k}_0$ is the updated nominal rotational stiffness (calculated in Section 1.3.2 by model updating). The new class of models is parameterized by $\theta = \{\theta_1,\theta_2,\theta_3,\theta_4\}^T$, where the superscript T represents the transpose. This class of models allows individual joints to have different rotational stiffness values, and the damage of a joint can be reflected by the reduction in its rotational stiffness. The new measure-of-fit function can be expressed as:

$$J^D(\theta) = \sum_{n=1}^{N} \left(\frac{f_n(\theta) - \hat{f}_n^D}{\hat{f}_n^D} \right)^2 \tag{1.3}$$

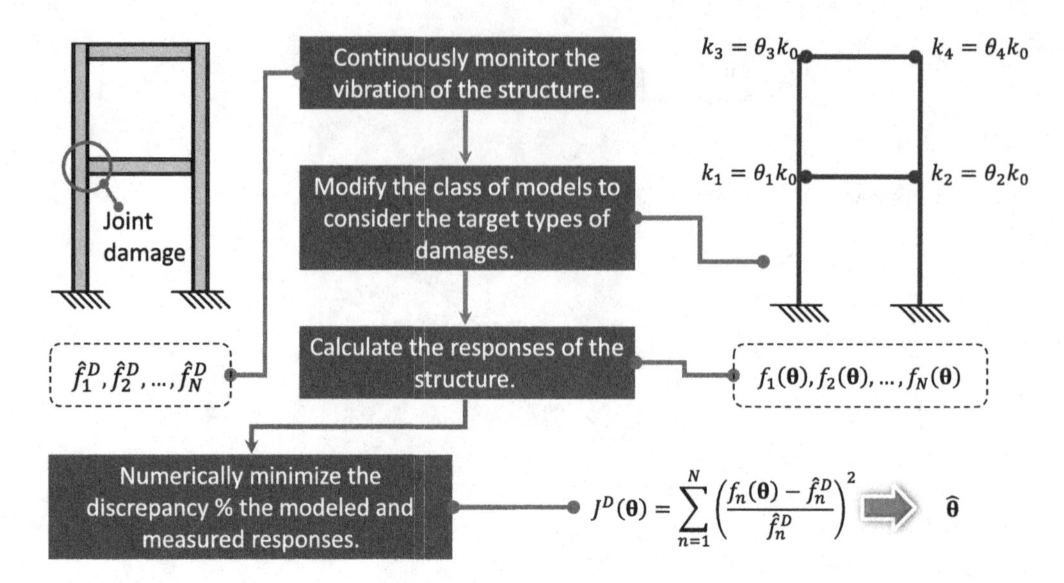

Figure 1.23 An example to illustrate the basic concept of model-based structural damage detection.

Model updating of the possibly damaged structure can be performed by numerically minimizing the J^D function to calculate the optimal scaling factors, $\hat{\theta} = \left\{\hat{\theta}_1, \hat{\theta}_2, \hat{\theta}_3, \hat{\theta}_4\right\}^T$. Consider the example in Figure 1.23: joint 1 is damaged (see the physical structure on the far left of the figure). It is expected that the identified $\hat{\theta}_1$ takes a value of less than unity (showing a reduction in rotational stiffness), while other optimal scaling factors should have a value very close to unity (implying no reduction in rotational stiffness). For example, if $\hat{\theta}_1 = 0.7$, it shows a 30% reduction in rotational stiffness at joint 1.

1.4 HOW TO USE THIS BOOK

This book is particularly designed for researchers who intend to work in the area of structural health monitoring through the model-based approach. The book can be divided into three parts: (1) the background knowledge in vibration test and structural dynamics in Chapters 1 and 2; (2) the techniques for modal analysis in Chapters 3, 4, and 5; and (3) the techniques for structural model updating and damage detection in Chapters 6 and 7. The main contents and target readers for different chapters of this book are discussed in the following and summarized in Table 1.1.

1.4.1 Part 1: Background knowledge

Chapter 1 is the introduction, and it covers the basic knowledge of vibration tests. It also provides beginners with basic concepts of system identification, modal analysis, model updating, and structural health monitoring. Chapter 2 contains the fundamentals of structural dynamics. It covers both the single-DOF and multi-DOF systems. This chapter is essential in preparing readers with the necessary theoretical backgrounds and skills needed to learn the more research-orientated topics, such as operational modal analysis

Table 1.1 Main Contents and Target Readers for Different Chapters of This Book

Chapter	*Description*	*Target readers*
1	• Basic knowledge about vibration tests • An introduction to structural model updating and structural health monitoring • The way to use this book	• Undergraduates • Postgraduates
2	• Structural dynamics • Time domain analysis of single-DOF system • Time domain analysis of multi-DOF system	• Undergraduates • Postgraduates
3	• Modal analysis by power spectral density (PSD)	• Postgraduates • Researchers
4	• Modal analysis by cross-correlation	• Postgraduates • Researchers
5	• System identification by vector ARMA	• Postgraduates • Researchers
6	• Model updating through numerical optimization	• Postgraduates • Researchers
7	• Bayesian model updating by Markov chain Monte Carlo (MCMC) simulation	• Researchers

and structural model updating in the latter part of this book. A comprehensive series of MATLAB codes are available for readers to learn and practice different methods for the dynamic analysis of various structural systems. Examples are available to ensure the learning objectives are achieved. These two chapters are suitable for both undergraduate and postgraduate students.

1.4.2 Part 2: Modal analysis

Chapter 3 steps into the area of system identification. This chapter introduces important techniques for modal analysis utilizing power spectral density (PSD). The concepts and formulations given in this chapter form the foundation for building up other modal analysis techniques. Chapter 4 extends the concept in Chapter 3 to develop modal analysis techniques by using cross-correlation. Chapter 5 puts forward a newly developed method for extracting modal parameters, such as natural frequencies, mode shapes, and damping ratios through the system identification of an autoregressive moving average (ARMA) model in vector form. Examples of civil engineering structures (through field tests) are given to illustrate the analysis procedures and demonstrate the applicability of the methods in real situations (not under laboratory conditions). These three chapters are suitable for postgraduate students but not undergraduate students.

1.4.3 Part 3: Model updating

Chapter 6 introduces the basic formulation of structural model updating through numerical minimization techniques (i.e., minimizing the discrepancy between the measured and model-predicted responses). For the completeness of this book, some numerical optimization algorithms are also introduced in this chapter. Methods in this chapter are mainly deterministic. A comprehensive series of MATLAB codes are provided for readers to implement various formulations. At the end of this chapter, several case studies of civil engineering structures (with measured data obtained from field tests) are provided to demonstrate the introduced methodologies in real applications. This chapter is not very difficult, and it is suitable for postgraduate students and researchers.

The final chapter, Chapter 7, is the most difficult one in this book. This chapter puts forward a computationally efficient and powerful model updating method following the Bayesian system identification framework (i.e., probabilistic approach). Unlike the deterministic approach that pinpoints a single model as the solution of model updating, the Bayesian framework aims to calculate the posterior (updated) probability density function (PDF) of uncertain model parameters. To cater for the high uncertainty nature of model updating problem for civil engineering structures, the Markov chain Monte Carlo (MCMC) simulation technique is employed in approximating the posterior PDF of the uncertain model parameters. Similar to other chapters, case studies utilizing field test data are provided to explain the complicated concepts. This chapter is suitable for researchers only but not students.

REFERENCES

Avci, O., Abdeljaber, O., Kiranyaz, S., Hussein, M., Gabbouj, M. and Inman, D.J., 2021. A review of vibration-based damage detection in civil structures: From traditional methods to machine learning and deep learning applications. *Mechanical Systems and Signal Processing*, 147, 107077. https://doi.org/10.1016/j.ymssp.2020.107077.

Beck, J.L. and Yuen, K.V., 2004. Model selection using response measurements: Bayesian probabilistic approach. *Journal of Engineering Mechanics*, 130, 192–203.

Chiu, P., 2018. Typhoon Mangkhut: Despite swaying high-rises, Hong Kong structures are built to handle strong winds, experts say. *South China Morning Post*. https://www.scmp.com/news/hong-kong/health-environment/article/2164448/typhoon-mangkhut-despite-swaying-high-rises-hong.

Dallard, P., Fitzpatrick, T., Flint, A. and Low, A., 2001. London Millennium Bridge: Pedestrian-induced lateral vibration. *Journal of Bridge Engineering*, 6(6). December 2001. https://doi.org/10.1061/(ASCE)1084-0702(2001)6:6(412).

Das, S., Saha, P. and Patro, S.K., 2016. Vibration-based damage detection techniques used for health monitoring of structures: A review. *Journal of Civil Structural Health Monitoring*, 6, 477–507. https://doi.org/10.1007/s13349-016-0168-5.

Hu, J., Lam, H.F. and Yang, J.H., 2017. Operational modal identification and finite element model updating of a coupled building following Bayesian approach. *Structural Control and Health Monitoring*, 25, e2089. https://doi.org/10.1002/stc.2089.

Hu, J. and Yang, J.H., 2019. Operational modal analysis and Bayesian model updating of a coupled building. *International Journal of Structural Stability and Dynamics*, 19(1), 1940012. https://doi.org/10.1142/S0219455419400121.

Huang, Y.H., Fu, J.Y., Wang, R.H., Gan, Q., Rao, R. and Liu, A.R., 2014. Practical formula to calculate tension of vertical cable with hinged-fixed conditions based on vibration method. *Journal of Vibroengineering*, 16(2), 997–1009.

Kim, B.H. and Park, T., 2007. Estimation of cable tension force using the frequency-based system identification method. *Journal of Sound and Vibration*, 304(3–5), 1067–1072.

Lam, H.F., Hu, J. and Adeagbo, M.O., 2019a. Bayesian model updating of a 20-story office building utilizing operational modal analysis results. *Advances in Structural Engineering*. https://doi.org/10.1177/1369433218825043.

Lam, H.F., Hu, J. and Yang, J.H., 2017. Bayesian operational modal analysis and Markov chain Monte Carlo-based model updating of a factor building. *Engineering Structures*, 132, 314–336. https://doi.org/10.1016/j.engstruct.2016.11.048.

Lam, H.F., Hu, J., Zhang, F.L. and Ni, Y.C., 2019. Markov chain Monte Carlo-based Bayesian model updating of a sailboat-shaped building using a parallel technique. *Engineering Structures*, 193, 12–27.
Lam, H.F., Zhang, F.L., Ni, Y.C. and Hu, J., 2017a. Operational modal identification of a boat-shaped building by a Bayesian approach. *Engineering Structures*, 138, 381–393.
Moller, P.W. and Friberg, O., 1998. An approach to the mode pairing problem. *Mechanical Systems and Signal Processing*, 12, 515–523.
National Instruments, 2003. LabView user manual, April 2003 Edition, part number 320999E-01.
Olson, L.D., 2005. Dynamic bridge substructure evaluation and monitoring, Publication No. FHWA-RD-03-089, U.S. Department of Transportation, Federal Highway Administration.
Ren, W.X., Chen, G. and Hu, W.H., 2005. Empirical formulas to estimate cable tension by cable fundamental frequency. *Structural Engineering and Mechanics*, 20(3), 363–380.
Zui, H., Shinke, T. and Namita, Y., 1996. Practical formulas for estimation of cable tension by vibration method. *Journal of Structural Engineering-ASCE*, 122(6), 651–656.

Chapter 2

Fundamentals of structural dynamics

2.1 SINGLE-DEGREE-OF-FREEDOM SYSTEMS

Most structures in the real world are too complicated to be considered as single-degree-of-freedom (DOF) systems. However, a complicated multi-DOF system can be decomposed into a series of single-DOF systems, which can then be analyzed separately. The vibration responses of the original complicated multi-DOF system can then be obtained by a combination of the series of calculated responses from the decomposed single-DOF systems. Therefore, the study of single-DOF systems is essential in the field of structural dynamics.

The behaviors of some complicated structural systems (not too many) under some special conditions can be approximated as an equivalent single-DOF system. For example, the bus stop shelter in Figure 2.1 can be assumed as a single-DOF system under small amplitude vibration. Another commonly used example is a single-story industrial building, as shown in Figure 2.2. Here, the rafter-column connection is rigid, and the column-base connection is pinned.

Figure 2.3 shows a typical single-DOF vibration system, which can be described by a single mass, m, connected to a spring with a spring constant, k, and a dashpot with a damping constant, c_v. Under the action of the external time-domain force, $f(t)$, the mass is allowed to travel along the spring elongation direction, x. Note that a DOF is an independent displacement or rotation at a given point of a structural system.

When the mass is moved along the positive x-direction, the spring will be elongated. The elongation is equal to the displacement x traveled by the mass. It is clear that the spring will exert a force on the mass in the negative direction of x. By Hooke's law, the magnitude of this spring force is proportional to the spring elongation and the string constant (i.e., kx). Note that the string constant, k, is also called the stiffness coefficient, as the stiffness can be considered as the force required to produce a unit displacement.

When an object is moving on the earth with a given velocity, the mechanical energy will be dissipated through various sources of damping effects. As a result, the velocity will be reduced, and the object will eventually be stopped if no additional force is applied (i.e., no additional energy to the system). The damping effect is considered in the vibration of single-DOF systems through the dashpot. In this chapter, the complicated damping effect is simplified to the simple viscous damping (and that is why the symbol of damping constant has a subscript of v). Under the assumption of viscous damping, the damping force is proportional to the damping constant and the velocity of the object (i.e., $c_v\dot{x}$, where $\dot{x} = dx / dt$ is the first derivative of the displacement, x, with respect to time, t).

DOI: 10.1201/9780429445866-2

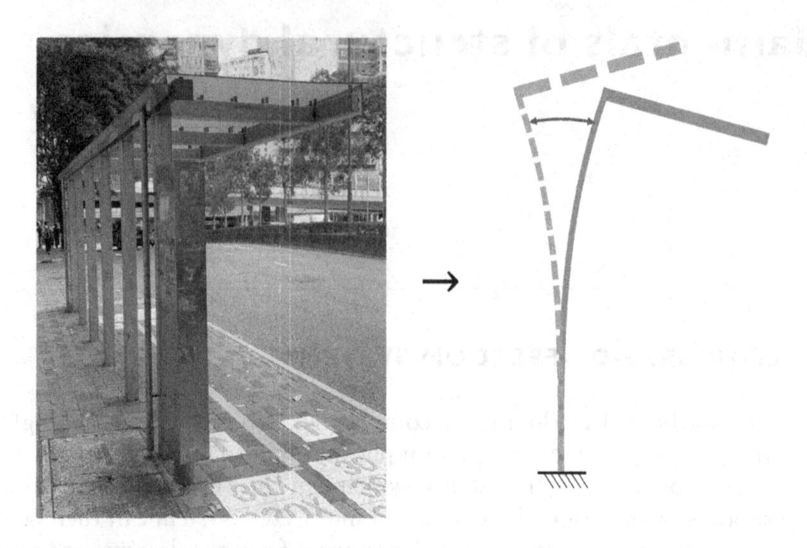

Figure 2.1 A bus stop shelter can be simplified as a single-DOF system.

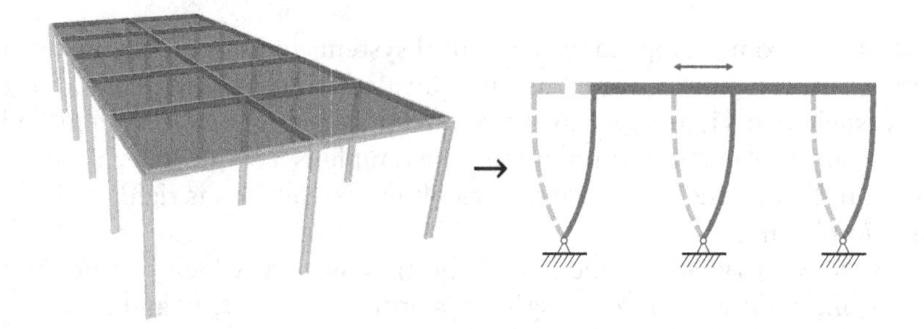

Figure 2.2 A single-story industrial building can be simplified as a single-DOF system.

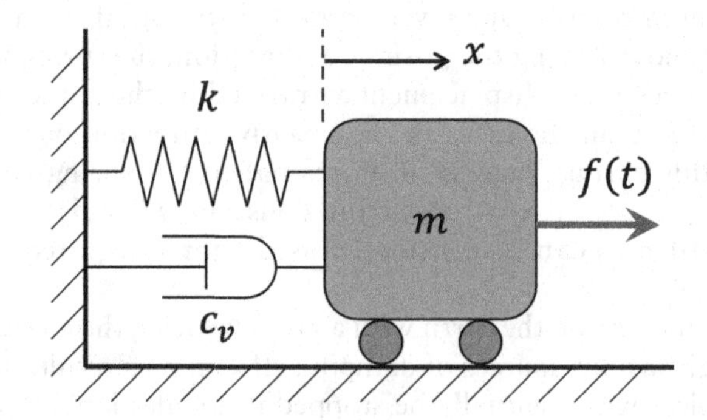

Figure 2.3 A typical model of a single-DOF system.

The vibration of a single-DOF system can be analyzed by Newton's second law of motion,

$$F = ma \tag{2.1}$$

where F is the net force on the object in N (i.e., the vibrating mass of the single-DOF system in this case); m is the mass of the system in kg; and a is the acceleration of the object

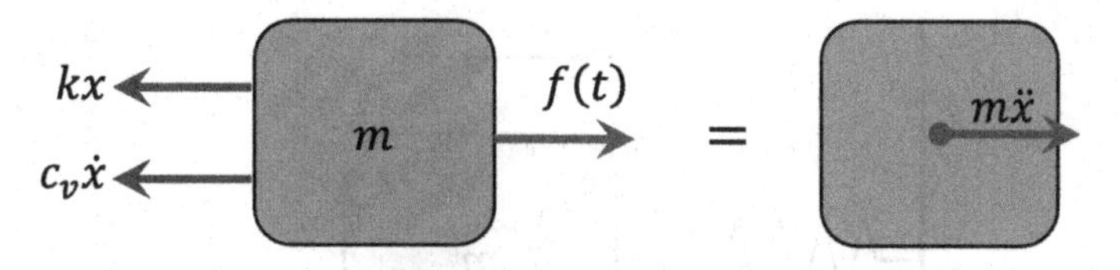

Figure 2.4 The free-body diagram of a single-DOF system.

in m/s². In this case, $a = \ddot{x}$, and $\ddot{x} = d^2x / dt^2$ is the second derivative of the displacement, x, with respect to time, t.

Consider the free-body diagram of the single-DOF system on the left-hand side of Figure 2.4, the net force on the object is equal to:

$$F = f(t) - kx - c_v\dot{x} \tag{2.2}$$

By Newton's second law of motion in Equation (2.1), one obtains:

$$F = f(t) - kx - c_v\dot{x} = m\ddot{x} \tag{2.3}$$

Rearranging to get the most important second-order non-homogeneous ordinary differential equation (ODE) as the governing equation for the vibration of a single-DOF system:

$$m\ddot{x} + c_v\dot{x} + kx = f(t) \tag{2.4}$$

where m is the mass of the system in kg, c_v is the damping constant of the system in N/(m/s), and k is the stiffness of the system in N/m. For a second-order differential equation of this type, two initial conditions are required to solve the two integration constants resulting from solving the ODE. There are many ways to provide the initial conditions, and the most commonly used one is to define the initial displacement, x_0, and initial velocity, $\dot{x}_0$ ($= v_0$, which is commonly used in the literature). With these two initial conditions, the ODE can be expressed as:

$$m\ddot{x} + c_v\dot{x} + kx = f(t), \quad \text{for} \quad x(0) = x_0 \quad \text{and} \quad \dot{x}(0) = \dot{x}_0 \tag{2.5}$$

The purpose here is to calculate the time-domain responses of the single-DOF system under a given excitation, that is, to solve the ODE and determine the displacement x as a function of t (i.e., $x(t)$). Under different damping and excitation conditions (i.e., under different assumptions), the vibration of a single-DOF system (i.e., the ODE in Equation (2.5)) will be solved separately in the following sections.

2.1.1 Undamped free vibration

In this section, the simplest situation, in which there is no external force (i.e., $f(t) = 0$) and no damping effect (i.e., $c_v = 0$), is considered as a background (see Figure 2.5). Under these two conditions, the ODE in Equation (2.5) is simplified as:

$$m\ddot{x} + kx = 0, \quad \text{for} \quad x(0) = x_0 \quad \text{and} \quad \dot{x}(0) = \dot{x}_0 \tag{2.6}$$

Equation (2.6) is a second-order homogeneous ODE. Without external force, the vibration of the single-DOF system is determined by the initial conditions. Furthermore, if

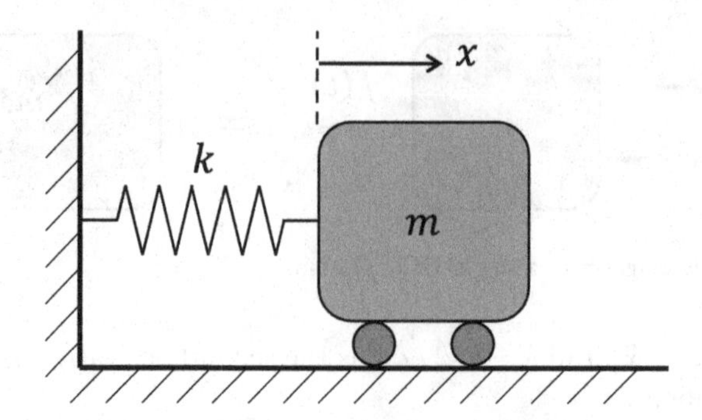

Figure 2.5 A model of an undamped single-DOF system under free vibration.

there is no resistance or damping in the system, the oscillatory motion will continue forever with a constant vibration amplitude. This is the characteristic of an undamped free vibration in the single-DOF system shown in Figure 2.5. Note that the two rollers under the object in the figure are used to show that the object can freely move on the surface of the ground without inducing any resistance.

For a second-order homogeneous ODE with constant coefficients (assuming that both the mass and stiffness are constant), the solution $x(t)$ is in the following form:

$$x(t) = Ae^{\lambda t} \tag{2.7}$$

where A and λ are two constants to be determined. The symbol e is the base of the natural logarithm, and it is approximately equal to 2.7183. Consider the derivative of a general exponential function:

$$\frac{d}{dx}\left(a^x\right) = \left(\ln a\right)a^x \tag{2.8}$$

where a is a positive constant. The special feature of e is that:

$$\frac{d}{dx}\left(e^x\right) = \left(\ln e\right)e^x = e^x \tag{2.9}$$

That is, the derivative of e^x is equal to e^x itself! Referring to the left-hand side of Equation (2.6), the second derivative of the solution must be in the same functional form as the solution (before taking the derivative). Otherwise, it is hopeless to cancel out the two terms and return zero on the right-hand side. Thus, the exponential function, e^x, is a perfect choice. The first and second derivatives of the solution in Equation (2.7) are:

$$\dot{x}(t) = A\lambda e^{\lambda t} \quad \text{and} \quad \ddot{x}(t) = A\lambda^2 e^{\lambda t} \tag{2.10}$$

By substituting Equations (2.7) and (2.10) into the ODE in Equation (2.6):

$$mA\lambda^2 e^{\lambda t} + kAe^{\lambda t} = \left(m\lambda^2 + k\right)Ae^{\lambda t} = 0$$

Considering a non-zero constant A, the following characteristic equation can be obtained:

$$m\lambda^2 + k = 0 \tag{2.11}$$

By solving this characteristic equation, the unknown constant λ can be calculated as:

$$\lambda = \pm i\sqrt{\frac{k}{m}} \tag{2.12}$$

where $i^2 = -1$ is the imaginary unit. To simplify the formulation, the following important substitution is employed:

$$\omega_n = \sqrt{\frac{k}{m}} \tag{2.13}$$

where ω_n is the natural angular frequency of the single-DOF system. As the characteristic equation returns two discrete complex roots (they are complex conjugates of each other), the solution of the ODE for undamped free vibration can be expressed as:

$$x(t) = A_1 e^{i\omega_n t} + A_2 e^{-i\omega_n t} \tag{2.14}$$

where the constant A in Equation (2.7) is extended into A_1 and A_2 here (one corresponding to each root of the characteristic equation). Referring to Euler's formula:

$$e^{i\theta} = \cos\theta + i\sin\theta \tag{2.15}$$

where θ in Euler's formula can be treated as the phase angle as shown in Figure 2.6. When θ increases, the exponential function, $e^{i\theta}$, describes an oscillator (i.e., vibration) in the form of a cosine curve in the real part and another oscillator in the form of a sine curve in the imaginary part.

With the expansion from Euler's formula, the solution in Equation (2.14) can be expressed as:

$$\begin{aligned} x(t) &= A_1\left(\cos(\omega_n t) + i\sin(\omega_n t)\right) + A_2\left(\cos(\omega_n t) - i\sin(\omega_n t)\right) \\ &= (A_1 + A_2)\cos(\omega_n t) + (A_1 - A_2)i\sin(\omega_n t) \\ &= B_1\cos(\omega_n t) + B_2\sin(\omega_n t) \end{aligned} \tag{2.16}$$

where B_1 and B_2 are two new constants (depending on A_1 and A_2) to be determined by the initial conditions $x(0) = x_0$ and $\dot{x}(0) = \dot{x}_0$. Considering the first initial condition, one obtains:

$$x(0) = B_1\cos(0) + B_2\sin(0) = B_1 = x_0 \tag{2.17}$$

With the second initial condition, one obtains:

$$\dot{x}(t) = -B_1\omega_n\sin(\omega_n t) + B_2\omega_n\cos(\omega_n t) \Rightarrow \dot{x}(0) = B_2\omega_n = \dot{x}_0 \Rightarrow B_2 = \frac{\dot{x}_0}{\omega_n} \tag{2.18}$$

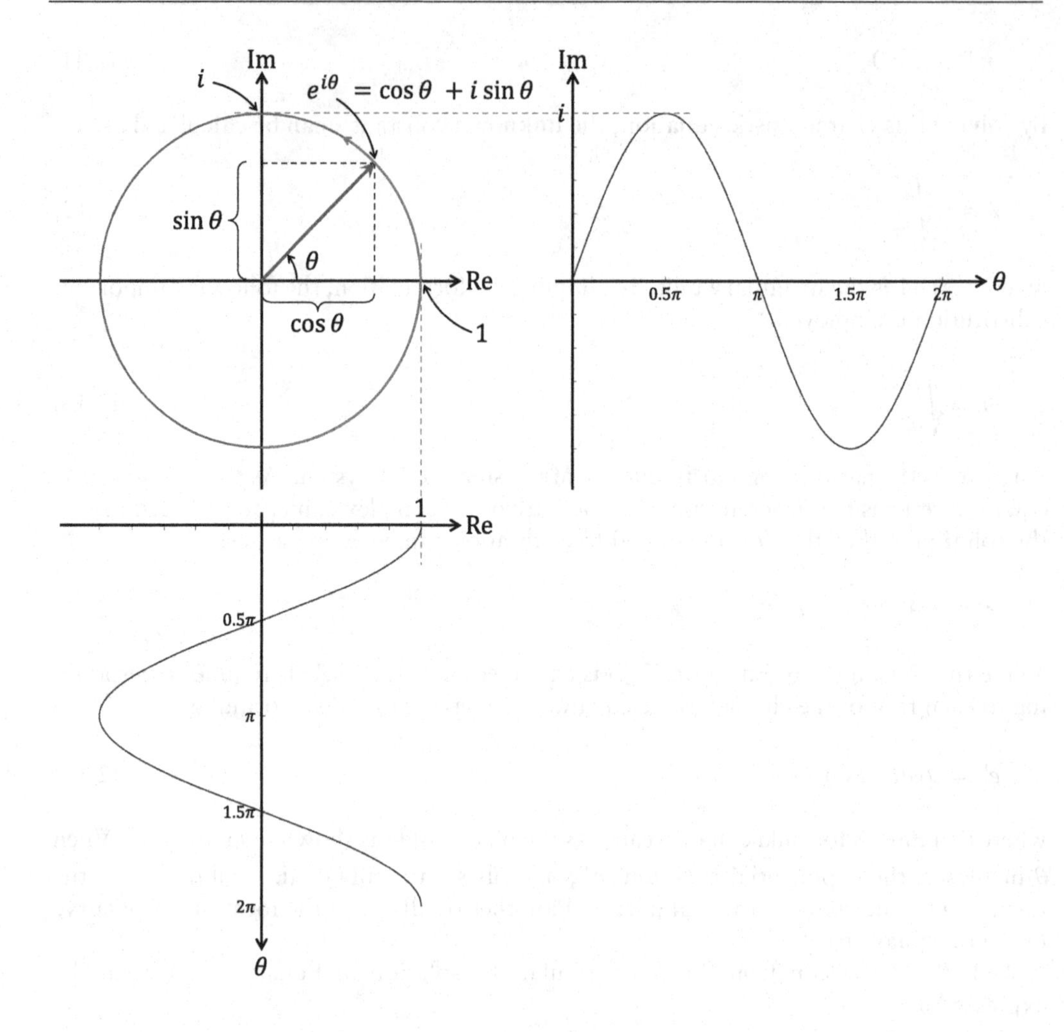

Figure 2.6 The representation of Euler's formula.

The solution of the ODE of undamped free vibration in Equation (2.6) is:

$$x(t) = x_0 \cos(\omega_n t) + \frac{\dot{x}_0}{\omega_n} \sin(\omega_n t) \tag{2.19}$$

Equation (2.19) consists of a cosine term and a sine term. It is clear that the solution represents an oscillation response. However, the vibration amplitude cannot be directly read from the expression. It is thus preferable to combine the two terms into a single term of cosine (or sine). This can be done by constructing a right triangle with the two legs of lengths equal to x_0 and $\dot{x}_0 / \omega_n$ as shown in Figure 2.7.

As shown in Figure 2.7, the angle φ_0 is fixed once the lengths of the two perpendicular legs are given. Furthermore, the length of the hypotenuse, A_0, can be calculated by the Pythagorean theorem. With this triangle, one obtains:

$$\cos\varphi_0 = \frac{x_0}{A_0} \quad \text{and} \quad \sin\varphi_0 = \frac{\dot{x}_0 / \omega_n}{A_0}$$

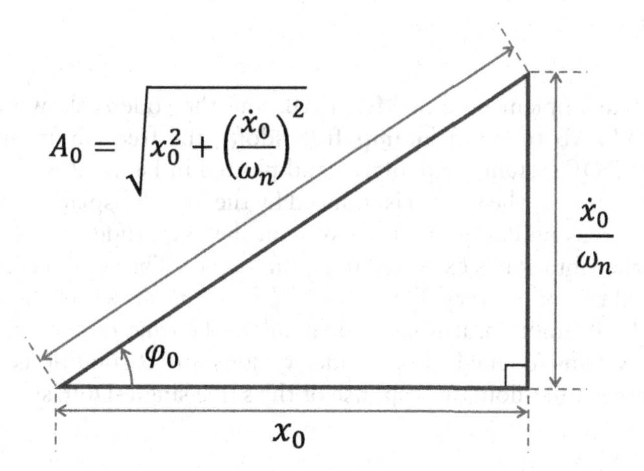

Figure 2.7 The right triangle constructed for transforming the formulation of $x(t)$.

Rearrange to get:

$$x_0 = A_0 \cos\varphi_0 \quad \text{and} \quad \frac{\dot{x}_0}{\omega_n} = A_0 \sin\varphi_0$$

Substitute these two expressions in Equation (2.19) to get:

$$x(t) = A_0 \cos\varphi_0 \cos(\omega_n t) + A_0 \sin\varphi_0 \sin(\omega_n t) \tag{2.20}$$

As $\cos A \cos B + \sin A \sin B = \cos(A - B) = \cos(B - A)$, the solution in Equation (2.19) can be expressed in the following equivalent form:

$$x(t) = A_0 \cos(\omega_n t - \varphi_0) \tag{2.21}$$

where the amplitude of vibration A_0 and the initial phase angle φ_0 are given by:

$$A_0 = \sqrt{x_0^2 + \left(\frac{\dot{x}_0}{\omega_n}\right)^2} \quad \text{and} \quad \varphi_0 = \tan^{-1}\left(\frac{\dot{x}_0}{x_0 \omega_n}\right) \tag{2.22}$$

It is now very clear from Equation (2.21) that the free vibration of an undamped single-DOF system is a cosine curve with a vibration frequency of $\omega_n = \sqrt{k/m}$ as given in Equation (2.13). As this vibration frequency is not induced by external force and is a natural property of the system, it is defined as the natural angular frequency of the single-DOF system in rad/s. For structural engineers, it is more popular to use the natural frequency, f_n, in Hz (i.e., the number of vibration circles per second):

$$f_n = \frac{\omega_n}{2\pi} \tag{2.23}$$

In the field of wind or seismic engineering, the natural period, T_n, in seconds is more popular:

$$T_n = \frac{1}{f_n} = \frac{2\pi}{\omega_n} \tag{2.24}$$

Equation (2.19) was implemented in MATLAB, and the code is shown in Program 2.1. By using the MATLAB function **undmp_free_sdof()**, the free vibration response of an undamped single-DOF system is calculated and plotted in Figure 2.8. It is clear from the figure that the vibration of the system is induced by the initial displacement, x_0, and initial velocity, $\dot{x}_0$. As there is no damping in the system, the amplitude of vibration A_0 will not be reduced, and the vibration is expected to go on forever. The system is expected to complete a circle of vibration in every T_n s (natural period) as shown in the figure. With the developed MATLAB function, it is easy to calculate the time-domain responses of various single-DOF systems (defined by ω_n) under various initial conditions. As an example, Figure 2.9 shows the time-domain response of the same single-DOF system in Figure 2.8

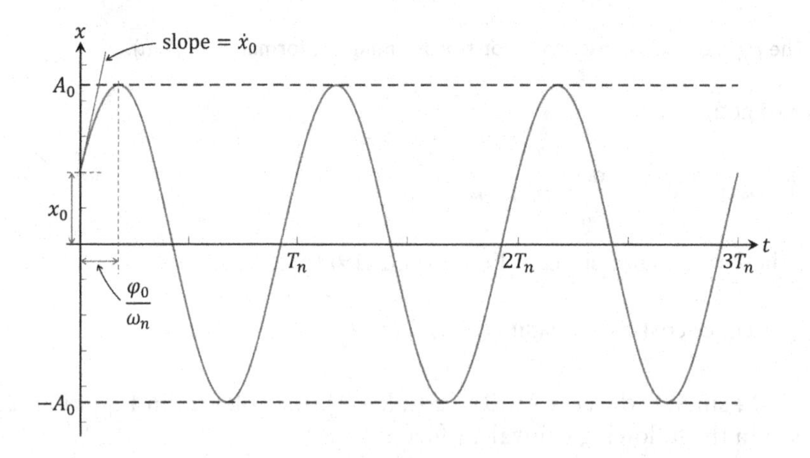

Figure 2.8 Time-domain response of an undamped single-DOF system under free vibration.

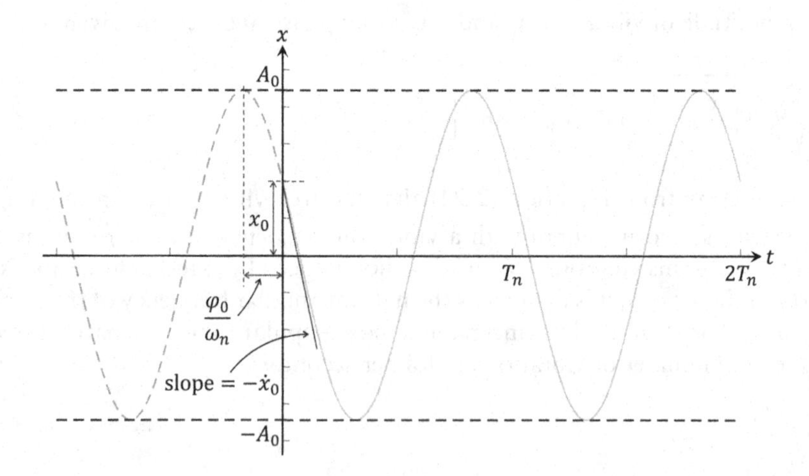

Figure 2.9 Time-domain response of an undamped single-DOF system under free vibration with negative initial velocity.

but with negative initial velocity (i.e., $\dot{x}_0$ is in the downward direction instead of upward). To show the negative phase angle φ_0, the response before t = 0 (it does not exist) is also plotted.

Program 2.1: **undmp_free_sdof()** for calculating the free vibration of an undamped single-DOF system.

```
function [x,tm_ax]=undmp_free_sdof(omg_n,x0,v0,dt,tm_tl)
tm_ax=0:dt:tm_tl;
x=x0.*cos(omg_n.*tm_ax)+v0/omg_n.*sin(omg_n.*tm_ax);
plot(tm_ax,x,'b-','LineWidth',2)
end
```

It must be pointed out that the ODE of the free vibration of an undamped single-DOF system (i.e., Equation (2.6)) is usually expressed in terms of the natural angular frequency as:

$$\ddot{x} + \omega_n^2 x = 0, \quad \text{for} \quad x(0) = x_0 \quad \text{and} \quad \dot{x}(0) = \dot{x}_0 \tag{2.25}$$

In general, an ODE is normalized to have unity coefficient for the term of the highest derivative. Note that the zero-damping assumption is not accurate. In reality, there must be some resistance during the vibration of a structural system. These resistances will dissipate the mechanical energy, and the vibration induced by the initial disturbance will be reduced and eventually stopped. Such a more realistic single-DOF system is studied in the next section.

2.1.2 Damped free vibration

This section considers the free vibration of a damped single-DOF system (see Figure 2.10), the governing equation can be obtained from Equation (2.5) by setting $f(t)$ = 0 as:

$$m\ddot{x} + c_v\dot{x} + kx = 0, \quad \text{for} \quad x(0) = x_0 \quad \text{and} \quad \dot{x}(0) = \dot{x}_0 \tag{2.26}$$

Similar to the ODE for undamped free vibration, this ODE is a second-order homogeneous ordinary differential equation with constant coefficients (as it is assumed that m, c_v, and k are constants).

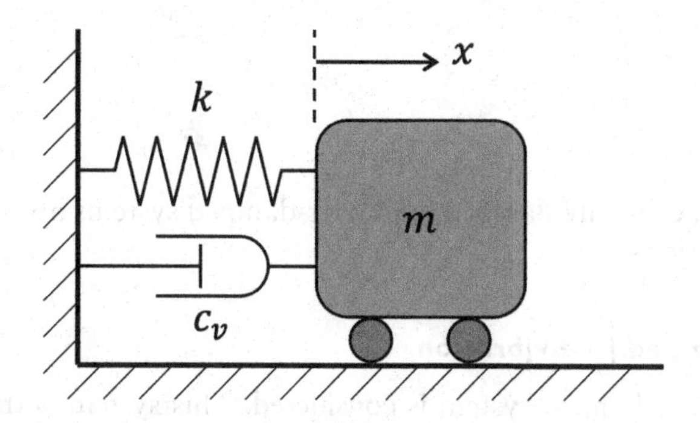

Figure 2.10 A model of a damped single-DOF system under free vibration.

For this type of ODE, the solution is in the form of $Ae^{\lambda t}$ as in Equation (2.7). By substituting the solution into the ODE in Equation (2.26), the characteristic equation can be obtained as:

$$m\lambda^2 + c_v\lambda + k = 0 \tag{2.27}$$

The roots of this quadratic equation are:

$$\lambda = \frac{-c_v \pm \sqrt{c_v^2 - 4mk}}{2m} \tag{2.28}$$

The characteristics of the solution can be classified into three types depending on the value of the discriminant of the quadratic equation:

1. If $c_v^2 - 4mk < 0$, the system is termed under-damped. The roots of the characteristic equation are complex conjugates. The system will vibrate with a given frequency, and the amplitude of vibration will decay exponentially.
2. If $c_v^2 - 4mk = 0$, the system is termed "critically damped." The roots of the characteristic equation are repeated. The initial disturbance of the system will decay to the equilibrium position with, at most, one overshoot.
3. If $c_v^2 - 4mk > 0$, the system is termed "over-damped." The roots of the characteristic equation are purely real and distinct. The initial disturbance of the system will decay to the equilibrium position without vibration. The time for the system to converge to the equilibrium position is longer than that of the critically damped system.

The damping constant of a critically damped system works as the boundary between the under-damped system (with vibration) and the over-damped system (without vibration). It is called the "critical damping," c_c, and is defined as:

$$c_c^2 - 4mk = 0 \Rightarrow c_c = 2\sqrt{mk} = 2m\sqrt{\frac{k}{m}} = 2m\omega_n \tag{2.29}$$

In general, the damping constant c_v of a system is extremely difficult to measure directly. To simplify the formulation (in the later part), it is more convenient to express the damping of a system in terms of its ratio to the critical damping. That is, to define the damping ratio, ζ, as:

$$\zeta = \frac{c_v}{c_c} \tag{2.30}$$

The under-damped, critically damped, and over-damped systems are studied one by one in the following sections.

2.1.2.1 Under-damped free vibration

First of all, the under-damped system is considered. This system is the most important for structural engineers, as the damping is small, and vibration is involved. The forced vibration of an under-damped system may fail if the rate of energy input is larger than the rate of energy dissipation through damping.

When $c_v^2 - 4mk < 0$ (equivalent to $\zeta < 1$ or $c_v < c_c$), the characteristic equation has a pair of complex conjugate roots:

$$\begin{aligned}\lambda &= -\frac{c_v}{2m} \pm i\sqrt{\frac{4mk}{4m^2} - \left(\frac{c_v}{2m}\right)^2} \\ &= -\frac{c_v}{2m}\frac{\omega_n}{\omega_n} \pm i\sqrt{\frac{k}{m} - \left(\frac{c_v}{2m}\frac{\omega_n}{\omega_n}\right)^2} \\ &= -\frac{c_v}{c_c}\omega_n \pm i\sqrt{\omega_n^2 - \left(\frac{c_v}{c_c}\omega_n\right)^2}\end{aligned} \tag{2.31}$$

Therefore:

$$\lambda = -\zeta\omega_n \pm i\sqrt{1-\zeta^2}\,\omega_n \tag{2.32}$$

To further simplify the expression, the damped angular frequency, ω_d, is defined as:

$$\omega_d = \sqrt{1-\zeta^2}\,\omega_n \tag{2.33}$$

The damped frequency $f_d = \omega_d / (2\pi)$ in Hz and damped period $T_d = 1 / f_d$ in seconds can be easily calculated from the damped angular frequency in rad/s. For real structures, the damping ratio ζ is a very small number with a value of 0.01 to 0.02 (i.e., 1% to 2% of critical damping, or 1% to 2% damping in short form) for steel structures, and a value of 0.02 to 0.03 (i.e., 2% to 3% damping) for concrete structures. As a result, the numerical value of damped angular frequency is very close to that of the natural angular frequency (the same for natural frequency and natural period). Consider a damping ratio $\zeta = 0.02$ as an example, where the damped angular frequency is $\sqrt{1-0.02^2} = 0.9996$ times the natural angular frequency (the factor is basically equal to unity).

The two complex roots of the characteristic equation can then be expressed as:

$$\lambda = -\zeta\omega_n \pm i\omega_d \tag{2.34}$$

The solution of the ODE in Equation (2.26) becomes:

$$\begin{aligned}x(t) &= A_1 e^{(-\zeta\omega_n + i\omega_d)t} + A_2 e^{(-\zeta\omega_n - i\omega_d)t} \\ &= e^{-\zeta\omega_n t}\left(A_1 e^{i\omega_d t} + A_2 e^{-i\omega_d t}\right)\end{aligned} \tag{2.35}$$

The solution consists of an exponential decay term, $e^{-\zeta\omega_n t}$, and a vibration term very similar to the responses of the undamped free vibration in Equation (2.14) (by replacing ω_n by ω_d). Following the same steps as in Equation (2.14), Equation (2.35) can be rearranged by Euler's formula as:

$$x(t) = e^{-\zeta\omega_n t}\left[B_1\cos(\omega_d t) + B_2\sin(\omega_d t)\right] \tag{2.36}$$

The unknown constants B_1 and B_2 can then be determined by the initial conditions $x(0) = x_0$ and $\dot{x}(0) = v_0$. Considering the first initial condition, one obtains:

$$x(0) = e^0\left[B_1\cos(0) + B_2\sin(0)\right] = B_1 = x_0 \tag{2.37}$$

To use the second initial condition, the first derivative of $x(t)$ is needed:

$$\begin{aligned}\dot{x}(t) &= -\zeta\omega_n e^{-\zeta\omega_n t}\left[B_1\cos(\omega_d t) + B_2\sin(\omega_d t)\right] \\ &\quad + e^{-\zeta\omega_n t}\left[-B_1\omega_d\sin(\omega_d t) + B_2\omega_d\cos(\omega_d t)\right]\end{aligned} \tag{2.38}$$

With the second initial condition, one obtains:

$$\begin{aligned}\dot{x}(0) &= -\zeta\omega_n e^0\left[B_1\cos(0) + B_2\sin(0)\right] + e^0\left[-B_1\omega_d\sin(0) + B_2\omega_d\cos(0)\right] \\ &= -\zeta\omega_n B_1 + B_2\omega_d = \dot{x}_0\end{aligned} \tag{2.39}$$

As $B_1 = x_0$, the other unknown constant can be calculated as:

$$B_2 = \frac{\dot{x}_0 + \zeta\omega_n x_0}{\omega_d} \tag{2.40}$$

The solution becomes:

$$x(t) = e^{-\zeta\omega_n t}\left[x_0\cos(\omega_d t) + \frac{\dot{x}_0 + \zeta\omega_n x_0}{\omega_d}\sin(\omega_d t)\right] \tag{2.41}$$

Similar to the undamped free vibration case, the sine and cosine terms can be combined by a simple transformation with a right triangle as:

$$x(t) = A_0 e^{-\zeta\omega_n t}\cos(\omega_d t - \varphi_0) \tag{2.42}$$

where:

$$A_0 = \sqrt{x_0^2 + \left(\frac{\dot{x}_0 + \zeta\omega_n x_0}{\omega_d}\right)^2} \quad \text{and} \quad \varphi_0 = \tan^{-1}\left(\frac{\dot{x}_0 + \zeta\omega_n x_0}{x_0\omega_d}\right) \tag{2.43}$$

Equation (2.41) was implemented in MATLAB, and the code is shown in Program 2.2. By using the MATLAB function **undcrtcl_dmp_free_sdof()**, the free vibration response of an under-damped single-DOF system with damping ratio $\zeta = 0.1$ (i.e., 10% damping) is calculated and plotted in Figure 2.11. Similar to the undamped free vibration case (in Figure 2.8), the vibration of the system is induced by the initial conditions. The effect of the damping is to reduce the amplitude of vibration in an exponential decay manner. The decay envelope is described by the exponential term, $A_0 e^{-\zeta\omega_n t}$, in Equation (2.42). It is clear from the figure that the time required for a cycle of vibration is damped period, T_d. Even though the damping in this example (10% damping) is relatively large (considering

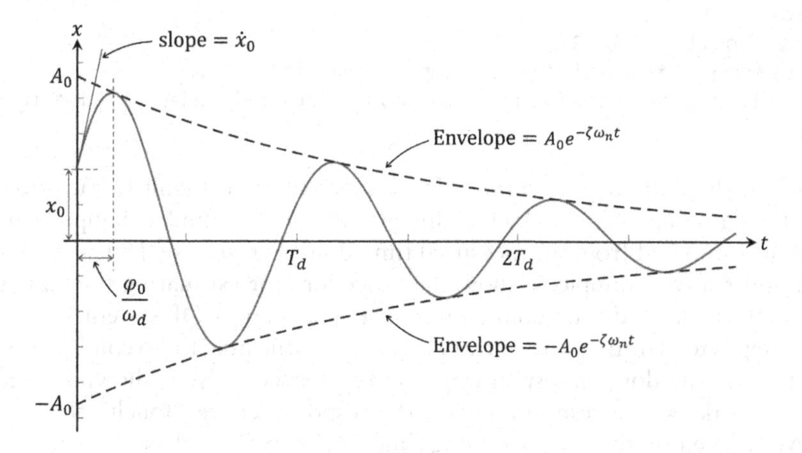

Figure 2.11 Time-domain response of an under-damped single-DOF system under free vibration.

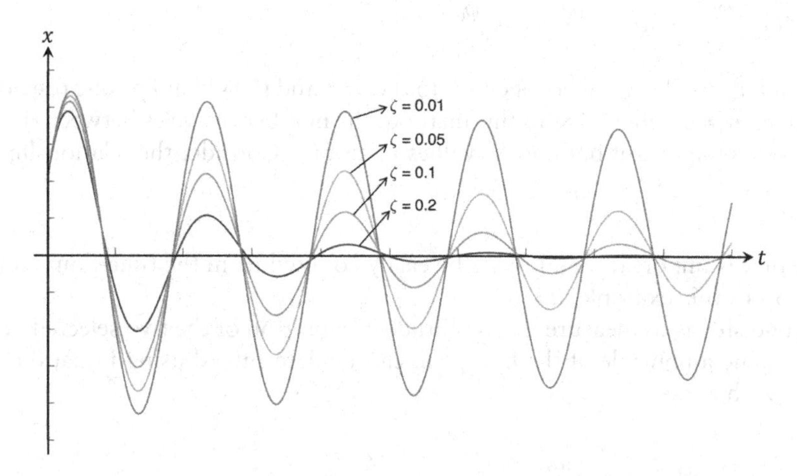

Figure 2.12 Effect of damping in the free vibration of a single-DOF system.

real structures), the damped period is still very close to the natural period ($T_d = 1.005T_n$). In Figure 2.11, the envelopes "touch" the response curve when the cosine term returns a value of unity (i.e., $\cos(\omega_d t - \varphi_0) = 1$) at the local peak of the curve. These local peaks are easy to identify, and they are useful in estimating the damping ratio of a single-DOF system through measured time-domain response (to be discussed later).

To observe the effect of various damping ratios on the vibration response of a single-DOF system, the responses of the same single-DOF system under different damping ($\zeta = 0.2, 0.1, 0.05, 0.01$) are calculated and plotted in Figure 2.12. The increase in damping reduces not only the amplitude of vibration but also the damped period (i.e., it takes a longer time for a higher damping system to complete a cycle of vibration) even if the increase is very small.

Program 2.2: **undcrtcl_dmp_free_sdof()** for calculating the free vibration of an under-damped single-DOF system.

```
function [x,tm_ax]=undcrtcl_dmp_free_sdof(omg_n,dmp,x0,v0,dt,tm_tl)
```

```
tm_ax=0:dt:tm_tl;
omg_d=omg_n*sqrt(1-dmp^2);
x=exp(-dmp*omg_n.*tm_ax).*(x0.*(dmp/(1-dmp^2)^0.5 ...
.*sin(omg_d.*tm_ax) +cos(omg_d.*tm_ax))+v0/omg_d.*sin(omg_d.*tm_ax));
end
```

As mentioned, the damping constant c_v in N/(m/s) is very difficult to measure directly. However, the damping ratio ζ (which is dimensionless) of an under-damped single-DOF system can be estimated from the measured time-domain response. Figure 2.13 is considered as an illustrative example to show the procedure for estimating the damping ratio simply from the measured time-domain response of a single-DOF system.

The first step is to identify two arbitrary local peaks (the first and second selected peaks) of the measured time-domain response at t_1 and t_2, respectively, as shown in Figure 2.13. At any local peak on the response curve, the response curve "touch" the exponential decay curve. Therefore, the corresponding time can be expressed as:

$$t_1 = n_1 T_d + \frac{\varphi_0}{\omega_d} \quad \text{and} \quad t_2 = n_2 T_d + \frac{\varphi_0}{\omega_d} \tag{2.44}$$

where n_1 and n_2 are the numbers of cycles to the first and the second peaks, respectively. In this example, $n_1 = 2$ and $n_2 = 8$. In this method, the number of cycles between the first and second peaks is important but not the values n_1 and n_2. Consider the relationship below:

$$n_2 = n_1 + n \tag{2.45}$$

The value of n is important, and it can be easily counted from the time-domain plot (i.e., Figure 2.13). In this example, $n = 6$.

The second step is to measure the amplitude of vibration of the two selected peaks (i.e., A_1 and A_2). The amplitude of the first peak can be determined using Equation (2.42) by replacing t with t_1 as:

$$A_1 = x\left(n_1 T_d + \frac{\varphi_0}{\omega_d}\right) = A_0 e^{-\zeta\omega_n\left(n_1 T_d + \frac{\varphi_0}{\omega_d}\right)} \cos\left(\omega_d\left(n_1 T_d + \frac{\varphi_0}{\omega_d}\right) - \varphi_0\right) \tag{2.46}$$

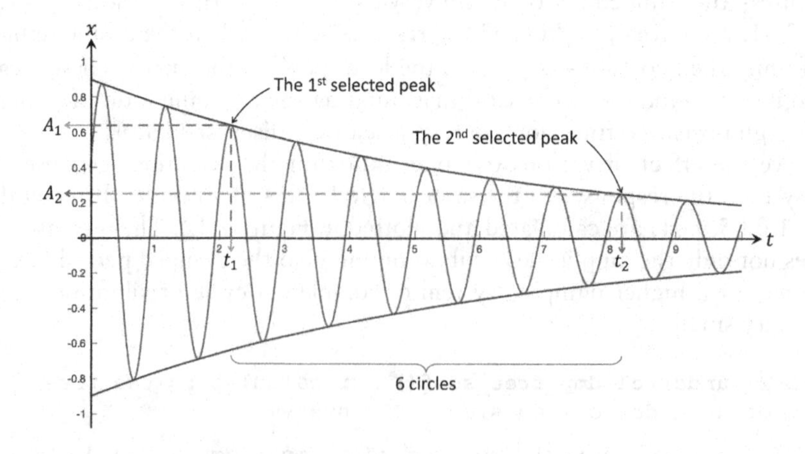

Figure 2.13 An illustrative example for damping ratio estimation.

It can be simplified as:

$$A_1 = A_0 e^{-\zeta\omega_n T_d n_1} e^{-\zeta\omega_n \frac{\varphi_0}{\omega_d}} \cos(n_1\omega_d T_d) = A_0 e^{-\zeta\omega_n T_d n_1} e^{-\zeta\omega_n \frac{\varphi_0}{\omega_d}} \cos(2\pi n_1) \tag{2.47}$$

As n_1 is an integer, the cosine term is equal to unity. The expression can then be simplified as:

$$A_1 = A_0 e^{-\zeta\omega_n T_d n_1} e^{-\zeta\omega_n \frac{\varphi_0}{\omega_d}} \tag{2.48}$$

Similarly, the amplitude of the second peak can be expressed as:

$$A_2 = A_0 e^{-\zeta\omega_n T_d n_2} e^{-\zeta\omega_n \frac{\varphi_0}{\omega_d}} \tag{2.49}$$

The ratio of the first amplitude to the second one can be expressed as:

$$\frac{A_1}{A_2} = \frac{e^{-\zeta\omega_n T_d n_1}}{e^{-\zeta\omega_n T_d n_2}} = e^{\zeta\omega_n T_d n_2 - \zeta\omega_n T_d n_1} = e^{\zeta\omega_n T_d n} \tag{2.50}$$

where T_d can be expressed as:

$$T_d = \frac{1}{f_d} = \frac{2\pi}{\omega_d} = \frac{2\pi}{\sqrt{1-\zeta^2}\,\omega_n} \tag{2.51}$$

Applying natural logarithm on both sides of Equation (2.50), one obtains:

$$\ln\frac{A_1}{A_2} = \zeta\omega_n T_d n = \frac{2\pi n\zeta}{\sqrt{1-\zeta^2}} \approx 2\pi n\zeta \tag{2.52}$$

As ζ is a small number, $\sqrt{1-\zeta^2} \approx 1$. Finally, the damping ratio can be calculated as:

$$\zeta \approx \frac{\ln(A_1 / A_2)}{2\pi n} \tag{2.53}$$

In this illustrative example, the values $A_1 = 0.62$ and $A_2 = 0.23$ can be easily read/measured from Figure 2.13. Therefore, the damping ratio in this example can be estimated as:

$$\zeta \approx \frac{\ln(0.62 / 0.23)}{2\pi(6)} = 0.0263 = 2.63\%$$

For a single-DOF system, this is one of the best ways to estimate the damping ratio of the system.

From the same figure, one can easily read the values of t_1 and t_2. The time difference $t_2 - t_1$ is the time required for the system to vibration n cycles. Therefore, the damped period of the system can be calculated as:

$$T_d = \frac{t_2 - t_1}{n} \tag{2.54}$$

As the damping ratio is estimated, the natural angular frequency in rad/s (and so as the natural frequency in Hz) can be easily calculated.

2.1.2.2 Critically damped free response

Referring to the characteristic equation in Equation (2.27), the system is critically damped if the discriminant of the quadratic characteristic equation is equal to zero, and therefore, $c_v^2 - 4mk = 0$ (equivalent to $\zeta = 1$ or $c_v = c_c$). Under such a situation, the characteristic equation has two repeated real roots.

$$\lambda = -\frac{c_v}{2m} = -\frac{2\sqrt{mk}}{2m} = -\sqrt{\frac{k}{m}} = -\omega_n \tag{2.55}$$

The solution of the ODE may be expressed as:

$$x(t) = Ae^{-\omega_n t} \tag{2.56}$$

However, this is not the general solution of the second-order ODE as there is only one unknown constant, A. It is necessary to search for another solution. Consider another form of solution:

$$x(t) = y(t)e^{-\omega_n t} \tag{2.57}$$

where $y(t)$ is a general function of t to be identified. The first and second derivatives of the new solution can be calculated as:

$$\dot{x}(t) = -\omega_n y(t)e^{-\omega_n t} + \dot{y}(t)e^{-\omega_n t} \tag{2.58}$$

and

$$\begin{aligned}\ddot{x}(t) &= \omega_n^2 y(t)e^{-\omega_n t} - \omega_n \dot{y}(t)e^{-\omega_n t} + \ddot{y}(t)e^{-\omega_n t} - \omega_n \dot{y}(t)e^{-\omega_n t} \\ &= \omega_n^2 y(t)e^{-\omega_n t} - 2\omega_n \dot{y}(t)e^{-\omega_n t} + \ddot{y}(t)e^{-\omega_n t}\end{aligned} \tag{2.59}$$

Substituting the first and second derivatives of the new solution into the ODE in Equation (2.26), one obtains:

$$m\left[\omega_n^2 y(t) - 2\omega_n \dot{y}(t) + \ddot{y}(t)\right] + c_v\left[-\omega_n y(t) + \dot{y}(t)\right] + ky(t) = 0 \tag{2.60}$$

Rearrange to get:

$$m\ddot{y}(t) + (c_v - 2\omega_n m)\dot{y}(t) + \left(m\omega_n^2 - c_v\omega_n + k\right)y(t) = 0 \tag{2.61}$$

As $-\omega_n$ is a root of the characteristic equation in Equation (2.27), one obtains $m\omega_n^2 - c_v\omega_n + k = 0$. Therefore, the $y(t)$ term in Equation (2.61) vanishes. With the root $-\omega_n$, the characteristic equation can be constructed as:

$$\left(\lambda+\omega_n\right)^2=\lambda^2+2\omega_n\lambda+\omega_n^2=0 \tag{2.62}$$

To compare with the constructed characteristic equation, the original characteristic equation in Equation (2.27) is normalized as:

$$\lambda^2+\frac{c_v}{m}\lambda+\frac{k}{m}=0 \tag{2.63}$$

Comparing the coefficient of λ in Equations (2.62) and (2.63), one obtains:

$$2\omega_n=\frac{c_v}{m}\Rightarrow c_v-2\omega_n m=0 \tag{2.64}$$

Therefore, the $\dot{y}(t)$ term in Equation (2.61) vanishes. Equation (2.61) becomes:

$$\ddot{y}(t)=0\Rightarrow\dot{y}(t)=A_1\Rightarrow y(t)=A_1t+A_2 \tag{2.65}$$

The new solution assumed in Equation (2.57) for the second-order ODE of a critically damped single-DOF system can be expressed as:

$$x(t)=\left(A_1t+A_2\right)e^{-\omega_nt} \tag{2.66}$$

where A_1 and A_2 are unknown constants to be determined by the initial conditions. With the first initial condition $x(0)=x_0$:

$$x(0)=A_2=x_0 \tag{2.67}$$

The first derivative of the new solution is:

$$\dot{x}(t)=A_1e^{-\omega_nt}-\omega_n\left(A_1t+A_2\right)e^{-\omega_nt} \tag{2.68}$$

With the second initial condition $\dot{x}(0)=\dot{x}_0$:

$$\dot{x}(0)=A_1-\omega_nA_2=\dot{x}_0\Rightarrow A_1=\dot{x}_0+\omega_nx_0 \tag{2.69}$$

Finally, the new solution becomes:

$$x(t)=\left[\left(\dot{x}_0+\omega_nx_0\right)t+x_0\right]e^{-\omega_nt} \tag{2.70}$$

The solution is the multiplication of a linear function of t to an exponential decay term. Unlike the under-damped case, there is no oscillation cosine term. Equation (2.70) was implemented in MATLAB, and the code is shown in Program 2.3. By the MATLAB function **crtcdmp_free_sdof()**, the response of a critically damped single-DOF system induced by initial conditions (both the initial displacement and velocity are positive) is calculated and plotted in Figure 2.14. It is clear that the response has no vibration. The initial disturbance decayed to the equilibrium position ($x = 0$) very fast. The system is almost at rest after one natural period, T_n.

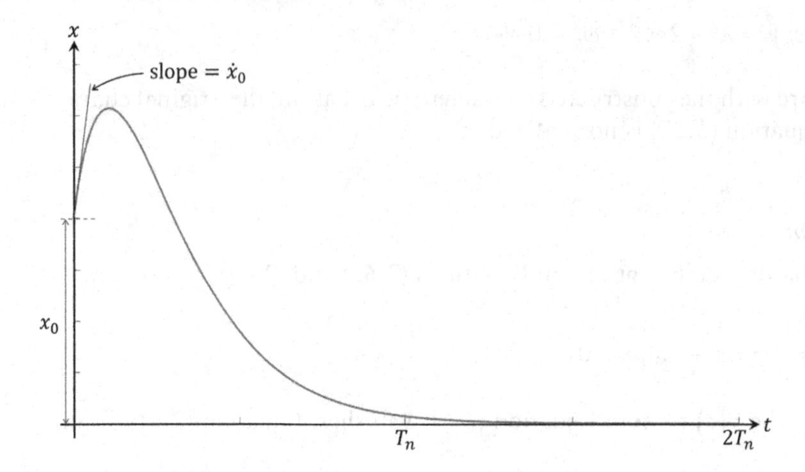

Figure 2.14 Time-domain response of a critically damped single-DOF system under free vibration.

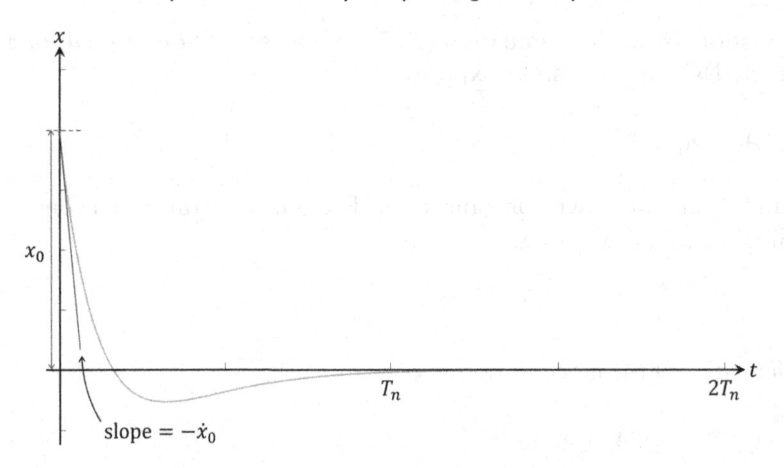

Figure 2.15 Time-domain response of a critically damped single-DOF system under free vibration with negative initial velocity.

Program 2.3: **crtcdmp_free_sdof()** for calculating the free vibration of a critically damped single-DOF system.

```
function [x,tm_ax]=crtcdmp_free_sdof(omg_n,x0,v0,dt,tm_tl)
tm_ax=0:dt:tm_tl;
x=(x0+(v0+x0*omg_n).*tm_ax).*exp(-omg_n.*tm_ax);
plot(tm_ax,x,'b-','LineWidth',2)
end
```

If the initial velocity is negative (i.e., downward), the calculated response is shown in Figure 2.15. There is one overshoot, and the response goes to the equilibrium position very fast. Again, the system almost returns back to the equilibrium position after one natural period of time.

2.1.2.3 Over-damped free response

Finally, the over-damped system is considered. This is when the discriminant of the quadratic characteristic equation in Equation (2.27) is greater than zero, and therefore,

$c_v^2 - 4mk > 0$ (equivalent to $\zeta > 1$ or $c_v > c_c$). Under such conditions, the characteristic equation has two distinct real roots:

$$\lambda = -\frac{c_v}{2m} \pm \sqrt{\left(\frac{c_v}{2m}\right)^2 - \frac{4mk}{4m^2}}$$

$$= -\frac{c_v}{2m}\frac{\omega_n}{\omega_n} \pm \sqrt{\left(\frac{c_v}{2m}\frac{\omega_n}{\omega_n}\right)^2 - \frac{k}{m}} \tag{2.71}$$

$$= -\frac{c_v}{c_c}\omega_n \pm \sqrt{\left(\frac{c_v}{c_c}\omega_n\right)^2 - \omega_n^2}$$

Therefore:

$$\lambda = -\zeta\omega_n \pm \sqrt{\zeta^2 - 1}\,\omega_n \tag{2.72}$$

The solution of the ODE becomes:

$$x(t) = A_1 e^{\left(-\zeta + \sqrt{\zeta^2 - 1}\right)\omega_n t} + A_2 e^{\left(-\zeta - \sqrt{\zeta^2 - 1}\right)\omega_n t} \tag{2.73}$$

The following substitution is used to simplified the expression:

$$\omega_D = \sqrt{\zeta^2 - 1}\,\omega_n \tag{2.74}$$

It must be pointed out that ω_D is not equal to ω_d in the under-damped system, and it is just a variable to simplify the expression (ζ can be a very large number if the system is heavily damped). The solution can be expressed as:

$$x(t) = A_1 e^{(-\zeta\omega_n + \omega_D)t} + A_2 e^{(-\zeta\omega_n - \omega_D)t} \tag{2.75}$$

By the first initial condition:

$$x(0) = A_1 + A_2 = x_0 \tag{2.76}$$

The first derivative of the solution is:

$$\dot{x}(t) = A_1(-\zeta\omega_n + \omega_D)e^{(-\zeta\omega_n + \omega_D)t} + A_2(-\zeta\omega_n - \omega_D)e^{(-\zeta\omega_n - \omega_D)t} \tag{2.77}$$

By the second initial condition:

$$\dot{x}(0) = (-\zeta\omega_n + \omega_D)A_1 + (-\zeta\omega_n - \omega_D)A_2 = \dot{x}_0 \tag{2.78}$$

The two unknown constants A_1 and A_2 can be solved by Equations (2.76) and (2.78) as:

$$A_1 = \frac{\dot{x}_0 + \zeta\omega_n x_0 + \omega_D x_0}{2\omega_D} \quad \text{and} \quad A_2 = \frac{-\dot{x}_0 - \zeta\omega_n x_0 + \omega_D x_0}{2\omega_D} \tag{2.79}$$

The solution becomes:

$$x(t)=\frac{\dot{x}_0+\zeta\omega_n x_0+\omega_D x_0}{2\omega_D}e^{(-\zeta\omega_n+\omega_D)t}+\frac{-\dot{x}_0-\zeta\omega_n x_0+\omega_D x_0}{2\omega_D}e^{(-\zeta\omega_n-\omega_D)t} \tag{2.80}$$

It can be simplified to:

$$x(t)=\frac{e^{-\zeta\omega_n t}}{2\omega_D}\left[\left(\dot{x}_0+\zeta\omega_n x_0\right)\left(e^{\omega_D t}-e^{-\omega_D t}\right)+\omega_D x_0\left(e^{\omega_D t}+e^{-\omega_D t}\right)\right] \tag{2.81}$$

If the system is heavily damped, the value of the damping ratio is much larger than unity (i.e., $\zeta \gg 1$). From Equation (2.74), one obtains:

$$\omega_D=\sqrt{\zeta^2-1}\,\omega_n \approx \zeta\omega_n \quad \text{for} \quad \zeta \gg 1 \tag{2.82}$$

Then, Equation (2.80) can be simplified when $\zeta \gg 1$:

$$x(t)\approx x_0+\frac{\dot{x}_0}{2\zeta\omega_n}\left(1-e^{-2\zeta\omega_n t}\right) \tag{2.83}$$

Equation (2.81) was implemented in MATLAB, and the code is shown in Program 2.4. By using the MATLAB function **ovrcrtc_dmp_free_sdof()**, the free responses of an over-damped single-DOF system with different damping ratios are calculated and plotted in Figure 2.16. It is clear from the figure that the higher the damping (i.e., the larger the value of damping ratio), the longer the time required to decay from the initial disturbance will be.

Program 2.4: **ovrcrtc_dmp_free_sdof()** for calculating the free response of an over-damped single-DOF system.

```
function [x,tm_ax]=ovrcrtc_dmp_free_sdof(omg_n,dmp,x0,v0,dt,tm_tl)
tm_ax=0:dt:tm_tl;
omg_do=omg_n*(dmp^2-1)^0.5;
x=exp(-dmp*omg_n.*tm_ax)/(2*omg_do).*((v0+dmp*omg_n*x0) ...
```

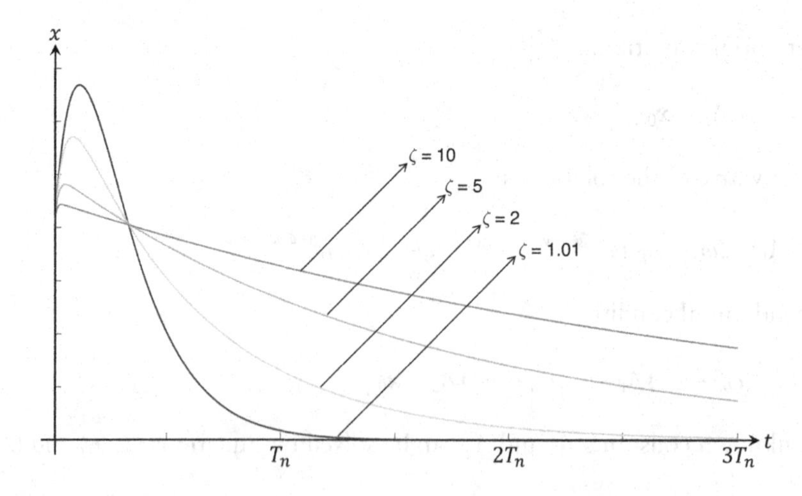

Figure 2.16 Time-domain response of over-damped single-DOF systems induced by initial conditions.

```
.*(exp(omg_do.*tm_ax)-exp(-omg_do.*tm_ax)) ...
+omg_do*x0.*(exp(omg_do.*tm_ax)+exp(-omg_do.*tm_ax)));
plot(tm_ax,x,'b-','LineWidth',2);
end
```

Note that the second-order ODE for the damped vibration of a single-DOF system in Equation (2.26) is usually normalized such that the coefficient of the second derivative term is equal to unity. That is:

$$\ddot{x}+\frac{c_v}{m}\dot{x}+\frac{k}{m}x=0 \Rightarrow \ddot{x}+2\zeta\omega_n\dot{x}+\omega_n^2x=0 \tag{2.84}$$

Together with the initial conditions, one obtains:

$$\ddot{x}+2\zeta\omega_n\dot{x}+\omega_n^2x=0, \quad \text{for} \quad x(0)=x_0 \quad \text{and} \quad \dot{x}(0)=\dot{x}_0 \tag{2.85}$$

The analysis of single-DOF systems under free vibration not only provides an understanding of the natural vibration characteristics of structural systems but also provides a foundation for developing analysis methods for the forced vibration of single-DOF systems. This will be studied in detail in the following sections.

2.1.3 Forced vibration: harmonic excitation

This section considers a more general situation where the external force, $f(t)$, in Equation (2.5) is not equal to zero but a harmonic function as shown in Figure 2.17. This kind of forcing function is essential in the understanding of the concepts of frequency-response-function (FRF) and resonance.

As almost all existing structural systems (e.g., buildings and bridges) have very low damping, only the under-damped single-DOF system is considered in this section. Note that the vibration of a system with very high damping (i.e., critically damped or over-damped) is usually not a problem as the mechanical energy can be easily dissipated.

Under such a forcing function, the ODE is a second-order non-homogeneous ordinary differential equation with constant coefficients. From Equation (2.5), the governing equation of a damped single-DOF system under harmonic excitation can be expressed as:

$$m\ddot{x}+c_v\dot{x}+kx=f_0e^{i\omega t}, \quad \text{for} \quad x(0)=x_0 \quad \text{and} \quad \dot{x}(0)=\dot{x}_0 \tag{2.86}$$

where ω is the excitation frequency in rad/s. Referring to Figure 2.6, this forcing function corresponds to a cosine excitation in the real part and the sine excitation in the imaginary

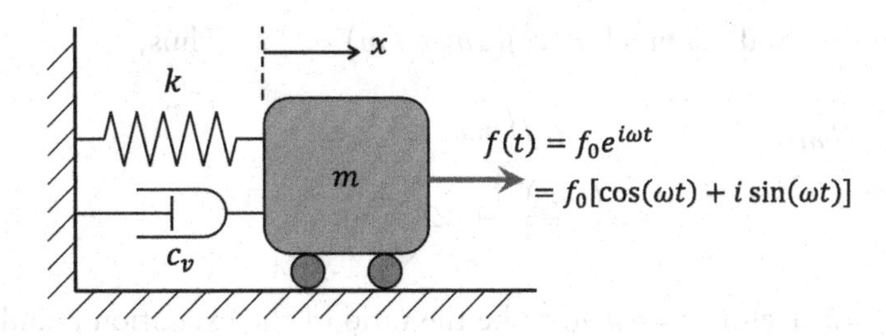

Figure 2.17 A model of a damped single-DOF system under the action of a harmonic force.

part. The force amplitude for both excitations is equal to f_0 in N. The forcing function of sine (or cosine) can be considered as a special case of the current forcing function. Although the formulation of using $f_0 e^{i\omega t}$ as the forcing function is more complicated, it is not necessary to repeat the derivation for both sine and cosine. The initial conditions considered in this section are the same as those in the free vibration cases.

The general solution of a non-homogeneous ODE is in the format of:

$$x(t) = x_c(t) + x_p(t) \tag{2.87}$$

where $x_c(t)$ is the complementary solution (i.e., the general solution of the corresponding homogeneous ODE), and $x_p(t)$ is the particular solution of the non-homogeneous ODE (it, of course, depends on the adopted forcing function). The determination of the general solution of a homogeneous ODE for free vibration of undamped (see Section 2.1.1), under-damped (see Section 2.1.2.1), critically damped (see Section 2.1.2.2), and over-damped (see Section 2.1.2.3) were covered in detail. This section focuses on the determination of the particular solution of a non-homogeneous ODE in the form of Equation (2.86).

The form of the particular solution follows the functional form of the forcing function. Thus, the following solution form is selected for harmonic excitation:

$$x_p(t) = Ae^{i\omega t} \tag{2.88}$$

where A is a constant to be determined. The first and second derivatives of $x_p(t)$ are:

$$\dot{x}_p(t) = i\omega A e^{i\omega t} \quad \text{and} \quad \ddot{x}_p(t) = -A\omega^2 e^{i\omega t} \tag{2.89}$$

By substituting $x_p(t)$ and its derivatives back into the ODE in Equation (2.86), one obtains:

$$-m\omega^2 A + c_v \omega i A + kA = f_0 \tag{2.90}$$

Thus,

$$A = \frac{f_0}{\left(k - m\omega^2\right) + c_v \omega i} \tag{2.91}$$

To clearly show the physical meaning of the expression, it is rearranged as:

$$A = \frac{f_0 / m}{\left(k / m - \omega^2\right) + \left(c_v / m\right)\omega i} \tag{2.92}$$

Note that $k / m = \omega_n^2$ and $c_v / m = \left(c_v / c_c\right)\left(2m\omega_n / m\right) = 2\zeta\omega_n$. Thus,

$$A = \frac{f_0 / m}{\left(\omega_n^2 - \omega^2\right) + 2\zeta\omega_n \omega i} = \frac{f_0 / \left(m\omega_n^2\right)}{\left[1 - \left(\frac{\omega}{\omega_n}\right)^2\right] + 2\zeta\left(\frac{\omega}{\omega_n}\right)i} \tag{2.93}$$

Note that $m\omega_n^2 = k$. Define $r = \omega / \omega_n$ to be the ratio of the excitation angular frequency to the natural angular frequency of the single-DOF system. The unknown constant can be expressed as:

$$A = \frac{f_0 / k}{(1-r^2) + 2r\zeta i} \tag{2.94}$$

where f_0 is the force amplitude and k is the stiffness of the system, and therefore, f_0 / k represents the static deformation of the system. That is, the displacement of the system when the force is applied very slowly to the system in such a way that the dynamic effect can be neglected. Assuming that this static displacement is equal to x_s, the particular solution of the ODE becomes:

$$x_p(t) = \frac{x_s}{(1-r^2) + 2r\zeta i} e^{i\omega t} \tag{2.95}$$

where $x_s = f_0 / k$ is the elongation of the spring under a static force f_0. The solution consists of the vibration amplitude and the oscillator, $e^{i\omega t}$. However, the complex number in the denominator makes it difficult to understand the real vibration behavior of the system. The complex number can be transformed to the numerator as:

$$x_p(t) = \frac{x_s}{(1-r^2) + 2r\zeta i}\left[\frac{(1-r^2) - 2r\zeta i}{(1-r^2) - 2r\zeta i}\right] e^{i\omega t} = \frac{(1-r^2) - 2r\zeta i}{(1-r^2)^2 + (2r\zeta)^2} x_s e^{i\omega t} \tag{2.96}$$

The formulation can be simplified by the following substitutions:

$$a = \frac{(1-r^2)x_s}{(1-r^2)^2 + (2r\zeta)^2} \quad \text{and} \quad b = \frac{2r\zeta x_s}{(1-r^2)^2 + (2r\zeta)^2} \tag{2.97}$$

Then, $x_p(t)$ can be expressed as:

$$\begin{aligned} x_p(t) &= (a - ib)e^{i\omega t} \\ &= (a - ib)\left[\cos(\omega t) + i\sin(\omega t)\right] \\ &= \left[a\cos(\omega t) + b\sin(\omega t)\right] + i\left[a\sin(\omega t) - b\cos(\omega t)\right] \end{aligned} \tag{2.98}$$

Consider a right triangle with leg lengths equal to a and b (see Figure 2.18), one obtains:

$$\sin\varphi_0 = \frac{b}{A_0} \quad \text{and} \quad \cos\varphi_0 = \frac{a}{A_0} \tag{2.99}$$

The particular solution can then be expressed in terms of A_0 and φ_0 as:

$$\begin{aligned} x_p(t) &= A_0\left[\cos\varphi_0 \cos(\omega t) + \sin\varphi_0 \sin(\omega t)\right] \\ &\quad + iA_0\left[\cos\varphi_0 \sin(\omega t) - \sin\varphi_0 \cos(\omega t)\right] \\ &= A_0 \cos(\omega t - \varphi_0) + iA_0 \sin(\omega t - \varphi_0) \\ &= A_0 e^{i(\omega t - \varphi_0)} \end{aligned} \tag{2.100}$$

Finally, the particular solution is:

$$x_p(t) = A_0 e^{i(\omega t - \varphi_0)} \tag{2.101}$$

where the vibration amplitude is defined in Figure 2.18 as:

$$A_0 = \sqrt{a^2 + b^2} = \sqrt{\left[\frac{(1-r^2)x_s}{(1-r^2)^2 + (2r\zeta)^2}\right]^2 + \left[\frac{2r\zeta x_s}{(1-r^2)^2 + (2r\zeta)^2}\right]^2}$$

$$= x_s \sqrt{\frac{(1-r^2)^2 + (2r\zeta)^2}{\left[(1-r^2)^2 + (2r\zeta)^2\right]^2}} \tag{2.102}$$

$$= \frac{x_s}{\sqrt{(1-r^2)^2 + (2r\zeta)^2}}$$

The phase angle is also defined in Figure 2.18 as:

$$\varphi_0 = \tan^{-1}\left(\frac{b}{a}\right) = \tan^{-1}\left(\frac{2r\zeta}{1-r^2}\right) \tag{2.103}$$

Note that the real part of the solution gives the response of the system under the cosine forcing function, and the imaginary part of the solution gives the response of the system under the sine forcing function.

From Equation (2.87), the final solution can be determined by combining the complementary solution (i.e., the general solution of the under-damped free vibration in Equation (2.36)) and the particular solution just obtained in Equation (2.101) as:

$$x(t) = x_c(t) + x_p(t) = e^{-\zeta\omega_n t}\left[B_1 \cos(\omega_d t) + B_2 \sin(\omega_d t)\right] + A_0 e^{i(\omega t - \varphi_0)} \tag{2.104}$$

Due to the exponential decay term, $e^{-\zeta\omega_n t}$, the response from the complementary solution will eventually vanish, leaving only the response from the particular solution. Thus, the

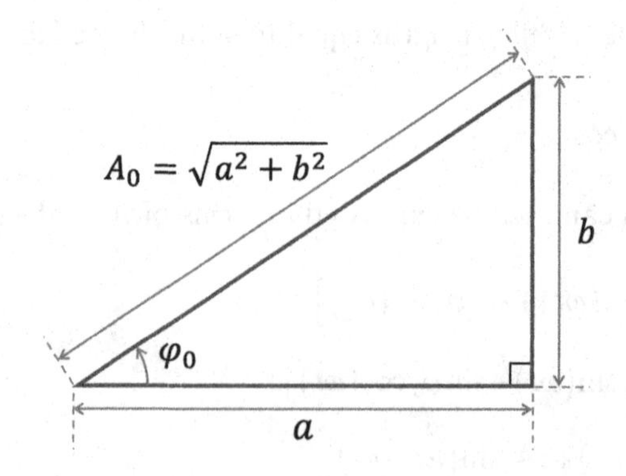

Figure 2.18 The right triangle constructed for transforming the formulation of $x_p(t)$.

complementary solution is considered as the transient response (it is less important in the forced vibration of a structural system), and the particular solution is considered as the steady-state response.

2.1.3.1 Dynamic multiplication factor

Without loss of generality, the steady-state response (i.e., $x_p(t)$) in Equation (2.101) is considered. It consists of two parts, the amplitude of vibration, A_0, as given in Equation (2.102), and the cosine oscillator, $e^{i(\omega t-\varphi_0)}$, which takes a value from +1 to −1. Thus, the maximum deflection of the system is dictated by the amplitude of vibration. It is clear from Equation (2.102) that the maximum deflection is equal to the static deformation, x_s, (the displacement of the single-DOF system when the force, f_0, is slowly applied without inducing any dynamic effect) multiplied by a factor, which is defined as the dynamic multiplication factor D:

$$D = \frac{1}{\sqrt{\left(1-r^2\right)^2 + \left(2r\zeta\right)^2}} \tag{2.105}$$

The dynamic multiplication factor of a single-DOF system depends on the frequency ratio $r = \omega / \omega_n$ and the critical damping ratio, ζ. From Equation (2.105), when the excitation frequency approaches the natural frequency of the system (i.e., $r \to 1$), the term $1 - r^2$ approaches zero. If the system is undamped (i.e., $\zeta = 0$), the value of D approaches infinity (i.e., $D \to \infty$). As most structural systems are under-damped (i.e., $\zeta < 1$), the value of D will become very large but not infinite. This phenomenon is known as "resonance." It is clear that the existence of damping (even if it is small as in most structural systems) stops the collapse of the structural system. From the structural engineering viewpoint, the damping effect is important as it provides a channel for the system to dissipate (or waste) the continuously incoming mechanical energy (from the applied force).

Figure 2.19 shows the dynamic multiplication factor calculated by Equation (2.105) under different damping ratios. Figure 2.19(a) shows curves which correspond to the relatively low damping ratios of 0.01 (e.g., for steel structures) and 0.03 (e.g., for concrete

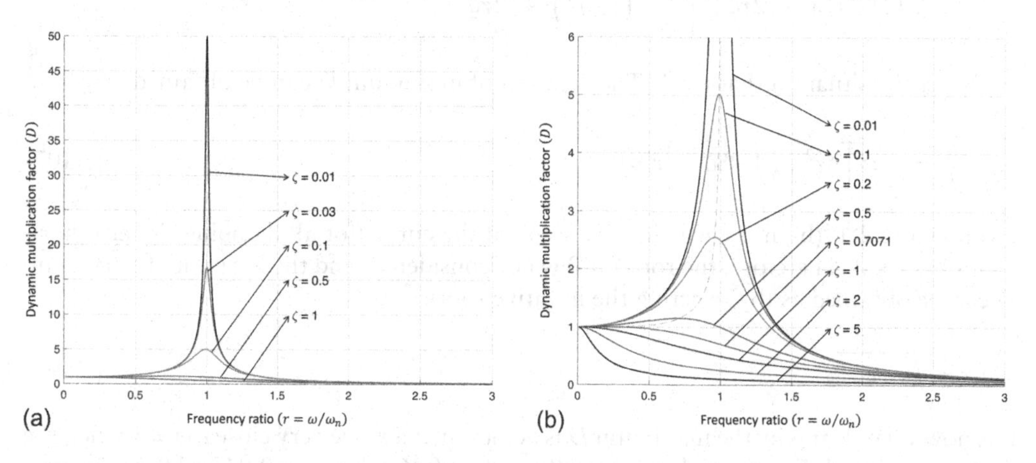

Figure 2.19 Dynamic multiplication factor under different damping. (a) Low and reference damping ratios; (b) High and reference damping ratios.

structures). Furthermore, the relative high damping ratios of 0.1 and 0.5 are also included for comparison. The curve for critical damping (i.e., $\zeta = 1$) is also plotted as a reference to link Figure 2.19(a) to Figure 2.19(b). In the general under-damped case, the curve starts at unity (at $r = 0$) corresponding to a slowly applied static force without dynamic effect. When the frequency ratio is increased (towards $r = 1$), the value of D is increased by the dynamic effect and achieves its maximum value near $r = 1$ for very low damping situation, as shown in Figure 2.19(a) (except when the damping is critical, $\zeta = 1$). Note that the location of maximum D is not exactly at $r = 1$ (to be discussed in Figure 2.19(b)). When the excitation frequency is further increased ($r > 1$), the value of D is decreased. When the excitation frequency is much higher than the natural frequency, the value of D drops below unity. When $r = 1$, the dynamic effects can increase the displacement of the system with a factor of 50 in the low damping situation (i.e., $\zeta = 0.01$). When the damping is increased from 1% to 3% and then to 10%, the dynamic multiplication factor is dropped from 50 to about 17 and then to only 5, as shown in the figure. This clearly shows the significance of the dynamic effect in structural safety.

Figure 2.19(a) clearly shows the curves at low damping but not at high damping. Therefore, Figure 2.19(b) is plotted with particular emphasis on the behavior of the system when the damping is relatively high (some low damping curves are also included for comparison purposes). It is now very clear from Figure 2.19(b) that the maximum D is achieved at $r < 1$. A dashed line is plotted in the figure to show the location of maximum D. When the damping becomes higher and higher, the location of maximum D moves farther and farther away from $r = 1$. Resonance is defined as the increase of vibration amplitude (related to the static displacement) when the excitation frequency is near the natural frequency of the system. From the figure, it seems that the resonance effect disappears when the damping ratio is increased to a value of about 0.7071. That is, the entire curve is below $D = 1$ when $\zeta \geq 0.7071$. To determine the value of r at which D achieve its maximum for a given damping ratio, one can calculate the first derivative of D with respect to r and assign it to zero. That is:

$$\frac{dD}{dr} = \frac{-4r\left(1-r^2\right)+8\zeta^2 r}{\left[\left(1-r^2\right)^2+\left(2r\zeta\right)^2\right]^{\frac{3}{2}}} = \frac{\left[4r^2-\left(4-8\zeta^2\right)\right]r}{\left[\left(1-r^2\right)^2+\left(2r\zeta\right)^2\right]^{\frac{3}{2}}} \tag{2.106}$$

Consider the situation that $r \neq 1$. The location of maximum D can be obtained as:

$$\frac{dD}{dr} = 0 \Rightarrow \left[4\hat{r}^2-\left(4-8\zeta^2\right)\right]\hat{r} = 0 \tag{2.107}$$

From Figure 2.19(b), it is clear that the slope of the curve (for all damping) is zero (horizontal) at $r = 0$. Therefore, the root $\hat{r} = 0$ is not considered, and the location of maximum D can be obtained as (by rejecting the negative root):

$$\hat{r} = \sqrt{1-2\zeta^2} \tag{2.108}$$

It is now very clear why the maximum D is achieved at a value very close to $r = 1$ when the damping is small. For example, $\hat{r} = 0.9999$ and 0.9996 when $\zeta = 0.01$ and 0.02, respectively. When the damping is increased, the difference between $\hat{r}$ and r is more observable.

For example, $\hat{r} = 0.7071$ when $\zeta = 0.5$. From Equation (2.108), $\hat{r}$ is well defined only if $1 - 2\zeta^2 \geq 0$. Therefore, the limited damping value for having a maximum D is:

$$\zeta \leq \frac{1}{\sqrt{2}} \approx 0.7071 \tag{2.109}$$

That is the reason for the disappearance of the resonance effect when the damping ratio exceeds 0.0701, as illustrated by the dashed line in Figure 2.19(b). At this limited damping value, the maximum $D = 1$ is at $r = 0$ (i.e., the static displacement).

The dynamic multiplication factor is important for understanding the vibration characteristics of a structural system. However, it is difficult to measure it directly as the static displacement, x_s, in Equation (2.102) is difficult to obtain for real structures. In practice, one usually considers measuring the steady-state vibration amplitude of a single-DOF system under a given excitation frequency expressed in terms of ω or $f = \omega / (2\pi)$. After testing the single-DOF system for a range of excitation frequencies, say, from f_a to f_b with n intervals, the results can be presented graphically as in Figure 2.20 (for excitation frequency between 1 Hz and 3 Hz, it is clear that the natural frequency of the system is at 2 Hz). This amplitude vs. frequency plot is part of the frequency response function (FRF). Note that the other part of the FRF is the phase vs. frequency plot. It must be pointed out that the amplitude vs. frequency plot looks very similar to the dynamic multiplication factor. One can rearrange Equation (2.102) to get the expression of this curve as:

$$A_0(f) = \frac{x_s}{\sqrt{\left[1 - \left(\frac{f}{f_n}\right)^2\right]^2 + \left[2\zeta\left(\frac{f}{f_n}\right)\right]^2}} \tag{2.110}$$

where f_n and ζ are the natural frequency and damping ratio, respectively, of the single-DOF system, and x_s is defined in Equation (2.95) as the static displacement of the system. This experimental measured plot can be used to estimate the damping ratio, which is very difficult to directly measure for real structures.

2.1.3.2 Estimation of damping ratio

If the static displacement of the system, x_s, is available, the damping ratio can be easily obtained at the peak of the amplitude vs. frequency plot. When the damping is low and the excitation frequency matches the natural frequency (i.e., $f = f_n$), the peak amplitude, A_{peak}, can be expressed as:

$$A_{\text{peak}} = A_0(f_n) = \frac{x_s}{2\zeta} \tag{2.111}$$

Therefore, the damping ratio is given by:

$$\zeta = \frac{x_s}{2A_{\text{peak}}} \tag{2.112}$$

However, the value of x_s is usually unavailable. Under such a situation, the half-power bandwidth method is commonly used to estimate the damping of a system through the measured amplitude vs. frequency plot.

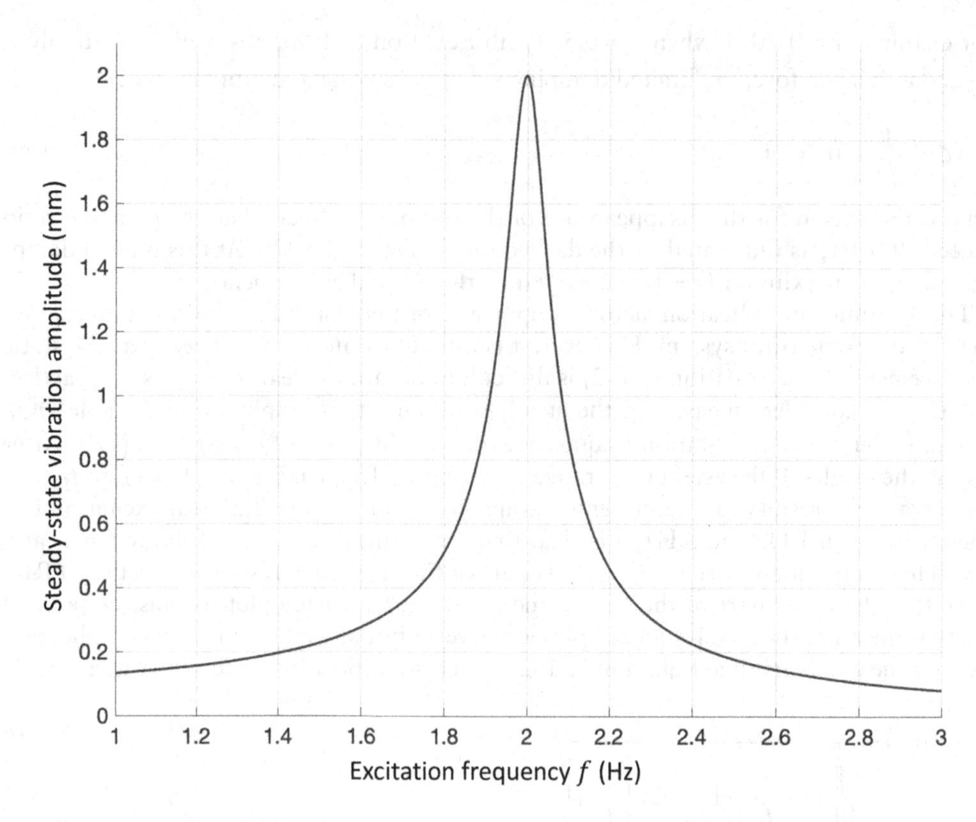

Figure 2.20 An illustrative example of a frequency response function (FRF) for a single-DOF system under harmonic excitation.

The procedure of the half-power bandwidth method is as follows (see Figure 2.21):

1. Identify the peak of the amplitude vs. frequency plot and measure the peak amplitude, A_{peak}.
2. Calculate the half-power amplitude $A_{\text{peak}} / \sqrt{2}$.
3. Mark the half-power points (i.e., the two points on the curve with the half-power amplitude), and get the corresponding frequencies, f_1 and f_2, where $f_1 < f_2$.
4. Approximate the damping ratio as $(f_2 - f_1)/(f_2 + f_1)$.

The verification is given below. The half-power amplitude is:

$$\frac{A_{\text{peak}}}{\sqrt{2}} = \frac{x_s}{2\sqrt{2}\zeta} \tag{2.113}$$

The frequencies corresponding to the half-power points can be obtained through Equation (2.110) as:

$$\frac{A_{\text{peak}}}{\sqrt{2}} = \frac{x_s}{2\sqrt{2}\zeta} = \frac{x_s}{\sqrt{\left(1-r^2\right)^2 + (2\zeta r)^2}} \tag{2.114}$$

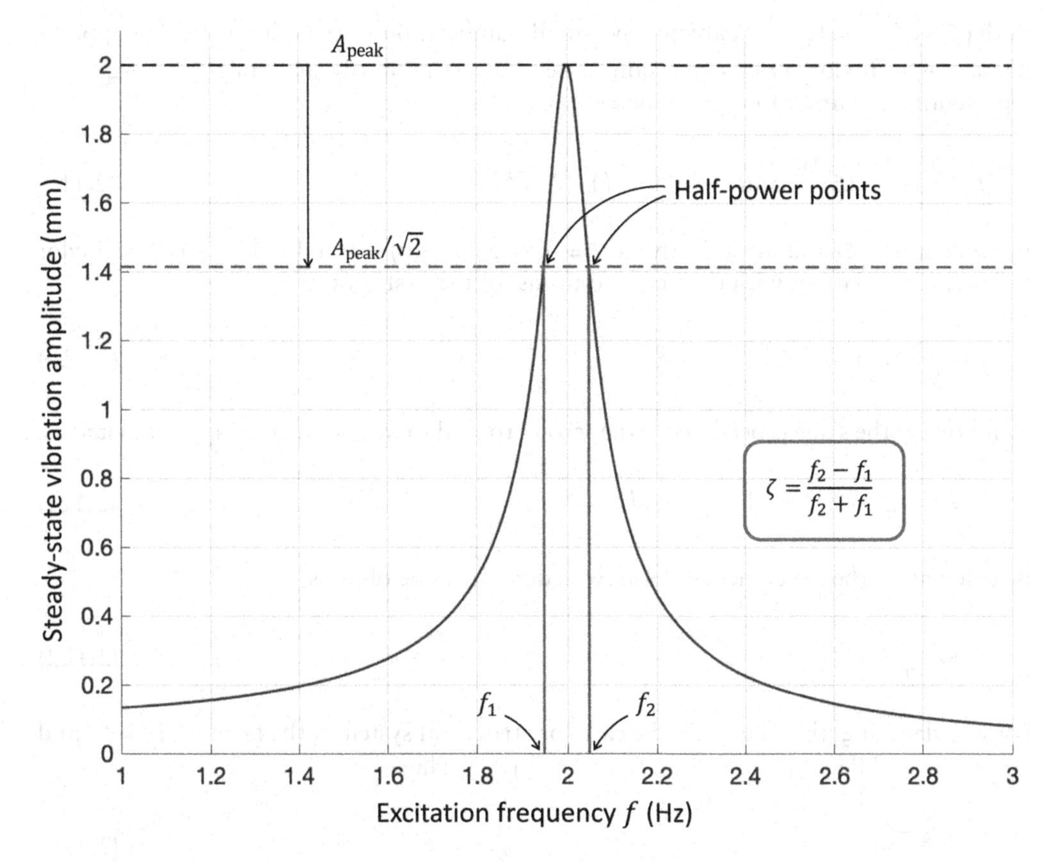

Figure 2.21 The estimation of damping ratio by the half-power bandwidth method.

Rearrange to get:

$$\sqrt{\left(1-r^2\right)^2+\left(2\zeta r\right)^2}=2\sqrt{2}\zeta \Rightarrow \left(1-r^2\right)^2+\left(2\zeta r\right)^2=8\zeta^2 \tag{2.115}$$

One can obtain a quadratic equation of r^2 as:

$$\left(r^2\right)^2+\left(4\zeta^2-2\right)r^2+1-8\zeta^2=0 \tag{2.116}$$

The two roots are:

$$r^2=1-2\zeta^2\pm 2\zeta\sqrt{1+\zeta^2} \tag{2.117}$$

As ζ^2 is very small, $\sqrt{1+\zeta^2}$ is very close to unity. The two roots (i.e., $r_1=f_1/f_n$ and $r_2=f_2/f_n$) can be approximated as:

$$r_1\approx\sqrt{1-2\left(\zeta+\zeta^2\right)} \quad \text{and} \quad r_2\approx\sqrt{1+2\left(\zeta-\zeta^2\right)} \tag{2.118}$$

Both $(\zeta + \zeta^2)$ and $(\zeta - \zeta^2)$ are positive small numbers. Both roots can be further approximated by a Taylor series expansion at zero. By temporarily assigning $\xi = \zeta + \zeta^2$ and representing the first root by $f(\xi)$, one obtains:

$$f(\xi) = (1 - 2\xi)^{\frac{1}{2}} \quad \text{and} \quad f'(\xi) = -(1 - 2\xi)^{-\frac{1}{2}} \tag{2.119}$$

By evaluating $f(\xi)$ and its derivative at $\xi = 0$, one gets $f(0) = 1$ and $f'(0) = -1$. The Taylor series approximation (with the first two terms) of the first root is:

$$f(\xi) = f(0) + f'^{(0)}\xi + \ldots \approx 1 - \xi \tag{2.120}$$

By following the same procedure for the second root, the two roots can be approximated as:

$$r_1 \approx 1 - (\zeta + \zeta^2) \quad \text{and} \quad r_2 \approx 1 + (\zeta - \zeta^2) \tag{2.121}$$

By calculating the difference of these two equations, one obtains:

$$\zeta \approx \frac{r_2 - r_1}{2} = \frac{f_2 - f_1}{2f_n} \tag{2.122}$$

For low damping (this is usually the case for structural systems), the peak is at $f = f_n$, and it is at the middle of f_1 and f_2 (i.e., $f_n = (f_1 + f_2)/2$). Thus:

$$\zeta \approx \frac{f_2 - f_1}{f_2 + f_1} \tag{2.123}$$

The half-power bandwidth method provides a convenient way to estimate the damping ratio based on the amplitude vs. frequency plot of a system.

Example 2.1

From a field test, the vertical vibration of a structural system is measured under the action of an actuator with different excitation frequencies. The resultant amplitude vs. excitation frequency plot is shown in Figure 2.22. Estimate the damping ratio of the structural system by the half-power bandwidth method.

SOLUTION

From Figure 2.23, the peak amplitude is 1.4 mm, and the half-power amplitude is $0.9899 = 1.4/\sqrt{2}$. The corresponding f_1 and f_2 are equal to 2.84 Hz and 3.14 Hz, respectively. The estimated damping ratio is:

$$\zeta \approx \frac{f_2 - f_1}{f_2 + f_1} = \frac{3.14 - 2.84}{3.14 + 2.84} = 0.0506 \tag{2.124}$$

Thus, the damping ratio of this structural system is about 5%.

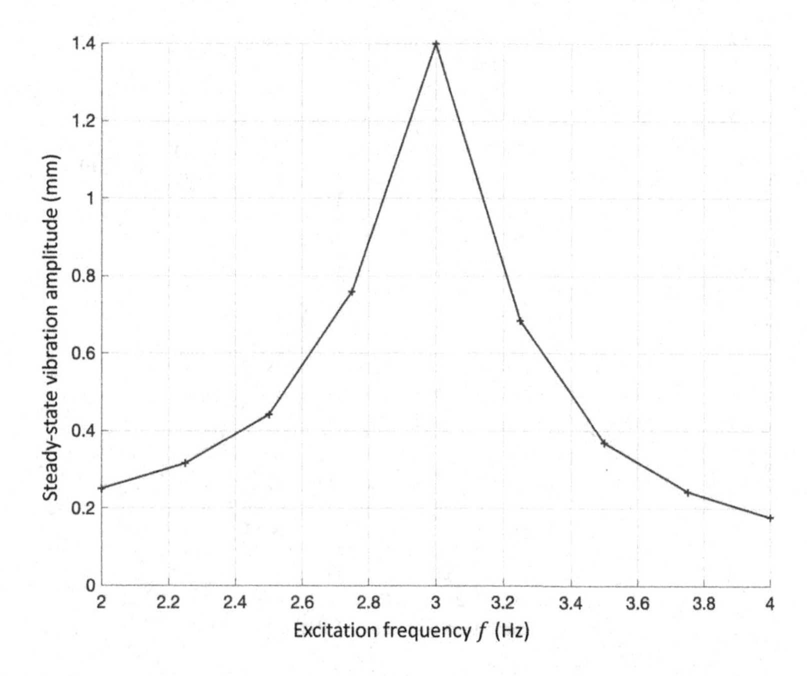

Figure 2.22 The measured vibration amplitude vs. excitation frequency plot in Example 2.1.

2.1.3.3 *Resonance*

To observe the effect of excitation frequency (with respect to the natural frequency) on the vibration response of a system, the solution in Equation (2.104) is simplified by reducing the damping to zero (i.e., it becomes an undamped system). The solution (of the undamped system) is expressed in terms of the frequency ratio r as:

$$x(t) = B_1 \cos(\omega_n t) + B_2 \sin(\omega_n t) + \frac{x_s}{1-r^2} e^{ir\omega_n t} \tag{2.125}$$

With the first initial condition, $x(0) = x_0$:

$$x(0) = B_1 \cos(0) + B_2 \sin(0) + \frac{x_s}{1-r^2} e^0 = B_1 + \frac{x_s}{1-r^2} = x_0$$

Thus,

$$\Rightarrow B_1 = x_0 - \frac{x_s}{1-r^2} \tag{2.126}$$

The first derivative of the solution is:

$$\dot{x}(t) = -\omega_n B_1 \sin(\omega_n t) + \omega_n B_2 \cos(\omega_n t) + \frac{ir\omega_n x_s}{1-r^2} e^{ir\omega_n t} \tag{2.127}$$

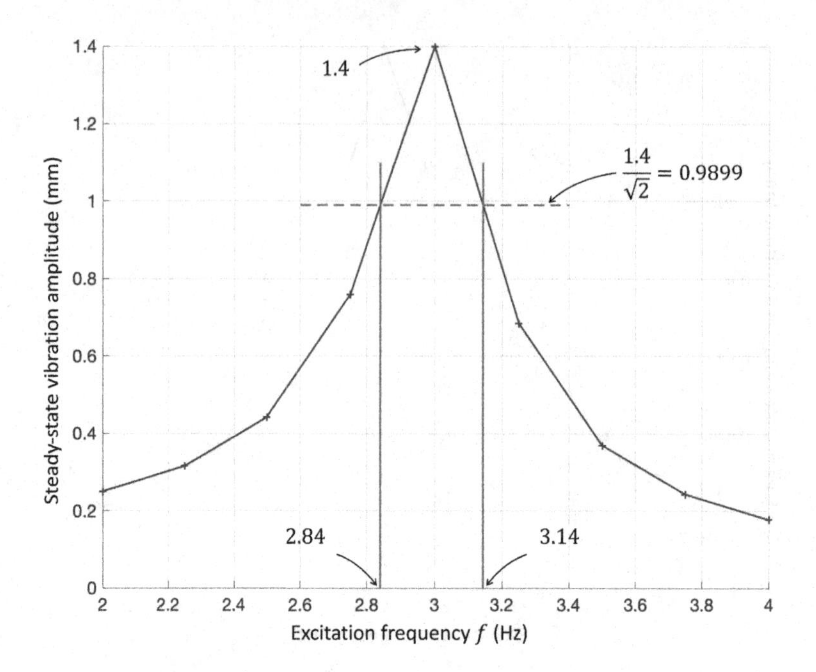

Figure 2.23 The estimation of damping ratio in Example 2.1.

With the second initial condition, $\dot{x}(0) = \dot{x}_0$:

$$\dot{x}(0) = \omega_n B_2 + \frac{ir\omega_n x_s}{1-r^2} = \dot{x}_0 \tag{2.128}$$

The second unknown constant is:

$$B_2 = \frac{1}{\omega_n}\left(\dot{x}_0 - \frac{ir\omega_n x_s}{1-r^2}\right) = \frac{\dot{x}_0}{\omega_n} - \frac{irx_s}{1-r^2} \tag{2.129}$$

The solution becomes:

$$x(t) = \left(x_0 - \frac{x_s}{1-r^2}\right)\cos(\omega_n t) + \left(\frac{\dot{x}_0}{\omega_n} - \frac{irx_s}{1-r^2}\right)\sin(\omega_n t) + \frac{x_s}{1-r^2}e^{ir\omega_n t} \tag{2.130}$$

When the excitation frequency is equal to the natural frequency, the frequency ratio $r = 1$. However, Equation (2.130) cannot be directly used, as some terms are divided by zero. To overcome this difficulty, L'Hôpital's rule is used. In this case, if $\lim_{r\to1} f(r) = \lim_{r\to1} g(r) = 0$ and both $f(r)$ and $g(r)$ are differentiable, then:

$$\lim_{r\to1}\frac{f(r)}{g(r)} = \lim_{r\to1}\frac{\dot{f}(r)}{\dot{g}(r)} \tag{2.131}$$

To use L'Hôpital's rule, Equation (2.130) is rearranged to:

$$x(t) = x_0 \cos(\omega_n t) + \frac{\dot{x}_0}{\omega_n} \sin(\omega_n t) + x_s \frac{e^{ir\omega_n t} - \cos(\omega_n t) - ir\sin(\omega_n t)}{1 - r^2} \tag{2.132}$$

where $f(r)$ and $g(r)$ are defined as:

$$f(r) = e^{ir\omega_n t} - \cos(\omega_n t) - ir\sin(\omega_n t) \quad \text{and} \quad g(r) = 1 - r^2 \tag{2.133}$$

It is clear that $\lim_{r \to 1} g(r) = 1 - 1^2 = 0$. For $f(r)$,

$$\lim_{r \to 1} f(r) = e^{i\omega_n t} - \cos(\omega_n t) - i\sin(\omega_n t) = e^{i\omega_n t} - e^{i\omega_n t} = 0 \tag{2.134}$$

The derivative of $f(r)$ with respect to r is:

$$\dot{f}(r) = i\omega_n t e^{ir\omega_n t} - i\sin(\omega_n t) \tag{2.135}$$

The derivative of $g(r)$ with respect to r is:

$$\dot{g}(r) = -2r \tag{2.136}$$

Therefore:

$$\begin{aligned} \lim_{r \to 1} \frac{\dot{f}(r)}{\dot{g}(r)} &= \lim_{r \to 1} \frac{i\omega_n t e^{ir\omega_n t} - i\sin(\omega_n t)}{-2r} = \frac{-i\omega_n t e^{i\omega_n t} + i\sin(\omega_n t)}{2} \\ &= \frac{-i\omega_n t(\cos\omega_n t + i\sin\omega_n t) + i\sin(\omega_n t)}{2} \\ &= \frac{\omega_n t}{2}\sin\omega_n t + i\frac{\sin\omega_n t - \omega_n t\cos\omega_n t}{2} \end{aligned} \tag{2.137}$$

As a result, when the excitation frequency is equal to the natural frequency, the response of the undamped single-DOF system is:

$$x(t) = x_0 \cos(\omega_n t) + \frac{\dot{x}_0}{\omega_n} \sin(\omega_n t) + x_s \left(\frac{\omega_n t}{2}\sin\omega_n t + i\frac{\sin\omega_n t - \omega_n t\cos\omega_n t}{2} \right) \tag{2.138}$$

When the excitation is $f_0 \cos\omega_n t$ (i.e., the real part of the current excitation), the solution becomes:

$$x(t) = x_0 \cos(\omega_n t) + \frac{\dot{x}_0}{\omega_n} \sin(\omega_n t) + \frac{x_s \omega_n t}{2} \sin\omega_n t \tag{2.139}$$

Similarly, when the excitation is $f_0 \sin\omega_n t$ (i.e., the imaginary part of the current excitation), the solution becomes:

$$x(t) = x_0 \cos(\omega_n t) + \frac{\dot{x}_0}{\omega_n} \sin(\omega_n t) + x_s \frac{\sin\omega_n t - \omega_n t\cos\omega_n t}{2} \tag{2.140}$$

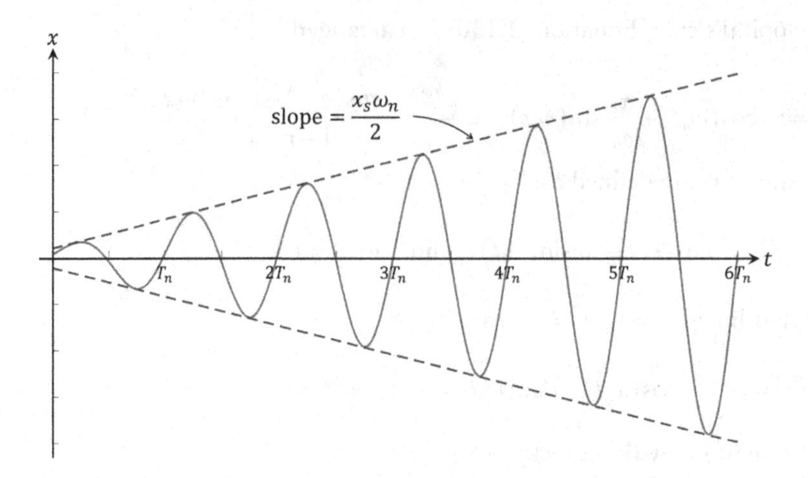

Figure 2.24 Forced vibration of an undamped single-DOF system at resonance.

The response of an undamped single-DOF system under the action of $f_0 \cos \omega_n t$ when $r = 1$ (i.e., resonance) is calculated by Equation (2.139) and plotted in Figure 2.24. It is very clear from the figure that the response is dominated by the last term in Equation (2.139). The amplitude of vibration increases linearly with time t with a slope $x_s \omega_n / 2$. As the force continuously pumps energy into the single-DOF system and there is no damping to dissipate the energy, the response of the system continuously increases. The curve is not started at zero (at $t = 0$) due to the initial conditions. However, the effect of the initial conditions (i.e., the first two terms in Equation (2.139)) is constant (independent of time). As the amplitude of vibration due to the excitation (i.e., the third term in Equation (2.139)) will increase with time, the effect of initial conditions will become relatively small that it can be neglected after a few cycles of vibration (as shown in Figure 2.24). Under a real situation, the system may become non-linear when the amplitude of vibration achieves a certain limited value (e.g., the yield strain). The non-linear effect will change the stiffness of the system, and therefore will alter the natural frequency (recall $\omega_n = \sqrt{k / m}$). This may stop the amplitude of vibration from continuously increasing as the frequency ratio r is not equal to unity. The best way to control the vibration of a structural system is to increase the damping (i.e., by installing dampers).

2.1.4 Forced vibration: general force

2.1.4.1 Response to unit impulse

Contrary to harmonic excitation, the forced vibration of an under-damped single-DOF system is studied under general excitation in the following. The response to the excitation in the basic form—a unit impulse, is first considered. Responses to other excitations in arbitrary forms can be extended using the unit impulse response. A unit impulse is an excitation that lasts an infinitely small period of time, ϵ; i.e., $\epsilon \to 0$, and the integration of a unit impulse, with respect to time, equals unity. Figure 2.25 shows a unit impulse at τ. The Dirac delta function is used to denote the unit impulse:

$$\delta(t-\tau) = \begin{cases} +\infty, t = \tau \\ 0, t \neq \tau \end{cases} \quad \text{and} \quad \int_{-\infty}^{\infty} \delta(t-\tau) dt = 1 \tag{2.141}$$

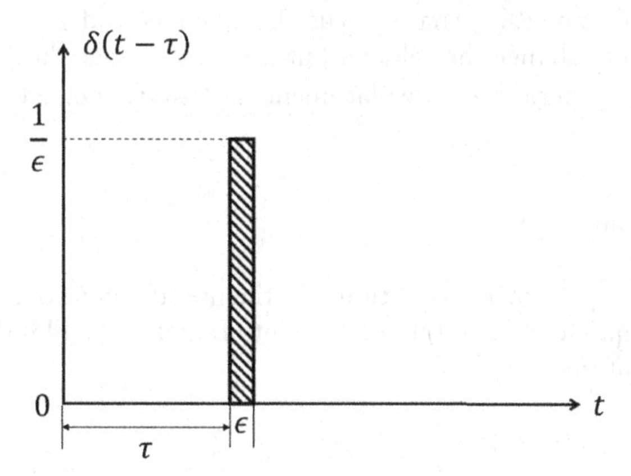

Figure 2.25 The definition of a unit impulse.

Equation (2.141) and Figure 2.25 indicate that the unit impulse only takes values around the neighborhood of τ, and is zero elsewhere. It also indicates that the unit impulse reaches a very large amplitude in a short time, with the constraint that its integration is unity, i.e., the shaded area under the unit impulse is unity. That is, when ϵ approaches zero (i.e., $\epsilon \to 0$), the amplitude approaches infinity (i.e., $1/\epsilon \to \infty$).

Two steps are considered for deriving the response to the unit impulse $\delta(t-\tau)$. The first step is to study the response during the infinitely small period of time $\epsilon \to 0$. The second step is to study the response after the time instance τ. The equation of motion can be obtained by modifying the ODE of a damped single-DOF system under external force in Equation (2.5) by replacing the forcing function $f(t)$ with the unit impulse applied at $\tau = 0$ (i.e., $\delta(t)$).

$$m\ddot{x} + c\dot{x} + kx = \delta(t), \quad \text{for} \quad x(0) = x_0 \quad \text{and} \quad \dot{x}(0) = \dot{x}_0 \tag{2.142}$$

Without loss of generality, the system is assumed to start from rest, and therefore, $x_0 = \dot{x}_0 = 0$. Integrating both sides of the ODE in Equation (2.142) over the duration of the unit impulse gives:

$$\int_0^\epsilon (m\ddot{x} + c\dot{x} + kx)\,dt = \int_0^\epsilon \delta(t)\,dt \tag{2.143}$$

where $\int_0^\epsilon \delta(t)\,dt = 1$, according to Equation (2.141). The three terms in the integral on the left-hand side of Equation (2.143) are evaluated by taking the limit when $\epsilon \to 0$. For the first term,

$$\lim_{\epsilon \to 0} \int_0^\epsilon m\ddot{x}(t)\,dt = \lim_{\epsilon \to 0} m\dot{x}(t)\Big|_0^\epsilon = \lim_{\epsilon \to 0} m\left[\dot{x}(\epsilon) - \dot{x}(0)\right] = m\dot{x}(0^+) \tag{2.144}$$

where the initial condition, $\dot{x}(0) = 0$, has been used; 0^+ is used to denote the moment right after the impulse, considering that $\epsilon \to 0$. For the second term,

$$\lim_{\epsilon \to 0} \int_0^\epsilon c\dot{x}(t)\,dt = \lim_{\epsilon \to 0} cx(t)\Big|_0^\epsilon = \lim_{\epsilon \to 0} c\left(x(\epsilon) - x(0)\right) = 0 \tag{2.145}$$

where the initial condition $x(0) = 0$ has been used; and it is considered that the impulse only provides momentum to change the velocity but does not change the mass location; so, in an infinitely small time period, the displacement is constant, i.e., $x(t) = x(\epsilon) = x(0) = 0$. For the third term,

$$\lim_{\epsilon \to 0} \int_0^{\epsilon} kx(t)dt = \lim_{\epsilon \to 0} kx(0)t\Big|_0^{\epsilon} = 0 \tag{2.146}$$

where the constant displacement condition and the initial condition, $x(0) = 0$, have been used. Substituting Equations (2.144) to (2.146) into Equation (2.143) then gives the velocity right after the unit impulse:

$$\dot{x}\left(0^+\right) = \frac{1}{m} \tag{2.147}$$

Therefore, the response for the second step is simply a free-decay response due to the initial velocity $\dot{x}\left(0^+\right)$ and the zero displacement. Using the solution of the free vibration of an under-damped single-DOF system (i.e., Equation (2.41)), the response to a unit impulse can be expressed as:

$$h(t) = \frac{1}{m\omega_d} e^{-\zeta\omega_n t} \sin\left(\omega_d t\right) \tag{2.148}$$

The response to a unit impulse will be extended in the following section to calculate the response to a general excitation by Duhamel's integral.

2.1.4.2 *Duhamel's integral*

In practice, a general excitation can be discretized into strips with equal time step Δt. When Δt is small, the excitation within one time step can be considered as a constant. Thus, a general excitation can be approximated by a sequence of strips (or rectangles) as shown in Figure 2.26. Comparing each rectangle in Figure 2.26 with the shaded rectangle in Figure 2.25, one can treat each rectangle in Figure 2.26 as an impulse, except that its area is not unity. For example, the force at t_i is an impulse with magnitude

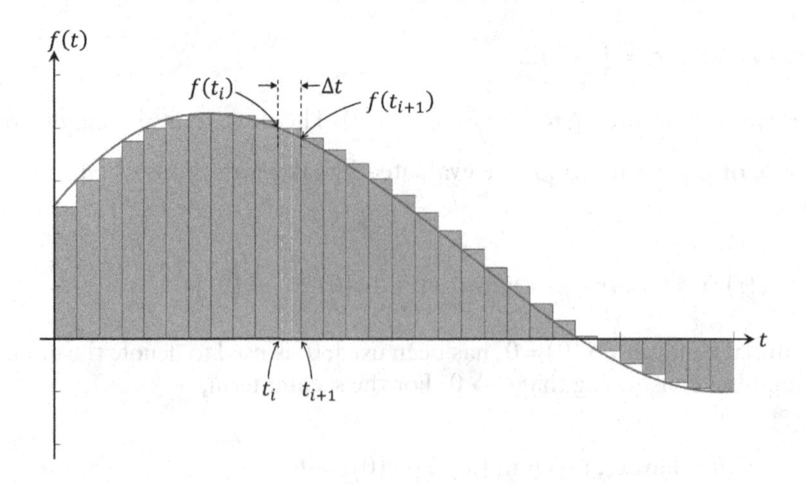

Figure 2.26 Representation of a general excitation based on impulses.

$f(t_i)\Delta t$ (i.e., the area of the rectangle). For a linear system, the response to this impulse alone can be obtained by the unit-impulse response, Equation (2.148), with a scaling factor $f(t_i)\Delta t$ as:

$$\Delta x(t|t_i) = f(t_i)\Delta t\, h(t-t_i) = \frac{f(t_i)}{m\omega_d} e^{-\zeta\omega_n(t-t_i)} \sin(\omega_d(t-t_i))\Delta t, \quad \text{for} \quad t > t_i \tag{2.149}$$

where $\Delta x(t|t_i)$ represents the response at t due to the impulse at t_i, and only the time after t_i is considered, as the impulse is not available before t_i. The total response to a general excitation can then be approximated by superposing the responses to the sequence of impulses represented by the rectangles in Figure 2.26:

$$\sum_{t>t_i} \Delta x(t|t_i) = \sum_{t>t_i} f(t_i) h(t-t_i)\Delta t \tag{2.150}$$

Assuming that the force is started at $t = 0$ and taking the limit as $\Delta t \to 0$ gives:

$$\begin{aligned} x(t) &= \lim_{\Delta t \to 0} \sum_{t>t_i} \Delta x(t|t_i) = \int_0^t f(\tau) h(t-\tau) d\tau \\ &= \int_0^t \frac{f(\tau)}{m\omega_d} e^{-\zeta\omega_n(t-\tau)} \sin(\omega_d(t-\tau)) d\tau \end{aligned} \tag{2.151}$$

Equation (2.151) is known as a Duhamel's integral, or convolution integral. Solving Duhamel's integral under different forcing function $f(\tau)$ is extremely important in the field of structural dynamics.

2.1.4.3 Numerical method for forced vibration

In the following, the main purpose is to formulate the numerical method for calculating the time domain response $x(t)$ of an under-damped single-DOF system by solving Duhamel's integral under a general forcing function. Without loss of generality, the acceleration excitation is considered in the formulation (preparing for the formulation in the modal superposition method of calculating the responses of a multi-DOF system) as:

$$x(t) = \frac{1}{\omega_d} \int_0^t a(\tau) e^{-\zeta\omega_n(t-\tau)} \sin(\omega_d(t-\tau)) d\tau \tag{2.152}$$

where $a(\tau) = f(\tau)/m$ is the acceleration excitation as a function of time with a unit of m/s^2 or g (i.e., gravitational acceleration).

The idea is:

(1) To divide the excitation duration of interest from t_a to t_b into n sub-intervals, where $\Delta t = (t_b - t_a)/n$.
(2) To solve Duhamel's integral within a general time interval from t_i to t_{i+1}, where $i = 0,1,\ldots,n$, with $t_a = t_0$ and $t_b = t_n$, so as to express the response (i.e., the displacement, velocity, and acceleration) at t_{i+1} in terms of the response at t_i. That is to formulate the recursive equations.

(3) To calculate the responses at all time steps starting from the given initial conditions (i.e., $x(0)$ and $\dot{x}(0)$) together with the forcing function at discrete time (i.e., the acceleration excitation in the current formulation).

Here, the most difficult part is in step (2). The response at t_{i+1} consists of two parts:

$$x_{i+1} = x_{a,i\to i+1} + x_{f,i\to i+1} \tag{2.153}$$

where $x_{a,i\to i+1}$ represents the response induced by the acceleration excitation. The subscript "$i \to i+1$" puts emphasis on the fact that the response at t_{i+1} is calculated based on response at t_i. $x_{f,i\to i+1}$ and represents the free-decay response at t_{i+1} due to the initial conditions at t_i.

First, the determination of $x_{a,i\to i+1}$ is addressed by solving Duhamel's integral within a given time step. To make the analytical solution of Duhamel's integral available, the forcing function within a given time step (i.e., from t_i to t_{i+1}) is approximated by a linear function (or linear interpolation). Figure 2.27 shows the linear approximation of the acceleration excitation at τ from t_i. That is:

$$a(\tau) = a_i + \frac{a_{i+1} - a_i}{\Delta t}\tau \tag{2.154}$$

Within this time step, Duhamel's integral can be expressed as:

$$\begin{aligned} x_{a,i\to i+1} &= \int_0^{\Delta t} f(\tau)h(\Delta t - \tau)d\tau \\ &= \frac{1}{\omega_d}\int_0^{\Delta t} a(\tau)e^{-\zeta\omega_n(\Delta t-\tau)}\sin\left(\omega_d(\Delta t - \tau)\right)d\tau \\ &= \frac{1}{\omega_d}\int_0^{\Delta t}\left(a_i + \frac{a_{i+1} - a_i}{\Delta t}\tau\right)e^{-\zeta\omega_n(\Delta t-\tau)}\sin\left(\omega_d(\Delta t - \tau)\right)d\tau \end{aligned} \tag{2.155}$$

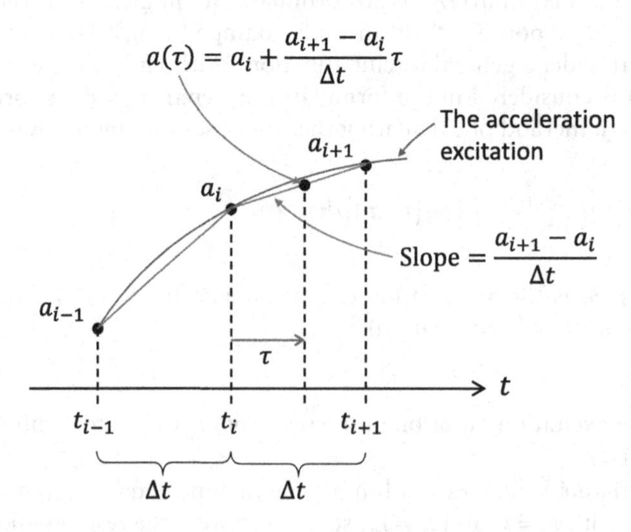

Figure 2.27 Linear approximation of the acceleration excitation in a given time step.

The integral in Equation (2.155) can be expanded to two integrals as:

$$x_{a,i\to i+1} = \frac{a_i}{\omega_d}\int_0^{\Delta t} e^{-\zeta\omega_n(\Delta t-\tau)}\sin\left(\omega_d\left(\Delta t-\tau\right)\right)d\tau + \frac{a_{i+1}-a_i}{\omega_d\Delta t}\int_0^{\Delta t}\tau e^{-\zeta\omega_n(\Delta t-\tau)}\sin\left(\omega_d\left(\Delta t-\tau\right)\right)d\tau \tag{2.156}$$

As $\sin(A-B)=\sin A\cos B-\cos A\sin B$, the two integrals in Equation (2.156) can be further expanded into four integrals as:

$$\begin{aligned} x_{a,i\to i+1} &= \frac{a_i}{\omega_d}e^{-\zeta\omega_n\Delta t}\sin(\omega_d\Delta t)\int_0^{\Delta t}e^{\zeta\omega_n\tau}\cos(\omega_d\tau)d\tau \\ &-\frac{a_i}{\omega_d}e^{-\zeta\omega_n\Delta t}\cos(\omega_d\Delta t)\int_0^{\Delta t}e^{\zeta\omega_n\tau}\sin(\omega_d\tau)d\tau \\ &+\frac{a_{i+1}-a_i}{\omega_d\Delta t}e^{-\zeta\omega_n\Delta t}\sin(\omega_d\Delta t)\int_0^{\Delta t}\tau e^{\zeta\omega_n\tau}\cos(\omega_d\tau)d\tau \\ &-\frac{a_{i+1}-a_i}{\omega_d\Delta t}e^{-\zeta\omega_n\Delta t}\cos(\omega_d\Delta t)\int_0^{\Delta t}\tau e^{\zeta\omega_n\tau}\sin(\omega_d\tau)d\tau \end{aligned} \tag{2.157}$$

The four terms in Equation (2.157) are represented by I_1, I_2, I_3, and I_4, respectively, as:

$$x_{a,i\to i+1} = I_1+I_2+I_3+I_4 \tag{2.158}$$

The four integrals in I_1 to I_4 are solved one by one in the following. The integral in I_1 can be solved by carrying out integration by parts twice. Integration by parts states that:

$$\int u\frac{dv}{d\tau}d\tau = uv-\int v\frac{du}{d\tau}d\tau \tag{2.159}$$

Firstly, consider:

$$u=\cos(\omega_d\tau)\Rightarrow\frac{du}{d\tau}=-\omega_d\sin(\omega_d\tau) \tag{2.160}$$

and:

$$\frac{dv}{d\tau}=e^{\zeta\omega_n\tau}\Rightarrow v=\frac{e^{\zeta\omega_n\tau}}{\zeta\omega_n} \tag{2.161}$$

Thus:

$$\int e^{\zeta\omega_n\tau}\cos(\omega_d\tau)d\tau=\frac{e^{\zeta\omega_n\tau}}{\zeta\omega_n}\cos(\omega_d\tau)+\int\frac{e^{\zeta\omega_n\tau}}{\zeta\omega_n}\omega_d\sin(\omega_d\tau)d\tau \tag{2.162}$$

Use integration by parts again and consider:

$$u=\omega_d\sin(\omega_d\tau)\Rightarrow\frac{du}{d\tau}=\omega_d^2\cos(\omega_d\tau) \tag{2.163}$$

and:

$$\frac{dv}{d\tau}=\frac{e^{\zeta\omega_n\tau}}{\zeta\omega_n}\Rightarrow v=\frac{e^{\zeta\omega_n\tau}}{(\zeta\omega_n)^2} \tag{2.164}$$

Equation (2.162) becomes:

$$\int e^{\zeta\omega_n\tau}\cos(\omega_d\tau)d\tau \\ =\frac{e^{\zeta\omega_n\tau}}{\zeta\omega_n}\cos(\omega_d\tau)+\frac{e^{\zeta\omega_n\tau}}{(\zeta\omega_n)^2}\omega_d\sin(\omega_d\tau)-\frac{\omega_d^2}{(\zeta\omega_n)^2}\int e^{\zeta\omega_n\tau}\cos(\omega_d\tau)d\tau \tag{2.165}$$

Note that the integrals on the left-hand side and the right-hand side of Equation (2.165) are the same, Thus, Equation (2.165) can be rearranged to:

$$\int e^{\zeta\omega_n\tau}\cos(\omega_d\tau)d\tau=\frac{e^{\zeta\omega_n\tau}}{(\zeta\omega_n)^2+\omega_d^2}\left[\zeta\omega_n\cos(\omega_d\tau)+\omega_d\sin(\omega_d\tau)\right]+c \tag{2.166}$$

where c is an integration constant. The integral in I_1 of Equation (2.157) can be expressed as:

$$\int_0^{\Delta t} e^{\zeta\omega_n\tau}\cos(\omega_d\tau)d\tau \\ =\frac{e^{\zeta\omega_n\Delta t}}{(\zeta\omega_n)^2+\omega_d^2}\left[\zeta\omega_n\cos(\omega_d\Delta t)+\omega_d\sin(\omega_d\Delta t)\right]-\frac{\zeta\omega_n}{(\zeta\omega_n)^2+\omega_d^2} \tag{2.167}$$

From Equation (2.33), $\omega_d=\sqrt{1-\zeta^2}\,\omega_n$. Therefore,

$$(\zeta\omega_n)^2+\omega_d^2=(\zeta\omega_n)^2+\omega_n^2(1-\zeta^2)=\omega_n^2 \tag{2.168}$$

Thus, Equation (2.167) can be simplified as:

$$\int_0^{\Delta t} e^{\zeta\omega_n\tau}\cos(\omega_d\tau)d\tau=e^{\zeta\omega_n\Delta t}\left[\frac{\zeta}{\omega_n}\cos(\omega_d\Delta t)+\frac{\sqrt{1-\zeta^2}}{\omega_n}\sin(\omega_d\Delta t)\right]-\frac{\zeta}{\omega_n} \tag{2.169}$$

The integral in I_2 of Equation (2.157) can be calculated in a similar manner (i.e., carrying out integration by parts twice) as:

$$\int_0^{\Delta t} e^{\zeta\omega_n\tau}\sin(\omega_d\tau)d\tau=e^{\zeta\omega_n\Delta t}\left[\frac{\zeta}{\omega_n}\sin(\omega_d\Delta t)-\frac{\sqrt{1-\zeta^2}}{\omega_n}\cos(\omega_d\Delta t)\right]+\frac{\sqrt{1-\zeta^2}}{\omega_n} \tag{2.170}$$

The integral in I_3 of Equation (2.157) is more complicated. But it can still be solved using integration by parts as:

$$\int \tau e^{\zeta\omega_n\tau}\cos(\omega_d\tau)d\tau$$

$$= \frac{e^{\zeta\omega_n\tau}}{\left((\zeta\omega_n)^2+\omega_d^2\right)^2}\left\{\left[\left((\zeta\omega_n)^2+\omega_d^2\right)\tau-2\zeta\omega_n\right]\omega_d\sin(\omega_d\tau)\right. \tag{2.171}$$

$$\left.+\left[\left((\zeta\omega_n)^2+\omega_d^2\right)\zeta\omega_n\tau+\omega_d^2-(\zeta\omega_n)^2\right]\cos(\omega_d\tau)\right\}$$

The corresponding definite integral is:

$$\int_0^{\Delta t}\tau e^{\zeta\omega_n\tau}\cos(\omega_d\tau)d\tau$$

$$= \frac{e^{\zeta\omega_n\Delta t}}{\left((\zeta\omega_n)^2+\omega_d^2\right)^2}\left\{\left[\left((\zeta\omega_n)^2+\omega_d^2\right)\Delta t-2\zeta\omega_n\right]\omega_d\sin(\omega_d\Delta t)\right. \tag{2.172}$$

$$\left.+\left[\left((\zeta\omega_n)^2+\omega_d^2\right)\zeta\omega_n\Delta t+\omega_d^2-(\zeta\omega_n)^2\right]\cos(\omega_d\Delta t)\right\}-\frac{\omega_d^2-(\zeta\omega_n)^2}{\left((\zeta\omega_n)^2+\omega_d^2\right)^2}$$

It can be simplified by Equations (2.33) and (2.168) as:

$$\int_0^{\Delta t}\tau e^{\zeta\omega_n\tau}\cos(\omega_d\tau)d\tau$$

$$= e^{\zeta\omega_n\Delta t}\left[\left(\frac{\sqrt{1-\zeta^2}}{\omega_n}\Delta t-2\frac{\zeta\sqrt{1-\zeta^2}}{\omega_n^2}\right)\sin(\omega_d\Delta t)\right. \tag{2.173}$$

$$\left.+\left(\frac{\zeta}{\omega_n}\Delta t+\frac{1-2\zeta^2}{\omega_n^2}\right)\cos(\omega_d\Delta t)\right]-\frac{1-2\zeta^2}{\omega_n^2}$$

Similarly, the integral in I_4 of Equation (2.157) can be solved using integration by parts as:

$$\int_0^{\Delta t}\tau e^{\zeta\omega_n\tau}\sin(\omega_d\tau)d\tau$$

$$= \frac{e^{\zeta\omega_n\Delta t}}{\left((\zeta\omega_n)^2+\omega_d^2\right)^2}\left\{\left[\left((\zeta\omega_n)^2+\omega_d^2\right)\zeta\omega_n\Delta t+\omega_d^2-(\zeta\omega_n)^2\right]\sin(\omega_d\Delta t)\right. \tag{2.174}$$

$$\left.-\left[\left((\zeta\omega_n)^2+\omega_d^2\right)\Delta t-2\zeta\omega_n\right]\omega_d\cos(\omega_d\Delta t)\right\}-\frac{2\zeta\omega_n\omega_d}{\left((\zeta\omega_n)^2+\omega_d^2\right)^2}$$

It can be simplified by Equations (2.33) and (2.168) as:

$$\int_0^{\Delta t} \tau e^{\zeta\omega_n\tau} \sin(\omega_d\tau)d\tau$$

$$= e^{\zeta\omega_n\Delta t}\left[\left(\frac{\zeta}{\omega_n}\Delta t + \frac{1-2\zeta^2}{\omega_n^2}\right)\sin(\omega_d\Delta t) - \left(\frac{\sqrt{1-\zeta^2}}{\omega_n}\Delta t - 2\frac{\zeta\sqrt{1-\zeta^2}}{\omega_n^2}\right)\cos(\omega_d\Delta t)\right] - 2\frac{\zeta\sqrt{1-\zeta^2}}{\omega_n^2} \tag{2.175}$$

The four integrals in Equation (2.157) are all solved. By Equation (2.169), one obtains:

$$I_1 = a_i\left(\frac{\zeta}{\omega_n\omega_d}\cos(\omega_d\Delta t)\sin(\omega_d\Delta t) + \frac{\sqrt{1-\zeta^2}}{\omega_n\omega_d}\sin^2(\omega_d\Delta t) - \frac{\zeta e^{-\zeta\omega_n\Delta t}}{\omega_n\omega_d}\sin(\omega_d\Delta t)\right) \tag{2.176}$$

By Equation (2.170), one obtains:

$$I_2 = -a_i\left(\frac{\zeta}{\omega_n\omega_d}\sin(\omega_d\Delta t)\cos(\omega_d\Delta t) - \frac{\sqrt{1-\zeta^2}}{\omega_n\omega_d}\cos^2(\omega_d\Delta t) + \frac{\sqrt{1-\zeta^2}e^{-\zeta\omega_n\Delta t}\cos(\omega_d\Delta t)}{\omega_n\omega_d}\right) \tag{2.177}$$

By Equation (2.173), one obtains:

$$I_3 = (a_{i+1} - a_i)\left(\left(\frac{\sqrt{1-\zeta^2}}{\omega_n}\Delta t - 2\frac{\zeta\sqrt{1-\zeta^2}}{\omega_n^2}\right)\frac{\sin^2(\omega_d\Delta t)}{\omega_d\Delta t} + \left(\frac{\zeta}{\omega_n}\Delta t + \frac{1-2\zeta^2}{\omega_n^2}\right)\frac{\cos(\omega_d\Delta t)\sin(\omega_d\Delta t)}{\omega_d\Delta t} - \left(\frac{1-2\zeta^2}{\omega_n^2}\right)\frac{e^{-\zeta\omega_n\Delta t}\sin(\omega_d\Delta t)}{\omega_d\Delta t}\right) \tag{2.178}$$

Finally, by Equation (2.175), one obtains the last term in Equation (2.157) as:

$$I_4 = -(a_{i+1} - a_i)\left(\left(\frac{\zeta}{\omega_n}\Delta t + \frac{1-2\zeta^2}{\omega_n^2}\right)\frac{\sin(\omega_d\Delta t)\cos(\omega_d\Delta t)}{\omega_d\Delta t} - \left(\frac{\sqrt{1-\zeta^2}}{\omega_n}\Delta t - 2\frac{\zeta\sqrt{1-\zeta^2}}{\omega_n^2}\right)\frac{\cos^2(\omega_d\Delta t)}{\omega_d\Delta t} - \left(2\frac{\zeta\sqrt{1-\zeta^2}}{\omega_n^2}\right)\frac{e^{-\zeta\omega_n\Delta t}\cos(\omega_d\Delta t)}{\omega_d\Delta t}\right) \tag{2.179}$$

Rearranging Equation (2.157) in terms of a_i and a_{i+1}, one obtains:

$$x_{a,i\to i+1} = Aa_i + Ba_{i+1} \tag{2.180}$$

where the factor A can be obtained by summing all the coefficients of a_i from I_1 to I_4 in Equations (2.176), (2.177), (2.178), and (2.179) and simplifying:

$$A = \left(\frac{1-2\zeta^2}{\omega_n\Delta t} - \zeta\right)\frac{e^{-\zeta\omega_n\Delta t}\sin(\omega_d\Delta t)}{\omega_n^2\sqrt{1-\zeta^2}} - \left(\frac{2\zeta}{\omega_n\Delta t}+1\right)\frac{e^{-\zeta\omega_n\Delta t}\cos(\omega_d\Delta t)}{\omega_n^2} + \frac{2\zeta}{\omega_n^3\Delta t} \tag{2.181}$$

Similarly, the factor B can be obtained by summing all the coefficients of a_{i+1} in I_2 and I_4 in Equations (2.178) and (2.179) and simplifying:

$$B = \left(\frac{2\zeta^2-1}{\omega_n^3\sqrt{1-\zeta^2}\Delta t}\right)e^{-\zeta\omega_n\Delta t}\sin(\omega_d\Delta t) + \left(\frac{2\zeta}{\omega_n^3\Delta t}\right)e^{-\zeta\omega_n\Delta t}\cos(\omega_d\Delta t) + \frac{1}{\omega_n^2} - \frac{2\zeta}{\omega_n^3\Delta t} \tag{2.182}$$

It is clear from Equation (2.180) that the response at t_{i+1} induced by the acceleration excitation depends on the excitation at both t_i and t_{i+1}. Next, the $x_{f,i\to i+1}$ (i.e., the free-decay response at t_{i+1} due to the initial conditions at t_i), is considered. From Equation (2.41), the free vibration response at t_{i+1} of an under-damped single-DOF system due to the initial displacement, x_i, and velocity, $\dot{x}_i$, at t_i can be expressed as:

$$x_{f,i\to i+1} = e^{-\zeta\omega_n\Delta t}\left[x_i\cos(\omega_d\Delta t) + \frac{\dot{x}_i + \zeta\omega_n x_i}{\omega_d}\sin(\omega_d\Delta t)\right] \tag{2.183}$$

Rearranging Equation (2.183) in terms of x_i and x_{i+1}, one obtains:

$$x_{f,i\to i+1} = Cx_i + D\dot{x}_i \tag{2.184}$$

where the factors C and D are given by:

$$C = e^{-\zeta\omega_n\Delta t}\left(\frac{\zeta}{\sqrt{1-\zeta^2}}\sin(\omega_d\Delta t) + \cos(\omega_d\Delta t)\right) \tag{2.185}$$

and

$$D = \frac{e^{-\zeta\omega_n\Delta t}}{\omega_d}\sin(\omega_d\Delta t) \tag{2.186}$$

From Equation (2.184), the response at t_{i+1} depends not only on the displacement x_i but also the velocity $\dot{x}_i$ at t_i. Therefore, the recursive equations for calculating $\dot{x}_{i+1}$ based on the response at t_i must be formulated. Similar to the displacement response, the velocity response also consists of two parts. They are the responses due to the acceleration excitation and the initial conditions:

$$\dot{x}_{i+1} = \dot{x}_{a,i\to i+1} + \dot{x}_{f,i\to i+1} \tag{2.187}$$

The response due to the acceleration excitation, $\dot{x}_{a,i\to i+1}$, can be determined from the derivative of the Duhamel's integral in Equation (2.152) as:

$$\dot{x}(t) = \frac{d}{dt}\left(\frac{1}{\omega_d}\int_0^t a(\tau)e^{-\zeta\omega_n(t-\tau)}\sin\left(\omega_d(t-\tau)\right)d\tau\right)$$
$$= \frac{1}{\omega_d}\int_0^t a(\tau)\frac{d}{dt}\left[e^{-\zeta\omega_n(t-\tau)}\sin\left(\omega_d(t-\tau)\right)\right]d\tau \tag{2.188}$$
$$= \int_0^t a(\tau)e^{-\zeta\omega_n(t-\tau)}\left[\cos\left(\omega_d(t-\tau)\right) - \frac{\zeta}{\sqrt{1-\zeta^2}}\sin\left(\omega_d(t-\tau)\right)\right]d\tau$$

Therefore, considering the time step from t_i to t_{i+1} and substituting $a(\tau)$ from Equation (2.154), one obtains:

$$\dot{x}_{a,i\to i+1} = \int_0^{\Delta t}\left(a_i + \frac{a_{i+1}-a_i}{\Delta t}\tau\right)e^{-\zeta\omega_n(\Delta t-\tau)}\left[\cos\left(\omega_d(\Delta t-\tau)\right) - \frac{\zeta}{\sqrt{1-\zeta^2}}\sin\left(\omega_d(\Delta t-\tau)\right)\right]d\tau \tag{2.189}$$

Similar to the situation of Equation (2.155), $\dot{x}_{a,i\to i+1}$ can be expressed in terms of a_i and a_{i+1} after solving the integrals as:

$$\dot{x}_{a,i\to i+1} = A'a_i + B'a_{i+1} \tag{2.190}$$

where:

$$A' = \left(\frac{1}{\omega_d} + \frac{\zeta}{\omega_n^2\Delta t\sqrt{1-\zeta^2}}\right)e^{-\zeta\omega_n\Delta t}\sin\left(\omega_d\Delta t\right) + \frac{1}{\omega_n^2\Delta t}\left(e^{-\zeta\omega_n\Delta t}\cos\left(\omega_d\Delta t\right) - 1\right) \tag{2.191}$$

and:

$$B' = \frac{1}{\omega_n^2\Delta t}\left(1 - \frac{\zeta}{\sqrt{1-\zeta^2}}e^{-\zeta\omega_n\Delta t}\sin\left(\omega_d\Delta t\right) - e^{-\zeta\omega_n\Delta t}\cos\left(\omega_d\Delta t\right)\right). \tag{2.192}$$

The free-decay term, $\dot{x}_{f,i\to i+1}$, can be obtained by calculating the first derivative of displacement response in Equation (2.41) and expressing it in terms of x_0 and $\dot{x}_0$:

$$\dot{x}(t) = -x_0\left(\frac{\omega_n}{\sqrt{1-\zeta^2}}\right)e^{-\zeta\omega_n t}\sin\left(\omega_d t\right) + \dot{x}_0 e^{-\zeta\omega_n t}\left(\cos\left(\omega_d t\right) - \frac{\zeta}{\sqrt{1-\zeta^2}}\sin\left(\omega_d t\right)\right) \tag{2.193}$$

Considering the time step from t_i to t_{i+1}, one obtains:

$$\dot{x}_{f,i\to i+1} = C'x_i + D'\dot{x}_i \tag{2.194}$$

where:

$$C' = -\left(\frac{\omega_n}{\sqrt{1-\zeta^2}}\right)e^{-\zeta\omega_n\Delta t}\sin\left(\omega_d\Delta t\right) \tag{2.195}$$

and:

$$D' = e^{-\zeta\omega_n\Delta t}\left(\cos(\omega_d\Delta t) - \frac{\zeta}{\sqrt{1-\zeta^2}}\sin(\omega_d\Delta t)\right) \tag{2.196}$$

Now, when the displacement and velocity at t_i are given, the displacement and velocity at t_{i+1} can be calculated. The recursive equation for calculating the acceleration of the system at t_{i+1} based on the given displacement and velocity at t_i can be obtained by normalizing the governing equation in Equation (2.4) as:

$$\ddot{x} + \frac{c_v}{m}\dot{x} + \frac{k}{m}x = \frac{f(t)}{m} \Rightarrow \ddot{x} + 2\zeta\omega_n\dot{x} + \omega_n^2 x = a(t) \tag{2.197}$$

Finally, the displacement response can be calculated by substituting Equations (2.180) and (2.184) into Equation (2.157); the velocity response can be calculated by substituting Equations (2.190) and (2.194) into (2.187); and the acceleration response can be calculated by considering Equation (2.197) at t_{i+1}. In summary:

$$\begin{aligned} x_{i+1} &= Aa_i + Ba_{i+1} + Cx_i + D\dot{x}_i \\ \dot{x}_{i+1} &= A'a_i + B'a_{i+1} + C'x_i + D'\dot{x}_i \\ \ddot{x}_{i+1} &= a_{i+1} - 2\zeta\omega_n\dot{x}_{i+1} - \omega_n^2 x_{i+1} \end{aligned} \tag{2.198}$$

where the factors A, B, C, and D are given in Equations (2.181), (2.182), (2.185), and (2.186), respectively, and the factors A', B', C', and D' are given in Equations (2.191), (2.192), (2.195), and (2.196), respectively. For given acceleration excitation (i.e., a_i for $i = 0,1,\ldots,n$) and initial conditions (i.e., x_0 and $\dot{x}_0$), one can use Equation (2.198) to calculate the displacement, velocity, and acceleration at all time steps.

The set of recursive formulas in Equation (2.198) is implemented in MATLAB to develop the function **sodf_duhamel()** as shown in Program 2.5. It is clear from Line 1 that the function requires seven input parameters. They are the natural circular frequency (**wn**) and the damping ratio (**zeta**) of the target single-DOF system; the time step size (**dt**) and the number of time steps considered (**Nt**); the forcing function (**fa**), which is a vector of dimensions **Nt+1** by 1 (note that when the time duration is divided into n sub-divisions, there will be $n + 1$ points); and the initial displacement (**x0**) and initial velocity (**v0**). The function returns three output parameters, and they are the displacement (**xx**), the velocity (**vv**), and acceleration (**aa**) vectors. They all have dimensions **Nt+1** by 1. From Lines 2 to 6, some variables are defined to shorten the lengths of the following equations. The time vector is defined on Line 7. The factors A, B, C, and D in Equations (2.181), (2.182), (2.185), and (2.186), respectively, and the factors A', B', C', and D' in Equations (2.191), (2.192), (2.195), and (2.196), respectively, are defined in Lines 8 to 15. The initial conditions are assigned (or calculated) to the response variables (i.e., **xx**, **vv**, and **aa**) on Line 16. Lines 17 to 21 contain the for-loop to implement the set of three recursive formulas in Equation (2.198).

Program 2.5: **sdof_duhamel()** for calculating the time-domain responses of an under-damped single-DOF system by solving Duhamel's integral.

```
Line 1.   function [xx,vv,aa]=sdof_duhamel(wn,zeta,dt,Nt,fa,x0,v0);
Line 2.   zz=(1-zeta^2)^0.5;
```

```
Line 3.   wd=wn*zz;
Line 4.   ee=exp(-zeta*wn*dt);
Line 5.   ss=sin(wd*dt);
Line 6.   cc=cos(wd*dt);
Line 7.   tt=[0:Nt]*dt;
Line 8.   A=((1-2*zeta^2)/wn/dt-zeta)*ee*ss/wn^2/zz-(2*zeta/wn/dt+1)*
          ee*cc/wn^2+2*zeta/wn^3/dt;
Line 9.   B=((2*zeta^2-1)/wn^3/zz/dt)*ee*ss+(2*zeta/wn^3/dt)*ee*cc+1/
          wn^2-2*zeta/wn^3/dt;
Line 10.  C=ee*(zeta/zz*ss+cc);
Line 11.  D=ee/wd*ss;
Line 12.  Ap=(1/wd+zeta/wn^2/dt/zz)*ee*ss+1/wn^2/dt*(ee*cc-1);
Line 13.  Bp=1/wn^2/dt*(1-zeta/zz*ee*ss-ee*cc);
Line 14.  Cp=-(wn/zz)*ee*ss;
Line 15.  Dp=ee*(cc-zeta/zz*ss);
Line 16.  xx(1)=x0; vv(1)=v0; aa(1)=fa(1)-2*zeta*wn*vv(1)-wn^2*xx(1);
Line 17.  for ii=1:Nt
Line 18.      xx(ii+1)=A*fa(ii)+B*fa(ii+1)+C*xx(ii)+D*vv(ii);
Line 19.      vv(ii+1)=Ap*fa(ii)+Bp*fa(ii+1)+Cp*xx(ii)+Dp*vv(ii);
Line 20.      aa(ii+1)=fa(ii+1)-2*zeta*wn*vv(ii+1)-wn^2*xx(ii+1);
Line 21.  end
Line 22.  end
```

As illustrated in Figure 2.27, one important assumption in formulating the numerical solution of Duhamel's integral is that the forcing function is assumed to be linear between two successive data points. A simple example is employed in the following to study if this assumption will seriously affect the accuracy of the numerical analysis result.

Consider an under-damped single-DOF system with natural frequency of 2 Hz and damping ratio of 0.03. The displacement response under the action of a sine (harmonic) excitation with excitation frequency of 1 Hz is calculated by solving Duhamel's integral (using **sdof_duhamel**() in Program 2.5) with $\Delta t = 0.2$, 0.1, and 0.01 s. The acceleration excitation with magnitude equal to 0.05 g is adopted in this study. The numerical results are then compared to the exact solution from Equation (2.104). It must be pointed out that the solution in Equation (2.104) considers both the cosine and sine excitations (when the exponential function with constant A_0 is expanded). In this study, only the sine part is considered. Therefore, Equation (2.104) is converted to:

$$x(t) = e^{-\zeta\omega_n t}\left[B_1\cos(\omega_d t) + B_2\sin(\omega_d t)\right] + A_0\sin(\omega t - \varphi_0) \tag{2.199}$$

To simplify the analysis, it is assumed that the system starts from rest. To calculate the exact solution, the unknown constants B_1 and B_2 in Equation (2.199) must be calculated from the initial conditions (i.e., $x_0 = \dot{x}_0 = 0$). Consider the first initial condition, $x(0) = x_0 = 0$, Equation (2.199) gives:

$$x(0) = B_1 - A_0\sin\varphi_0 = 0 \Rightarrow B_1 = A_0\sin\varphi_0 \tag{2.200}$$

The first derivative of Equation (2.199) gives:

$$\begin{aligned}\dot{x}(t) = &-\zeta\omega_n e^{-\zeta\omega_n t}\left[B_1\cos(\omega_d t) + B_2\sin(\omega_d t)\right] \\ &+ e^{-\zeta\omega_n t}\left[-\omega_d B_1\sin(\omega_d t) + \omega_d B_2\cos(\omega_d t)\right] + \omega A_0\cos(\omega t - \varphi_0)\end{aligned} \tag{2.201}$$

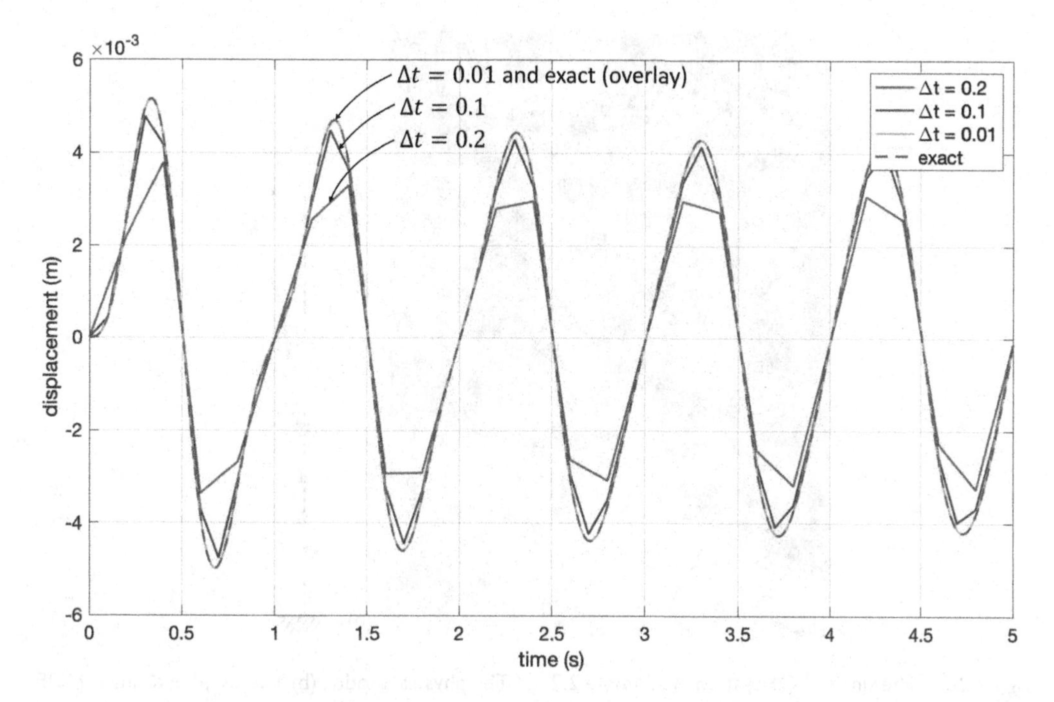

Figure 2.28 Testing the accuracy of the formulated numerical method (Duhamel's integral).

Considering the second initial condition, $\dot{x}(0) = \dot{x}_0 = 0$, it becomes:

$$\dot{x}(0) = -\zeta\omega_n B_1 + \omega_d B_2 + \omega A_0 \cos\varphi_0 = 0 \tag{2.202}$$

Substitute Equation (2.200) into Equation (2.202) and rearrange:

$$B_2 = \frac{A_0}{\omega_d}\left(\zeta\omega_n \sin\varphi_0 - \omega\cos\varphi_0\right) \tag{2.203}$$

As A_0 and φ_0 can be calculated by Equations (2.102) and (2.103), respectively, the exact solution of the displacement response can be calculated by Equation (2.199) by using B_1 in Equation (2.200) and B_2 in Equation (2.203).

The numerical solutions with various Δt are plotted together with the exact solution in Figure 2.28. When Δt is large (i.e., 0.2 in this case), the approximation of the solution near the peaks of the curve is very bad. However, the approximated solution can capture the trend of the solution, and there is no shift between the exact and approximated solutions even when $\Delta t = 0.2$s. when the time step size is reduced to 0.1 s, significant improvement on the approximation is observed. The approximated solution is very close to the exact one even near the peaks. When the time step size is further reduced to 0.01s, the approximate solution is basically overlayed on the exact solution.

Example 2.2

Under small amplitude vibration, the vertical cantilever with a concentrated mass at the free tip (as shown in Figure 2.29) can be considered as a single-DOF system with the lateral displacement at the tip (corresponding to bending about the minor axis of

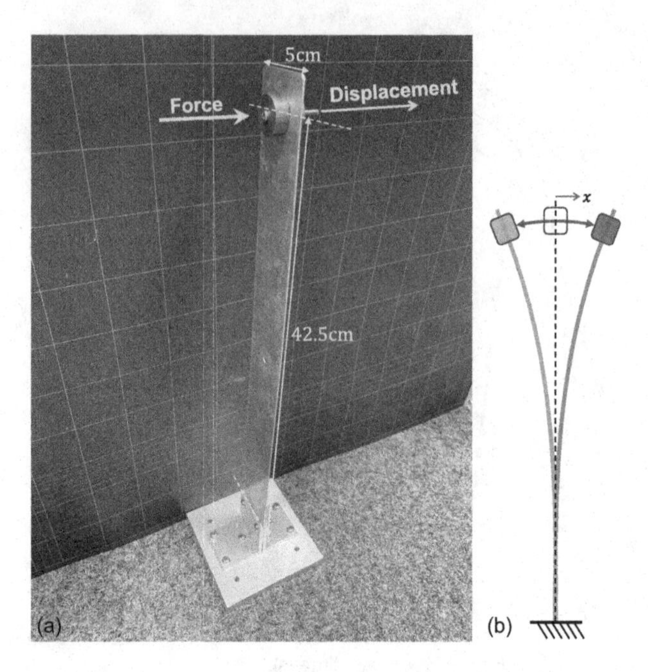

Figure 2.29 The single-DOF system in Example 2.2. (a) The physical model; (b) The simplified single-DOF system.

the cross-section) as the degree-of-freedom. The vertical cantilever is made of aluminum (with modulus of elasticity, $E = 70\,\text{GPa}$), the height of the plate from the fixed support to the mass center of the free tip mass is 42.5 cm. The width and the thickness of the aluminum plate are 5 cm and 1 mm, respectively. The mass at the free tip is equal to 90 g (neglect the mass of the aluminum plate). It is assumed that the damping ratio of this system is 2%, and the system starts from rest.

a) With a time step of $\Delta t = 0.01\,\text{s}$, determine the first ten seconds of displacement response of this single-DOF system under the action of a rectangular force with magnitude 0.1 N applied suddenly to the system with a duration of 5 s.
b) Determine the displacement response again if the force duration is reduced to 1 s.
c) Determine the displacement response again if the force duration is further reduced to 0.1 s (like an impulse).

SOLUTION

First, the stiffness of this single-DOF system is calculated. The stiffness is defined as the force required to produce a unit displacement. In this case, the force is the lateral force applied at the free tip of the cantilever and the displacement is in the same direction and location of the force as shown in Figure 2.29(a). Consider the cantilever beam with a displacement Δ at the free tip induced by the force F as shown in Figure 2.30, it is a statically determinate system, and the support reaction force and moment can be easily calculated by statics. Virtually cutting the beam at x from the left support and considering the left beam segment as shown in Figure 2.30, the sum of all moments at x from the left support must be equal to zero for the system in equilibrium. Thus:

$$\sum M_x = 0 \Rightarrow M(x) + FL - Fx = 0 \tag{2.204}$$

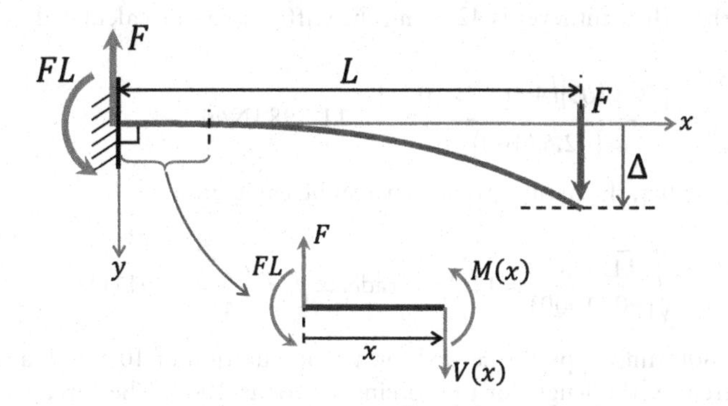

Figure 2.30 Determine the stiffness of the single-DOF system.

Therefore, the bending moment of the cantilever beam as a function of x can be expressed as:

$$M(x) = F(x - L) \tag{2.205}$$

Consider the second-order ordinary differential equation (ODE) for beam bending:

$$y'' = -\frac{M(x)}{EI} \Rightarrow EIy'' = -M(x) = F(L - x) \tag{2.206}$$

Integrating Equation (2.206) to obtain:

$$EIy' = F\left(Lx - \frac{x^2}{2}\right) \tag{2.207}$$

where the integration constant is equal to zero as the slope (i.e., the first derivative of the displacement) at the cantilever support must be zero. Integration Equation (2.207), one obtains:

$$EIy = F\left(L\frac{x^2}{2} - \frac{x^3}{6}\right) \tag{2.208}$$

where the integration constant is equal to zero again as the displacement at the support must vanish. At the free tip (i.e., when $x = L$), the displacement is equal to Δ. Thus:

$$EI\Delta = FL^3\left(\frac{1}{2} - \frac{1}{6}\right) = \frac{FL^3}{3} \tag{2.209}$$

Therefore, the stiffness can be calculated as:

$$k = \frac{F}{\Delta} = \frac{3EI}{L^3} \tag{2.210}$$

The cross-section of the aluminum cantilever is a rectangle with breadth, $b = 500$ mm, and thickness, $t = 1$ mm. The second moment of area is then:

$$I = \frac{bt^3}{12} = \frac{(500/1000)(1/1000)^3}{12} = 4.1667 \times 10^{-12}\ \text{m}^4 \tag{2.211}$$

As the length of the cantilever is 42.5 cm, the stiffness can be calculated as:

$$k = \frac{3EI}{L^3} = \frac{3\left(70\times10^9\right)\left(4.1667\times10^{-12}\right)}{\left(42.5/100\right)^3} = 11.3983\,\text{N/m} \tag{2.212}$$

As the mass is 90 g, the natural frequency can be calculated as:

$$\omega_n = \sqrt{\frac{k}{m}} = \sqrt{\frac{11.3983}{(90/1000)}} = 11.2538\ \text{rad/s} \Rightarrow f_n = \frac{\omega_n}{2\pi} = 1.7911\,\text{Hz} \tag{2.213}$$

In part a), 1000 time steps are needed for a time duration of 10 s with a step size of 0.01 s. Therefore, the length of the forcing vector is 1001. The force magnitude is $f_0 = 0.1\,\text{N}$, and therefore, the acceleration excitation is $1.1111\,\text{m/s}^2$ $(= f_0 / m)$, where $m = 0.09\,\text{kg}$. The force duration is 5 s, and therefore, only the first 501 elements in the acceleration excitation vector are equal to 1.1111, and all other elements are equal to zero. By using the function **sdof_duhamel**() in Program 2.5, the displacement response can be calculated and plotted together with the acceleration excitation in Figure 2.31. As there is no initial displacement and velocity, the effect of the acceleration excitation is clearly observed in the figure. When the force is suddenly applied at $t = 0$, the system is displaced in the positive direction and vibrating with a positive mean position. The amplitude of vibration is decaying due to the damping effect (i.e., the 2% damping). At $t = 5\text{s}$, the force suddenly disappears, the effect is the same as a suddenly applied force in the negative direction. Thus, the system is set into vibration again with the mean position at zero (as the force is already vanished).

In part b), everything is the same as that in part a), except the force duration is reduced to 1 s. Therefore, the acceleration excitation vector becomes zero after the 101th element. Figure 2.32 shows the calculated displacement response and the

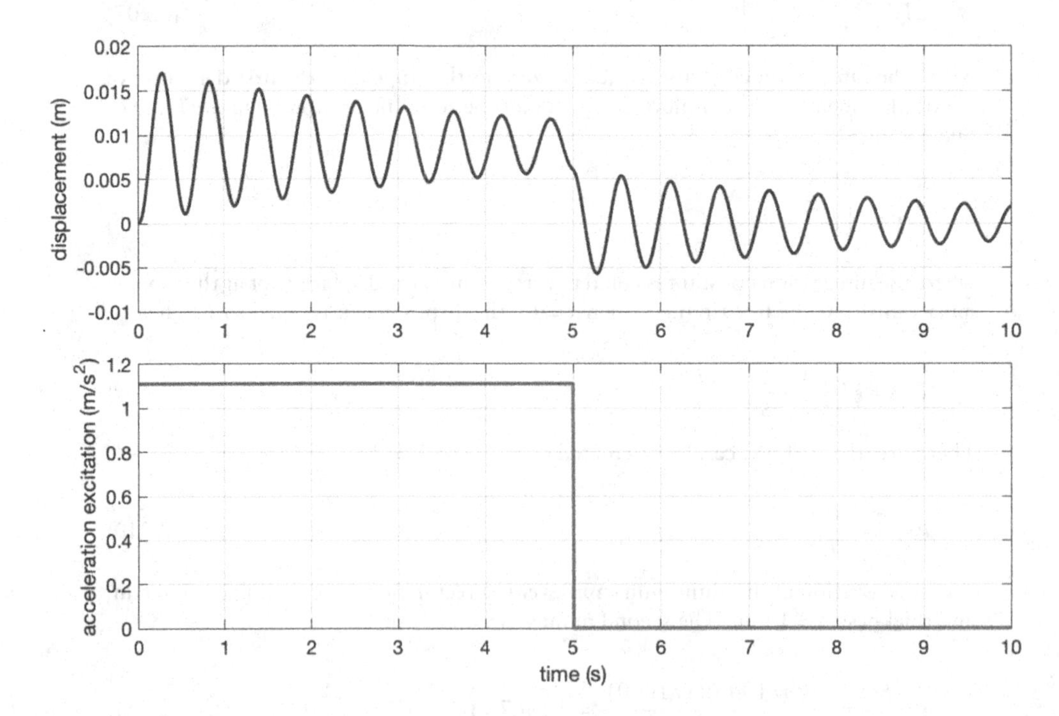

Figure 2.31 Part a) force duration = 5s.

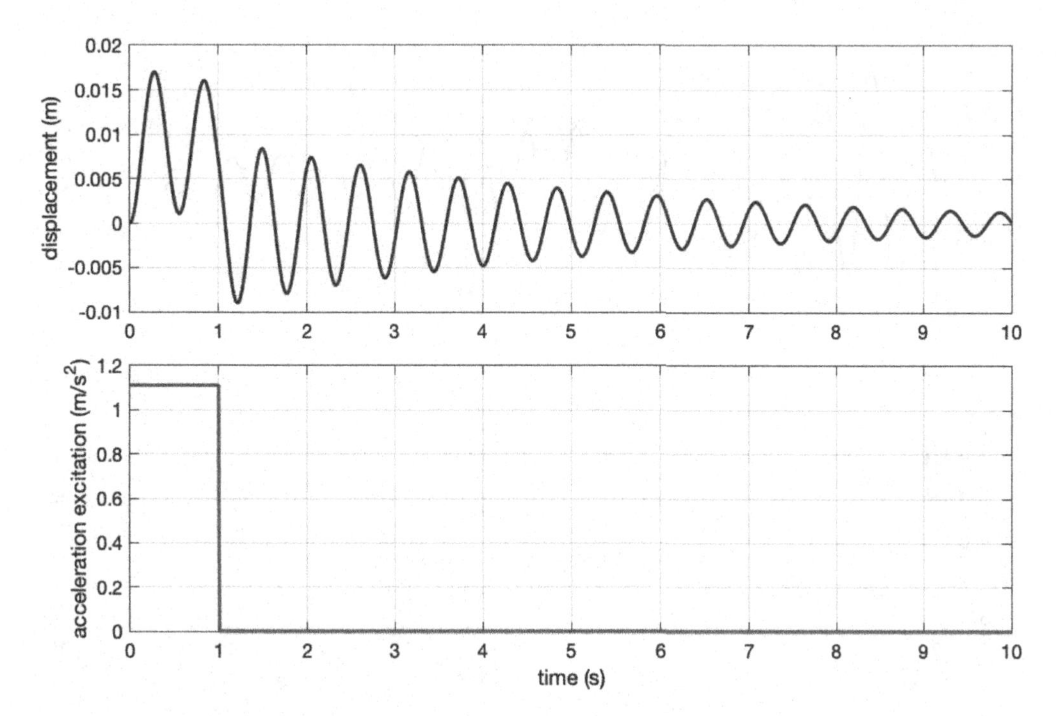

Figure 2.32 Part b) force duration = 1s.

acceleration excitation. The vibration behavior in this case is very similar to that obtained in the previous case. The vibration can be easily divided into two phases. The mean position is positive in the first phase, and the mean position becomes zero in the second phase, when the force disappears.

In part c), the force duration is further reduced to 0.1 s, and therefore, only the first 11 elements in the acceleration excitation vector are non-zero. Figure 2.33 shows the analysis result. As the force duration is too short, the two-phase vibration phenomenon cannot be observed. The displacement response looks very similar to the free vibration of an under-damped single-DOF system (see Figure 2.11 as an example). In fact, a rectangular force with such a short force duration can be treated as an impulse. The effect of an impulse can be replaced by an initial velocity. Consider Newton's second law:

$$F = ma = m\frac{v-0}{\Delta t} \tag{2.214}$$

where F is the impulse force applied on the mass m, and a is the acceleration. Consider a system which starts from rest, the velocity of the system changes from 0 to v in a short time Δt by the impulse force. As the impulse force in this case has a magnitude of 0.1N and a duration of 0.1s, the corresponding initial velocity (to achieve the same effect under free vibration) can be estimated as:

$$v = \frac{F\Delta t}{m} = \frac{0.1\times 0.1}{0.09} = 0.1111\,\text{m/s} \tag{2.215}$$

By evaluating Equation (2.41) with $x_0 = 0$ and $\dot{x}_0 = 0.1111$, the free-vibration response of the same single-DOF system is calculated and plotted in Figure 2.34. Comparing Figure 2.34 and Figure 2.33, the displacement responses in both situations are almost the same except for a small discrepancy at the beginning (near $t = 0$).

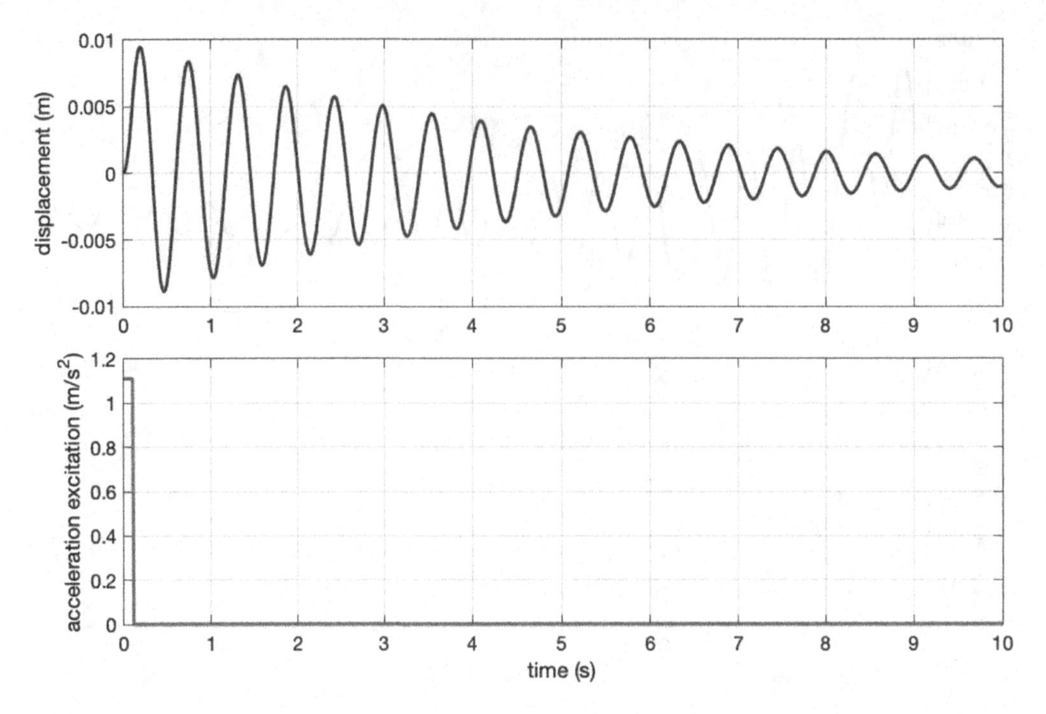

Figure 2.33 Part c) force duration = 0.1 s.

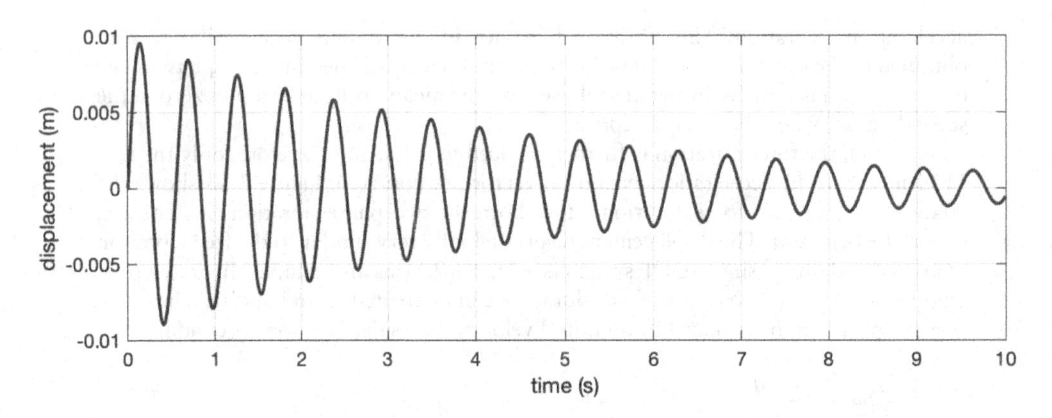

Figure 2.34 Free vibration response of the same system under the action of an initial velocity.

Example 2.3

Determine the displacement responses of a single-DOF system with natural frequency $f_n = 1\text{Hz}$ and damping ratio $\zeta = 0.02$ (i.e., 2% damping) under the action of a harmonic excitation, and the acceleration excitation is given by:

$$a(t) = \frac{f_0}{m}\sin\omega t = a_0 \sin 2\pi f t \tag{2.216}$$

where ω is the excitation circular frequency in rad/s and f is the excitation frequency in Hz, f_0 is the force magnitude, and m is the mass of the single-DOF system, a_0 is the amplitude of the acceleration excitation, and it is given as $a_0 = 0.05\text{g}$. Assuming that the initial displacement and velocity are both equal to zero, determine the displacement response of the system when the excitation frequency is equal to 0.8, 0.9, 1.0, 1.1, and 1.2 Hz.

SOLUTION

Consider a time step size of $\Delta t = 0.01\text{s}$ and the number of time steps n = 2000 (i.e., t = 0 to 20 s), the displacement response can be calculated by the function **sdof_duhamel()** in Program 2.5.

The excitation frequency of 0.8 Hz is handled first. The amplitude of acceleration excitation is $0.4905\,\text{m/s}^2$ $(= 0.05 \times 9.81)$. The acceleration excitation is calculated and plotted in Figure 2.35. It is then input into **sdof_duhamel()** for calculating the time-domain responses. The calculated displacement response is plotted in Figure 2.36. The vibration response seems to consist of two frequencies. The higher frequency is very clear to be the excitation frequency in 0.8 Hz. It corresponds to a period of 1.25 s (= 1/0.8). This frequency can be easily identified by matching the peaks in Figure 2.36 to those in Figure 2.35. The lower frequency seems to have a period of about 5 s. It corresponds to the beat frequency (due to the superposition of two similar but not exactly the same frequencies). The beat frequency is equal to the difference between the two similar frequencies. In this case, the two frequencies are the excitation frequency of 0.8 Hz and the natural frequency of 1 Hz. The beat frequency is equal to 0.2 Hz (= 1 – 0.8), and the corresponding period is 5 s $(= 1/0.2)$. Due to the effect of damping, the maximum amplitude of vibration within a "beat" reduces progressively. Therefore, the maximum absolute displacement of the single-DOF system can be measured from this plot as 54.3 mm.

The displacement response of the system with an excitation frequency of 0.9 Hz is next calculated and plotted in Figure 2.37. The beat frequency in this case is 0.1 Hz $(= 1 - 0.9)$, and the corresponding period is 10 s $(= 1/0.1)$. The excitation frequency and the beat frequency can be clearly observed from the figure. The maximum absolute displacement of the single-DOF system can be measured from the plot as 95.6 mm, which is larger than that in the previous case (when the excitation frequency is equal to 0.8 Hz). This is expected as the excitation frequency in this case is closer to the natural frequency of the system.

Next, the situation when the excitation frequency is equal to the natural frequency (i.e., $f = f_n = 1\text{Hz}$) of the system is considered. The calculated displacement response

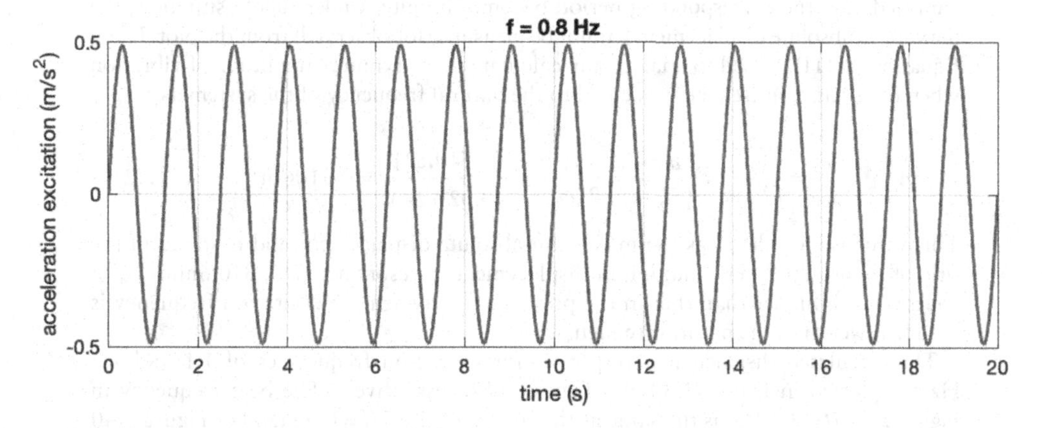

Figure 2.35 The acceleration excitation when the excitation frequency is 0.8 Hz.

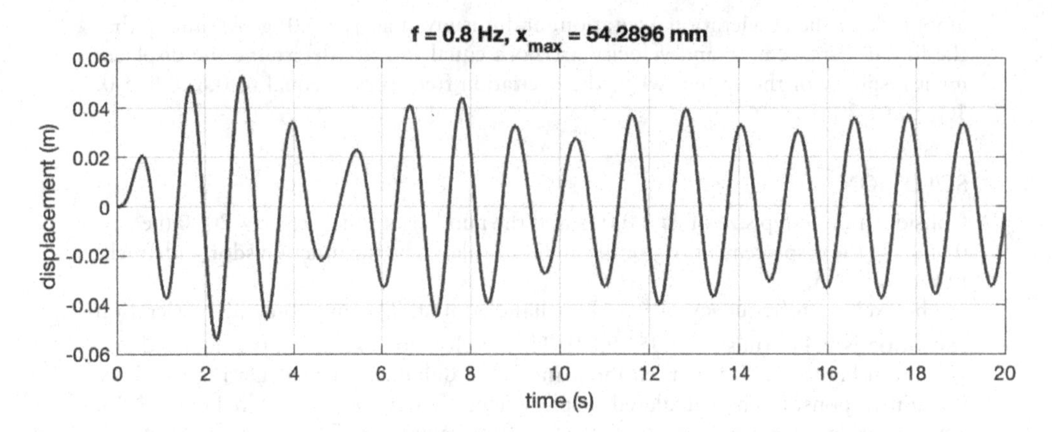

Figure 2.36 The displacement response when the excitation frequency is 0.8 Hz.

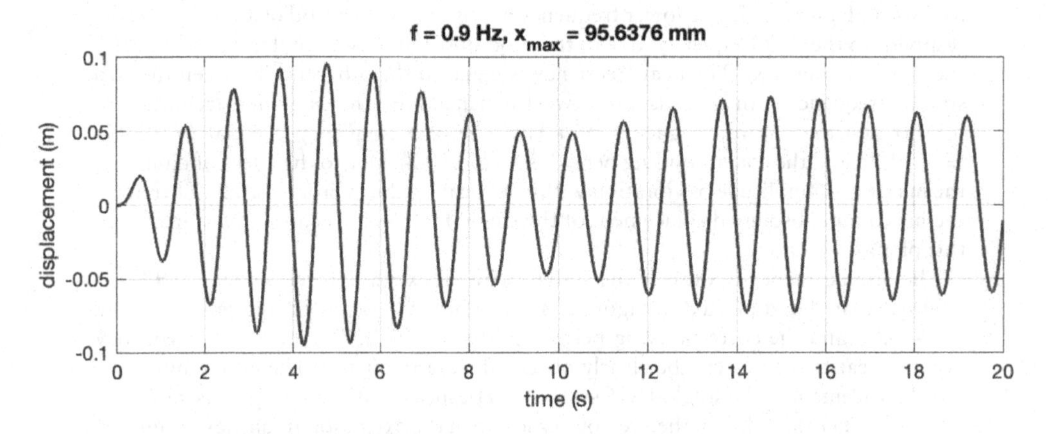

Figure 2.37 The displacement response when the excitation frequency is 0.9 Hz.

is shown in Figure 2.38. It is clear from the figure that the amplitude of vibration continuously increases. If one considers the concept of "beats," the beat frequency now is vanished, and the corresponding period becomes infinite. Under such a situation, the maximum absolute displacement cannot be measured (or observed) from the plot. Here, Equation (2.111) is used to analytically calculate the maximum amplitude of vibration when the excitation frequency is equal to the natural frequency of the system as:

$$A_{\text{peak}} = \frac{x_s}{2\zeta} = \frac{f_0}{(2\zeta)k} = \frac{a_0 m}{(2\zeta)k} = \frac{a_0}{2\zeta\omega_n^2} = \frac{0.05(9.81)}{2(0.02)(2\pi)^2} = 0.3106\,\text{m} \tag{2.217}$$

This value is considered as the maximum absolute displacement and is stated in the title of Figure 2.38. The maximum displacement in resonance (i.e., 310.6mm) is, of course, much higher than that in the previous case (when the excitation frequency is 0.1 Hz lower than the natural frequency).

The calculated displacement responses for excitation frequencies of 1.1 and 1.2 Hz are plotted in Figure 2.39 and Figure 2.40, respectively. The beat frequency in Figure 2.39 (f = 1.1 Hz) is the same as that in Figure 2.37 (f = 0.9 Hz). For Figure 2.40 (f = 1.2 Hz), the beat frequency is the same as that in Figure 2.36 (f = 0.8 Hz).

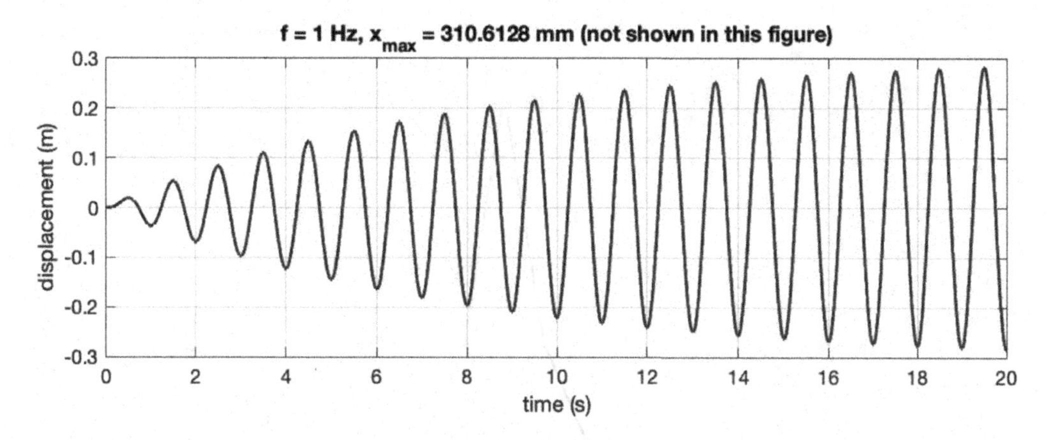

Figure 2.38 The displacement response when the excitation frequency is 1.0 Hz.

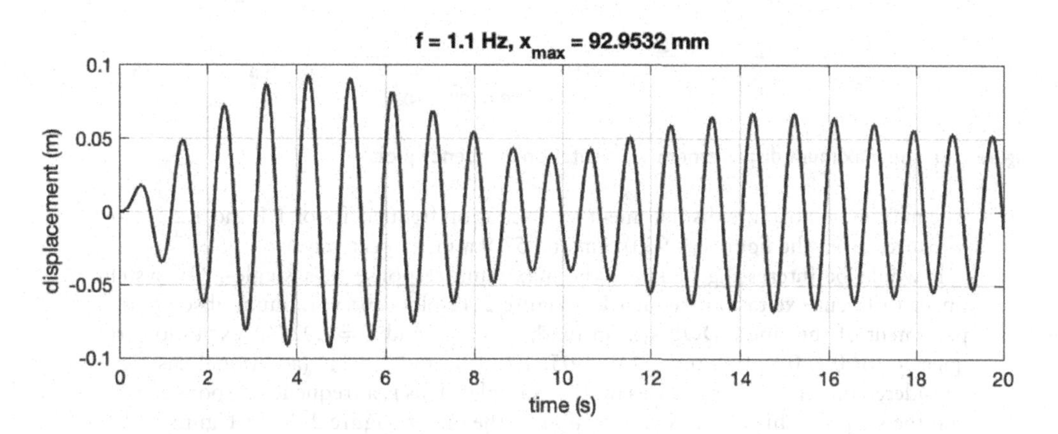

Figure 2.39 The displacement response when the excitation frequency is 1.1 Hz.

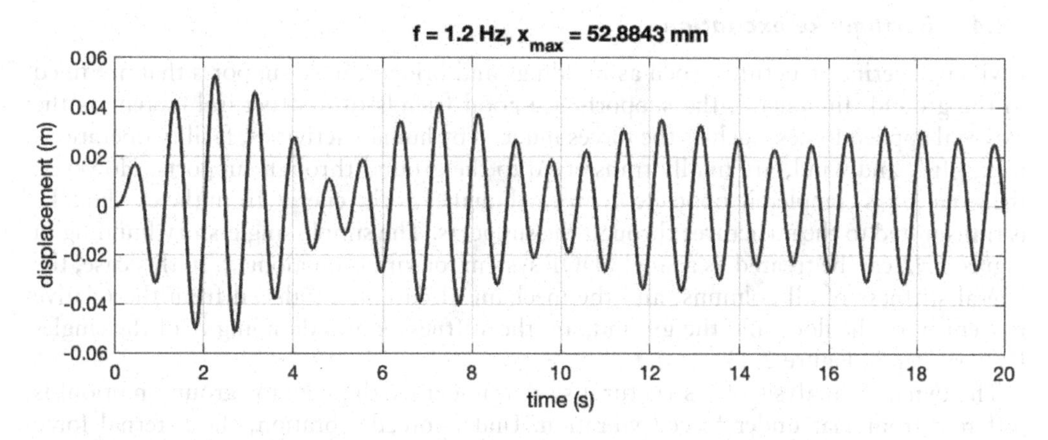

Figure 2.40 The displacement response when the excitation frequency is 1.2 Hz.

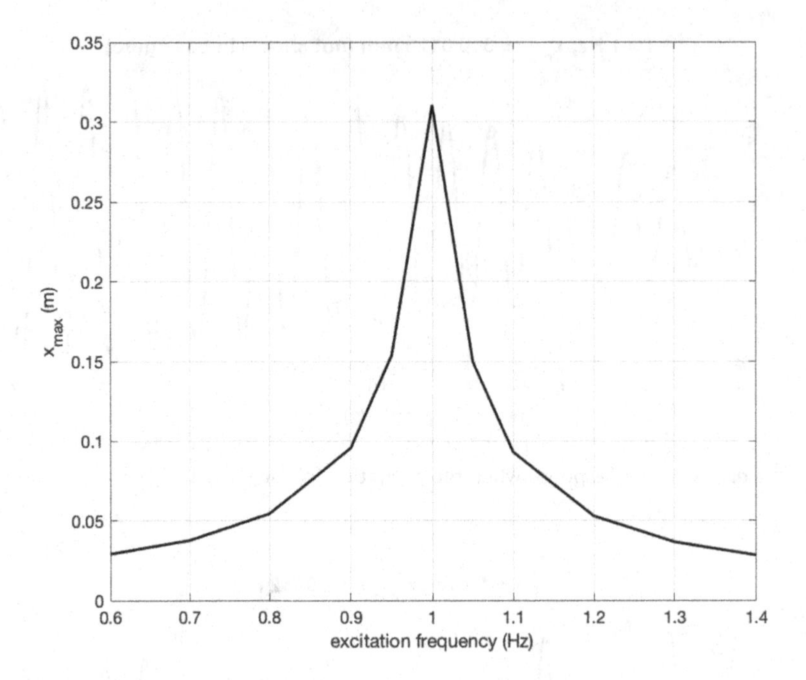

Figure 2.41 the maximum displacement vs. excitation frequency plot.

The maximum absolute displacement for excitation frequencies of 1.1 and 1.2 Hz are measured from the figures as 93.0 mm and 52.9 mm, respectively.

It would be interesting to know the maximum response of a single-DOF system under different excitation frequencies. Figure 2.41 plots the maximum absolute displacement of the single-DOF system (with $f_n = 1\text{Hz}$ and $\zeta = 0.02$) at excitation frequencies of 0.7, 0.8, 0.9, 0.95, 1.0, 1.05, 1.1, 1.2, and 1.3 Hz (additional cases are considered on top of the five cases in this example). This is a frequency response curve, and the shape of this curve is very similar to the one in Figure 2.20. If Figure 2.41 is obtained from a vibration test of the targeted system, the damping ratio of the system can be estimated by the half-power bandwidth method (see Section 2.1.3.2).

2.1.4.4 Earthquake excitation

Civil engineering structures, such as buildings and bridges, have supports that are fixed to the ground. In general, the supports are good friends of a structural system as the external applied forces, such as the forces induced by human activities, facility operations (e.g., lifts), and wind, are finally transferred to the ground through supports. However, the situation is completely opposite in an earthquake, as the energy from the earthquake is transmitted to the structures through the supports. The simple single-story building in Figure 2.42 can be treated as a single-DOF system for vibration along x. In this case, the lateral stiffness of all columns, and the mechanical damping induced from the relative movement of the floor and the ground, are the stiffness k and damping c_v of the single-DOF system in Figure 2.43.

The dynamic analysis of a structural system under earthquake (or ground motion) is different from that under forced vibration. Under forced vibration, the external force is applied at the degree-of-freedom of the system (see x in Figure 2.43). However, an

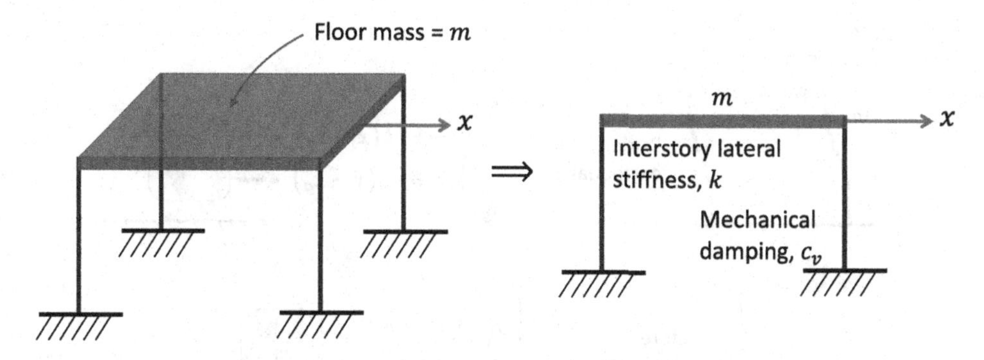

Figure 2.42 A single-story building model.

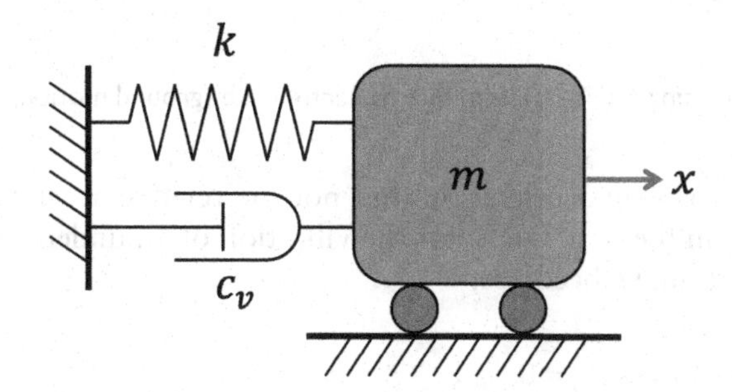

Figure 2.43 A typical model of a single-DOF system.

earthquake induces excitation at the support of a structural system. In the model, the support is not supposed to have any motion (consider the rigid wall on the left of Figure 2.43). An earthquake is usually recorded as ground acceleration. The question now is how to convert the ground acceleration to the equivalent external applied force to the single-DOF system.

The left part of Figure 2.44 shows the single-story building model, while the right part shows the corresponding single-DOF system. The lower part of the figure shows the condition before an earthquake, and the upper part shows the scenario during an earthquake. For the building model, consider the end of the left column as the reference point. During an earthquake, the ground moves and the ground displacement is represented by x_g in the figure, and x is the absolute displacement of the floor mass. As the stiffness force on the floor mass is directly proportional to the lateral deflection of the column, it is clear from the figure that the stiffness force is not directly proportional to the absolute displacement x (unlike in the forced vibration situation) but the relative displacement x_r:

$$x_r = x - x_g \tag{2.218}$$

Evaluating the first and second derivatives of Equation (2.218), one obtains:

$$\dot{x}_r = \dot{x} - \dot{x}_g \quad \text{and} \quad \ddot{x}_r = \ddot{x} - \ddot{x}_g \tag{2.219}$$

Referring to the single-DOF model on the right of Figure 2.44, the damping force is proportional to the relative velocity. However, the inertia force on the mass is still

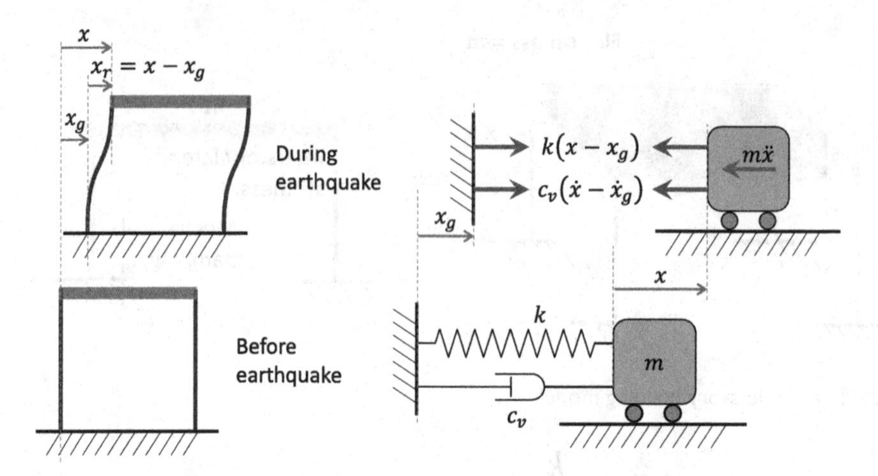

Figure 2.44 Vibration of a single-DOF system due to earthquake (ground motion).

proportional to the absolute acceleration (but not the relative acceleration). As a result, the equation of motion (i.e., the ODE) for the vibration of an under-damped single-DOF system under an earthquake becomes:

$$m\ddot{x} + c_v\left(\dot{x} - \dot{x}_g\right) + k\left(x - x_g\right) = 0 \tag{2.220}$$

Subtract both sides of Equation (2.220) by $m\ddot{x}_g$, one obtains:

$$m\left(\ddot{x} - \ddot{x}_g\right) + c_v\left(\dot{x} - \dot{x}_g\right) + k\left(x - x_g\right) = -m\ddot{x}_g \tag{2.221}$$

It becomes:

$$m\ddot{x}_r + c_v\dot{x}_r + kx_r = -m\ddot{x}_g \tag{2.222}$$

It is very clear from Equation (2.222) that the equation of motion for earthquake vibration is in the same form as for the forced vibration. However, the calculated displacement, velocity, and acceleration are all relative to the ground. Under the action of earthquakes, the equivalent forcing function is equal to the product of the negative ground acceleration and the system mass (i.e., $-m\ddot{x}_g$). This is good news, as all formulations in the forced vibration analysis can then be used to calculate the relative responses of the system under earthquake. If the absolute acceleration is needed, it can be calculated by summing the calculated relative acceleration to the given ground acceleration as shown in Equation (2.219).

Example 2.4

Consider a single-story building model with natural circular frequency $\omega_n = \sqrt{k/m} = 2\pi$ and damping ratio $\zeta = 0.05$ (i.e., 5% damping). Determine the relative displacement (x_r), relative velocity ($\dot{x}_r$), and absolute acceleration ($\ddot{x}$) responses of this single-DOF system under the action of ground motion given by the El Centro earthquake record as shown in Figure 2.45 with $\Delta t = 0.02\,\text{s}$.

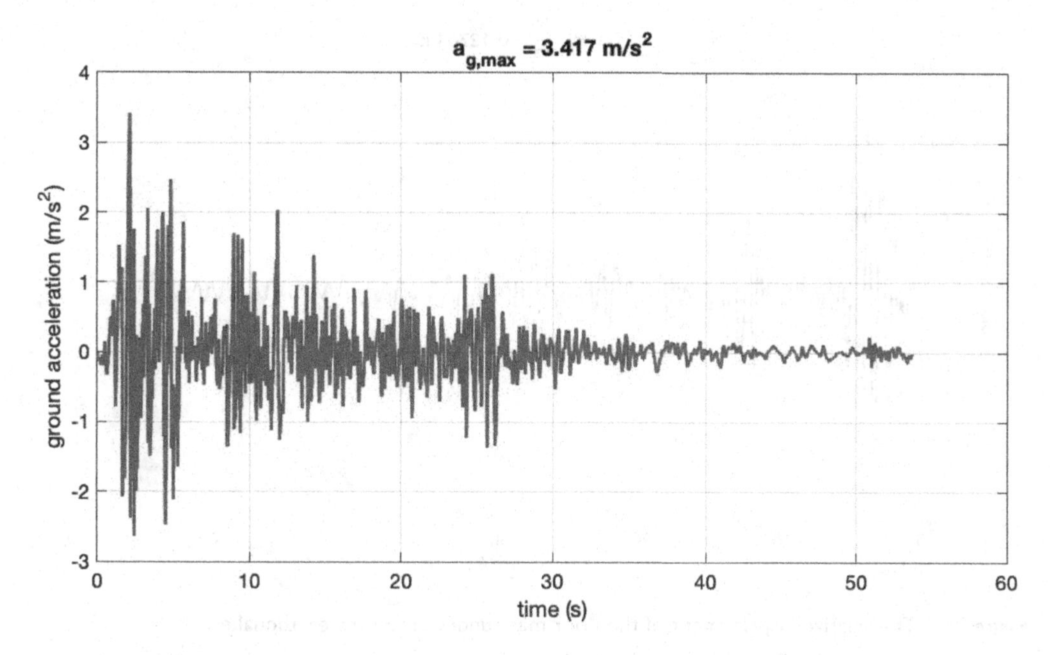

Figure 2.45 Ground acceleration record of the 1940 El Centro Earthquake.

SOLUTION

To use the self-developed MATLAB function **sdof_duhamel()**, Equation (2.222) has to be normalized as:

$$\ddot{x}_r + 2\zeta\omega_n\dot{x}_r + \omega_n^2 x_r = -\ddot{x}_g \tag{2.223}$$

Therefore, the acceleration excitation is $-\ddot{x}_g$. The relative displacement (x_r), relative velocity ($\dot{x}_r$), and relative acceleration ($\ddot{x}_r$) are calculated and plotted in Figure 2.46, Figure 2.47, and Figure 2.48, respectively. The absolute acceleration ($\ddot{x}$) can be calculated as:

$$\ddot{x} = \ddot{x}_r + \ddot{x}_g \tag{2.224}$$

The calculated absolute acceleration is plotted in Figure 2.49.

2.1.5 Fast response calculation based on Duhamel's integral

Having formulated the responses to a general excitation in the form of Duhamel's integral, fast algorithms for calculating responses when excitations are measured in practice are developed in this section. These algorithms were proposed by Beck and Dowling (1988). The idea is to assume that the excitation within each time step is linear, so that the Duhamel's integral for each time step can be evaluated analytically. The calculation is progressed in a recursive way. The algorithm for calculating accelerations is first discussed because accelerations are measured in practice most of the time.

2.1.5.1 Acceleration algorithm

In field tests, accelerations are usually measured. Duhamel's integral for accelerations is derived in the following. This is done by first differentiating the displacement response Equation (2.151), to get the velocity response, and then differentiating the velocity

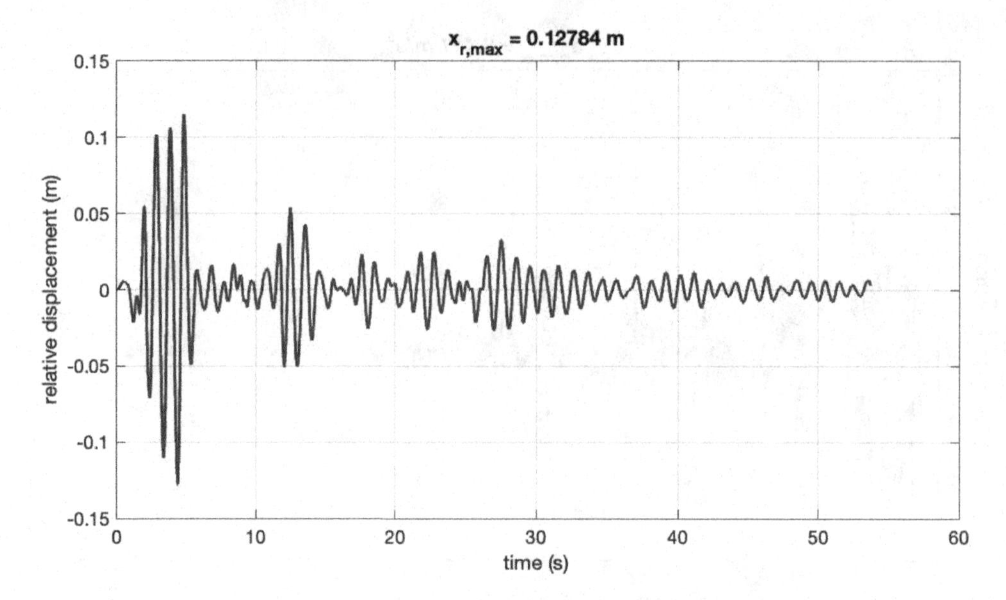

Figure 2.46 The relative displacement of the floor mass under El Centro earthquake.

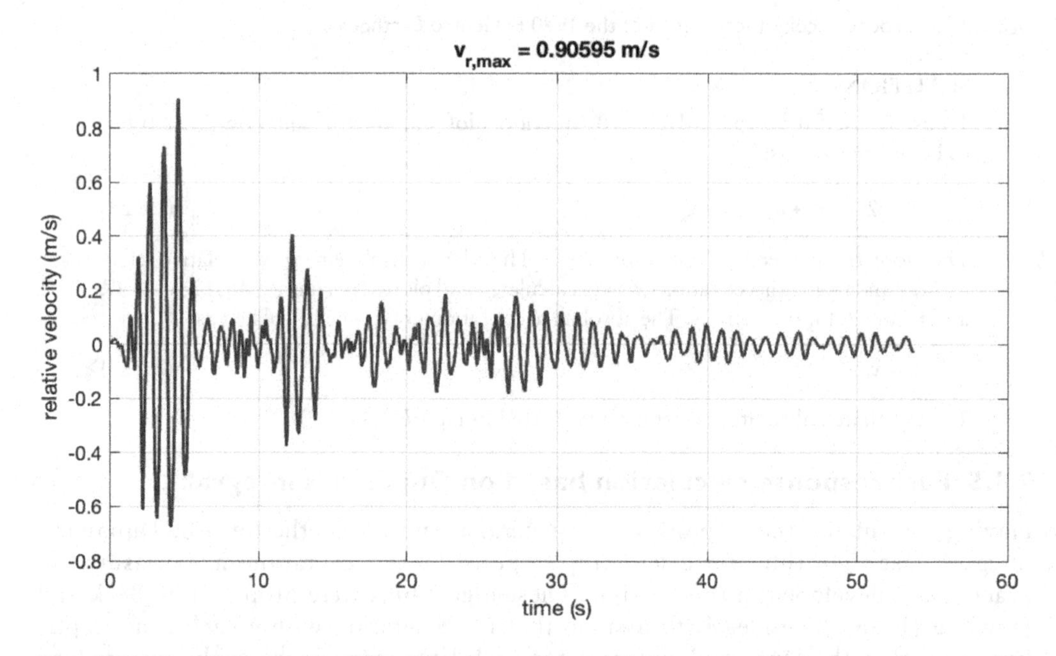

Figure 2.47 The relative velocity of the floor mass under the El Centro earthquake.

response to get the acceleration response. The "Leibniz rule of differentiation" under integral sign is needed (Au 2018). For $y(\tau, t)$ with lower and upper integration limits equal to $l(t)$ and $u(t)$, one obtains:

$$\frac{d}{dt}\int_{l(t)}^{u(t)} y(\tau,t)\,d\tau = y(u(t),t)\dot{u}(t) - y(l(t),t)\dot{l}(t) + \int_{l(t)}^{u(t)} \frac{\partial y(\tau,t)}{\partial t}\,d\tau \tag{2.225}$$

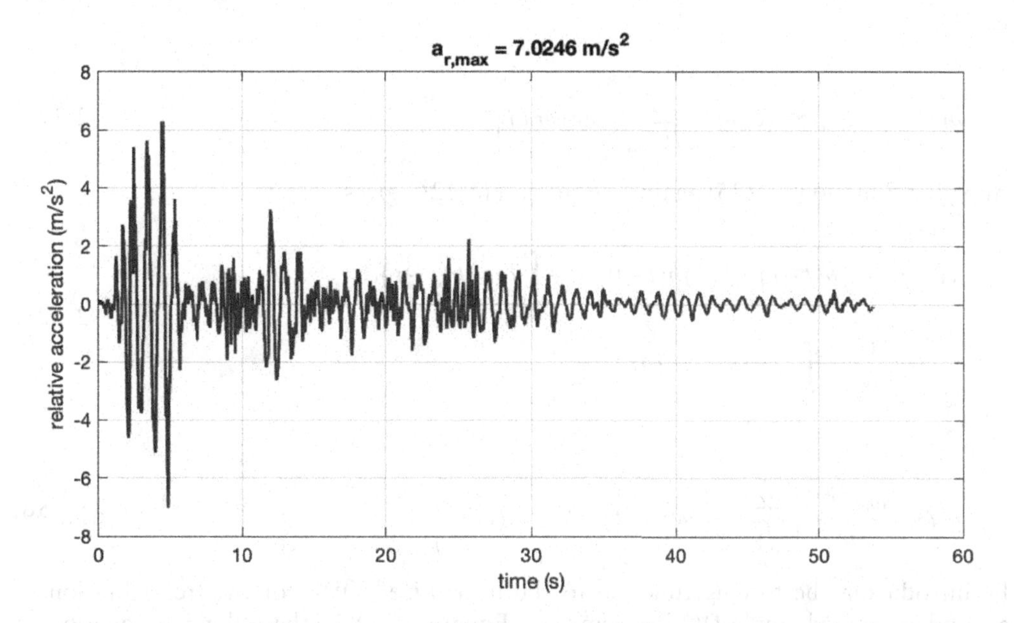

Figure 2.48 The relative acceleration of the floor mass under the El Centro earthquake.

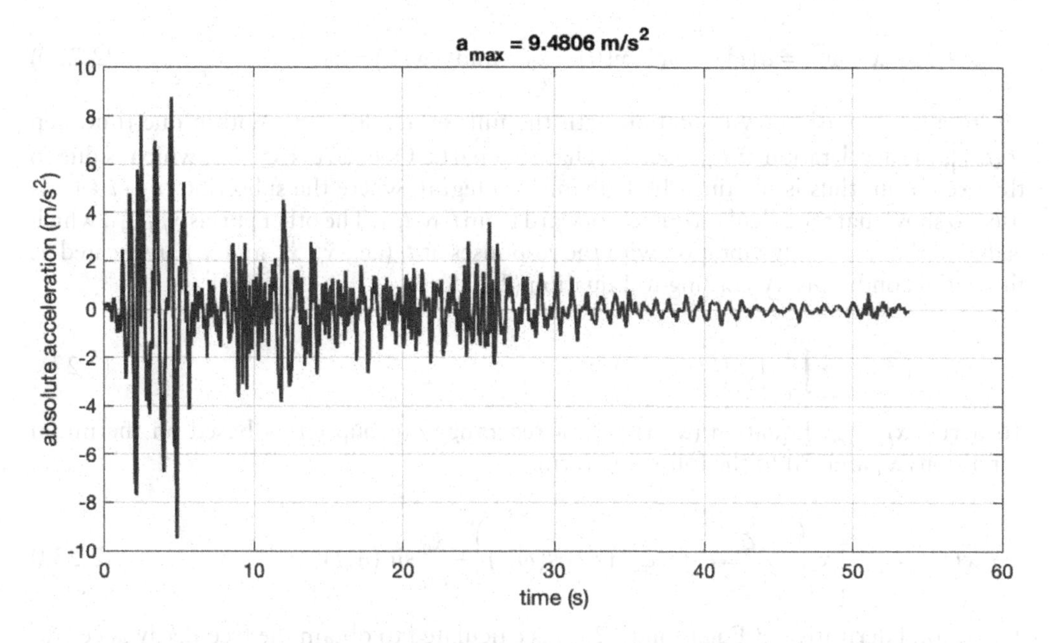

Figure 2.49 The absolute acceleration of the floor mass under the El Centro earthquake.

Applying Equation (2.225) to Equation (2.151) gives:

$$
\begin{aligned}
\dot{x}(t) &= f(t)h(t-t) - f(0)h(t-0)\cdot 0 + \int_0^t f(\tau)\dot{h}(t-\tau)d\tau \\
&= \int_0^t f(\tau)\dot{h}(t-\tau)d\tau
\end{aligned}
\tag{2.226}
$$

where

$$\dot{h}(t)=\frac{e^{-\zeta\omega_n t}}{m}\left(\cos(\omega_d t)-\frac{\zeta}{\sqrt{1-\zeta^2}}\sin(\omega_d t)\right) \tag{2.227}$$

Applying Equation (2.225) again to Equation (2.226) gives:

$$\begin{aligned}\ddot{x}(t)&=f(t)\dot{h}(t-t)-f(0)\dot{h}(t-0)\cdot 0+\int_0^t f(\tau)\ddot{h}(t-\tau)d\tau\\&=\frac{f(t)}{m}+\int_0^t f(\tau)\ddot{h}(t-\tau)d\tau\end{aligned} \tag{2.228}$$

where

$$\ddot{h}(t)=\frac{\omega_n e^{-\zeta\omega_n t}}{m}\left(\frac{2\zeta^2-1}{\sqrt{1-\zeta^2}}\sin(\omega_d t)-2\zeta\cos(\omega_d t)\right) \tag{2.229}$$

By introducing the forcing function to the normalized ODE for the free vibration of an under-damped single-DOF system, i.e., Equation (2.85), the following equation of motion can be obtained:

$$\ddot{x}+2\zeta\omega_n\dot{x}+\omega_n^2 x=a(t),\quad \text{for}\quad x(0)=x_0\quad\text{and}\quad \dot{x}(0)=\dot{x}_0 \tag{2.230}$$

where $a(t)$ is an arbitrary excitation with the unit of acceleration. Within one time step $[t_i,t_{i+1}]$, the acceleration at t_{i+1} (i.e., $\ddot{x}_{i+1}$) has two parts. One part is $\ddot{x}_{a,i\to i+1}$, which is due to the excitation, thus is obtained by Duhamel's integral, where the subscript "$i\to i+1$" is used to show that the calculation goes forward from t_i to t_{i+1}. The other part is $\ddot{x}_{f,i\to i+1}$, which is due to the free-decay vibration with the responses at t_i (i.e., $\ddot{x}_i$, $\dot{x}_i$, and x_i) are treated as the initial conditions. According to Equation (2.228), $\ddot{x}_{a,i\to i+1}$ is expressed as:

$$\ddot{x}_{a,i\to i+1}=a_{i+1}+\int_{t_i}^{t_{i+1}} a(\tau)\ddot{h}(t_{i+1}-\tau)d\tau \tag{2.231}$$

To derive $\ddot{x}_{f,\,i\to i+1}$, Equation (2.41) is first rearranged (group terms based on the initial conditions x_0 and $\dot{x}_0$) to the following form:

$$x(t)=\mathrm{e}^{-\zeta\omega_n t}\left[x_0\left(\frac{\zeta}{\sqrt{1-\zeta^2}}\sin(\omega_d t)+\cos(\omega_d t)\right)+\frac{\dot{x}_0}{\omega_d}\sin(\omega_d t)\right] \tag{2.232}$$

The second derivative of Equation (2.232) is calculated to obtain the free-decay acceleration due to initial conditions:

$$\begin{aligned}\ddot{x}(t)&=e^{-\zeta\omega_n t}\left(2\zeta\omega_n\dot{x}_0+\omega_n^2 x_0\right)\left(\frac{\zeta}{\sqrt{1-\zeta^2}}\sin(\omega_d t)-\cos(\omega_d t)\right)\\&\quad-\frac{e^{-\zeta\omega_n t}\sin(\omega_d t)}{\sqrt{1-\zeta^2}}\omega_n\dot{x}_0\end{aligned} \tag{2.233}$$

Notice that the initial conditions satisfy the equation of motion:

$$\ddot{x}_0 + 2\zeta\omega_n\dot{x}_0 + \omega_n^2 x_0 = 0 \tag{2.234}$$

By writing $2\zeta\omega_n\dot{x}_0 + \omega_n^2 x_0 = -\ddot{x}_0$ using Equation (2.234), and substituting it into (2.233), x_0 can be canceled, so Equation (2.233) becomes:

$$\ddot{x}(t) = \ddot{x}_0 e^{-\zeta\omega_n t}\left(-\frac{\zeta}{\sqrt{1-\zeta^2}}\sin(\omega_d t) + \cos(\omega_d t)\right) - \frac{e^{-\zeta\omega_n t}\sin(\omega_d t)}{\sqrt{1-\zeta^2}}\omega_n\dot{x}_0 \tag{2.235}$$

$\ddot{x}_{f,\, i\to i+1}$ is obtained by Equation (2.235):

$$\ddot{x}_{f,\, i\to i+1} = \ddot{x}_i e^{-\zeta\omega_n \Delta t}\left(-\frac{\zeta}{\sqrt{1-\zeta^2}}\sin(\omega_d \Delta t) + \cos(\omega_d \Delta t)\right) - \frac{e^{-\zeta\omega_n t}\sin(\omega_d \Delta t)}{\sqrt{1-\zeta^2}}\omega_n\dot{x}_i \tag{2.236}$$

where $\Delta t = t_{i+1} - t_i$ is the length of the time step. $\ddot{x}_{i+1}$ is thus obtained by Equations (2.231) and (2.236):

$$\begin{aligned}\ddot{x}_{i+1} &= \ddot{x}_{a,i\to i+1} + \ddot{x}_{f,\, i\to i+1} \\ &= a_{i+1} + \int_{t_i}^{t_{i+1}} a(\tau)\ddot{h}(t_{i+1} - \tau)\,d\tau \\ &\quad + \ddot{x}_i e^{-\zeta\omega_n \Delta t}\left(-\frac{\zeta}{\sqrt{1-\zeta^2}}\sin(\omega_d \Delta t) + \cos(\omega_d \Delta t)\right) - \frac{e^{-\zeta\omega_n \Delta t}\sin(\omega_d \Delta t)}{\sqrt{1-\zeta^2}}\omega_n\dot{x}_i\end{aligned} \tag{2.237}$$

It can be seen from Equation (2.237) that the acceleration at the current time instant depends not only on the acceleration at the previous time instant, but also on the velocity at the previous time instant (see the last term in Equation (2.237)). It means that to use Equation (2.237), both accelerations and velocities are required to be calculated. This could be a waste of computational time if only accelerations are needed. In the following, we try to eliminate the velocity term from the formulation, so that accelerations can be calculated without the need to calculate velocities. The computational efficiency will thereby be improved.

Equation (2.237) is derived by applying Duhamel's integral forward from t_i to t_{i+1}. Similarly, we can express $\ddot{x}_i$ in terms of $\ddot{x}_{i-1}$ and $\dot{x}_i$ by applying Duhamel's integral backward from t_i to t_{i-1}. First, consider the displacement of Duhamel's integral in the backward direction within one time step from Δt to t:

$$x(t) = \int_t^{\Delta t} \left(-f(\tau)\right)h(t-\tau)\,d\tau \tag{2.238}$$

Differentiating Equation (2.238) according to Equation (2.225) gives:

$$\begin{aligned}\dot{x}(t) &= \left(-f(\Delta t)\right)h(t-\Delta t)\cdot 0 - \left(-f(t)\right)h(t-t) + \int_t^{\Delta t}\left(-f(\tau)\right)\dot{h}(t-\tau)\,d\tau \\ &= \int_t^{\Delta t}\left(-f(\tau)\right)\dot{h}(t-\tau)\,d\tau\end{aligned} \tag{2.239}$$

Differentiating Equation (2.239) according to Equation (2.225) gives:

$$\begin{aligned}\ddot{x}(t) &= \left(-f(\Delta t)\right)\dot{h}(t-\Delta t)\cdot 0-\left(-f(t)\right)\dot{h}(t-t)+\int_t^{\Delta t}\left(-f(\tau)\right)\ddot{h}(t-\tau)d\tau \\ &= \frac{f(t)}{m}+\int_t^{\Delta t}\left(-f(\tau)\right)\ddot{h}(t-\tau)d\tau\end{aligned} \tag{2.240}$$

By applying Equation (2.240), the backward acceleration Duhamel's integral from t_i to t_{i-1} is obtained:

$$\ddot{x}_{a,i\to i-1} = a_{i-1}+\int_{t_{i-1}}^{t_i}\left(-a(\tau)\right)\ddot{h}(t_{i-1}-\tau)d\tau \tag{2.241}$$

The free-decay response in the backward direction with $\ddot{x}_i$ and $\dot{x}_i$ as the initial conditions is written as

$$\begin{aligned}\ddot{x}_{f,\,i\to i-1} &= \ddot{x}_i e^{-\zeta\omega_n(-\Delta t)}\left(-\frac{\zeta}{\sqrt{1-\zeta^2}}\sin\left(\omega_d(-\Delta t)\right)+\cos\left(\omega_d(-\Delta t)\right)\right) \\ &\quad -\frac{e^{-\zeta\omega_n(-\Delta t)}\sin\left(\omega_d(-\Delta t)\right)}{\sqrt{1-\zeta^2}}\omega_n\dot{x}_i\end{aligned} \tag{2.242}$$

Using Equations (2.241) and (2.242), $\ddot{x}_{i-1}$ can be written in terms of $\ddot{x}_i$ and $\dot{x}_i$:

$$\begin{aligned}\ddot{x}_{i-1} &= \ddot{x}_{a,i\to i-1}+\ddot{x}_{f,\,i\to i-1} \\ &= a_{i-1}-\int_{t_{i-1}}^{t_i} a(\tau)\ddot{h}(t_{i-1}-\tau)d\tau \\ &\quad +\ddot{x}_i e^{\zeta\omega_n\Delta t}\left(\frac{\zeta}{\sqrt{1-\zeta^2}}\sin(\omega_d\Delta t)+\cos(\omega_d\Delta t)\right)+\frac{e^{\zeta\omega_n\Delta t}\sin(\omega_d\Delta t)}{\sqrt{1-\zeta^2}}\omega_n\dot{x}_i\end{aligned} \tag{2.243}$$

With Equations (2.237) and (2.241), we can eliminate the velocity term. Multiplying $e^{-2\xi\omega_n\Delta t}$ to Equation (2.243) and adding it to Equation (2.237) gives:

$$\begin{aligned}\ddot{x}_{i+1} &= a_{i+1}+e^{-2\zeta\omega_n\Delta t}a_{i-1}-e^{-2\zeta\omega_n\Delta t}\ddot{x}_{i-1}+\int_{t_i}^{t_{i+1}} a(\tau)\ddot{h}(t_{i+1}-\tau)d\tau \\ &\quad -e^{-2\zeta\omega_n\Delta t}\int_{t_{i-1}}^{t_i} a(\tau)\ddot{h}(t_{i-1}-\tau)d\tau+2\ddot{x}_i e^{-\zeta\omega_n\Delta t}\cos(\omega_d\Delta t)\end{aligned} \tag{2.244}$$

By assuming that the excitation is linear within each time step, $a(\tau)$ in the above equation has the following form:

$$a(\tau)=\begin{cases} a_{i-1}+\dfrac{a_i-a_{i-1}}{\Delta t}(\tau-t_{i-1}),\tau\in[t_{i-1},t_i] \\ a_i+\dfrac{a_{i+1}-a_i}{\Delta t}(\tau-t_i),\tau\in[t_i,t_{i+1}]\end{cases} \tag{2.245}$$

Given the excitation expressed in the linear form and Equation (2.230), the two Duhamel's integrals can be evaluated analytically:

$$\int_{t_i}^{t_{i+1}} a(\tau)\ddot{h}(t_{i+1}-\tau)d\tau = \int_0^{\Delta t}\left(a_i + \frac{a_{i+1}-a_i}{\Delta t}\tau\right)\ddot{h}(\Delta t-\tau)d\tau$$

$$= -a_{i+1} + \frac{e^{-\zeta\omega_n\Delta t}\sin(\omega_d\Delta t)}{\omega_d\Delta t}a_{i+1} + \left(e^{-\zeta\omega_n\Delta t}\cos(\omega_d\Delta t)\right)a_i \tag{2.246}$$

$$-\frac{e^{-\zeta\omega_n\Delta t}\sin(\omega_d\Delta t)}{\omega_d\Delta t}a_i - \frac{\zeta e^{-\zeta\omega_n\Delta t}\sin(\omega_d\Delta t)}{\sqrt{1-\zeta^2}}a_i$$

$$\int_{t_{i-1}}^{t_i} a(\tau)\ddot{h}(t_{i-1}-\tau)d\tau = \int_0^{\Delta t}\left(a_{i-1} + \frac{a_i-a_{i-1}}{\Delta t}\tau\right)\ddot{h}(0-\tau)d\tau$$

$$= \frac{\zeta e^{\zeta\omega_n\Delta t}\sin(\omega_d\Delta t)}{\sqrt{1-\zeta^2}}a_i - \frac{e^{\zeta\omega_n\Delta t}\sin(\omega_d\Delta t)}{\omega_d\Delta t}a_i \tag{2.247}$$

$$+\left(e^{\zeta\omega_n\Delta t}\cos(\omega_d\Delta t)\right)a_i - a_{i-1} + \frac{e^{\zeta\omega_n\Delta t}\sin(\omega_d\Delta t)}{\omega_d\Delta t}a_{i-1}$$

Substituting Equations (2.246) and (2.247) into Equation (2.244) and simplifying gives:

$$\ddot{x}_{i+1} = b_1\ddot{x}_i + b_2\ddot{x}_{i-1} + b_3a_{i+1} + b_4a_i + b_5a_{i-1} \tag{2.248}$$

where

$$b_1 = 2e^{-\zeta\omega_n\Delta t}\cos(\omega_d\Delta t)$$

$$b_2 = -e^{-2\zeta\omega_n\Delta t}$$

$$b_3 = \frac{e^{-\zeta\omega_n\Delta t}\sin(\omega_d\Delta t)}{\omega_d\Delta t} \tag{2.249}$$

$$b_4 = 2e^{-\zeta\omega_n\Delta t}\left(\cos(\omega_d\Delta t) - \frac{\sin(\omega_d\Delta t)}{\omega_d\Delta t}\right)$$

$$b_5 = \frac{e^{-\zeta\omega_n\Delta t}\sin(\omega_d\Delta t)}{\omega_d\Delta t}$$

It is observed that the calculation of accelerations in Equation (2.258) only involves accelerations and excitations without the need to calculate displacements and velocities. This improves the efficiency of the computational process. An initializing equation from t_0 to t_1 is needed to calculate $\ddot{x}_1$ as the acceleration in each time instance depends on accelerations in two previous time instances. Similarly, $\ddot{x}_1$ consists of two parts, one due to the excitation $\ddot{x}_{a,\,0\to1}$ and the other due to the free-decay response $\ddot{x}_{f,\,0\to1}$. $\ddot{x}_{f,\,0\to1}$ is obtained using Equation (2.233):

$$\ddot{x}_{f,\,0\to1} = b_6x_0 + b_7\dot{x}_0 \tag{2.250}$$

where

$$b_6 = \omega_n^2 e^{-\zeta\omega_n\Delta t}\left(\frac{\zeta}{\sqrt{1-\zeta^2}}\sin(\omega_d\Delta t) - \cos(\omega_d\Delta t)\right)$$
$$b_7 = \omega_n e^{-\zeta\omega_n\Delta t}\left(\frac{2\zeta^2-1}{\sqrt{1-\zeta^2}}\sin(\omega_d\Delta t) - 2\zeta\cos(\omega_d\Delta t)\right) \tag{2.251}$$

$\ddot{x}_{a,\,0\to1}$ is obtained by the forward Duhamel's integral based on Equations (2.231) and (2.246):

$$\ddot{x}_{a,0\to1} = b_8 a_1 + b_9 a_0 \tag{2.252}$$

where

$$b_8 = \frac{e^{-\zeta\omega_n\Delta t}\sin(\omega_d\Delta t)}{\omega_d\Delta t}$$
$$b_9 = e^{-\zeta\omega_n\Delta t}\left(\cos(\omega_d\Delta t) - \frac{1+\zeta\omega_n\Delta t}{\omega_d\Delta t}\sin(\omega_d\Delta t)\right) \tag{2.253}$$

Therefore, the initializing equation for calculating $\ddot{x}_1$ is:

$$\begin{aligned}\ddot{x}_1 &= \ddot{x}_{f,\,0\to1} + \ddot{x}_{a,0\to1}\\ &= b_6 x_0 + b_7\dot{x}_0 + b_8 a_1 + b_9 a_0\end{aligned} \tag{2.254}$$

By using Equations (2.248), (2.249), (2.254), (2.251), and (2.253), accelerations under a general excitation can be efficiently calculated.

2.2 MULTI-DEGREE-OF-FREEDOM SYSTEMS

All structures in the real world are continuous systems with an infinite number of degrees-of-freedom (DOFs). Depending on the amount of information that one intends to obtain from the analysis, a multi-degree of freedom model (multi-DOF) can be used to approximate the behavior of an infinity-degree of freedom system. The purpose of this section is to introduce the basic knowledge about the dynamic analysis of multi-DOF systems. This section will first introduce the easiest type of multi-DOF system—multi-story shear buildings, which work as examples to demonstrate several important characteristics of a general multi-DOF structural system. Then, the concept of natural frequencies and mode shapes will be covered. It turns out that the analysis of a multi-degree of freedom system can be converted into the analyses of a series of single-DOF systems by using the method of modal superposition. Following this approach, the single-DOF response calculation techniques can be employed in solving multi-DOF system problems.

2.2.1 Shear building model—a multi-DOF system

The shear building model is not only an example of multi-DOF systems but also a very good tool to study some important structural behaviors of a complicated building structure.

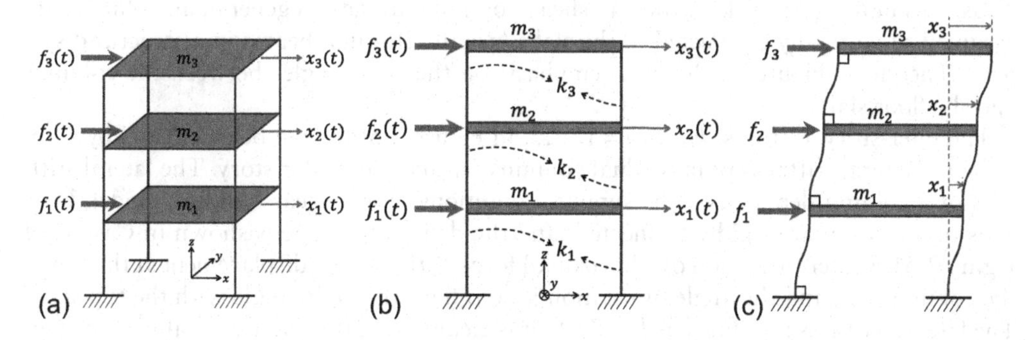

Figure 2.50 Shear building model. (a) Typical shear building model in 3D. (b) 2D projection of the shear building model. (c) Deflection shape of a shear building model.

Figure 2.50(a) shows a typical shear building model with three stories in three dimensions (3D). As shown in the figure, the origin of the coordinate system is on the ground and directly under the centroid of the floor slab. In the figure, m_i, for $i = 1,\ldots,3$, are the total floor mass at the i-th floor; $f_i(t)$, for $i = 1,\ldots,3$, are the external forces applied at the i-th floor; $x_i(t)$, for $i = 1,\ldots,3$, are the displacements of the i-th floor. For explanation purposes, each floor is supported by four columns. More columns can be considered in real situations. To simplify the situation, only the motion along the x-direction is considered. After understanding the method, the motion along the y-direction can be considered in the same manner. When the structural system is symmetrical about the xz-plane at $y = 0$, a force applied through the centroid of the floor in the x-direction will only induce displacement in the x-direction. Under such a situation, the 3D shear building model can be considered as a 2D model, as shown in Figure 2.50(b), where k_i, for $i = 1,\ldots,3$, is the inter-story stiffness of the i-th story. Note that the inter-story stiffness is the sum of lateral stiffness of all columns in a given story. In this example, there are four columns in each story.

The detailed assumptions of a shear building model (as shown in Figure 2.50(b)) are given below:

1. The total mass of the structure is assumed to lump at the floor levels. It implies that the mass of columns is either neglected (the mass of columns is usually small when compared to the mass of slabs and beams) or equally distributed to the upper and lower floors.
2. The bending and axial stiffnesses of the floor slabs (together with the beams, if any) are assumed to be infinitely rigid when compared to those of the columns. Thus, the lateral deflections of the building (at the DOFs) are due to the bending of columns. To graphically show this property, the depth of slabs is much larger than the depth of the columns in the figure.
3. The (vertical) axial deformation of the columns is independent of the (horizontal) lateral deformation of the building. Thus, only the lateral displacement (or vibration) is considered for each floor.

With these assumptions, a building structure with an infinite number of DOFs (as a distributed system) is converted to an N-DOF system, where N is the number of stories (or floors). N is equal to 3 for the example given in Figure 2.50(b). The lateral deflection of this building model follows the deflection shape of a shear cantilever (see Figure 2.50(c)).

Thus, this building model is called a "shear" building model. In general, all columns are assumed to be rigidly connected to the slabs (usually through beams in a structural system). Therefore, Figure 2.50(c) puts emphasis on the right angles between the columns and the floor slabs.

The inter-story stiffness, k_i, for $i = 1,\ldots,3$, in Figure 2.50(b) can be calculated by summing the lateral stiffness of individual columns on that particular story. The lateral stiffness of a column depends on the support conditions. For a shear building model, both ends of a column are rigidly connected, and the deflection shape is shown in Case C of Figure 2.51. Under the action of a horizontal force F, the lateral displacement at the top of the column is Δ, and the angle of rotation is zero (i.e., at a right angle with the top slab). The lateral stiffness is defined as $k = F/\Delta$. It is clear that Case C is a statically indeterminate system. Many methods are available for calculating the required force F to induce the displacement Δ for given column properties, such as the modulus of elasticity, E, the second moment of area of the column about the bending axis, I, and the length of the column, L. Here, the feasibility method (or force method) is employed to solve the problem following the idea of superposition. In Figure 2.51, the statically indeterminate system in Case C can be considered as the superposition of the two systems in Cases A and B; both cantilever columns (statically determinate). The cantilever in Case A is subjected to a unit load that induced a lateral displacement of δ_A with a rotation of α_A, while the cantilever in Case B is subjected to unit moment that induced a lateral displacement of δ_B with a rotation of α_B. All δs and αs can be easily calculated (the corresponding systems are statically determinate). For a linear system, the lateral displacement and rotation in both Cases A and B can be adjusted by applying (or multiplying) a force F to Case A and a moment M to Case B. The requirement for the superposition of Cases A and B to return Case C is that the sum of the rotations in Cases A and B must vanish, as there is no rotation in Case C. Under such conditions, the sum of lateral displacements in both Cases A and B returns Δ in Case C. Thus, one obtains the compatibility conditions (or equations) as:

$$\begin{aligned} &\delta_A F + \delta_B M = \Delta \\ &\alpha_A F + \alpha_B M = 0 \end{aligned} \tag{2.255}$$

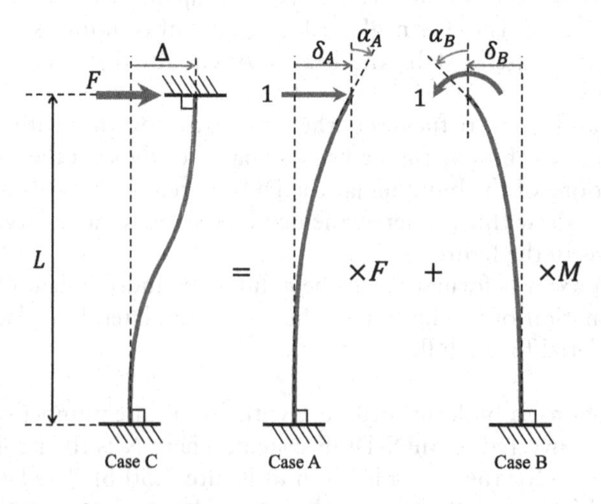

Figure 2.51 The calculation of lateral stiffness of a column.

where F and M are the unknowns, and δ_A, δ_B, α_A, and α_B are the feasibility coefficients (i.e., the displacement/rotation induced by unit force/moment). Thus, this method is called the feasibility method. As the main purpose of Equation (2.255) is to solve for the unknown forces F and M, this method is also called the force method.

To construct the bending moment diagram (and so as to calculate the deflection of the system), the column in Case A is rotated through 90° as shown in Figure 2.52(a). Consider a virtual cut on the beam at x from the left support (the origin of the coordinate system) and focus on the free-body diagram of the left beam segment. Calculating the sum of moments at x from the left end of the beam segment, one obtains:

$$\Sigma M_x = 0 \Rightarrow M(x) = x - L \tag{2.256}$$

Considering the second-order ODE for beam bending:

$$y'' = -\frac{M(x)}{EI} \Rightarrow EIy'' = -M(x) = L - x \tag{2.257}$$

where $y'' = d^2y / dx^2$ is the second derivative of the deflection, y, and EI is the bending rigidity of the beam (both E and I were previously defined). The deflection of the beam, y, can be obtained by the double integration of the negative bending moment as a function of x. Integrating Equation (2.257):

$$EIy' = Lx - \frac{x^2}{2} + c_1 \tag{2.258}$$

where c_1 is the integration constant, and it is equal to zero due to the boundary condition that the slope of the beam is zero at the left end (i.e., $y'|_{x=0} = 0$). Integrating Equation (2.258):

$$EIy = \frac{Lx^2}{2} - \frac{x^3}{6} + c_2 \tag{2.259}$$

where c_2 is another integration constant, and it is also equal to zero due to the boundary condition that the deflection of the beam is zero at the left end (i.e., $y|_{x=0} = 0$). The slope (or the rotation) and deflection of the beam in Case A can be expressed by Equations (2.258) and (2.259) as:

$$y' = \frac{x}{EI}\left(L - \frac{x}{2}\right) \quad \text{and} \quad y = \frac{x^2}{2EI}\left(L - \frac{x}{3}\right) \tag{2.260}$$

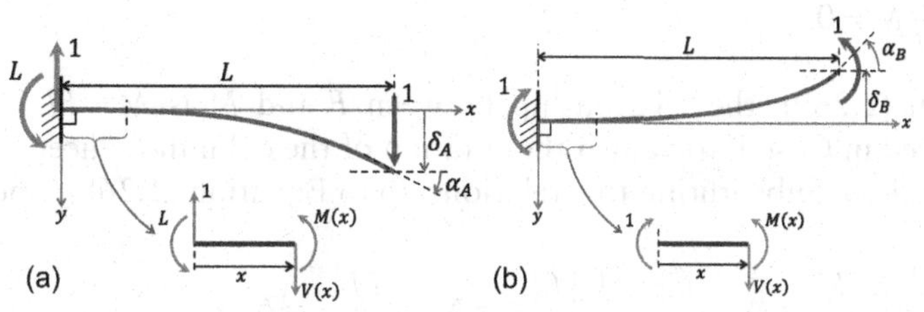

Figure 2.52 The Calculation of feasibility coefficients in Cases A and B. (a) Case A. (b) Case B.

Thus, the feasibility coefficients for Case A can be calculated as:

$$\alpha_A = y'\big|_{x=L} = \frac{L^2}{2EI} \quad \text{and} \quad \delta_A = y\big|_{x=L} = \frac{L^3}{3EI} \tag{2.261}$$

Referring to Figure 2.52(b) for the free-body diagram of the left beam segment, the bending moment in Case B can be expressed as:

$$M(x) = 1 \tag{2.262}$$

The second-order ODE becomes:

$$EIy'' = -M(x) = -1 \tag{2.263}$$

Integrating Equation (2.263), one obtains:

$$EIy' = -x + c_1, \quad \text{where } c_1 = 0 \text{ as } y'\big|_{x=0} = 0 \tag{2.264}$$

and

$$EIy = -\frac{x^2}{2} + c_2, \quad \text{where } c_2 = 0 \text{ as } y\big|_{x=0} = 0 \tag{2.265}$$

The slope (or rotation) and deflection of the beam in Case B can be calculated as:

$$y' = -\frac{x}{EI} \quad \text{and} \quad y = -\frac{x^2}{2EI} \tag{2.266}$$

Thus, the feasibility coefficients for Case B can be calculated as:

$$\alpha_B = y'\big|_{x=L} = -\frac{L}{EI} \quad \text{and} \quad \delta_B = y\big|_{x=L} = -\frac{L^2}{2EI} \tag{2.267}$$

Substitute the feasibility coefficients in Equations (2.261) and (2.267) into Equation (2.255), one obtains:

$$\frac{L^3}{3EI}F - \frac{L^2}{2EI}M = \Delta \tag{2.268}$$

$$\frac{L^2}{2EI}F - \frac{L}{EI}M = 0 \tag{2.269}$$

From Equation (2.269), the relationship between F and M is $M = FL/2$. This is the moment required in Case B to vanish the rotation of the column induced by F (i.e., compatibility condition). Substituting this relationship to Equation (2.268), one obtains:

$$\frac{L^3}{3EI}F - \frac{L^2}{2EI}\left(\frac{FL}{2}\right) = \Delta \Rightarrow \left(\frac{1}{3} - \frac{1}{4}\right)\frac{FL^3}{EI} = \Delta \Rightarrow \frac{FL^3}{12EI} = \Delta \tag{2.270}$$

The lateral stiffness of the column can then be calculated as:

$$k = \frac{F}{\Delta} = \frac{12EI}{L^3} \tag{2.271}$$

For the shear building model in Figure 2.50, the inter-story stiffness of the i-th story can be calculated by summing the lateral stiffness of individual columns at that story.

For given values of the inter-story stiffnesses and masses, the shear building model in Figure 2.50 can be represented by a simple spring-mass model as shown in Figure 2.53. The elongation of the spring of the i-th story is the inter-story drift of that story. It must be pointed out that the damping effect is neglected here to simplify the formulation. The effect of damping will be considered later after introducing the important concept of natural frequencies and mode shapes. The equation of motion can be obtained by considering the free-body diagram of each floor mass, as shown in Figure 2.54. Considering the equilibrium of m_1, m_2, and m_3, one obtains:

$$\begin{aligned} &f_1(t) - k_1 x_1 + k_2(x_2 - x_1) = m_1 \ddot{x}_1 \\ &f_2(t) - k_2(x_2 - x_1) + k_3(x_3 - x_2) = m_2 \ddot{x}_2 \\ &f_3(t) - k_3(x_3 - x_2) = m_3 \ddot{x}_3 \end{aligned} \tag{2.272}$$

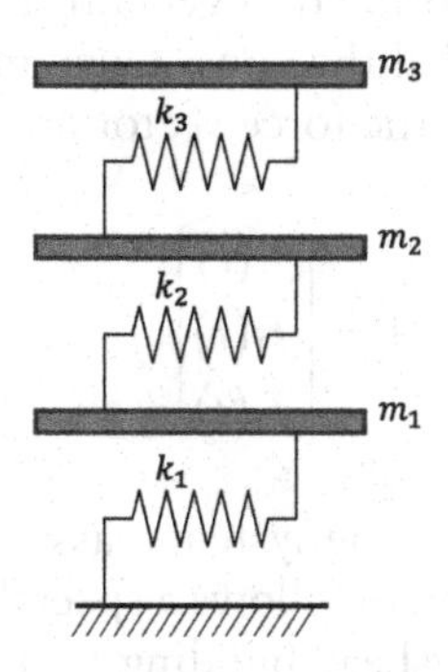

Figure 2.53 The spring-mass representation of the shear building model.

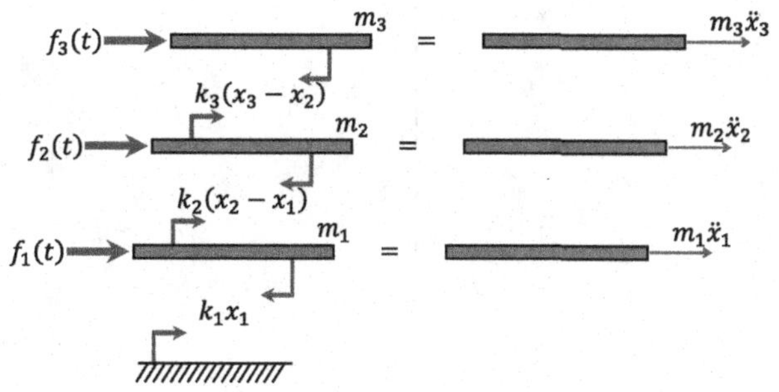

Figure 2.54 The free-body diagrams for each floor of the shear building model.

Rearranging them to get the set of governing equations for the undamped forced vibration of a three-story shear building model:

$$\begin{aligned} m_1\ddot{x}_1 + (k_1 + k_2)x_1 - k_2x_2 &= f_1(t) \\ m_2\ddot{x}_2 - k_2x_1 + (k_2 + k_3)x_2 - k_3x_3 &= f_2(t) \\ m_3\ddot{x}_3 - k_3x_2 + k_3x_3 &= f_3(t) \end{aligned} \tag{2.273}$$

This set of governing equations is convenient to be presented in matrix form as:

$$\mathbf{M}\ddot{\mathbf{X}} + \mathbf{K}\mathbf{X} = \mathbf{F} \tag{2.274}$$

where **M** and **K** are the system mass matrix and system stiffness matrix of the multi-DOF system (i.e., the three-story shear building model in this case) defined as:

$$\mathbf{M} = \begin{bmatrix} m_1 & 0 & 0 \\ 0 & m_2 & 0 \\ 0 & 0 & m_3 \end{bmatrix} \quad \text{and} \quad \mathbf{K} = \begin{bmatrix} k_1 + k_2 & -k_2 & 0 \\ -k_2 & k_2 + k_3 & -k_3 \\ 0 & -k_3 & k_3 \end{bmatrix} \tag{2.275}$$

Note that the form of **M** defined here is, in general, denoted as the lumped mass matrix (as the masses are lumped as each DOFs). For a general structural system, another common form of system mass matrix is called the consistency mass matrix. The acceleration and displacement vectors together with the force vector are defined as:

$$\ddot{\mathbf{X}} = \begin{Bmatrix} \ddot{x}_1 \\ \ddot{x}_2 \\ \ddot{x}_3 \end{Bmatrix} \quad \text{and} \quad \mathbf{X} = \begin{Bmatrix} x_1 \\ x_2 \\ x_3 \end{Bmatrix} \quad \text{and} \quad \mathbf{F} = \begin{Bmatrix} f_1(t) \\ f_2(t) \\ f_3(t) \end{Bmatrix} \tag{2.276}$$

It is clear from Equation (2.275) that the system mass matrix is a diagonal matrix of floor masses, and the system stiffness matrix follows a special pattern. Thus, it is easy to extend the formulation of the three-story shear building to a general N-story one as shown in Figure 2.55.

The system mass matrix becomes:

$$\mathbf{M} = \begin{bmatrix} m_1 & & & & & \\ & m_2 & & & & \\ & & \ddots & & & \\ & & & m_i & & \\ & & & & \ddots & \\ & & & & & m_N \end{bmatrix} \tag{2.277}$$

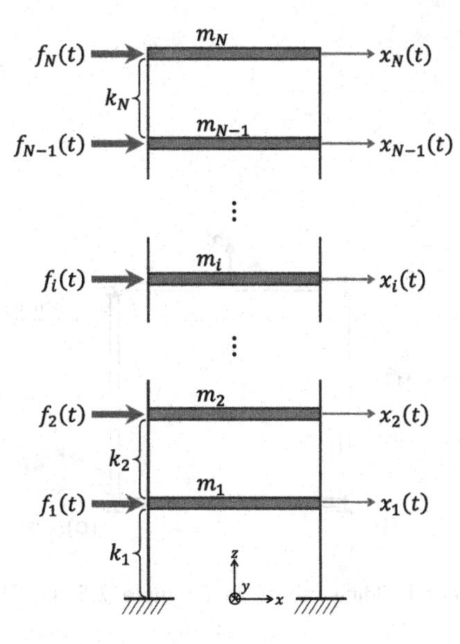

Figure 2.55 The shear building model with *N* stories.

where all empty elements are equal to zero. The system stiffness matrix becomes:

$$\mathbf{K} = \begin{bmatrix} k_1 + k_2 & -k_2 & & & & \\ -k_2 & k_2 + k_3 & \ddots & & & \\ & \ddots & \ddots & -k_i & & \\ & & -k_i & k_i + k_{i+1} & \ddots & \\ & & & \ddots & \ddots & -k_N \\ & & & & -k_N & k_N \end{bmatrix} \tag{2.278}$$

and the acceleration, displacement, and force vectors become:

$$\ddot{\mathbf{X}} = \begin{Bmatrix} \ddot{x}_1 \\ \ddot{x}_2 \\ \vdots \\ \ddot{x}_i \\ \vdots \\ \ddot{x}_N \end{Bmatrix} \quad \text{and} \quad \mathbf{X} = \begin{Bmatrix} x_1 \\ x_2 \\ \vdots \\ x_i \\ \vdots \\ x_N \end{Bmatrix} \quad \text{and} \quad \mathbf{F} = \begin{Bmatrix} f_1(t) \\ f_2(t) \\ \vdots \\ f_i(t) \\ \vdots \\ f_N(t) \end{Bmatrix} \tag{2.279}$$

Example 2.5

Determine the system stiffness and mass matrices for the four-story shear building shown in Figure 2.56(a) (consider vibration along the x-direction only). It is assumed that the four floors have the same floor plan as the one given in Figure 2.56(b). All columns are made of steel with the same cross-section as shown in Figure 2.56(c), where h = 600 mm (section height), b = 400 mm (section breadth), t_f = 40mm (flange

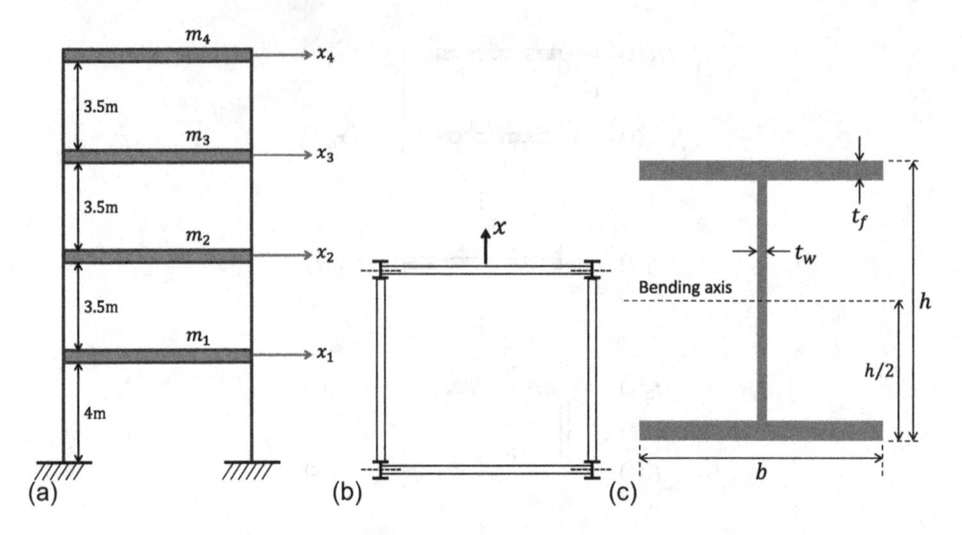

Figure 2.56 The four-story shear building model in Example 2.5. (a) Elevation view. (b) Plan view. (c) Column section.

thickness), and $t_w = 20\text{mm}$ (web thickness). It is also assumed that the column section is bending about the centroidal axis. The modulus of elasticity of steel is 200 GPa, and the total mass of each floor is equal to 18,000 kg.

SOLUTION

The second moment of area, I, of the column section about the bending axis is:

$$I = \frac{h^3 b}{12} - \frac{(h - 2t_f)^3 (b - t_w)}{12} = 2.7474 \times 10^9 \text{ mm}^4 = 2.7474 \times 10^{-3} \text{ m}^4 \tag{2.280}$$

The inter-story stiffness of the first story (i.e., the sum of the lateral stiffness of all four fix-fix columns on the first story) is (see Equation (2.271) for the lateral stiffness of a fix-fix column):

$$k_1 = 4 \times \frac{12EI}{L_1^3} = 4 \times \frac{12(200 \times 10^9)(2.7474 \times 10^{-3})}{4^3} = 4.1211 \times 10^8 \text{ N/m} \tag{2.281}$$

Similarly for the inter-story stiffness of other stories:

$$k_2 = k_3 = k_4 = 4 \times \frac{12(200 \times 10^9)(2.7474 \times 10^{-3})}{3.5^3} = 6.1516 \times 10^8 \text{ N/m} \tag{2.282}$$

The system stiffness matrix can be expressed by assigning $N = 4$ in Equation (2.278) to get:

$$\mathbf{K} = \begin{bmatrix} k_1 + k_2 & -k_2 & 0 & 0 \\ -k_2 & k_2 + k_3 & -k_3 & 0 \\ 0 & -k_3 & k_3 + k_4 & -k_4 \\ 0 & 0 & -k_4 & k_4 \end{bmatrix}$$
$$= \begin{bmatrix} 10.2728 & -6.1516 & 0 & 0 \\ -6.1516 & 12.3033 & -6.1516 & 0 \\ 0 & -6.1516 & 12.3033 & -6.1516 \\ 0 & 0 & -6.1516 & 6.1516 \end{bmatrix} \times 10^8 \text{ N/m} \tag{2.283}$$

The system mass matric is (assigning $N = 4$ in Equation (2.277)):

$$\mathbf{M} = \begin{bmatrix} m_1 & 0 & 0 & 0 \\ 0 & m_2 & 0 & 0 \\ 0 & 0 & m_3 & 0 \\ 0 & 0 & 0 & m_4 \end{bmatrix} = \begin{bmatrix} 1 & 0 & 0 & 0 \\ 0 & 1 & 0 & 0 \\ 0 & 0 & 1 & 0 \\ 0 & 0 & 0 & 1 \end{bmatrix} \times 1.8 \times 10^4 \text{ kg} \tag{2.284}$$

Example 2.6

Determine the system stiffness matrix of the two-story shear building as shown in Figure 2.57. Note that all column-base connections are pinned. It is given that the second moment of area for all columns is $I = 3.7 \times 10^{-5}\ \text{m}^4$, and the modulus of elasticity is $E = 200$ GPa.

SOLUTION

The columns on the first story are pinned at the bottom and fixed at the top. By flipping these columns upside down, they will be the same as the one in Case A of Figure 2.51. Thus, the column deflection shape can be represented by the one in Figure 2.52(a), and the feasibility coefficient (i.e., the displacement induced by a unit force) is given by δ_A is Equation (2.261). The corresponding lateral stiffness (i.e., the force required to produce a unit displacement) can be calculated as:

$$k = \frac{1}{\delta_A} = \frac{3EI}{L^3} \tag{2.285}$$

The inter-story stiffness of the first story is:

$$k_1 = 4 \times \frac{3EI}{L_1^3} = 4 \times \frac{3\left(200 \times 10^9\right)\left(3.7 \times 10^{-5}\right)}{4^3} = 1387500\,\text{N/m} \tag{2.286}$$

The inter-story stiffness of the second story is:

$$k_2 = 4 \times \frac{12EI}{L_2^3} = 4 \times \frac{12\left(200 \times 10^9\right)\left(3.7 \times 10^{-5}\right)}{4^3} = 5550000\,\text{N/m} \tag{2.287}$$

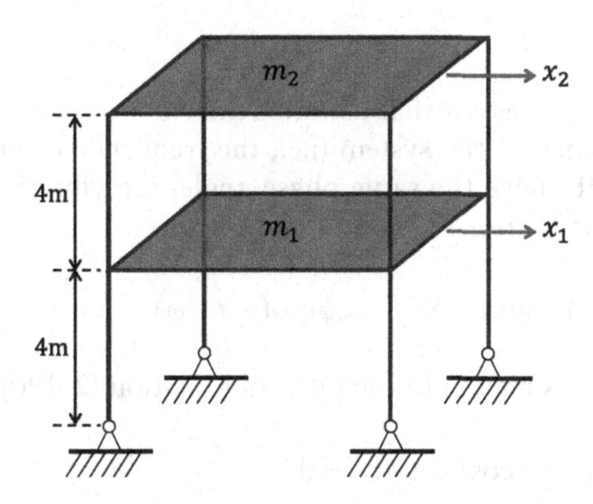

Figure 2.57 The two-story shear building model in Example 2.6.

The system stiffness matrix can be obtained by assigning $N = 2$ in Equation (2.278) to get:

$$\mathbf{K} = \begin{bmatrix} k_1 + k_2 & -k_2 \\ -k_2 & k_2 \end{bmatrix} = \begin{bmatrix} 6937500 & -5550000 \\ -5550000 & 5550000 \end{bmatrix} \text{N/m} \tag{2.288}$$

2.2.2 Undamped free vibration of a multi-DOF system

2.2.2.1 *Natural frequencies and mode shapes*

The free vibration of a multi-DOF system induced by initial disturbance is extremely important in the structural dynamic analysis of the system. In this section, the very important concept of natural frequencies and mode shapes of a multi-DOF system is introduced. For explanation purposes, the shear building model defined in the previous section may be employed as examples in this section.

It must be pointed out that Equation (2.274) is for a general multi-DOF system (not only for shear building models). By removing the external excitation (i.e., $\mathbf{F} = 0$), one obtains the governing equation for the free vibration of an N-DOF system:

$$\mathbf{M}\ddot{\mathbf{X}} + \mathbf{K}\mathbf{X} = 0 \tag{2.289}$$

where $\mathbf{M}$ and $\mathbf{K}$ are the system mass matrix and system stiffness matrix, respectively, of the target structure with dimensions $N \times N$; $\mathbf{0}$ is a zero vector with dimension $N \times 1$; and $\ddot{\mathbf{X}}$ and $\mathbf{X}$ are the acceleration vector and displacement vector, respectively, with dimensions $N \times 1$.

Figure 2.58 shows an example of a two-DOF system in free vibration (this is not a shear building model). Under an initial disturbance (e.g., an initial displacement or initial velocity induced by a horizontal impulse at one of the two masses), the free vibration of the system follows a given pattern. First, the top mass, m_2, is vibrating with a larger amplitude when compared to that of the bottom mass, m_1. The ratio between the two amplitudes of vibration is constant. Second, the vibrations of the two masses are in phase. That is, when the top mass is on the left, the bottom mass is also on the left (and vice versa). With these observations, the solution form of $\mathbf{X}$ can be expressed by extending the one for the free vibration of an undamped single-DOF system (see Equation (2.21)) as:

$$\mathbf{X} = \boldsymbol{\phi}\cos(\omega_n t - \varphi) \tag{2.290}$$

where $\boldsymbol{\phi} = \{\phi_1, \phi_2, \ldots, \phi_N\}^T$ is a vector that contains the vibration amplitudes of all N DOFs; ω_n is the natural frequency of the system (i.e., the frequency of vibration); and φ is the phase angle. As all DOFs have the same phase angle, the vibrations of all DOFs are in phase. The derivatives of $\mathbf{X}$ are:

$$\dot{\mathbf{X}} = -\omega_n \boldsymbol{\phi}\sin(\omega_n t - \varphi) \quad \text{and} \quad \ddot{\mathbf{X}} = -\omega_n^2 \boldsymbol{\phi}\cos(\omega_n t - \varphi) \tag{2.291}$$

Substituting the derivatives back into the ODE in Equation (2.290), one obtains:

$$-\omega_n^2 \mathbf{M}\boldsymbol{\phi}\cos(\omega_n t - \varphi) + \mathbf{K}\boldsymbol{\phi}\cos(\omega_n t - \varphi) = 0 \tag{2.292}$$

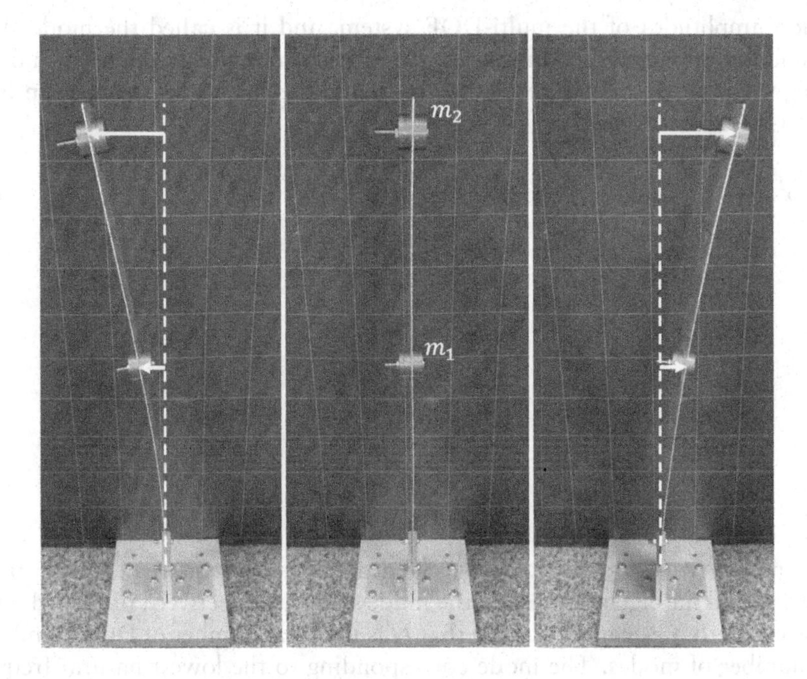

Figure 2.58 The vibration shape of a two-DOF system (under small amplitude vibration).

As the cosine function is not always equal to zero, one obtains:

$$\left(\mathbf{K}-\omega_n^2\mathbf{M}\right)\boldsymbol{\phi}=0 \tag{2.293}$$

This is a set of N homogeneous linear equations with N unknown amplitudes ϕ_i, for $i=1,\ldots,N$, and an unknown parameter, ω_n^2. Equation (2.293) is an important mathematical problem known as an eigenproblem or eigenvalue problem. $\phi_i=0$, for $i=1,\ldots,N$, is a solution of this set of equations. The non-trivial solution requires the determinant of the coefficient matrix to be equal to zero. That is:

$$\left|\mathbf{K}-\omega_n^2\mathbf{M}\right|=0 \tag{2.294}$$

Expanding the determinant in Equation (2.294) gives an N-th order polynomial of ω_n^2. By solving this polynomial, one obtains roots, ω_i^2, for $i=1,\ldots,N$, where ω_i is the natural circular frequency of the i-th mode of vibration in rad/s. Thus, Equation (2.293) becomes:

$$\left(\mathbf{K}-\omega_i^2\mathbf{M}\right)\boldsymbol{\phi}_i=0 \quad \text{for} \quad i=1,\ldots,N. \tag{2.295}$$

where $\boldsymbol{\phi}_i$ is the vector of unknown amplitudes corresponding to the i-th mode of vibration. To solve $\boldsymbol{\phi}_i$, one needs to substitute ω_i back into Equation (2.293) one by one. For each ω_i, one obtains a set of N homogeneous linear equations, for which the exact vibration amplitude at each DOF cannot be determined but only the "relative" amplitudes. That is, if $\boldsymbol{\phi}_i$ satisfies the set of N homogeneous linear equations in Equation (2.295), $a\boldsymbol{\phi}_i$ also satisfies, where a is a constant (it can be positive or negative). Thus, $\boldsymbol{\phi}_i$ only gives the "shape" of vibration amplitudes of the multi-DOF system, and it is called the mode shapes of the i-th mode of vibration. To summarize, the natural circular frequencies and the corresponding mode shapes as the results from solving the eigenvalue problem in Equation (2.293) can be expressed as:

$$\Omega = \begin{bmatrix} \omega_1^2 & & & \\ & \omega_2^2 & & \\ & & \ddots & \\ & & & \omega_N^2 \end{bmatrix} \quad \text{and} \quad \Phi = [\phi_1, \phi_2, \cdots, \phi_N] = \begin{bmatrix} \phi_{11} & \phi_{12} & \cdots & \phi_{1N} \\ \phi_{21} & \phi_{22} & \cdots & \phi_{2N} \\ \vdots & \vdots & \ddots & \vdots \\ \phi_{N1} & \phi_{N2} & \cdots & \phi_{NN} \end{bmatrix} \tag{2.296}$$

where ω_i^2 and $\phi_i = \{\phi_{1i}, \phi_{2i}, \ldots, \phi_{Ni}\}^T$ are the eigenvalue (i.e., the square of the natural circular frequency) and the eigenvector (i.e., the mode shape), respectively, of the i-th vibration mode. It is clear from Equation (2.296) that N is the total number of DOFs, and it is also the total number of modes. The mode corresponding to the lowest natural frequency is called the fundamental mode, which is of particular importance. It is believed that the fundamental mode (or the first mode) of a structural system is the most easily excited one. Furthermore, it is generally the easiest one to be measured in a field test (or vibration test under laboratory conditions). Other modes are usually called the higher modes (or higher harmonics). In general, the mode shapes for lower modes are relatively simple (without fluctuations in space). However, the mode shapes for higher modes become more and more complicated (with more and more fluctuations in space), as shown in Example 2.10.

Example 2.7

Determine the natural frequencies and mode shapes of the two-story shear building model in Figure 2.59.

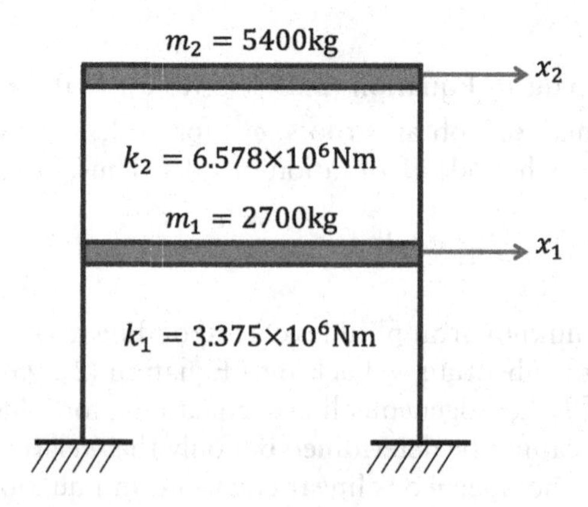

Figure 2.59 The two-story shear building model in Example 2.7.

SOLUTION

The system stiffness matrix is:

$$\mathbf{K}=\begin{bmatrix} k_1+k_2 & -k_2 \\ -k_2 & k_2 \end{bmatrix}=\begin{bmatrix} 9.953 & -6.578 \\ -6.578 & 6.578 \end{bmatrix}\times 10^6 \text{ N/m} \tag{2.297}$$

The system mass matrix is:

$$\mathbf{M}=\begin{bmatrix} m_1 & 0 \\ 0 & m_2 \end{bmatrix}=\begin{bmatrix} 2700 & 0 \\ 0 & 5400 \end{bmatrix}\text{kg} \tag{2.298}$$

Consider the eigenvalue problem from Equation (2.293):

$$\begin{gathered}\left(\mathbf{K}-\omega_n^2\mathbf{M}\right)\boldsymbol{\phi}=\begin{bmatrix} k_1+k_2-\omega_n^2 m_1 & -k_2 \\ -k_2 & k_2-\omega_n^2 m_2 \end{bmatrix}\begin{Bmatrix} \phi_1 \\ \phi_2 \end{Bmatrix}=\begin{Bmatrix} 0 \\ 0 \end{Bmatrix} \\ \Rightarrow\begin{cases} \left(k_1+k_2-\omega_n^2 m_1\right)\phi_1-k_2\phi_2=0 \\ -k_2\phi_1+\left(k_2-\omega_n^2 m_2\right)\phi_2=0 \end{cases}\end{gathered} \tag{2.299}$$

This is a set of two homogeneous linear equations. The condition for this set of equations to have non-trivial solutions is:

$$\begin{vmatrix} k_1+k_2-\omega_n^2 m_1 & -k_2 \\ -k_2 & k_2-\omega_n^2 m_2 \end{vmatrix}=0 \tag{2.300}$$

Rearrange to get the second-order polynomial of ω_n^2 as:

$$m_1 m_2\left(\omega_n^2\right)^2-\left[\left(k_1+k_2\right)m_2+m_1 k_2\right]\omega_n^2+k_1 k_2=0 \tag{2.301}$$

Substituting the numerical values of system parameters, one obtains:

$$1.4580\times 10^6\left(\omega_n^2\right)^2-7.1507\times 10^{10}\omega_n^2+2.2201\times 10^{13}=0 \tag{2.302}$$

By solving this quadratic equation, one obtains the two roots (i.e., the eigenvalues) as:

$$\omega_1^2=333.0930 \quad \text{and} \quad \omega_2^2=4571.3514 \tag{2.303}$$

The natural circular frequencies (in rad/s) of the two-story shear building are:

$$\omega_1=18.2508\,\text{rad/s} \quad \text{and} \quad \omega_2=67.6118\,\text{rad/s} \tag{2.304}$$

It must be pointed out that the order of natural frequencies depends on their numerical values. The smaller the numerical value, the lower the mode will be. Thus, 18.25 rad/s must be the first mode and not 67.61 rad/s. The natural frequencies in Hz can be calculated as:

$$f_1=2.9047\,\text{Hz} \quad \text{and} \quad f_2=10.7607\,\text{Hz} \tag{2.305}$$

Finally, the natural periods can be obtained:

$$T_1 = 0.3443\,\text{s} \quad \text{and} \quad T_2 = 0.0929\,\text{s} \tag{2.306}$$

Next, the calculated eigenvalues in Equation (2.303) are substituted back to the set of homogeneous linear equations in Equation (2.299) one by one to solve for the corresponding eigenvectors. Considering the first mode, one obtains:

$$\begin{aligned}(\mathbf{K} - \omega_1^2\mathbf{M})\boldsymbol{\phi}_1 &= \begin{bmatrix} k_1 + k_2 - \omega_1^2 m_1 & -k_2 \\ -k_2 & k_2 - \omega_1^2 m_2 \end{bmatrix}\begin{Bmatrix}\phi_{11}\\ \phi_{21}\end{Bmatrix} = \begin{Bmatrix}0\\0\end{Bmatrix} \\ &\Rightarrow \begin{cases} 9.0536\phi_{11} - 6.5780\phi_{21} = 0 \\ -6.5780\phi_{11} + 4.7793\phi_{21} = 0 \end{cases}\end{aligned} \tag{2.307}$$

Since this set of linear equations is homogeneous, there are only N – 1 independent equations. Both linear equations in Equation (2.299) are normalized, and only a single equation is returned:

$$\phi_{11} - 0.7266\phi_{21} = 0 \tag{2.308}$$

It is clear that the exact values of ϕ_{11} and ϕ_{21} cannot be obtained, but only the relative values. If it is assumed that $\phi_{21} = 1$, then $\phi_{11} = -0.7266$. Thus, one obtains the mode shape of the first vibration mode as:

$$\boldsymbol{\phi}_1 = \begin{Bmatrix}\phi_{11}\\ \phi_{21}\end{Bmatrix} = \begin{Bmatrix}0.7266\\ 1\end{Bmatrix} \tag{2.309}$$

The mode shape of the second vibration mode can be obtained in a similar way by substituting the second eigenvalue in Equation (2.303) to Equation (2.299) to obtain:

$$\begin{aligned}(\mathbf{K} - \omega_2^2\mathbf{M})\boldsymbol{\phi}_2 &= \begin{bmatrix} k_1 + k_2 - \omega_2^2 m_1 & -k_2 \\ -k_2 & k_2 - \omega_2^2 m_2 \end{bmatrix}\begin{Bmatrix}\phi_{12}\\ \phi_{22}\end{Bmatrix} = \begin{Bmatrix}0\\0\end{Bmatrix} \\ &\Rightarrow \begin{cases} 0.2390\phi_{12} + 0.6578\phi_{22} = 0 \\ 0.6578\phi_{12} + 1.8107\phi_{22} = 0 \end{cases}\end{aligned} \tag{2.310}$$

Both linear equations in Equation (2.310) are normalized to a single independent equation as:

$$\phi_{12} + 2.7527\phi_{22} = 0 \tag{2.311}$$

If it is assumed that $\phi_{22} = 1$, then $\phi_{12} = -2.7527$. The mode shape of the second vibration mode is:

$$\boldsymbol{\phi}_2 = \begin{Bmatrix}\phi_{12}\\ \phi_{22}\end{Bmatrix} = \begin{Bmatrix}-2.7527\\ 1\end{Bmatrix} \tag{2.312}$$

Finally, the mode shape matrix of the two-story shear building model can be expressed as:

$$\boldsymbol{\Phi} = [\boldsymbol{\phi}_1, \boldsymbol{\phi}_2] = \begin{bmatrix} 0.7266 & -2.7527 \\ 1 & 1 \end{bmatrix} \tag{2.313}$$

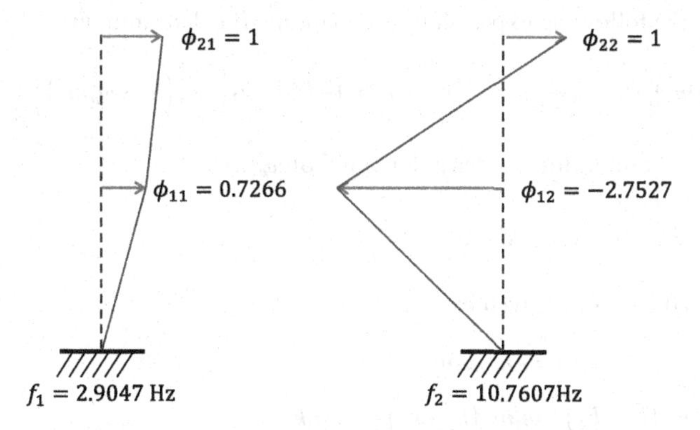

Figure 2.60 The natural frequencies and mode shapes of the two-story shear building model in Example 2.7.

The mode shapes of a structural system are best presented graphically, as in Figure 2.60. It is clear that the vibrations of the two DOFs are in-phase in mode 1 (i.e., the first vibration mode). That is, the vibrations are in the same direction with a zero-degree phase angle. However, the two DOFs are vibrating out-of-phase (i.e., with 180° phase angle) in mode 2.

Example 2.8

Determine the natural frequencies and mode shapes of a three-story shear building with $m_1 = m_2 = m_3 = 1\,\text{kg}$ and $k_1 = k_2 = k_3 = 10\,\text{N/m}$.

SOLUTION

The system mass matric can be expressed based on Equations (2.277) as:

$$\mathbf{M} = \begin{bmatrix} m_1 & 0 & 0 \\ 0 & m_2 & 0 \\ 0 & 0 & m_3 \end{bmatrix} = \begin{bmatrix} 1 & 0 & 0 \\ 0 & 1 & 0 \\ 0 & 0 & 1 \end{bmatrix} \text{kg} \tag{2.314}$$

And the system stiffness matric can be expressed based on Equation (2.278) as:

$$\mathbf{K} = \begin{bmatrix} k_1 + k_2 & -k_2 & 0 \\ -k_2 & k_2 + k_3 & -k_3 \\ 0 & -k_3 & k_3 \end{bmatrix} = \begin{bmatrix} 20 & -10 & 0 \\ -10 & 20 & -10 \\ 0 & -10 & 10 \end{bmatrix} \text{N/m} \tag{2.315}$$

The corresponding eigenvalue problem is:

$$\left(\mathbf{K} - \omega_n^2 \mathbf{M}\right)\boldsymbol{\phi} = \begin{bmatrix} k_1 + k_2 - \omega_n^2 m_1 & -k_2 & \\ -k_2 & k_2 + k_3 - \omega_n^2 m_2 & -k_3 \\ & -k_3 & k_3 - \omega_n^2 m_3 \end{bmatrix} \begin{Bmatrix} \phi_1 \\ \phi_2 \\ \phi_3 \end{Bmatrix} = \begin{Bmatrix} 0 \\ 0 \\ 0 \end{Bmatrix} \tag{2.316}$$

The condition for the set of homogeneous equations to have non-trivial solutions is:

$$\begin{vmatrix} k_1 + k_2 - \omega_n^2 m_1 & -k_2 & \\ -k_2 & k_2 + k_3 - \omega_n^2 m_2 & -k_3 \\ & -k_3 & k_3 - \omega_n^2 m_3 \end{vmatrix} = 0 \tag{2.317}$$

One obtains the following expression by expanding the determinant:

$$\left(k_1 + k_2 - \omega_n^2 m_1\right)\left[\left(k_2 + k_3 - \omega_n^2 m_2\right)\left(k_3 - \omega_n^2 m_3\right) - k_3^2\right] + k_2\left[-k_2\left(k_3 - \omega_n^2 m_3\right)\right] = 0 \quad (2.318)$$

Rearrange to obtain a third-order polynomial of ω_n^2 as:

$$a\left(\omega_n^2\right)^3 + b\left(\omega_n^2\right)^2 + c\omega_n^2 + d = 0 \quad (2.319)$$

where the coefficients are given by:

$$\begin{aligned} a &= -m_1 m_2 m_3 \\ b &= m_2 m_3\left(k_1 + k_2\right) + m_1 m_2\left(k_2 + k_3\right) + m_1 m_3 k_3 \\ c &= k_2\left(k_2 m_2 - k_1 m_1\right) - \left(k_1 + k_2\right)\left(k_2 m_3 + k_3 m_3 + k_3 m_2\right) \\ d &= k_1 k_2 k_3 \end{aligned} \quad (2.320)$$

By substituting the numerical values of inter-story stiffnesses and masses, one obtains:

$$-\left(\omega_n^2\right)^3 + 50\left(\omega_n^2\right)^2 - 600\omega_n^2 + 1000 = 0 \quad (2.321)$$

Unlike the resultant quadratic equation in the two-DOF cases, solving a cubic equation is more difficult. Here, Cardano's Method (Brilliant.org 2021) is employed. Consider the general cubic equation as:

$$ax^3 + bx^2 + cx + d = 0 \quad (2.322)$$

One of the three roots can be calculated as:

$$x_1 = \sqrt[3]{\alpha + \sqrt{\alpha^2 + \beta^3}} + \sqrt[3]{\alpha - \sqrt{\alpha^2 + \beta^3}} - \frac{b}{3a} \quad (2.323)$$

where:

$$\begin{aligned} \alpha &= \frac{9abc - 27a^2 d - 2b^3}{54a^3} \\ \beta &= \frac{3ac - b^2}{9a^2} \end{aligned} \quad (2.324)$$

After solving for the first root, x_1, the cubic equation in Equation (2.322) can be rewritten as:

$$\left(x - x_1\right)\left(Ax^2 + Bx + C\right) = 0 \quad (2.325)$$

Expanding to get:

$$Ax^3 + \left(B - x_1 A\right)x^2 + \left(C - x_1 B\right)x - x_1 C = 0 \quad (2.326)$$

By comparing the coefficients to Equation (2.322), one can solve for $A = a$, $B = b + x_1 A$ and $C = c + x_1 B = -d / x_1$. They are the coefficients of the quadratic equation for calculating the other two roots of the original cubic equation.

The cubic equation in Equation (2.321) can be solved now. Based on Cardano's Method, the two constants in Equation (2.324) are $\alpha = 129.6296$ and $\beta = -77.7778$. The first root can be calculated by Equation (2.323) as $x_1 = 32.4698$. The coefficients of the quadratic equation in Equation (2.325) can be calculated as $A = -1.0000$, $B = 17.5302$ and $C = -30.7979$. The three eigenvalues (sorted in ascending order) are:

$$\omega_1^2 = 1.9806, \omega_2^2 = 15.5496, \omega_3^2 = 32.4698 \quad (2.327)$$

The corresponding natural frequencies are:

$$f_1 = 0.2240\,\text{Hz}, f_2 = 0.6276\,\text{Hz}, f_3 = 0.9069\,\text{Hz} \quad (2.328)$$

Next, the first eigenvalue is substituted back into Equation (2.316) for calculating the corresponding mode shape:

$$\left(\mathbf{K} - \omega_1^2\mathbf{M}\right)\boldsymbol{\phi}_1 = \begin{bmatrix} 18.0194 & -10.0000 & 0 \\ -10.0000 & 18.0194 & -10.0000 \\ 0 & -10.0000 & 8.0194 \end{bmatrix} \begin{Bmatrix} \phi_{11} \\ \phi_{21} \\ \phi_{31} \end{Bmatrix} = \begin{Bmatrix} 0 \\ 0 \\ 0 \end{Bmatrix} \quad (2.329)$$

The concept of Gauss elimination can be followed to solve this set of three homogeneous equations. First, the first equation (row 1) is normalized such that the coefficient for ϕ_{11} is equal to unity:

$$\begin{bmatrix} 1.0000 & -0.5550 & 0 \\ -10.0000 & 18.0194 & -10.0000 \\ 0 & -10.0000 & 8.0194 \end{bmatrix} \begin{Bmatrix} \phi_{11} \\ \phi_{21} \\ \phi_{31} \end{Bmatrix} = \begin{Bmatrix} 0 \\ 0 \\ 0 \end{Bmatrix} \quad (2.330)$$

The coefficient of ϕ_{11} in the second equation (row 2) is eliminated by subtracting the first equation multiplied by -10 (the coefficient of ϕ_{11} in the second equation):

$$\begin{bmatrix} 1.0000 & -0.5550 & 0 \\ 0 & 12.4698 & -10.0000 \\ 0 & -10.0000 & 8.0194 \end{bmatrix} \begin{Bmatrix} \phi_{11} \\ \phi_{21} \\ \phi_{31} \end{Bmatrix} = \begin{Bmatrix} 0 \\ 0 \\ 0 \end{Bmatrix} \quad (2.331)$$

The coefficient of ϕ_{21} in the second equation (row 2) is then normalized to unity, and this equation (row) is used to eliminate the coefficient of ϕ_{21} in the third equation (row 3):

$$\begin{bmatrix} 1.0000 & -0.5550 & 0 \\ 0 & 1.0000 & -0.8019 \\ 0 & 0 & 0 \end{bmatrix} \begin{Bmatrix} \phi_{11} \\ \phi_{21} \\ \phi_{31} \end{Bmatrix} = \begin{Bmatrix} 0 \\ 0 \\ 0 \end{Bmatrix} \quad (2.332)$$

As this set of equations is not independent and with a free variable (here ϕ_{31} is the free variable), the third equation is eliminated, leaving two independent equations for calculating ϕ_{11} and ϕ_{21} for a given value of ϕ_{31} (usually assumed to be unity for shear building models). To complete the process, the coefficient of ϕ_{21} in the first equation (row 1) is eliminated by using the second equation (row 2) as:

$$\begin{bmatrix} 1.0000 & 0 & -0.4450 \\ 0 & 1.0000 & -0.8019 \\ 0 & 0 & 0 \end{bmatrix} \begin{Bmatrix} \phi_{11} \\ \phi_{21} \\ \phi_{31} \end{Bmatrix} = \begin{Bmatrix} 0 \\ 0 \\ 0 \end{Bmatrix} \quad (2.333)$$

By assuming $\phi_{31}=1$, $\phi_{11}=0.4450$ by the first equation (row 1), and $\phi_{21}=0.8019$ by the second equation (row 2). Thus, the mode shape for the first vibration mode is:

$$\phi_1=\begin{Bmatrix}\phi_{11}\\\phi_{21}\\\phi_{31}\end{Bmatrix}=\begin{Bmatrix}0.4450\\0.8019\\1\end{Bmatrix} \tag{2.334}$$

Next, the second eigenvalue is substituted back into Equation (2.316):

$$\left(\mathbf{K}-\omega_2^2\mathbf{M}\right)\phi_2=\begin{bmatrix}4.4504 & -10.0000 & 0\\-10.0000 & 4.4504 & -10.0000\\0 & -10.0000 & -5.5496\end{bmatrix}\begin{Bmatrix}\phi_{12}\\\phi_{22}\\\phi_{32}\end{Bmatrix}=\begin{Bmatrix}0\\0\\0\end{Bmatrix} \tag{2.335}$$

By following the idea of Gauss elimination again, the coefficient matrix can be rearranged as:

$$\begin{bmatrix}1.0000 & 0 & 1.2470\\0 & 1.0000 & 0.5550\\0 & 0 & 0\end{bmatrix}\begin{Bmatrix}\phi_{12}\\\phi_{22}\\\phi_{32}\end{Bmatrix}=\begin{Bmatrix}0\\0\\0\end{Bmatrix} \tag{2.336}$$

By assuming $\phi_{32}=1$, the mode shape of the second vibration mode is:

$$\phi_2=\begin{Bmatrix}\phi_{12}\\\phi_{22}\\\phi_{32}\end{Bmatrix}=\begin{Bmatrix}-1.2470\\-0.5550\\1\end{Bmatrix} \tag{2.337}$$

The last eigenvalue is then substituted back into Equation (2.316):

$$\left(\mathbf{K}-\omega_3^2\mathbf{M}\right)\phi_3=\begin{bmatrix}-12.4698 & -10.0000 & 0\\-10.0000 & -12.4698 & -10.0000\\0 & -10.0000 & -22.4698\end{bmatrix}\begin{Bmatrix}\phi_{13}\\\phi_{23}\\\phi_{33}\end{Bmatrix}=\begin{Bmatrix}0\\0\\0\end{Bmatrix} \tag{2.338}$$

By following the idea of Gauss elimination again, the coefficient matrix can be rearranged as:

$$\begin{bmatrix}1.0000 & 0 & -1.8019\\0 & 1.0000 & 2.2470\\0 & 0 & 0\end{bmatrix}\begin{Bmatrix}\phi_{13}\\\phi_{23}\\\phi_{33}\end{Bmatrix}=\begin{Bmatrix}0\\0\\0\end{Bmatrix} \tag{2.339}$$

By assuming $\phi_{33}=1$, the mode shape of the second vibration mode is:

$$\phi_3=\begin{Bmatrix}\phi_{13}\\\phi_{23}\\\phi_{33}\end{Bmatrix}=\begin{Bmatrix}1.8019\\-2.2470\\1\end{Bmatrix} \tag{2.340}$$

The mode shape matrix of the three-story shear building model in this example can then be formed:

$$\Phi=\left[\phi_1,\phi_2,\phi_3\right]=\begin{bmatrix}0.4450 & -1.2470 & 1.8019\\0.8019 & -0.5550 & -2.2470\\1 & 1 & 1\end{bmatrix} \tag{2.341}$$

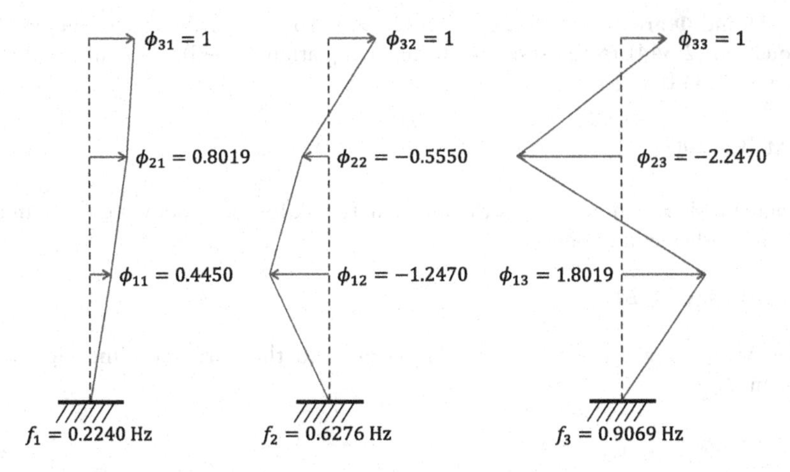

Figure 2.61 The natural frequencies and mode shapes of the three-story shear building model in Example 2.8.

The three mode shapes (i.e., the three columns in the matrix) are graphically presented in Figure 2.61. In mode 1, the vibrations of all three DOFs are in-phase. In mode 2, the vibrations of the first two DOFs are in-phase, while that of the third DOF is out-of-phase. In mode 3, the vibrations of the first and third DOFs are in-phase, while that of the second DOF is out-of-phase.

This example clearly shows that the calculation of modal parameters, such as natural frequencies and mode shapes, by solving the polynomial resulting from the eigenvalue problem is very time-consuming for structures with many DOFs. It is relatively easy for a two-DOF system (solving a quadratic equation), but not easy for a three-DOF system (solving a cubic equation). For systems with four or more DOFs, there is no analytical formula available for calculating the roots of a polynomial with an order higher than three. One may consider using other algorithms in solving the eigenvalue problem (e.g., subspace iteration). Here, the use of MATLAB function **eig()** is introduced. Depending on the input parameters, eig() uses different algorithms in solving the eigenvalue problem. The syntax for solving the standard eigenvalue problem is:

```
>> [V,D]=eig(A)
```

where **A** is the input matrix with dimensions $N \times N$, **D** is the diagonal matrix of eigenvalues, and **V** is the matrix of eigenvectors:

$$\mathbf{D} = \begin{bmatrix} d_1 & & & \\ & d_2 & & \\ & & \ddots & \\ & & & d_N \end{bmatrix} \quad \text{and} \quad \mathbf{V} = \left[\mathbf{v}_1, \mathbf{v}_2, \cdots, \mathbf{v}_N\right] = \begin{bmatrix} v_{11} & v_{12} & \cdots & v_{1N} \\ v_{21} & v_{22} & \cdots & v_{2N} \\ \vdots & \vdots & \ddots & \vdots \\ v_{N1} & v_{N2} & \cdots & v_{NN} \end{bmatrix} \tag{2.342}$$

The corresponding standard eigenvalue problem is:

$$\mathbf{AV} = \mathbf{VD} \tag{2.343}$$

Consider the eigenvalue problem for a particular mode (depending on the substituted eigenvalue) in Equation (2.293), it can be generalized to all N modes as:

$$\mathbf{K\Phi} = \mathbf{M\Phi\Omega} \tag{2.344}$$

where $\boldsymbol{\Omega}$ and $\boldsymbol{\Phi}$ are defined in Equation (2.296). To convert the eigenvalue problem in Equation (2.344) to the standard form in Equation (2.343), pre-multiply $\mathbf{M}^{-1}$ to Equation (2.344) to get:

$$\mathbf{M}^{-1}\mathbf{K}\boldsymbol{\Phi} = \boldsymbol{\Phi}\boldsymbol{\Omega} \tag{2.345}$$

The matrix $\mathbf{M}^{-1}\mathbf{K}$ can be used as the input matrix **A** for **eig**() following the standard syntax. Another syntax of eig() is:

```
>> [V,D]=eig(A,B)
```

where **A** and **B** are the two input matrices, and the corresponding eigenvalue problem is:

$$\mathbf{AV} = \mathbf{BVD} \tag{2.346}$$

By comparing Equation (2.346) with Equation (2.344), it is clear that the two input matrices are $\mathbf{A}=\mathbf{K}$ and $\mathbf{B}=\mathbf{M}$.

Example 2.9

Repeat Example 2.8 by using the MATLAB function **eig**().

SOLUTION

In this example, a MATLAB function is first developed for assembling the system stiffness matrix of a shear building model based on the provided inter-story stiffnesses using Equation (2.278). The developed MATLAB function **shear_building**() is presented in Program 2.6. This function gets the input parameter **k**, which is a vector of inter-story stiffness values. The first element is the inter-story stiffness of the first story, the second element is the inter-story stiffness of the second floor, and so on.

Program 2.7 shows the MATLAB script for calculating the natural frequencies and mode shapes of a shear building model with inter-story stiffnesses as defined by the **k** vector on Line 1 and floor mass as defined by the **m** vector on Line 2. By replacing the **k** and **m** vectors on these two lines, the script can be used for other shear building models. The length of the **k** vector is calculated on Line 3, and it defines the number of stories (i.e., the number of DOFs). The system mass matrix, **M**, is generated from the **m** vector on Line 4. Line 5 calculates the system stiffness matrix, **K**, by calling the **shear_building**() function in Program 2.7. Line 6 calls the MATLAB function **eig**() to solve the standard eigenvalue problem with an input matrix $\mathbf{A} = \mathbf{M}^{-1}\mathbf{K}$. The natural frequencies in Hz are calculated from the **D** matrix and assigned to the variable **nf** on Line 7. Finally, the calculated eigenvectors in **V** are normalized to have unity at the highest DOF ($N = 3$ in this example) on Line 8.

Program 2.6: **shear_building()** for calculating the system stiffness matrix of a shear building model based on the provided inter-story stiffness.

```
function K=shear_building(k)
if nargin<1, error('ERROR! k is missing'); end
N=length(k);
K=zeros(N,N);
```

```
for ii=1:N
  if ii==1
    K(ii,ii)=k(ii)+k(ii+1);
    K(ii,ii+1)=-k(ii+1);
  elseif ii==N
    K(ii,ii-1)=-k(ii);
    K(ii,ii)=k(ii);
  else
    K(ii,ii-1)=-k(ii);
    K(ii,ii)=k(ii)+k(ii+1);
    K(ii,ii+1)=-k(ii+1);
  end
end
end
```

```
Program 2.7: The MATLAB script for calculating the natural frequencies
and mode shapes of the three-story shear building in Example 2.9.

Line 1.   k=[10 10 10];
Line 2.   m=[1 1 1];
Line 3.   N=length(k);
Line 4.   M=diag(m);
Line 5.   K=shear_building(k);
Line 6.   [V,D]=eig(inv(M)*K);
Line 7.   nf=diag(D).^0.5/2/pi;
Line 8.   V=V./V(N,:);
```

Running the script in MATLAB, and show the calculated variables f and V:

```
>> f

f =
     0.2240
     0.6276
     0.9069
>> V
V =
     0.4450    -1.2470    1.8019
     0.8019    -0.5550   -2.2470
     1.0000     1.0000    1.0000
```

The calculated natural frequencies in f and the calculated mode shapes in V are exactly the same as those in Equation (2.328) and Equation (2.341), respectively, in Example 2.8. The **shear_building**() MATLAB function developed in this example can be used in other examples.

Example 2.10

Calculate the natural frequencies and mode shapes of an eight-story shear building model (a scaled-down model) with constant inter-story stiffness of 1000 N/m and constant floor mass of 1 kg through the entire building. Plot the mode shapes in MATLAB.

SOLUTION

Program 2.8 shows the MATLAB script to calculate and plot the natural frequencies and mode shapes of the targeted eight-story shear building model with constant inter-story stiffness of 1000 N/m and floor mass of 1 kg. Lines 1, 4, and 10 are comments to divide the scripts into three parts (i.e., the parts for defining the input parameters, for calculating the modal parameters, and for plotting the calculated results). Lines 2 and 3 are used to define the three input parameters **N**, **k**, and **m** for the number of stories, inter-story stiffness, and mass, respectively. The script can be used for other shear building models by replacing these three input parameters. Line 5 is used to generate the system mass matrix, and the function **shear_building()** is called on Line 6 to generate the system stiffness matrix. The natural frequencies and mode shapes are calculated on Lines 7, 8, and 9. Line 11 generates the y-axis (i.e., the story numbers) for plotting the mode shape. The for-loop from Line 12 to Line 23 plots the mode shapes of all **N** modes one by one.

In the for-loop (with the loop counter **ii**), a figure is created on Line 13, and the handle of this figure (i.e., **hf**) is defined. This handle is used later to modify the figure. The mode shape of mode **ii** is plotted on Line 14, where [0 **V**(:,**ii**)'] adds a zero at the beginning of the calculated mode shape (in **V**). The added zero corresponds to the mode shape value at the support of the shear building model. On the same line, the undeformed shape is plotted as a dashed line. Line 15 shows the natural frequency as the label on the x-axis of the plot. Line 16 shows the label of the y-axis. Line 17 increases the width of the lines on the plot, while Line 18 sets the position and size of the figure. Line 19 gets the handle of the axes, so the font size for both axes can be increased on Line 20. The title of the figure is defined on Line 21, and the generated figure is printed to a PNG file (for use in other applications) on Line 22.

Figure 2.62 shows the calculated natural frequencies and mode shapes of the eight-story shear building model in this example. The mode shape value at the roof of the shear building is normalized to unity in all modes. The vibrations of all DOFs are in-phase in mode 1, while the vibrations of adjacent DOFs are all out-of-phase in the last mode (i.e., mode 8).

Program 2.8: The MATLAB script for calculating and plotting the natural frequencies and mode shapes of the eight-story shear building model in Example 2.10.

```
Line 1.  % input parameters
Line 2.  N=8;
Line 3.  k=1000; m=1;
Line 4.  % calculation starts
Line 5.  M=diag(ones(N,1)*m);
Line 6.  K=shear_building(ones(N,1)*k);
Line 7.  [V,D]=eig(inv(M)*K);
Line 8.  nf=diag(D).^0.5/2/pi;
Line 9.  V=V./V(N,:);
Line 10. % plotting the mode shapes
Line 11. y=[0:N];
Line 12. for ii=1:N
Line 13.   hf=figure;
Line 14.   hp=plot([0 V(:,ii)'],y,'bo-',[0 0],[0 N],'k--'); grid on
Line 15.   xlabel(['f_' num2str(ii) ' = ' num2str(nf(ii)) ' Hz']);
Line 16.   ylabel('story');
Line 17.   set(hp,'LineWidth',1.5);
Line 18.   set(hf,'Position',[100 100 300 800]);
```

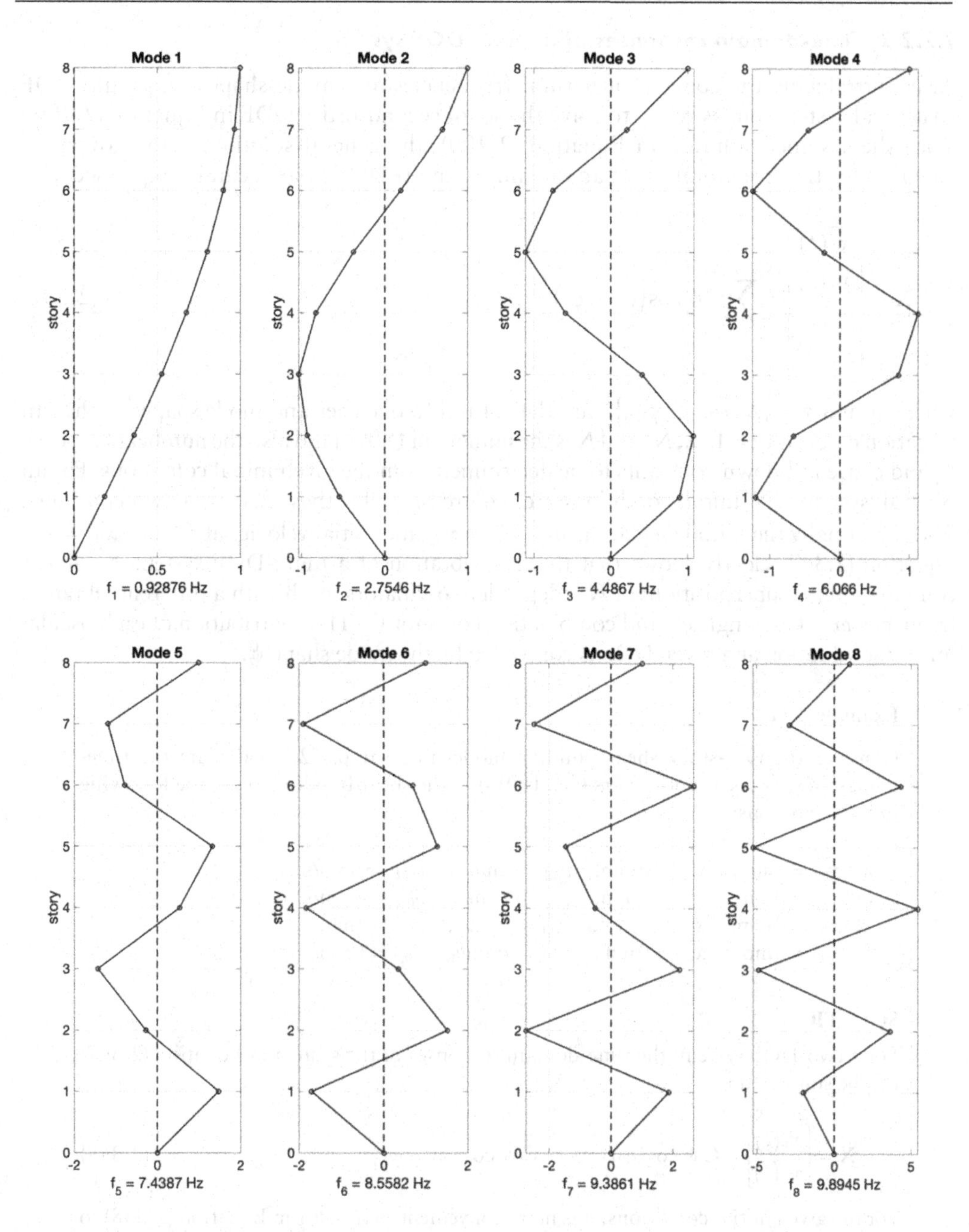

Figure 2.62 The natural frequencies and mode shapes of the eight-story shear building model in Example 2.10.

```
Line 19.    ax = gca;
Line 20.    set(ax,'FontSize',18);
Line 21.    ht=title(['Mode ' num2str(ii)]);
Line 22.    print(['modehsape' num2str(ii)],'-dpng');
Line 23. end
```

2.2.2.2 *Time-domain responses of a multi-DOF system*

After introducing the concept of natural frequencies and mode shapes of a multi-DOF structural system, one is ready to solve the set of second-order ODE in Equation (2.289). With the assumed solution in Equation (2.290), the general solution of the governing equation for the free vibration of an undamped multi-DOF system can be expressed as:

$$\mathbf{X} = \begin{Bmatrix} x_1(t) \\ x_2(t) \\ \vdots \\ x_N(t) \end{Bmatrix} = \sum_{i=1}^{N} C_i \boldsymbol{\phi}_i \cos(\omega_i t - \varphi_i) \tag{2.347}$$

where ω_i and $\boldsymbol{\phi}_i = \{\phi_{1i}, \phi_{2i}, \ldots, \phi_{Ni},\}^T$ are the natural frequencies and mode shapes of the i-th vibration mode, for $i = 1, \ldots, N$, and N is the number of DOFs (and also the number of modes), C_i and φ_i are unknown constants to be determined from the given initial conditions. For an N-DOF system, $2N$ initial conditions are required to solve these $2N$ unknown constants. They are usually the initial displacement $x_i(0) = x_{0,i}$ and initial velocity at $\dot{x}_i(0) = \dot{x}_{0,i} = v_{0,i}$. Equation (2.347) clearly shows that the free vibration of a multi-DOF system is a linear combination (or superposition) of N independent oscillators, each with a different vibration frequency ω_i, phase angle φ_i, and combination constant C_i. The contribution of each oscillator to the vibration at a given DOF is controlled by the mode shape $\boldsymbol{\phi}_i$.

Example 2.11

Consider the two-story shear building model in Example 2.7, calculate the time-domain responses at both floors (i.e., DOFs) of the shear building under the following initial conditions:

a) $x_{0,1} = 1\,\text{mm}$, $x_{0,2} = 0\,\text{mm}$, $v_{0,1} = 0\,\text{mm/s}$, $v_{0,2} = 0\,\text{mm/s}$.
b) $x_{0,1} = 0\,\text{mm}$, $x_{0,2} = 1\,\text{mm}$, $v_{0,1} = 0\,\text{mm/s}$, $v_{0,2} = 0\,\text{mm/s}$.
c) $x_{0,1} = 0\,\text{mm}$, $x_{0,2} = 0\,\text{mm}$, $v_{0,1} = 1\,\text{mm/s}$, $v_{0,2} = 0\,\text{mm/s}$.
d) $x_{0,1} = 0\,\text{mm}$, $x_{0,2} = 0\,\text{mm}$, $v_{0,1} = 0\,\text{mm/s}$, $v_{0,2} = 1\,\text{mm/s}$

SOLUTION

For a two-DOF system, the time-domain responses of the system in Equation (2.347) can be simplified to:

$$\mathbf{X} = \begin{Bmatrix} x_1(t) \\ x_2(t) \end{Bmatrix} = C_1 \boldsymbol{\phi}_1 \cos(\omega_1 t - \varphi_1) + C_2 \boldsymbol{\phi}_2 \cos(\omega_2 t - \varphi_2) \tag{2.348}$$

For the given initial conditions, it is more convenient to transform Equation (2.348) to the following form based on the right triangle as shown in Figure 2.63.

$$\mathbf{X} = a_1 \boldsymbol{\phi}_1 \cos(\omega_1 t) + b_1 \boldsymbol{\phi}_1 \sin(\omega_1 t) + a_2 \boldsymbol{\phi}_2 \cos(\omega_2 t) + b_2 \boldsymbol{\phi}_2 \sin(\omega_2 t) \tag{2.349}$$

With the initial displacement conditions, one obtains:

$$\mathbf{X}(0) = \begin{Bmatrix} x_{0,1} \\ x_{0,2} \end{Bmatrix} = a_1 \boldsymbol{\phi}_1 + a_2 \boldsymbol{\phi}_2 \tag{2.350}$$

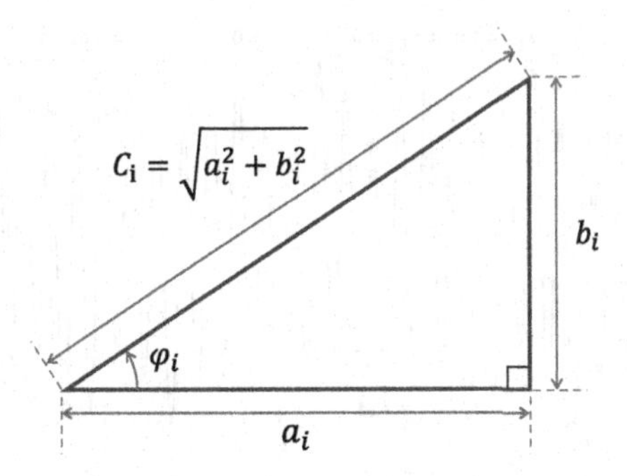

Figure 2.63 The right triangle constructed for transforming the formulation of **X**.

Resulting in the following two linear equations:

$$\begin{cases} \phi_{11}a_1 + \phi_{12}a_2 = x_{0,1} \\ \phi_{21}a_1 + \phi_{22}a_2 = x_{0,2} \end{cases} \tag{2.351}$$

To consider the initial velocity conditions, the first derivative of **X** is calculated:

$$\dot{\mathbf{X}} = -a_1\omega_1\boldsymbol{\phi}_1 \sin(\omega_1 t) + b_1\omega_1\boldsymbol{\phi}_1 \cos(\omega_1 t) - a_2\omega_2\boldsymbol{\phi}_2 \sin(\omega_2 t) + b_2\omega_2\boldsymbol{\phi}_2 \cos(\omega_2 t) \tag{2.352}$$

With the initial velocity conditions, one obtains:

$$\dot{\mathbf{X}}(0) = \begin{Bmatrix} v_{0,1} \\ v_{0,2} \end{Bmatrix} = b_1\omega_1\boldsymbol{\phi}_1 + b_2\omega_2\boldsymbol{\phi}_2 \tag{2.353}$$

Resulting in the following two linear equations:

$$\begin{cases} \omega_1\phi_{11}b_1 + \omega_2\phi_{12}b_2 = v_{0,1} \\ \omega_1\phi_{21}b_1 + \omega_2\phi_{22}b_2 = v_{0,2} \end{cases} \tag{2.354}$$

The numerical values of natural circular frequencies and mode shapes are available in Equation (2.304) and Equation (2.313), respectively. Considering the initial conditions in (a), the system of four linear equations becomes:

$$\begin{cases} 0.7266\,a_1 - 2.7527\,a_2 = 1 \\ a_1 + a_2 = 0 \end{cases}$$
$$\begin{cases} 13.2603b_1 - 186.1153b_2 = 0 \\ 18.2508b_1 + 67.6118b_2 = 0 \end{cases} \tag{2.355}$$

The four unknowns can be calculated from these four equations as:

$$a_1 = 0.2874, \quad a_2 = -0.2874, \quad b_1 = 0, \quad b_2 = 0 \tag{2.356}$$

The time-domain responses (for the first 3 seconds) of the two-story shear building can be calculated by Equation (2.349) and plotted in Figure 2.64(a). By following the same steps, the unknowns corresponding to initial conditions (b), (c), and (d) can be calculated and summarized below. For initial conditions (b), the unknowns are:

$$a_1 = 0.7912, \quad a_2 = 0.2088, \quad b_1 = 0, \quad b_2 = 0 \tag{2.357}$$

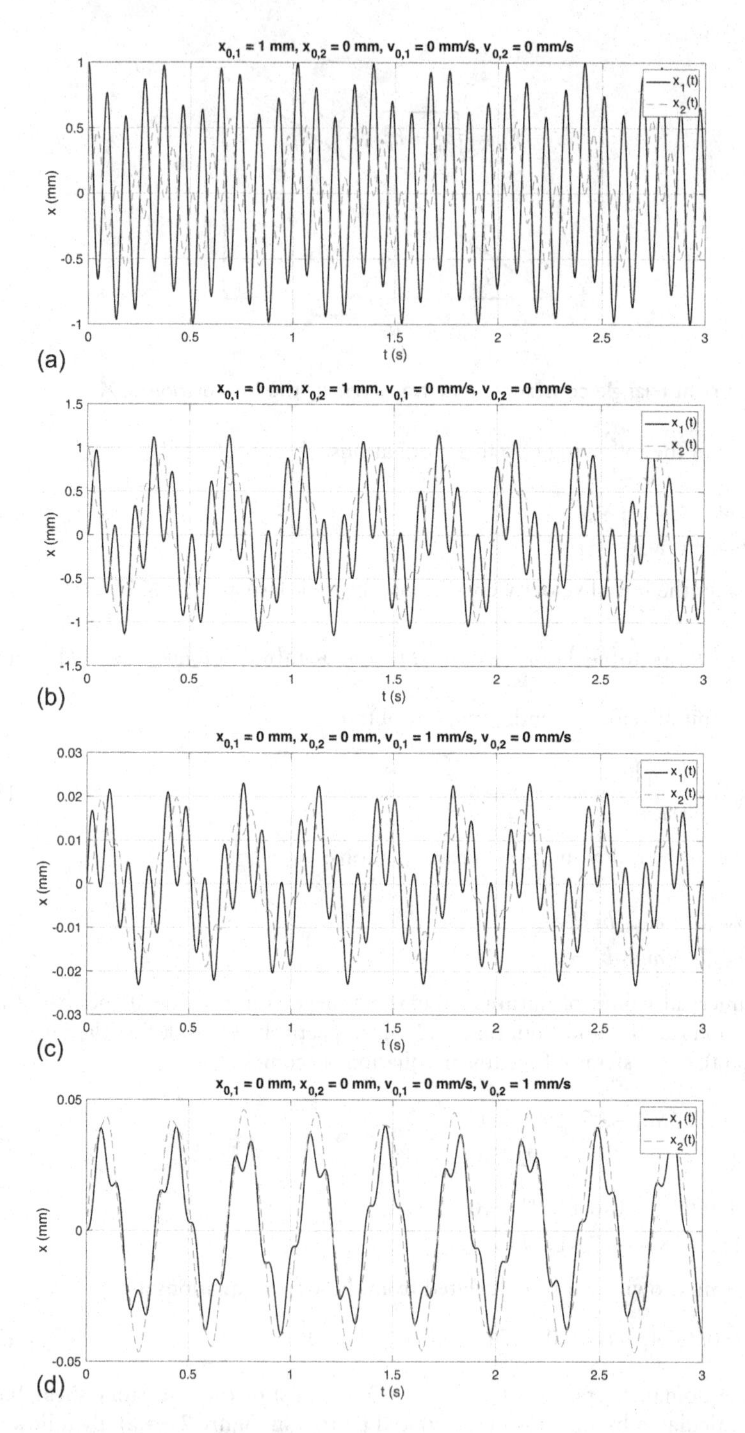

Figure 2.64 Calculated time-domain responses of the two-story shear building model under different initial conditions in Example 2.11.

and the corresponding calculated time-domain responses are presented in Figure 2.64(b). For initial conditions (c), the unknowns are:

$$a_1 = 0, \quad a_2 = 0, \quad b_1 = 0.0157, \quad b_2 = -0.0043 \tag{2.358}$$

and the corresponding calculated time-domain responses are presented in Figure 2.64(c). For the last initial conditions, the unknowns are:

$$a_1 = 0, \quad a_2 = 0, \quad b_1 = 0.0434, \quad b_2 = 0.0031 \tag{2.359}$$

and the corresponding calculated time-domain responses are presented in Figure 2.64(d).

2.2.2.3 Orthogonal property of mode shapes

The orthogonal property of modes (or mode shapes) is a very important property in the analysis of multi-DOF structural systems. It is introduced in this section. Considering the a-th mode of vibration in Equation (2.295), one obtains:

$$\mathbf{K}\phi_a = \omega_a^2 \mathbf{M}\phi_a \tag{2.360}$$

Pre-multiply Equation (2.360) with the transpose of the mode shape for the b-th vibration mode to get:

$$\phi_b^T \mathbf{K}\phi_a = \omega_a^2 \phi_b^T \mathbf{M}\phi_a \tag{2.361}$$

Similarly, one can consider the b-th mode in Equation (2.295) and pre-multiply it with the mode shape of the a-th mode to get:

$$\phi_a^T \mathbf{K}\phi_b = \omega_b^2 \phi_a^T \mathbf{M}\phi_b \tag{2.362}$$

Equation (2.361) is transposed to get:

$$\left(\phi_b^T \mathbf{K}\phi_a\right)^T = \omega_a^2 \left(\phi_b^T \mathbf{M}\phi_a\right)^T \Rightarrow \phi_a^T \mathbf{K}^T \phi_b = \omega_a^2 \phi_a^T \mathbf{M}^T \phi_b \tag{2.363}$$

As both $\mathbf{K}$ and $\mathbf{M}$ are symmetrical, one obtains:

$$\phi_a^T \mathbf{K}\phi_b = \omega_a^2 \phi_a^T \mathbf{M}\phi_b \tag{2.364}$$

Calculating the difference between Equation (2.364) and (2.362), one obtains:

$$0 = \left(\omega_a^2 - \omega_b^2\right) \phi_a^T \mathbf{M}\phi_b \tag{2.365}$$

From Equation (2.365), when a is not equal to b, $\left(\omega_a^2 - \omega_b^2\right)$ is not equal to zero and $\phi_a^T \mathbf{M}\phi_b$ must equal to equal. When a is equal to b, $\left(\omega_a^2 - \omega_b^2\right)$ is equal to zero and it is not necessary for $\phi_a^T \mathbf{M}\phi_b$ to be vanished. This can be presented as:

$$\phi_a^T \mathbf{M}\phi_b = 0, \quad \text{for } a \neq b \tag{2.366}$$

This is the orthogonal property to the system mass matrix, and it can be easily extended to the system stiffness matrix through Equation (2.364) as ω_a^2 is not equal to zero. In summary, the orthogonal property can be presented as:

$$\phi_a^T \mathbf{K} \phi_b = 0 \quad \text{and} \quad \phi_a^T \mathbf{M} \phi_b = 0, \quad \text{for } a \neq b \tag{2.367}$$

2.2.2.4 *Mode shape normalization*

Temporarily assign $\tilde{\mathbf{K}} = \Phi^T \mathbf{K} \Phi$ and $\tilde{\mathbf{M}} = \Phi^T \mathbf{M} \Phi$. The orthogonal property of mode shapes implies that both $\tilde{\mathbf{K}}$ and $\tilde{\mathbf{M}}$ are diagonal matrices as:

$$\tilde{\mathbf{K}} = \Phi^T \mathbf{K} \Phi = \begin{bmatrix} \tilde{k}_1 & & & \\ & \tilde{k}_2 & & \\ & & \ddots & \\ & & & \tilde{k}_N \end{bmatrix} \quad \text{and} \quad \tilde{\mathbf{M}} = \Phi^T \mathbf{M} \Phi = \begin{bmatrix} \tilde{m}_1 & & & \\ & \tilde{m}_2 & & \\ & & \ddots & \\ & & & \tilde{m}_N \end{bmatrix} \tag{2.368}$$

Recall the fact that a mode shape is not a vector of absolute amplitudes but only the relative amplitudes of vibration at various DOFs. Thus, if ϕ_i is a mode shape, then $a\phi_i$ is also a mode shape, where a is a constant (it can be positive or negative). As a result, the scaling of mode shape is an issue. For a shear building model, it is convenient to scale (or normalize) the mode shape such that the value at the roof is equal to unity. However, this normalization method is not applicable for general structures (other than shear building models). The orthogonal property to the system mass matrix provides a general method to normalize mode shapes. When the system mass matrix and the mode shape matrix are available, one can calculate the $\tilde{\mathbf{M}}$ matrix as shown in Equation (2.368). Then, one can normalize all mode shapes as:

$$\psi_i = \frac{1}{\sqrt{\tilde{m}_i}} \phi_i = \frac{1}{\sqrt{\phi_i^T \mathbf{M} \phi_i}} \phi_i, \quad \text{for } i = 1, \ldots, N. \tag{2.369}$$

where ψ_i is the mass normalized mode shape of the i-th mode of vibration. Similar to the mode shape matrix, it is convenient to define the mass normalized mode shape matrix as:

$$\Psi = \left[\psi_1, \psi_2, \cdots, \psi_N\right] = \begin{bmatrix} \psi_{11} & \psi_{12} & \cdots & \psi_{1N} \\ \psi_{21} & \psi_{22} & \cdots & \psi_{2N} \\ \vdots & \vdots & \ddots & \vdots \\ \psi_{N1} & \psi_{N2} & \cdots & \psi_{NN} \end{bmatrix} \tag{2.370}$$

The orthogonal property of mass normalized mode shapes to the system mass matrix becomes:

$$\psi_i^T \mathbf{M} \psi_j = \begin{cases} 0, & \text{for } i \neq j \\ 1, & \text{for } i = j \end{cases} \tag{2.371}$$

Replacing the mode shapes in Equation (2.361) with the mass normalized mode shapes, it becomes:

$$\psi_i^T \mathbf{K} \psi_j = \omega_j^2 \psi_i^T \mathbf{M} \psi_j \tag{2.372}$$

With Equation (2.371), one obtains:

$$\psi_i^T \mathbf{K} \psi_j = \begin{cases} 0, & \text{for } i \neq j \\ \omega_i^2, & \text{for } i = j \end{cases} \tag{2.373}$$

The orthogonal property of mass normalized mode shapes in Equations (2.371) and (2.373) can be presented in the matrix form as:

$$\boldsymbol{\Psi}^T \mathbf{M} \boldsymbol{\Psi} = \mathbf{I} \quad \text{and} \quad \boldsymbol{\Psi}^T \mathbf{K} \boldsymbol{\Psi} = \boldsymbol{\Omega} \tag{2.374}$$

where both **I** (a unit matrix) and **Ω** are diagonal matrices with dimensions $N \times N$:

$$\mathbf{I} = \begin{bmatrix} 1 & & & \\ & 1 & & \\ & & \ddots & \\ & & & 1 \end{bmatrix} \quad \text{and} \quad \boldsymbol{\Omega} = \begin{bmatrix} \omega_1^2 & & & \\ & \omega_2^2 & & \\ & & \ddots & \\ & & & \omega_N^2 \end{bmatrix} \tag{2.375}$$

For general structural systems, mass normalization is employed by default.

Example 2.12

Consider the two-story shear building model in Example 2.7, determine the mass normalized mode shapes, and verify the orthogonal property.

SOLUTION

The mode shape matrix and the system mass matrix are available in Equation (2.313) and Equation (2.298), respectively. Pre- and post-multiply the system mass matrix with the transpose of the mode shape matrix, $\boldsymbol{\Phi}^T$, and the mode shape matrix, $\boldsymbol{\Phi}$, respectively, to get:

$$\begin{aligned} \boldsymbol{\Phi}^T \mathbf{M} \boldsymbol{\Phi} &= \begin{bmatrix} 0.7266 & 1 \\ -2.7527 & 1 \end{bmatrix} \begin{bmatrix} 2700 & 0 \\ 0 & 5400 \end{bmatrix} \begin{bmatrix} 0.7266 & -2.7527 \\ 1 & 1 \end{bmatrix} \\ &= \begin{bmatrix} 6.8253 & 0 \\ 0 & 25.8589 \end{bmatrix} \times 10^3 \text{ kg} \end{aligned} \tag{2.376}$$

The mass normalized mode shape of the first mode can then be obtained by:

$$\psi_1 = \frac{1}{\sqrt{6.8253 \times 10^3}} \phi_1 = \begin{Bmatrix} 0.0088 \\ 0.0121 \end{Bmatrix} \tag{2.377}$$

The mass normalized mode shape of the second mode is:

$$\psi_2 = \frac{1}{\sqrt{25.8589 \times 10^3}} \phi_2 = \begin{Bmatrix} -0.0171 \\ 0.0062 \end{Bmatrix} \tag{2.378}$$

Thus, the mass normalized mode shape matrix of the two-story shear building model in this example is:

$$\Psi = [\psi_1, \psi_2] = \begin{bmatrix} 0.0088 & -0.0171 \\ 0.0121 & 0.0062 \end{bmatrix} \tag{2.379}$$

To verify the orthogonal property of mode shapes using the system mass matrix, one calculates:

$$\begin{aligned} \Psi^T \mathbf{M} \Psi &= \begin{bmatrix} 0.0088 & 0.0121 \\ -0.0171 & 0.0062 \end{bmatrix} \begin{bmatrix} 2700 & 0 \\ 0 & 5400 \end{bmatrix} \begin{bmatrix} 0.0088 & -0.0171 \\ 0.0121 & 0.0062 \end{bmatrix} \\ &= \begin{bmatrix} 1 & 0 \\ 0 & 1 \end{bmatrix} = \mathbf{I} \end{aligned} \tag{2.380}$$

Considering the system stiffness matrix in Equation (2.297), one calculates:

$$\begin{aligned} \Psi^T \mathbf{K} \Psi &= \begin{bmatrix} 0.0088 & 0.0121 \\ -0.0171 & 0.0062 \end{bmatrix} \begin{bmatrix} 9.953 & -6.578 \\ -6.578 & 6.578 \end{bmatrix} \begin{bmatrix} 0.0088 & -0.0171 \\ 0.0121 & 0.0062 \end{bmatrix} \times 10^6 \\ &= \begin{bmatrix} 333.0930 & 0 \\ 0 & 4571.3514 \end{bmatrix} \end{aligned} \tag{2.381}$$

Referring to Equation (2.303), one obtains:

$$\Psi^T \mathbf{K} \Psi = \begin{bmatrix} 333.0930 & 0 \\ 0 & 4571.3514 \end{bmatrix} = \begin{bmatrix} \omega_1^2 & 0 \\ 0 & \omega_2^2 \end{bmatrix} = \Omega \tag{2.382}$$

The orthogonal property is verified.

2.2.3 Forced vibration of an undamped multi-DOF system

The main problem for solving the set of second-order ODEs of a multi-DOF system is that these ODEs are coupled in general. Consider the two-story shear building model in Example 2.7 as an example, it is a two-DOF system and the set of ODEs (i.e., the governing equations) is:

$$\mathbf{M}\ddot{\mathbf{X}} + \mathbf{K}\mathbf{X} = \mathbf{F} \Rightarrow \begin{cases} m_1\ddot{x}_1 + (k_1 + k_2)x_1 - k_2x_2 = f_1 \\ m_2\ddot{x}_2 - k_2x_1 + k_2x_2 = f_2 \end{cases} \tag{2.383}$$

Both second-order ODEs contain the dependent variables x_1 and x_2. Thus, they are dependent, and they must be solved simultaneously. The main objective of this section is to transform the set of coupled ODEs into a system of independent (or uncoupled) ODEs. This can be done by using the modal parameters, such as the natural frequencies and mode shapes, of the multi-DOF system. Consider a general N-DOF system and introduce a set of new dependent variables $\mathbf{Y} = \{y_1(t), y_2(t), \ldots, y_N(t)\}^T$ that satisfies:

$$\mathbf{X} = \Psi\mathbf{Y} \tag{2.384}$$

With this transformation, the set of ODEs for the forced vibration of an undamped multi-DOF system becomes:

$$\mathbf{M\Psi\ddot{Y}} + \mathbf{K\Psi Y} = \mathbf{F} \tag{2.385}$$

Pre-multiplying the transpose of the mass normalized mode shape matrix to Equation (2.385), one obtains:

$$\mathbf{\Psi}^T\mathbf{M\Psi\ddot{Y}} + \mathbf{\Psi}^T\mathbf{K\Psi Y} = \mathbf{\Psi}^T\mathbf{F} \tag{2.386}$$

According to the orthogonal property of mode shapes, Equation (2.386) becomes:

$$\mathbf{I\ddot{Y}} + \mathbf{\Omega Y} = \mathbf{P} \tag{2.387}$$

The new force vector **P** is defined as:

$$\mathbf{P} = \begin{Bmatrix} p_1(t) \\ p_2(t) \\ \vdots \\ p_N(t) \end{Bmatrix} = \mathbf{\Psi}^T\mathbf{F} = \begin{Bmatrix} \boldsymbol{\psi}_1^T\mathbf{F} \\ \boldsymbol{\psi}_2^T\mathbf{F} \\ \vdots \\ \boldsymbol{\psi}_N^T\mathbf{F} \end{Bmatrix} \tag{2.388}$$

or it can be expressed as:

$$p_i(t) = \sum_{j=1}^{N} \psi_{ji} f_j(t), \quad \text{for } i = 1, \ldots, N \tag{2.389}$$

The transformed ODE of **Y** becomes:

$$\begin{gathered} \ddot{y}_1(t) + \omega_1^2 y_1(t) = p_1(t) \\ \ddot{y}_2(t) + \omega_2^2 y_2(t) = p_2(t) \\ \vdots \\ \ddot{y}_N(t) + \omega_N^2 y_N(t) = p_N(t) \end{gathered} \tag{2.390}$$

Equation (2.390) is a set of N second-order ODEs and each corresponds to the forced vibration of an undamped single-DOF system. Here, the original N-DOF system is transformed into N single-DOF systems. In other words, the original N coupled ODEs are transformed into N uncoupled ODEs that can be solved individually. The forced vibration of an under-damped single-DOF system under a general forcing function can be solved by the numerical method in Section 2.1.4 (by solving the Duhamel's integral). After solving **Y**, the time-domain response **X** of the original N-DOF system can be calculated by the mode shape (again) using Equation (2.384). As the method can be treated as the superposition of the responses of N single-DOF systems to get the responses of an N-DOF system, it is called the modal superposition method. In general, the initial conditions of a multi-DOF system are given as:

$$\mathbf{X}_0 = \begin{Bmatrix} x_{0,1} \\ x_{0,2} \\ \vdots \\ x_{0,N} \end{Bmatrix} \quad \text{and} \quad \dot{\mathbf{X}}_0 = \begin{Bmatrix} \dot{x}_{0,1} \\ \dot{x}_{0,2} \\ \vdots \\ \dot{x}_{0,N} \end{Bmatrix} \tag{2.391}$$

where $x_{0,i}$ and $\dot{x}_{0,i}$ are the initial displacement and velocity, respectively, at the i-th DOF of the structural system, for $i = 1,\ldots,N$. They must be transformed to the initial conditions for $\mathbf{Y}$:

$$\mathbf{Y}_0 = \begin{Bmatrix} y_{0,1} \\ y_{0,2} \\ \vdots \\ y_{0,N} \end{Bmatrix} \quad \text{and} \quad \dot{\mathbf{Y}}_0 = \begin{Bmatrix} \dot{y}_{0,1} \\ \dot{y}_{0,2} \\ \vdots \\ \dot{y}_{0,N} \end{Bmatrix} \tag{2.392}$$

where $y_{0,i}$ and $\dot{y}_{0,i}$ are the initial displacement and velocity, respectively, for the i-th mode of the structural system, for $i = 1,\ldots,N$. This can be done by the transformation in Equation (2.384) as:

$$\mathbf{X}_0 = \mathbf{\Psi}\mathbf{Y}_0 \tag{2.393}$$

Thus:

$$\mathbf{Y}_0 = \mathbf{\Psi}^{-1}\mathbf{X}_0 \quad \text{and} \quad \dot{\mathbf{Y}}_0 = \mathbf{\Psi}^{-1}\dot{\mathbf{X}}_0 \tag{2.394}$$

In general, the mass-normalized mode shape matrix $\mathbf{\Psi}$ is a full matrix (and so is the mode shape matrix $\mathbf{\Phi}$). The calculation of its inverse is time-consuming. Usually, the inverse of $\mathbf{\Psi}$ can be avoided by pre-multiplying Equation (2.393) by $\mathbf{\Psi}^T\mathbf{M}$ as:

$$\mathbf{\Psi}^T\mathbf{M}\mathbf{X}_0 = \mathbf{\Psi}^T\mathbf{M}\mathbf{\Psi}\mathbf{Y}_0 \tag{2.395}$$

With the orthogonal property of the mass normalized mode shape matrix (as in Equation (2.374)), $\mathbf{\Psi}^T\mathbf{M}\mathbf{\Psi} = \mathbf{I}$. Thus:

$$\mathbf{Y}_0 = \mathbf{\Psi}^T\mathbf{M}\mathbf{X}_0 \tag{2.396}$$

By comparing Equation (2.394) to Equation (2.396), the inverse of the mass normalized mode shape matric can be calculated as:

$$\mathbf{\Psi}^{-1} = \mathbf{\Psi}^T\mathbf{M} \tag{2.397}$$

As a result, the initial conditions of $\mathbf{Y}$ can be calculated as:

$$\mathbf{Y}_0 = \mathbf{\Psi}^T\mathbf{M}\mathbf{X}_0 \quad \text{and} \quad \dot{\mathbf{Y}}_0 = \mathbf{\Psi}^T\mathbf{M}\dot{\mathbf{X}}_0 \tag{2.398}$$

In Equation (2.390), each ODE corresponds to a single-DOF system with a different natural circular frequency. In general, the contribution of higher modes in the dynamic responses of a multi-DOF system is very small. Thus, it is possible to use only the first

10 to 20 modes for the dynamic analysis in practice. This will significantly reduce the computational power required.

2.2.4 Forced vibration of an under-damped multi-DOF system

The set of governing equations of an undamped multi-DOF system can be modified to consider an under-damped *N*-DOF system by adding the damping force term as:

$$\mathbf{M}\ddot{\mathbf{X}} + \mathbf{C}\dot{\mathbf{X}} + \mathbf{K}\mathbf{X} = \mathbf{F} \tag{2.399}$$

where **C** is the damping matrix, which is very difficult to be directly measured or calculated for real structures. Equation (2.399) is a set of coupled second-order ODEs. Similar to the undamped vibration situation, one may try to use the mode shape to uncouple the set of *N* ODEs in Equation (2.399) to a series of *N* single-DOF systems. With the transformation in Equation (2.384), Equation (2.399) becomes:

$$\mathbf{M\Psi}\ddot{\mathbf{Y}} + \mathbf{C\Psi}\dot{\mathbf{Y}} + \mathbf{K\Psi Y} = \mathbf{F} \tag{2.400}$$

Pre-multiplying the transpose of the mass normalized mode shape matrix, one obtains:

$$\mathbf{\Psi}^T\mathbf{M\Psi}\ddot{\mathbf{Y}} + \mathbf{\Psi}^T\mathbf{C\Psi}\dot{\mathbf{Y}} + \mathbf{\Psi}^T\mathbf{K\Psi Y} = \mathbf{\Psi}^T\mathbf{F} \tag{2.401}$$

Based on the orthogonal property of mode shapes, it becomes:

$$\mathbf{I}\ddot{\mathbf{Y}} + \mathbf{D}\dot{\mathbf{Y}} + \mathbf{\Omega Y} = \mathbf{P} \tag{2.402}$$

where **I**, **Ω**, and **P** are defined in Equations (2.375) and (2.388), and **D** is defined as:

$$\mathbf{D} = \mathbf{\Psi}^T\mathbf{C\Psi} \tag{2.403}$$

It is clear that both **I** and **Ω** are diagonal matrices. However, **D** is in general not a diagonal matrix as the orthogonal condition is only applicable for the system mass and stiffness matrices, but not the damping matrix. To uncouple Equation (2.402), it is assumed that the orthogonal property is also appliable to the damping matrix. That is, to enforce that **D** is also a diagonal matrix. For a classically damped structural system, the damping matrix can be defined as:

$$\psi_i^T\mathbf{C}\psi_j = \begin{cases} 0 & \text{for } i \neq j \\ 2\zeta_i\omega_i & \text{for } i = j \end{cases} \tag{2.404}$$

where ζ_i is the critical damping ratio of the *i*-th mode of the structural system. With this condition, the set of ODEs in Equation (2.402) can be uncoupled into:

$$\begin{gathered} \ddot{y}_1(t) + 2\zeta_1\omega_1\dot{y}_1(t) + \omega_1^2 y_1(t) = p_1(t) \\ \ddot{y}_2(t) + 2\zeta_2\omega_2\dot{y}_2(t) + \omega_2^2 y_2(t) = p_2(t) \\ \vdots \\ \ddot{y}_N(t) + 2\zeta_N\omega_N\dot{y}_N(t) + \omega_N^2 y_N(t) = p_N(t) \end{gathered} \tag{2.405}$$

Similar to the undamped situation, the set of uncoupled ODEs in Equation (2.405) can be solved individually, and the responses of the original multi-DOF system can be obtained by using the transformation in Equation (2.384).

In the calculation of the vibration responses of an under-damped multi-DOF system by modal superposition method, the damping effect is usually specified by defining the damping ratios ζ_i, for $i = 1,\ldots,N$, instead of defining the damping matrix C. It is clear from Equation (2.405) that only damping ratios from various modes are needed, and the damping matrix is not directly required.

When the damping ratios for all modes are available, one can calculate the damping matrix by Equation (2.401) as:

$$\mathbf{C} = \left(\Psi^T\right)^{-1} \mathbf{D}\Psi^{-1} \tag{2.406}$$

To reduce the computational power required in calculating the inverse of the full matrix $\boldsymbol{\Psi}$, Equation (2.407) is first rearranged as:

$$\mathbf{C} = \left(\Psi^T\right)^{-1} \mathbf{D}\Psi^{-1} = \left(\Psi^{-1}\right)^T \mathbf{D}\Psi^{-1} \tag{2.407}$$

Substituting in Equation (2.397) yields:

$$\mathbf{C} = \left(\Psi^T\mathbf{M}\right)^T \mathbf{D}\left(\Psi^T\mathbf{M}\right) = \left(\mathbf{M}\Psi\right)\mathbf{D}\left(\Psi^T\mathbf{M}\right) \tag{2.408}$$

The C matrix will be needed if the analysis is not carried out by the modal superposition method but by other numerical methods. Based on experience, the identified damping ratios of different modes from field tests are usually very close to each other. Thus, one may assume the damping ratios for all modes are the same to simplify the analysis procedure.

2.2.4.1 Rayleigh damping

Another popular way to construct the damping matrix of a classically damped system is the Rayleigh damping. In this method, the damping matrix is calculated as a linear combination of the system stiffness and mass matrices to ensure its orthogonal property as:

$$\mathbf{C} = a_\mathrm{M}\mathbf{M} + a_\mathrm{K}\mathbf{K} \tag{2.409}$$

where a_M and a_K are two proportional constants for the system mass and stiffness matrices, respectively. If $a_\mathrm{M} \neq 0$ and $a_\mathrm{K} = 0$, it is called mass-proportional damping. If $a_\mathrm{M} = 0$ and $a_\mathrm{K} \neq 0$, it is called stiffness-proportional damping. Pre- and post-multiplying Equation (2.409) by Ψ^T and $\boldsymbol{\Psi}$, respectively, one obtains the following based on the orthogonal property:

$$2\zeta_i\omega_i = a_\mathrm{M} + a_\mathrm{K}\omega_i^2, \quad \text{for } i = 1,\ldots,N. \tag{2.410}$$

The two unknown constants can then be calculated by specifying the damping ratios of two different modes. Assuming that the damping ratios for modes i and j are employed, one obtains a system of two equations for calculating the two unknown constants as:

$$\frac{1}{2\omega_i}a_{\mathrm{M}}+\frac{\omega_i}{2}a_{\mathrm{K}}=\zeta_i \quad \text{and} \quad \frac{1}{2\omega_j}a_{\mathrm{M}}+\frac{\omega_j}{2}a_{\mathrm{K}}=\zeta_j \tag{2.411}$$

One may express this set of equations in matrix form as:

$$\frac{1}{2}\begin{bmatrix}1/\omega_i & \omega_i\\ 1/\omega_j & \omega_j\end{bmatrix}\begin{Bmatrix}a_{\mathrm{M}}\\ a_{\mathrm{K}}\end{Bmatrix}=\begin{Bmatrix}\zeta_i\\ \zeta_j\end{Bmatrix} \tag{2.412}$$

After solving for a_{M} and a_{K}, they can be used to construct the damping matrix by Equation (2.409).

2.2.4.2 *Caughey damping*

Rayleigh damping allows one to fit the damping ratios of two modes only. This limitation is due to the use of **K** and **M** as matrices with orthogonal property, such that two unknown proportional constants need two conditions for the equation to be solved (they are the specified damping ratios of the two modes). If one can find an additional matrix with orthogonal property, the damping for one additional mode can be fitted. Consider pre-multiplying Equation (2.360) with $\phi_b^T\mathbf{K}\mathbf{M}^{-1}$, one obtains:

$$\phi_b^T\mathbf{K}\mathbf{M}^{-1}\mathbf{K}\phi_a=\omega_a^2\phi_b^T\mathbf{K}\mathbf{M}^{-1}\mathbf{M}\phi_a=\omega_a^2\phi_b^T\mathbf{K}\phi_a \tag{2.413}$$

As **K** has the orthogonal property, the matrix $\mathbf{K}\mathbf{M}^{-1}\mathbf{K}$ must also have. Next, pre-multiplying Equation (2.360) with $\phi_b^T\left(\mathbf{K}\mathbf{M}^{-1}\right)^2$, one obtains:

$$\phi_b^T\left(\mathbf{K}\mathbf{M}^{-1}\right)^2\mathbf{K}\phi_a=\omega_a^2\phi_b^T\left(\mathbf{K}\mathbf{M}^{-1}\right)\left(\mathbf{K}\mathbf{M}^{-1}\right)\mathbf{M}\phi_a=\omega_a^2\phi_b^T\mathbf{K}\mathbf{M}^{-1}\mathbf{K}\phi_a \tag{2.414}$$

As $\mathbf{K}\mathbf{M}^{-1}\mathbf{K}$ has the orthogonal property, the matrix $\left(\mathbf{K}\mathbf{M}^{-1}\right)^2\mathbf{K}$ must also have. By mathematical induction, all matrices $\left(\mathbf{K}\mathbf{M}^{-1}\right)^l\mathbf{K}$, for any positive integer l, have the orthogonal property. There is an infinite number of matrices from this series. Thus, one can fit the damping ratios of as many modes as one wants. To include the original two (i.e., **M** and **K**) in the series, the series of matrices can be expressed as $\mathbf{M}\left(\mathbf{M}^{-1}\mathbf{K}\right)^l$, for any positive integer l (including zero). When $l = 0$, $\mathbf{M}\left(\mathbf{M}^{-1}\mathbf{K}\right)^0=\mathbf{M}$, and it returns the system mass matrix. When $l = 1$, $\mathbf{M}\left(\mathbf{M}^{-1}\mathbf{K}\right)^1=\mathbf{K}$, and it returns the system stiffness matrix. When $l = 2$, $\mathbf{M}\left(\mathbf{M}^{-1}\mathbf{K}\right)\left(\mathbf{M}^{-1}\mathbf{K}\right)=\mathbf{K}\mathbf{M}^{-1}\mathbf{K}$. This is the matrix from Equation (2.413). When $l = 3$, $\mathbf{M}\left(\mathbf{M}^{-1}\mathbf{K}\right)\left(\mathbf{M}^{-1}\mathbf{K}\right)\left(\mathbf{M}^{-1}\mathbf{K}\right)=\left(\mathbf{K}\mathbf{M}^{-1}\right)^2\mathbf{K}$. This is the matrix from Equation (2.414). By assuming the system follows Caughey damping, the system damping matrix can be constructed as:

$$\mathbf{C}=\mathbf{M}\sum_{l=0}^{N_\zeta-1}a_l\left(\mathbf{M}^{-1}\mathbf{K}\right)^l \tag{2.415}$$

where N_ζ is the number of damping ratios to be fitted. To calculate the proportional constants a_i, for $l = 0,\ldots,N_\zeta - 1$, One needs to pre- and post-multiply Equation (2.415) by Ψ_b^T and Ψ_a, respectively, to get:

$$\Psi_b^T\mathbf{C}\Psi_a = \Psi_b^T\mathbf{M}\Big(a_0\Psi_a + a_1\left(\mathbf{M}^{-1}\mathbf{K}\right)\Psi_a + a_2\left(\mathbf{M}^{-1}\mathbf{K}\right)^2\Psi_a + \cdots + a_l\left(\mathbf{M}^{-1}\mathbf{K}\right)^l\Psi_a + \cdots + a_{N_\zeta-1}\left(\mathbf{M}^{-1}\mathbf{K}\right)^{N_\zeta-1}\Psi_a\Big) \tag{2.416}$$

Due to orthogonal property, the left-hand side is equal to $2\zeta_a\omega_a$ if $a = b$, otherwise it vanishes. The first term on the right-hand side is $a_0\Psi_b^T\mathbf{M}\Psi_a$. Due to the orthogonal property, this term will vanish if $a \neq b$. For $a = b$, it is equal to a_0. The second term on the right-hand side is $a_1\Psi_b^T\mathbf{M}\mathbf{M}^{-1}\mathbf{K}\Psi_a$. From Equation (2.360), $\mathbf{K}\Psi_a = \omega_a^2\mathbf{M}\Psi_a$. Therefore, the second term becomes $a_1\omega_a^2\Psi_b^T\mathbf{M}\Psi_a$. Due to orthogonal property, it is equal to $a_1\omega_a^2$. The third term on the right-hand side is $a_2\Psi_b^T\mathbf{M}\mathbf{M}^{-1}\mathbf{K}\mathbf{M}^{-1}\mathbf{K}\Psi_a$. By Equation (2.360), it becomes $a_2\omega_a^2\Psi_b^T\mathbf{K}\Psi_a$. By Equation (2.360) again, it becomes $a_2\omega_a^{2\times2}\Psi_b^T\mathbf{M}\Psi_a = a_2\omega_a^{2\times2}$. By mathematical induction, the fourth term can be expressed as $a_3\omega_a^{2\times3}$. Similarly, Equation (2.416) becomes:

$$2\zeta_a\omega_a = a_0 + a_1\omega_a^2 + a_2\omega_a^4 + \cdots + a_l\omega_a^{2l} + \cdots + a_{N_\zeta-1}\omega_a^{2(N_\zeta-1)} \tag{2.417}$$

Rearranging and replacing the counter a by i:

$$\zeta_i = \frac{1}{2}\left(a_0\omega_i^{-1} + a_1\omega_i + a_2\omega_i^3 + \cdots + a_l\omega_i^{2l-1} + \cdots + a_{N_\zeta-1}\omega_i^{2(N_\zeta-1)-1}\right) \tag{2.418}$$

Assuming that the damping ratios of the first N_ζ modes are given, Equation (2.418) returns N_ζ equations with N_ζ unknowns, and it can be expressed in matrix form as:

$$\frac{1}{2}\begin{bmatrix} 1/\omega_1 & \omega_1 & \omega_1^3 & \cdots & \omega_1^{2l-1} & \cdots & \omega_1^{2N_\zeta-3} \\ 1/\omega_2 & \omega_2 & \omega_2^3 & \cdots & \omega_2^{2l-1} & \cdots & \omega_2^{2N_\zeta-3} \\ \vdots & \vdots & \vdots & & \vdots & & \vdots \\ 1/\omega_{N_\zeta} & \omega_{N_\zeta} & \omega_{N_\zeta}^3 & \cdots & \omega_{N_\zeta}^{2l-1} & \cdots & \omega_{N_\zeta}^{2N_\zeta-3} \end{bmatrix}\begin{Bmatrix} a_0 \\ a_1 \\ a_2 \\ \vdots \\ a_l \\ \vdots \\ a_{N_\zeta-1} \end{Bmatrix} = \begin{Bmatrix} \zeta_1 \\ \zeta_2 \\ \vdots \\ \zeta_{N_\zeta} \end{Bmatrix} \tag{2.419}$$

After solving a_0 to $a_{N_\zeta-1}$, Equation (2.415) can be used to construct the system damping matrix. When Equation (2.418) of Caughey damping is compared to Equation (2.411) of Rayleigh damping, Rayleigh damping can be treated as a special case of Caughey damping with a_M and a_K in Equation (2.411) being the same as a_0 and a_1 in Equation (2.418).

Program 2.9 shows the MATLAB function **mdof_msuper()** developed to implement the modal superposition method for calculating the time-domain responses of a structural system under the action of both externally applied forces (i.e., the acceleration excitation) at all DOFs and initial conditions at all DOFs. The structural system is defined by the system stiffness (**K**) and mass (**M**) matrices with dimensions $N \times N$. The damping of the system is defined by a vector of damping ratios (**zeta**) with dimensions $N \times 1$ (i.e., damping ratios for all modes are required). The analysis duration is defined by the time step size (**dt**) and the number of time steps (**Nt**=n). The forcing matrix (**Fa**) with dimensions $N \times n$ defines the acceleration excitation at all N DOFs and all n time steps. The first row of **Fa** is the acceleration excitation for DOF 1, the second row of **Fa** is the acceleration excitation for DOF 2, and so on. The initial displacements and velocities are defined by **X0** and **V0**, respectively, with dimensions $N \times 1$. On Line 1, the number of DOF N is determined from the size of the system stiffness matrix. Line 2 calls the MATLAB function **eig**() to solve the eigenvalue problem and calculate the eigenvalues (**D**) and eigenvectors (**V**). To ensure the order of eigenvalues is ascending, the MATLAB function **sort**() is employed on Line 4. The natural circular frequencies are calculated on Line 5, and the order of eigenvectors (i.e., the mode shapes) is arranged as the one sorted in Line 4. The mass normalization is carried out from Line 7 to 10 by Equation (2.369), where **U** is the mass normalized mode shape matrix with dimensions $N \times N$. Both initial displacements and velocities are transformed to the initial modal displacements (**Y0**) and velocities (**U0**) by Equation (2.398). On Line 13, the acceleration excitation at all DOF is transformed to the modal forces for all modes (**Pa**) by Equation (2.388). Line 14 to 19 is a for-loop to call the self-developed **sdof_duhamel**() MATLAB function (see Program 2.5) to calculate the responses of a series of single-DOF systems (each represents a mode) as given in Equation (2.405). The calculated modal displacement, velocity, and acceleration are stored in the matrices **My**, **Mu** and **Ma**, respectively. The responses of these N single-DOF systems were transformed back to the responses of the targeted N-DOF system on Lines 20 to 22, where **XX**, **VV**, and **AA** are the displacement, velocity, and acceleration (with dimensions $N \times n$) at all DOFs. Finally, the natural frequencies (**nf**) and mass normalized mode shapes (**ms**) are stored in the object variable **info**. It is clear from Line 1 that all **XX**, **VV**, **AA**, **tt** (the time axis for plotting the responses), and **info** are output parameters.

Program 2.9: **mdof_msuper()** to calculate the time-domain responses of an under-damped multi-DOF system under the action of acceleration excitation and initial conditions.

```
Line 1.   function [XX,VV,AA,tt,info]=mdof_msuper(K, M, zeta, dt, Nt,
          Fa, X0, V0);
Line 2.   N=length(K);
Line 3.   [V,D]=eig(inv(M)*K);
Line 4.   [w2,indx]=sort(diag(D));
Line 5.   wn=w2.^0.5;
Line 6.   V=V(:,indx);
Line 7.   MM=V'*M*V;
Line 8.   for ii=1:N
Line 9.       U(:,ii)=V(:,ii)/MM(ii,ii).^0.5;
Line 10.  end
Line 11.  Y0=U'*M*X0;
Line 12.  U0=U'*M*V0;
Line 13.  Pa=U'*Fa;
Line 14.  for ii=1:N
Line 15.      [yy,uu,aa,tt]=sdof_duhamel(wn(ii), zeta(ii), dt, Nt,
          Pa(ii,:), Y0(ii), U0(ii));
Line 16.      My(ii,:)=yy;
Line 17.      Mu(ii,:)=uu;
```

```
Line 18.     Ma(ii,:)=aa;
Line 19. end
Line 20. XX=U*My;
Line 21. VV=U*Mu;
Line 22. AA=U*Ma;
Line 23. info.nf=wn/2/pi;
Line 24. info.ms=U;
Line 25. end
```

Example 2.13

Consider a scaled-down three-story shear building model with constant inter-story stiffness of 2000 N/m and floor mass 2 kg, determine the time-domain responses under the action of the specified initial conditions and excitation. Assume that the damping ratios for all modes are the same and are equal to 0.02.

- The initial displacements at all floors are the same and are equal to 1 mm, and the initial velocities at all floors are the same and are equal to 1 cm/s.
- The sine acceleration excitation is only applied at the top floor with an amplitude of 0.5 g and excitation frequency of 1 Hz.

SOLUTION

The script in Program 2.10 is developed to prepare the input parameters for running the self-developed MATLAB function **mdof_msuper()** in Program 2.9. The information of the shear building model is defined on Line 1. The number of stories (i.e., the number of DOFs) is defined by **N**. The inter-story stiffness and floor mass are stored in **k** and **m**, respectively. The vector of damping ratios is defined on Line 2 (**zeta**). The initial displacements (**X0**) and velocity (**V0**) are defined on Lines 3 and 4, respectively. The time information is defined on Line 5. It must be pointed out that the time step size (**dt**) is set to 0.01 s, and the number of time steps (**Nt**) is defined as 1500 (i.e., the time duration for analysis is 15 s). The acceleration amplitude (**A**) and excitation frequency (**fext**) are defined on Line 6. The time vector (**tt**) and the acceleration excitation for the top floor (**fa**) are generated on Lines 7 and 8, respectively. The acceleration excitation for all DOFs is generated on Lines 9 and 10. The sine excitation is only applied at the N-th DOF (i.e., the top floor). The system stiffness matrix is determined by the self-developed MATLAB function **shear_building()** in Program 2.6 on Line 12. The time-domain responses are calculated on Line 13 by calling the function **mdof_msuper()**. Lines 14 to 21 are used to plot the calculated displacement responses for each floor. Note that one may simply change the value of N on Line 1 for calculating the responses of a shear building model with other number of stories (and also for plotting the calculated displacement responses). The calculated displacement, velocity, and acceleration responses for all floors are shown in Figure 2.65, Figure 2.66, and Figure 2.67, respectively. It is clear from the figures that the transient responses due to the initial conditions are damped out at about 10 s, and the steady-state responses dominate the system responses thereafter.

Some of the important intermediate analysis results are shown below for readers to check their calculations. The natural frequencies are:

$$f_1 = 2.2399\,\text{Hz}, \quad f_2 = 6.2760\,\text{Hz}, \quad \text{and} \quad f_3 = 9.0690\,\text{Hz}. \tag{2.420}$$

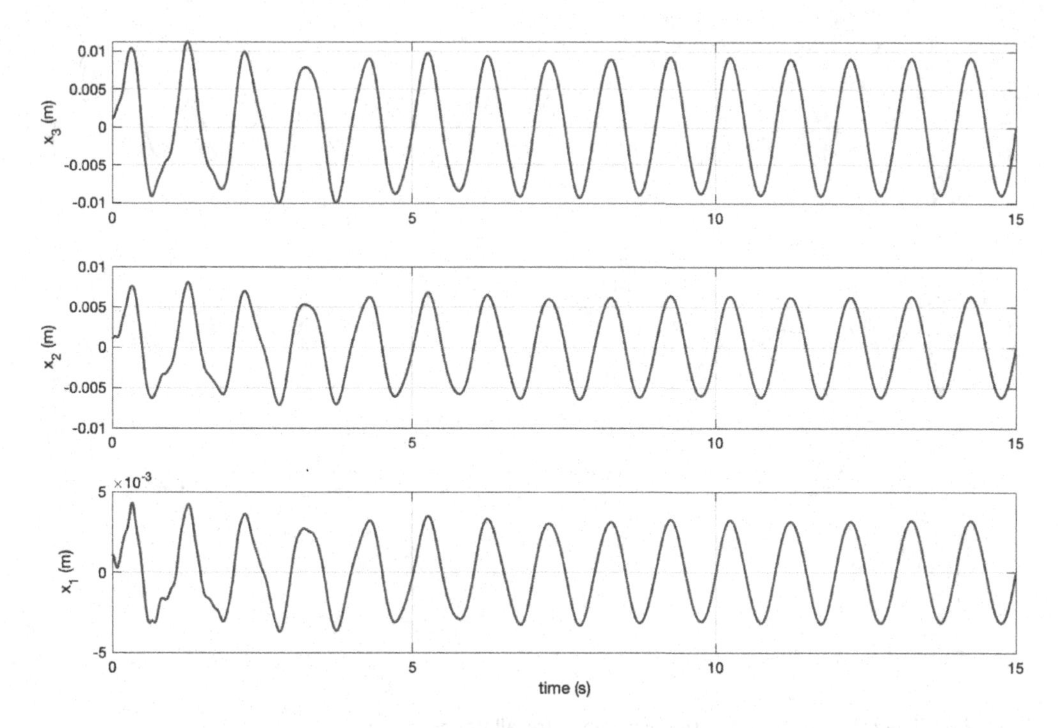

Figure 2.65 The calculated displacement responses for all floors.

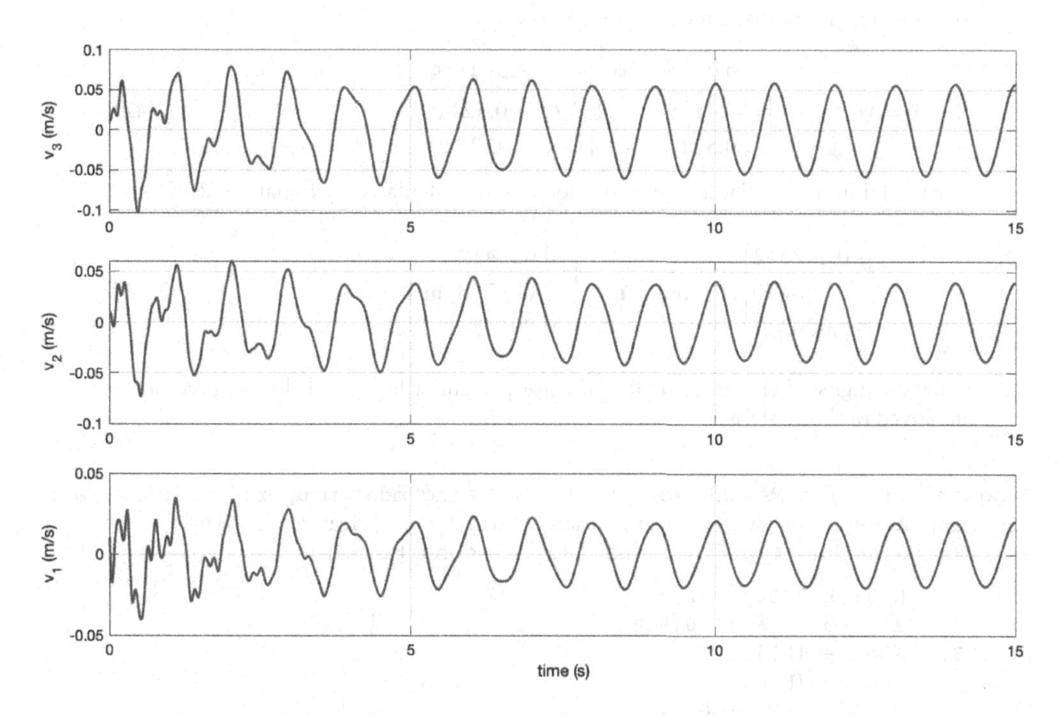

Figure 2.66 The calculated velocity responses for all floors.

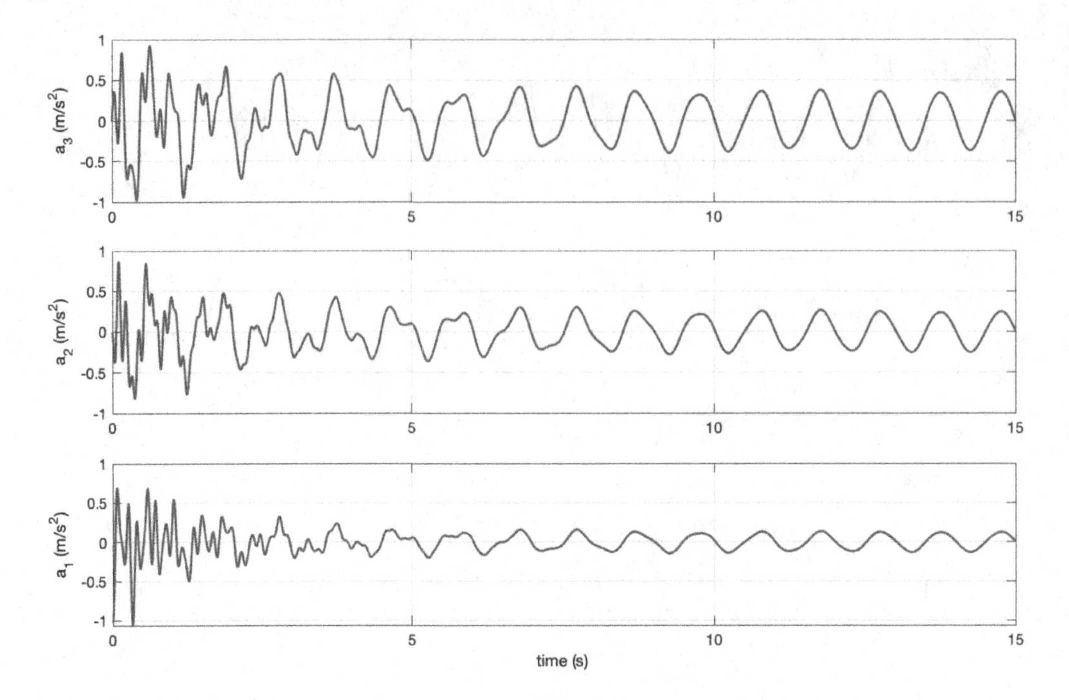

Figure 2.67 The calculated acceleration responses for all floors.

and the mass normalized mode shape matrix is:

$$\Psi = \left[\psi_1, \psi_2, \psi_3\right] = \begin{bmatrix} 0.2319 & 0.5211 & -0.4179 \\ 0.4179 & 0.2319 & 0.5211 \\ 0.5211 & -0.4179 & -0.2319 \end{bmatrix} \tag{2.421}$$

The modal initial displacements and velocities are calculated by Equation (2.398) as:

$$\mathbf{Y}_0 = \begin{Bmatrix} -0.002342 \\ 0.000670 \\ -0.000257 \end{Bmatrix} \text{m} \quad \text{and} \quad \dot{\mathbf{Y}}_0 = \begin{Bmatrix} -0.023419 \\ 0.006703 \\ -0.002574 \end{Bmatrix} \text{m/s} \tag{2.422}$$

Only six digits after the decimal place are presented here, and double precision is employed in the analysis.

Program 2.10: The MATLAB scripts for calling **mdof_msuper()** to calculate the responses of a multi-story shear building under both sine excitation at the top floor and initial conditions.

```
Line 1.   N=3; k=2000; m=2;
Line 2.   zeta=ones(N,1)*0.02;
Line 3.   X0=ones(N,1)./1000;
Line 4.   V0=ones(N,1)./100;
Line 5.   dt=0.01; Nt=1500;
Line 6.   A=0.5*9.81; fext=1;
Line 7.   tt=[0:Nt]*dt;
```

```
Line 8.   fa=A.*sin(2*pi*fext.*tt);
Line 9.   Fa=zeros(N,length(tt));
Line 10.  Fa(N,:)=fa;
Line 11.  M=diag(ones(N,1)*m);
Line 12.  K=shear_building(ones(N,1)*k);
Line 13.  [XX,VV,AA,tt,info]=mdof_msuper(K,M,zeta,dt,Nt,Fa,X0,V0);
Line 14.  figure;
Line 15.  for ii=1:N
Line 16.      subplot(N,1,ii);
Line 17.     thisN=N-ii+1;
Line 18.     plot(tt,XX(thisN,:),'LineWidth',1.5);
Line 19.     ylabel(['x_' num2str(thisN) ' (m)']);
Line 20.     grid on
Line 21.  end
Line 22.  xlabel('time (s)');
```

Example 2.14

Consider a five-story shear building model (i.e., $N = 5$) with inter-story stiffness 2000 N/m and floor mass 2 kg. Construct the system damping matrix in the following two cases.

1. Assuming a damping ratio of 0.02 for all 5 modes.
2. Assuming Rayleigh damping with the damping ratios for modes 1 and 2 equal to 0.02.

In both cases, determine the displacement responses under the action of a sine acceleration excitation with an amplitude of 0.5 g and excitation frequency of 1 Hz at the top floor (no excitation at other floors). Compare the two displacement responses of the system.

3. Construct the damping matrix again by assuming Rayleigh damping with damping ratios for modes 4 and 5 equal to 0.02.

Determine the displacement responses under the same acceleration excitation as before and compare the results with case 1.

SOLUTION

The system stiffness and mass matrices are:

$$\mathbf{K} = \begin{bmatrix} 4000 & -2000 & 0 & 0 & 0 \\ -2000 & 4000 & -2000 & 0 & 0 \\ 0 & -2000 & 4000 & -2000 & 0 \\ 0 & 0 & -2000 & 4000 & -2000 \\ 0 & 0 & 0 & -2000 & 2000 \end{bmatrix} \text{ and } \mathbf{M} = \begin{bmatrix} 2 & 0 & 0 & 0 & 0 \\ 0 & 2 & 0 & 0 & 0 \\ 0 & 0 & 2 & 0 & 0 \\ 0 & 0 & 0 & 2 & 0 \\ 0 & 0 & 0 & 0 & 2 \end{bmatrix} \quad (2.423)$$

Note that the units for stiffness and mass are N/m and kg, respectively. The eigenvalues and eigenvectors are determined by solving the eigenvalue problem of the system stiffness and mass matrices. The natural circular frequencies of the five-story shear building model are presented in vector form as:

$$\omega_n = \begin{Bmatrix} \omega_1 \\ \omega_2 \\ \vdots \\ \omega_{N-1} \\ \omega_N \end{Bmatrix} = \begin{Bmatrix} 9.0008 \\ 26.2732 \\ 41.4170 \\ 53.2055 \\ 60.6837 \end{Bmatrix} \text{rad/s} \tag{2.424}$$

The mass-normalized mode shapes matrix is:

$$\Psi = \begin{bmatrix} 0.1201 & -0.3223 & 0.4221 & 0.3879 & -0.2305 \\ 0.2305 & -0.4221 & 0.1201 & -0.3223 & 0.3879 \\ 0.3223 & -0.2305 & -0.3879 & -0.1201 & -0.4221 \\ 0.3879 & 0.1201 & -0.2305 & 0.4221 & 0.3223 \\ 0.4221 & 0.3879 & 0.3223 & -0.2305 & -0.1201 \end{bmatrix} \tag{2.425}$$

In Case 1, the diagonal damping matric $\mathbf{D}^{(1)}$ as defined by Equations (2.403) and (2.404) can be constructed from the damping ratios and natural circular frequencies for all five modes as:

$$\mathbf{D}^{(1)} = \begin{bmatrix} 2\zeta_1^{(1)}\omega_1 & & & & \\ & 2\zeta_2^{(1)}\omega_2 & & & \\ & & \ddots & & \\ & & & 2\zeta_{N-1}^{(1)}\omega_{N-1} & \\ & & & & 2\zeta_N^{(1)}\omega_N \end{bmatrix} = \begin{bmatrix} 0.3600 & 0 & 0 & 0 & 0 \\ 0 & 1.0509 & 0 & 0 & 0 \\ 0 & 0 & 1.6567 & 0 & 0 \\ 0 & 0 & 0 & 2.1282 & 0 \\ 0 & 0 & 0 & 0 & 2.4273 \end{bmatrix} \tag{2.426}$$

where the superscript (1) signifies that this is for Case 1. The system damping matrix $\mathbf{C}^{(1)}$ can then be constructed by Equation (2.408) as:

$$\mathbf{C}^{(1)} = (\mathbf{M}\Psi)\mathbf{D}^{(1)}(\Psi^T\mathbf{M}) = \begin{bmatrix} 3.4345 & -0.9846 & -0.1687 & -0.0681 & -0.0434 \\ -0.9846 & 3.2657 & -1.0527 & -0.2121 & -0.1115 \\ -0.1687 & -1.0527 & 3.2223 & -1.0961 & -0.2802 \\ -0.0681 & -0.2121 & -1.0961 & 3.1542 & -1.2648 \\ -0.0434 & -0.1115 & -0.2802 & -1.2648 & 2.1696 \end{bmatrix} \tag{2.427}$$

It is usually a fully symmetrical matrix. To save the required memory in computational analysis, one may store only the upper (or lower) triangular matrix. This matrix certainly has the orthogonal property.

In Case 2, the system damping matrix is constructed when the system is assumed to follow Rayleigh damping with $\zeta_1^{(2)} = \zeta_2^{(2)} = 0.02$. By Equation (2.412), one obtains:

$$\frac{1}{2}\begin{bmatrix} 1/\omega_1 & \omega_1 \\ 1/\omega_2 & \omega_2 \end{bmatrix} \begin{Bmatrix} a_M^{(2)} \\ a_K^{(2)} \end{Bmatrix} = \begin{Bmatrix} \zeta_1^{(2)} \\ \zeta_2^{(2)} \end{Bmatrix} \Rightarrow \begin{bmatrix} 0.0556 & 4.5004 \\ 0.0190 & 13.1366 \end{bmatrix} \begin{Bmatrix} a_M^{(2)} \\ a_K^{(2)} \end{Bmatrix} = \begin{Bmatrix} 0.02 \\ 0.02 \end{Bmatrix} \tag{2.428}$$

After solving this set of linear equations, the proportional constants for the mass and stiffness matrices can be obtained as $a_M^{(2)} = 0.2682$ and $a_K^{(2)} = 0.0011$. The system damping matrix $\mathbf{C}^{(2)}$ can then be calculated by Equation (2.409) as:

$$\mathbf{C}^{(2)} = \begin{bmatrix} 5.0723 & -2.2680 & 0 & 0 & 0 \\ -2.2680 & 5.0723 & -2.2680 & 0 & 0 \\ 0 & -2.2680 & 5.0723 & -2.2680 & 0 \\ 0 & 0 & -2.2680 & 5.0723 & -2.2680 \\ 0 & 0 & 0 & -2.2680 & 2.8043 \end{bmatrix} \tag{2.429}$$

This is also a symmetrical matrix. But it is usually not a full matrix as the system stiffness, and mass matrices are usually not full matrices. Other damping ratios (for modes 3, 4, and 5) can be determined by extracting the diagonal of the $\mathbf{D}^{(2)}$ matrix, which can be calculated from the system damping matrix by Equation (2.403):

$$\mathbf{D}^{(2)} = \Psi^T \mathbf{C}^{(2)} \Psi = \begin{bmatrix} 0.3600 & 0 & 0 & 0 & 0 \\ 0 & 1.0509 & 0 & 0 & 0 \\ 0 & 0 & 2.2134 & 0 & 0 \\ 0 & 0 & 0 & 3.4783 & 0 \\ 0 & 0 & 0 & 0 & 4.4441 \end{bmatrix} \tag{2.430}$$

From the diagonal elements of the $\mathbf{D}^{(2)}$ matrix, one can calculate the damping ratios and present them in a vector form as:

$$\zeta^{(2)} = \begin{Bmatrix} \zeta_1^{(2)} \\ \zeta_2^{(2)} \\ \vdots \\ \zeta_{N-1}^{(2)} \\ \zeta_N^{(2)} \end{Bmatrix} = \begin{Bmatrix} 0.0200 \\ 0.0200 \\ 0.0267 \\ 0.0327 \\ 0.0366 \end{Bmatrix} \tag{2.431}$$

As expected, the calculated damping ratios for modes 1 and 2 are equal to 0.02 (the assumed values for constructing the system damping matrix). The damping ratios for all other modes are calculated. In this case, the damping ratios for higher modes are larger in value (the increases are from 2% for the first and second modes to about 3.7% for the fifth mode).

The displacement responses for Cases 1 and 2 are calculated by the MATLAB function **mdof_msuper()** in Program 2.9 and plotted in Figure 2.68. The displacement responses for the two cases are overlaying on each other and the discrepancy is so minimal that it can be neglected. The excitation frequency of 1 Hz (= 6.2832 rad/s) is relatively close to the natural circular frequency of mode 1 (9.0008 rad/s) as shown in Equation (2.424). It is believed that mode 1 of the system dominates the response, and the damping ratio for the first mode is important. In both cases, the first mode damping ratios are the same, and therefore, the similarity of the displacement responses is reasonable.

In Case 3, the system damping matrix is constructed again by assuming that the system follows Rayleigh damping with damping ratios for modes 4 and 5 equal to 0.02. By following similar steps, the proportional constants can be calculated as

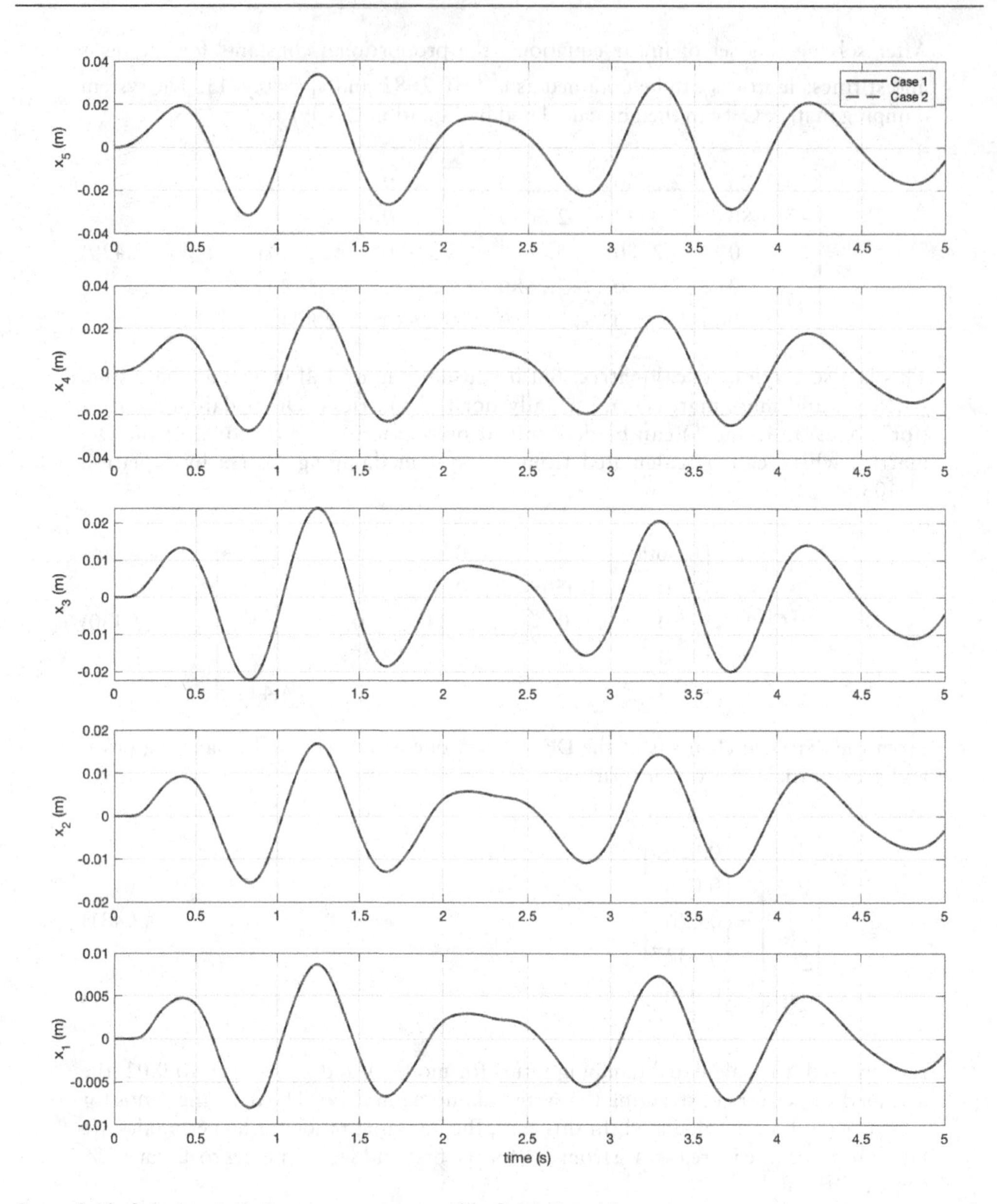

Figure 2.68 **Calculated displacement responses for Cases 1 and 2.**

$a_M^{(3)} = 1.1340$ and $a_K^{(3)} = 0.0004$. The system damping matrix $C^{(3)}$ can then be calculated by Equation (2.409) as:

$$C^{(3)} = \begin{bmatrix} 3.6728 & -0.7024 & 0 & 0 & 0 \\ -0.7024 & 3.6728 & -0.7024 & 0 & 0 \\ 0 & -0.7024 & 3.6728 & -0.7024 & 0 \\ 0 & 0 & -0.7024 & 3.6728 & -0.7024 \\ 0 & 0 & 0 & -0.7024 & 2.9704 \end{bmatrix} \quad (2.432)$$

The diagonal damping matrix $\mathbf{D}^{(3)}$ can be calculated as:

$$\mathbf{D}^{(3)} = \Psi^T \mathbf{C}^{(3)} \Psi = \begin{bmatrix} 1.1624 & 0 & 0 & 0 & 0 \\ 0 & 1.3764 & 0 & 0 & 0 \\ 0 & 0 & 1.7365 & 0 & 0 \\ 0 & 0 & 0 & 2.1282 & 0 \\ 0 & 0 & 0 & 0 & 2.4273 \end{bmatrix} \quad (2.433)$$

The damping ratios can then be calculated from the diagonal elements of the $\mathbf{D}^{(3)}$ matrix as:

$$\zeta^{(3)} = \begin{Bmatrix} \zeta_1^{(3)} \\ \zeta_2^{(3)} \\ \vdots \\ \zeta_{N-1}^{(3)} \\ \zeta_N^{(3)} \end{Bmatrix} = \begin{Bmatrix} 0.0646 \\ 0.0262 \\ 0.0210 \\ 0.0200 \\ 0.0200 \end{Bmatrix} \quad (2.434)$$

Unlike the previous case, the damping ratios for lower modes are larger in value. As the damping ratio for the first mode is very different from that in Case 1, it is believed that the calculated displacement responses for Cases 1 and 3 are not overlaying each other. By using **mdof_msuper()**, the displacement responses for Case 3 are calculated and presented together with the responses for Case 1 in Figure 2.69. As expected, the discrepancies between the responses for Cases 1 and 3 are not small.

In Rayleigh damping, the damping ratio of a given mode depends on the value of the corresponding natural circular frequency. From Equation (2.411), one obtains:

$$\zeta_i = \frac{1}{2\omega_i} a_{\mathrm{M}} + \frac{\omega_i}{2} a_{\mathrm{K}} \quad (2.435)$$

For Case 2 (the damping ratios for both modes 1 and 2 are equal to 0.02), the Rayleigh damping curve (i.e., Equation (2.435)) is illustrated in Figure 2.70(a). After fixing the curve by modes 1 and 2, it is clear from the figure that the damping ratios for all higher modes must increase. For Case 3 (damping ratios for both modes 4 and 5 are equal to 0.02), Figure 2.70(b) illustrates the corresponding Rayleigh damping curve. It is clear that the damping ratios for lower modes must increase once the curve is fixed by modes 4 and 5. By following this concept, if the damping ratios for modes 1 and 5 are enforced to be the same (i.e., $\zeta_1 = \zeta_5$), then all the intermediate modes (i.e., modes 2, 3, and 4) must have damping ratios that are lower than ζ_1. Readers may verify this by following the abovementioned steps.

In real applications, only damping ratios of lower modes can be measured with acceptable accuracy. By following the Rayleigh damping curve, the higher the mode, the larger the natural circular frequency, and the higher the damping ratio will be. This analytical interpretation may not agree with most field test results, in which the damping ratios for various modes have similar numerical values.

Example 2.15

Consider the same five-story shear building model in Example 2.14. Construct the system damping matrix under Caughey damping in the following cases.

1. The damping ratios of the first three modes are equal to 0.02.
2. The damping ratios of the first, third, and fifth modes are equal to 0.02.
3. The damping ratios of all five modes are equal to 0.02.

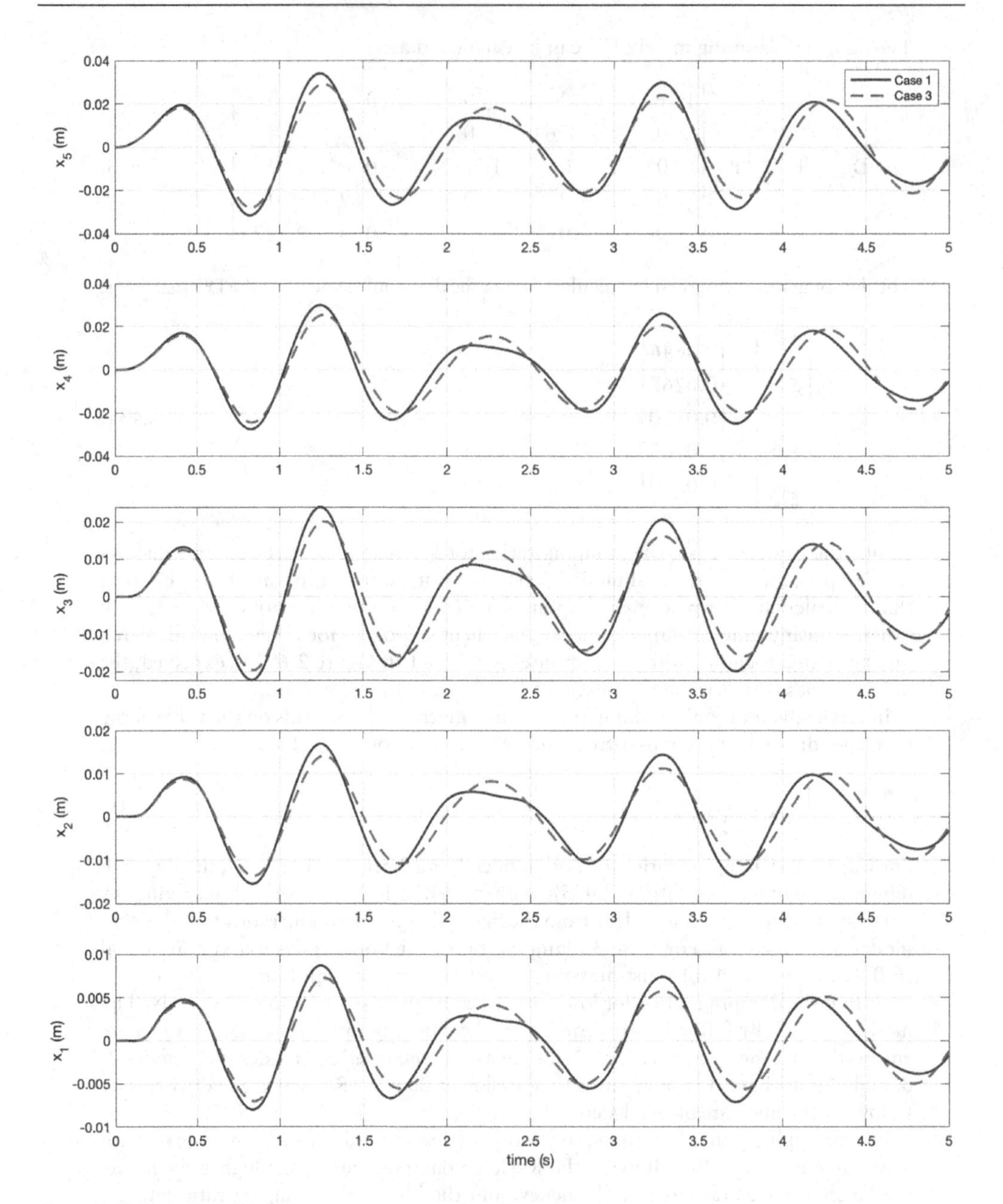

Figure 2.69 Calculated displacement responses for Cases 1 and 3.

SOLUTION

In Case 1, the damping ratios of the first three modes are fixed at 0.02 for constructing the system damping matrix (i.e., for calculating the damping ratios of other modes). As only three modes are considered, the set of equations for calculating the unknown proportional constants in Equation (2.419) can be simplified with $N_\zeta = 3$ as:

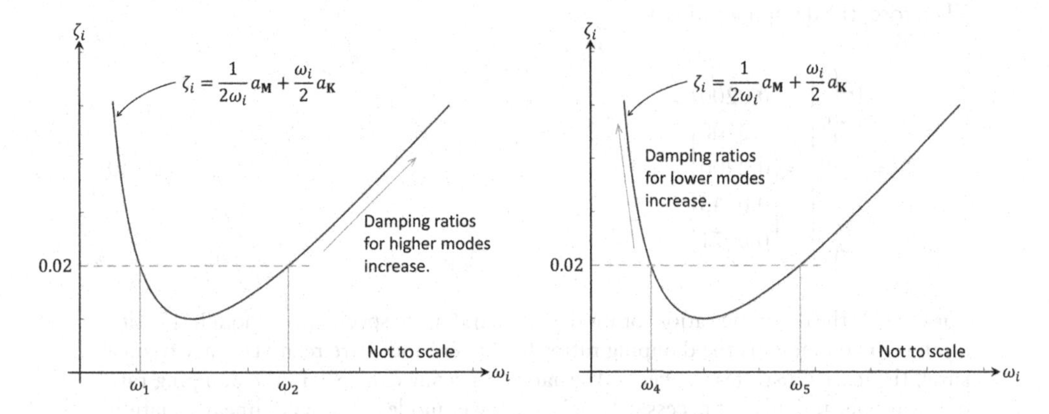

Figure 2.70 **The Rayleigh damping curves for Cases 2 and 3. (a) Case 2. (b) Case 3.**

$$\frac{1}{2}\begin{bmatrix} 1/\omega_1 & \omega_1 & \omega_1^3 \\ 1/\omega_2 & \omega_2 & \omega_2^3 \\ 1/\omega_3 & \omega_3 & \omega_3^3 \end{bmatrix} \begin{Bmatrix} a_0^{(1)} \\ a_1^{(1)} \\ a_2^{(1)} \end{Bmatrix} = \begin{Bmatrix} \zeta_1^{(1)} \\ \zeta_2^{(1)} \\ \zeta_3^{(1)} \end{Bmatrix} \tag{2.436}$$

Here the superscript of (1) is to put emphasis on the case identity (i.e., Case 1). With the calculated natural circular frequencies in Equation (2.424), Equation (2.436) becomes:

$$\begin{bmatrix} 0.0556 & 4.5004 & 364.5949 \\ 0.0190 & 13.1366 & 9067.8965 \\ 0.0121 & 20.7085 & 35522.7716 \end{bmatrix} \begin{Bmatrix} a_0^{(1)} \\ a_1^{(1)} \\ a_2^{(1)} \end{Bmatrix} = \begin{Bmatrix} 0.02 \\ 0.02 \\ 0.02 \end{Bmatrix} \tag{2.437}$$

After solving this set of linear equations, one obtains $a_0^{(1)} = 0.2496$, $a_1^{(1)} = 0.1390 \times 10^{-2}$, and $a_2^{(1)} = -0.33227 \times 10^{-6}$. The system damping matrix can then be constructed by Equation (2.415) as:

$$\mathbf{C}^{(1)} = a_0^{(1)}\mathbf{M} + a_1^{(1)}\mathbf{K} + a_2^{(1)}\mathbf{K}\mathbf{M}^{-1}\mathbf{K}$$
$$= \begin{bmatrix} 2.7375 & -0.1223 & -0.6645 & 0 & 0 \\ -0.1223 & 2.0729 & -0.1223 & -0.6645 & 0 \\ -0.6645 & -0.1223 & 2.0729 & -0.1223 & -0.6645 \\ 0 & -0.6645 & -0.1223 & 2.0729 & -0.7869 \\ 0 & 0 & -0.6645 & -0.7869 & 1.9506 \end{bmatrix} \tag{2.438}$$

The damping ratios for other modes (i.e., modes 4 and 5) can be calculated from the diagonal of the $\mathbf{D}^{(1)}$ matrix as:

$$\mathbf{D}^{(1)} = \mathbf{\Psi}^T\mathbf{C}^{(1)}\mathbf{\Psi} = \begin{bmatrix} 0.3600 & 0 & 0 & 0 & 0 \\ 0 & 1.0509 & 0 & 0 & 0 \\ 0 & 0 & 1.6567 & 0 & 0 \\ 0 & 0 & 0 & 1.5225 & 0 \\ 0 & 0 & 0 & 0 & 0.8633 \end{bmatrix} \tag{2.439}$$

Therefore, the damping ratios are:

$$\zeta^{(1)} = \begin{Bmatrix} \zeta_1^{(1)} \\ \zeta_2^{(1)} \\ \zeta_3^{(1)} \\ \zeta_4^{(1)} \\ \zeta_5^{(1)} \end{Bmatrix} = \begin{Bmatrix} 0.0200 \\ 0.0200 \\ 0.0200 \\ 0.0143 \\ 0.0071 \end{Bmatrix} \tag{2.440}$$

For Case 2, the damping ratios for modes 1, 3, and 5 are specified. Although this situation is uncommon (as the damping ratios for lower modes are relatively easy to measure), this case illustrates the general situation that one can specify the damping ratio of any mode, and it is not necessarily only the lower modes. The set of linear equations for solving the unknown proportional constants becomes:

$$\frac{1}{2}\begin{bmatrix} 1/\omega_1 & \omega_1 & \omega_1^3 \\ 1/\omega_3 & \omega_3 & \omega_3^3 \\ 1/\omega_5 & \omega_5 & \omega_5^3 \end{bmatrix} \begin{Bmatrix} a_0^{(2)} \\ a_1^{(2)} \\ a_2^{(2)} \end{Bmatrix} = \begin{Bmatrix} \zeta_1^{(2)} \\ \zeta_3^{(2)} \\ \zeta_5^{(2)} \end{Bmatrix} \tag{2.441}$$

The numerical values are:

$$\begin{bmatrix} 0.0556 & 4.5004 & 364.5949 \\ 0.0121 & 20.7085 & 35522.7716 \\ 0.0082 & 30.3418 & 111734.0106 \end{bmatrix} \begin{Bmatrix} a_0^{(2)} \\ a_1^{(2)} \\ a_2^{(2)} \end{Bmatrix} = \begin{Bmatrix} 0.02 \\ 0.02 \\ 0.02 \end{Bmatrix} \tag{2.442}$$

The unknown constants are $a_0^{(2)} = 0.2803, a_1^{(2)} = 9.9368 \times 10^{-4}$, and $a_2^{(2)} = -1.1151 \times 10^{-7}$. The system damping matrix can then be constructed by Equation (2.415) as:

$$\mathbf{C}^{(2)} = a_0^{(2)}\mathbf{M} + a_1^{(2)}\mathbf{K} + a_2^{(2)}\mathbf{K}\mathbf{M}^{-1}\mathbf{K}$$
$$= \begin{bmatrix} 3.4202 & -1.0953 & -0.2230 & 0 & 0 \\ -1.0953 & 3.1971 & -1.0953 & -0.2230 & 0 \\ -0.2230 & -1.0953 & 3.1971 & -1.0953 & -0.2230 \\ 0 & -0.2230 & -1.0953 & 3.1971 & -1.3183 \\ 0 & 0 & -0.2230 & -1.3183 & 2.1019 \end{bmatrix} \tag{2.443}$$

The damping ratios for other modes (i.e., modes 2 and 4) can be calculated from the diagonal of the $\mathbf{D}^{(2)}$ matrix as:

$$\mathbf{D}^{(2)} = \Psi^T\mathbf{C}^{(2)}\Psi = \begin{bmatrix} 0.3600 & 0 & 0 & 0 & 0 \\ 0 & 0.9130 & 0 & 0 & 0 \\ 0 & 0 & 1.6567 & 0 & 0 \\ 0 & 0 & 0 & 2.1996 & 0 \\ 0 & 0 & 0 & 0 & 2.4273 \end{bmatrix} \tag{2.444}$$

Therefore, the damping ratios are:

$$\zeta^{(2)} = \begin{Bmatrix} \zeta_1^{(2)} \\ \zeta_2^{(2)} \\ \zeta_3^{(2)} \\ \zeta_4^{(2)} \\ \zeta_5^{(2)} \end{Bmatrix} = \begin{Bmatrix} 0.0200 \\ 0.0174 \\ 0.0200 \\ 0.0207 \\ 0.0200 \end{Bmatrix} \tag{2.445}$$

Case 3 is used to verify that the Caughey damping formulation can return the same system damping matrix as Equation (2.408), which calculates **C** directly from **D**, when the damping ratios for all modes are specified. From Equation (2.419), the system of linear equations for solving unknown constants becomes:

$$\frac{1}{2}\begin{bmatrix} 1/\omega_1 & \omega_1 & \omega_1^3 & \omega_1^5 & \omega_1^7 \\ 1/\omega_2 & \omega_2 & \omega_2^3 & \omega_2^5 & \omega_2^7 \\ 1/\omega_3 & \omega_3 & \omega_3^3 & \omega_3^5 & \omega_3^7 \\ 1/\omega_4 & \omega_4 & \omega_4^3 & \omega_4^5 & \omega_4^7 \\ 1/\omega_5 & \omega_5 & \omega_5^3 & \omega_5^5 & \omega_5^7 \end{bmatrix} \begin{Bmatrix} a_0^{(3)} \\ a_1^{(3)} \\ a_2^{(3)} \\ a_3^{(3)} \\ a_4^{(3)} \end{Bmatrix} = \begin{Bmatrix} \zeta_1^{(3)} \\ \zeta_2^{(3)} \\ \zeta_3^{(3)} \\ \zeta_4^{(3)} \\ \zeta_5^{(3)} \end{Bmatrix} \tag{2.446}$$

The calculated proportional constants are:

$$\begin{Bmatrix} a_0^{(3)} \\ a_1^{(3)} \\ a_2^{(3)} \\ a_3^{(3)} \\ a_4^{(3)} \end{Bmatrix} = \begin{Bmatrix} 0.2348 \\ 1.6043\times10^{-3} \\ -7.4436\times10^{-7} \\ 2.0765\times10^{-10} \\ -2.1699\times10^{-14} \end{Bmatrix} \tag{2.447}$$

Finally, one can calculate the system damping matrix by Equation (2.415) as:

$$\mathbf{C}^{(3)} = a_0^{(3)}\mathbf{M} + a_1^{(3)}\mathbf{K} + a_2^{(3)}\mathbf{K}\mathbf{M}^{-1}\mathbf{K} + a_3^{(3)}\left(\mathbf{K}\mathbf{M}^{-1}\right)^2\mathbf{K} + a_4^{(3)}\left(\mathbf{K}\mathbf{M}^{-1}\right)^3\mathbf{K}$$
$$= \begin{bmatrix} 3.4345 & -0.9846 & -0.1687 & -0.0681 & -0.0434 \\ -0.9846 & 3.2657 & -1.0527 & -0.2121 & -0.1115 \\ -0.1687 & -1.0527 & 3.2223 & -1.0961 & -0.2802 \\ -0.0681 & -0.2121 & -1.0961 & 3.1542 & -1.2648 \\ -0.0434 & -0.1115 & -0.2802 & -1.2648 & 2.1696 \end{bmatrix} \tag{2.448}$$

This system damping matrix is exactly the same as the one in Equation (2.427).

Although only shear building models are employed in this section, the theories and formulations presented here are appliable to general multi-DOF systems for given **K** and **M** (the system damping matrix **C** can be calculated from the specified damping ratios as demonstrated in Example 2.14 and Example 2.15).

REFERENCES

Au, S.K., 2018. *Operational modal analysis: Modeling, Bayesian inference, uncertainty laws*. Springer.

Beck, J.L. and Dowling, M.J., 1988. Quick algorithms for computing either displacement, velocity or acceleration of an oscillator. *Earthquake Engineering & Structural Dynamics*, 16(2), 245–253.

Brilliant.org, 2021. Cardano's method. Retrieved 19:50, April 5, 2021, from https://brilliant.org/wiki/cardano-method/.

Chapter 3

Modal analysis based on power spectral density data

3.1 POWER SPECTRAL DENSITY

When dealing with full-scale structures under ambient excitations, usually measured structural responses are not directly used for modal analysis. The reason is that ambient excitations cannot be measured, so model-predicted responses cannot be calculated for fitting to the measured ones. Due to this reason, modal parameters cannot be extracted by minimizing differences between measured and model-predicted responses. One way to bypass this problem is to use power spectral density (PSD) as data. Fourier transform data can be also used and very efficient methods for modal analysis have been developed (Au 2011b; Au 2018). This section introduces how to obtain experimental PSD data from measured structural responses.

In practice, PSD matrices are usually estimated based on measured accelerations with a finite duration T:

$$\hat{\mathbf{S}}_k = \hat{\mathbf{X}}_{k,T}\hat{\mathbf{X}}_{k,T}^* \tag{3.1}$$

where $\hat{\mathbf{X}}_{k,T} \in R^{N_d}$ is the scaled discrete Fourier transform of measured accelerations $\left\{\hat{\ddot{\mathbf{x}}}_j \in R^{N_d}: j=0,1,\ldots,N_T-1\right\}$ at the discrete frequency $\omega_k = \dfrac{2\pi k}{T}$; N_d is the number of measured DOFs; N_T is the number of data points; $\hat{\mathbf{S}}_k \in R^{N_d \times N_d}$ is the PSD matrix at ω_k; $*$ is the conjugate transpose operator; $T = N_T\Delta t$; Δt is the sampling time step; subscript "T" is used to remind that a scaling factor $\dfrac{1}{\sqrt{T}}$ is multiplied to the discrete Fourier transform $\hat{\mathbf{X}}_k$, i.e.,

$$\begin{aligned}\hat{\mathbf{X}}_{k,T} &= \frac{1}{\sqrt{T}}\hat{\mathbf{X}}_k \\ &= \frac{1}{\sqrt{N_T\Delta t}}\sum_{j=0}^{N_T}\hat{\ddot{\mathbf{x}}}_j \exp\left(-ik\frac{2\pi}{N_T\Delta t}j\Delta t\right)\Delta t \\ &= \sqrt{\frac{\Delta t}{N_T}}\sum_{j=0}^{N_T}\hat{\ddot{\mathbf{x}}}_j \exp\left(-i2\pi jk/N_T\right)\end{aligned} \tag{3.2}$$

The scaling factor is used because Fourier transform does not exist for a stationary stochastic process, and measured structural responses under ambient excitations are stationary stochastic processes. This can be seen by inspecting the definition of Fourier

DOI: 10.1201/9780429445866-3

transform, $\mathbf{X}(\omega)=\int_{-\infty}^{\infty}\mathbf{x}(t)\exp(-i\omega t)dt$. If $\mathbf{x}(t)$ is stationary (non-decaying), the integration (summation) will be infinite when the integration limits go to infinity, i.e., $\mathbf{X}(\omega)\rightarrow\infty$ if $\mathbf{x}(t)$ is stationary. However, note that the scaling factor is $\frac{1}{\sqrt{T}}$ but not $\frac{1}{T}$, because $\frac{1}{T}$ is related to signal energy.

Here, only the diagonal of $\hat{\mathbf{S}}_k$ is used as data:

$$\hat{\mathbf{s}}_k=\text{diag}\left(\hat{\mathbf{X}}_{k,T}\hat{\mathbf{X}}_{k,T}^{*}\right) \tag{3.3}$$

where "diag" extracts the diagonal of a matrix; each element of $\hat{\mathbf{s}}_k$ contains the information of only one measured DOF, so it is called auto-PSD. On the other hand, each non-diagonal element of $\hat{\mathbf{S}}_k$ contains information of two different measured DOFs, so it is called cross-PSD.

The auto-PSD estimated by Equation (3.3) contains original information in data for modal analysis, but it stochastically revolves around the theoretical auto-PSD, making it inconvenient for visualizing spectral peaks in PSD spectra. In order to reduce statistical variances of PSDs for better visualization, stochastic averaging is done using multiple samples of auto-PSDs calculated by Equation (3.3). To calculate the sample average of PSD, the whole measurement $\left\{\hat{\ddot{\mathbf{x}}}_j: j=0,1,\ldots,N_T-1\right\}$ is usually divided into L non-overlapping segments, each of which has equal length N_T/L (assume that N_T/L is an integer), $\left\{\hat{\ddot{\mathbf{x}}}_j^l: j=0,1,\ldots,N_T/L-1; l=1,2,\ldots,L\right\}$, where the superscript l is the segment index. Each segment $\left\{\hat{\ddot{\mathbf{x}}}_j^l\right\}$ is used to calculate one sample of auto-PSD with Equation (3.3), $\hat{\mathbf{s}}_k^l=\text{diag}\left(\hat{\mathbf{X}}_{k,T}^l\hat{\mathbf{X}}_{k,T}^{l*}\right)$, where $\hat{\mathbf{X}}_{k,T}^l$ is the Fourier transform of $\left\{\hat{\ddot{\mathbf{x}}}_j^l\right\}$ at the k-th frequency. The averaged (smooth) auto-PSD is then obtained:

$$\hat{\mathbf{u}}_k=\frac{1}{L}\sum_{l=1}^{L}\hat{\mathbf{s}}_k^l \tag{3.4}$$

The variance of $\hat{\mathbf{u}}_k$ is smaller than $\hat{\mathbf{s}}_k$, and $\hat{\mathbf{u}}_k$ converges to the theoretical auto-PSD when $L\rightarrow\infty$. This can be reasoned by Central Limit Theorem. This theorem says that if the samples of auto-PSDs are independent and identically distributed (i.i.d.) random vectors (which is a reasonable assumption for our case) with theoretical mean $\text{E}\left(\hat{\mathbf{s}}_k^l\right)=\bar{\mathbf{u}}_k$ and covariance matrix $\text{COV}\left(\hat{\mathbf{s}}_k^l\right)=\mathbf{C}_k$, then as L approaches infinity, the probability distribution of the random vector $\sqrt{L}\left(\frac{1}{L}\sum_{l=1}^{L}\hat{\mathbf{s}}_k^l-\bar{\mathbf{u}}_k\right)$ converges to a normal distribution, i.e.,

$$\sqrt{L}\left(\hat{\mathbf{u}}_k-\bar{\mathbf{u}}_k\right)\sim N\left(0,\mathbf{C}_k\right) \tag{3.5}$$

where $N\left(0,\mathbf{C}_k\right)$ denotes a normal distribution with mean zero and covariance matrix $\mathbf{C}_k$. Therefore, it can be inferred that

$$\hat{\mathbf{u}}_k\sim N\left(\bar{\mathbf{u}}_k,\frac{\mathbf{C}_k}{L}\right) \tag{3.6}$$

Equation (3.6) indicates that the sample average $\hat{\mathbf{u}}_k$ has the mean equal to the theoretical auto-PSD and the variance of $\hat{\mathbf{u}}_k$ is $1/L$ of the variance of $\hat{\mathbf{s}}_k^l$. Therefore, by calculating the sample average, the statistical variance of auto-PSD is reduced. However, the reduction of variance comes at the expense of reducing the frequency resolution of PSD spectra because the length of measured responses is shortened by diving the whole measurement into multiple segments.

Note that this sample average of auto-PSD (Equation (3.4)) is only for getting smooth auto-PSD spectra, so that spectral peaks can be conveniently visualized, but it is not used as data for modal analysis. The auto-PSD that contains the most information (Equation (3.3)) is used as data. The following example shows auto-PSD spectra obtained from measured accelerations of a laboratory shear building and compares auto-PSD spectra without and with averaging.

Example 3.1

The four-story shear building was excited by a big fan to simulate stochastic wind excitation (see Figure 3.1). Each story was instrumented with one accelerometer to measure accelerations. The number of measured DOF N_d was thus four. The accelerations were measured for 10 minutes with sampling frequency 2048 Hz, and then downsampled to 128 Hz for analysis. Figure 3.2 shows the accelerations of stories 1 to 4 from the top to the bottom. The auto-PSD spectra without averaging were first constructed. Fourier transforms of the whole measurement were calculated, and auto-PSDs were obtained at different frequencies $f_k = k\Delta f$ using Equation (3.3), where Δf is the frequency resolution, which is $1/600\text{s} \approx 0.0017$ Hz in this case. The auto-PSD spectra were then constructed by plotting the auto-PSDs versus different f_ks and shown in Figure 3.3. The four lines in Figure 3.3 correspond to the four stories, respectively. Variations are observed for the auto-PSD spectra without averaging, but spectral peaks can be still identified.

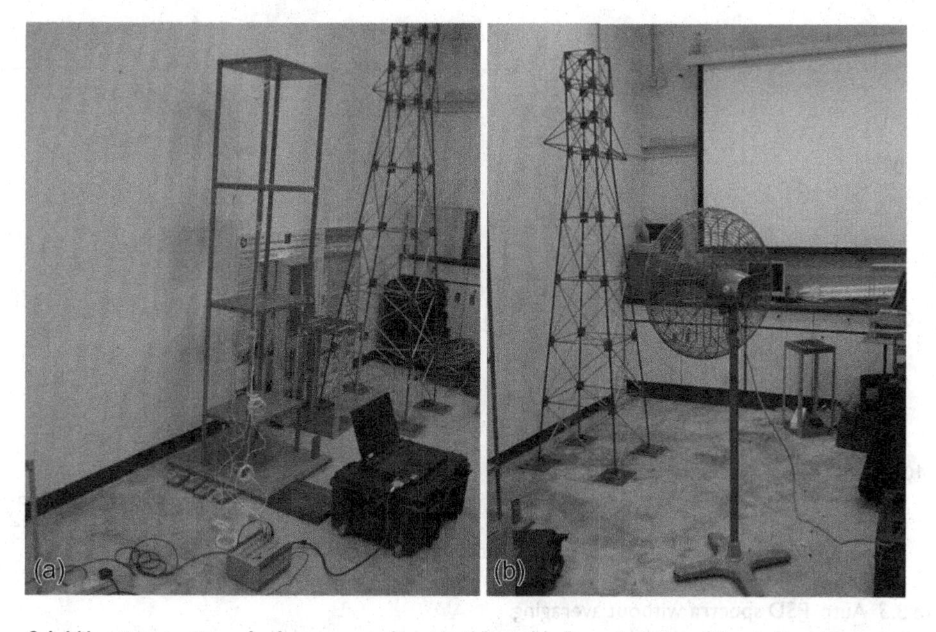

Figure 3.1 Vibration testing of a four-story shear building: (a) shear building; (b) wind excitation.

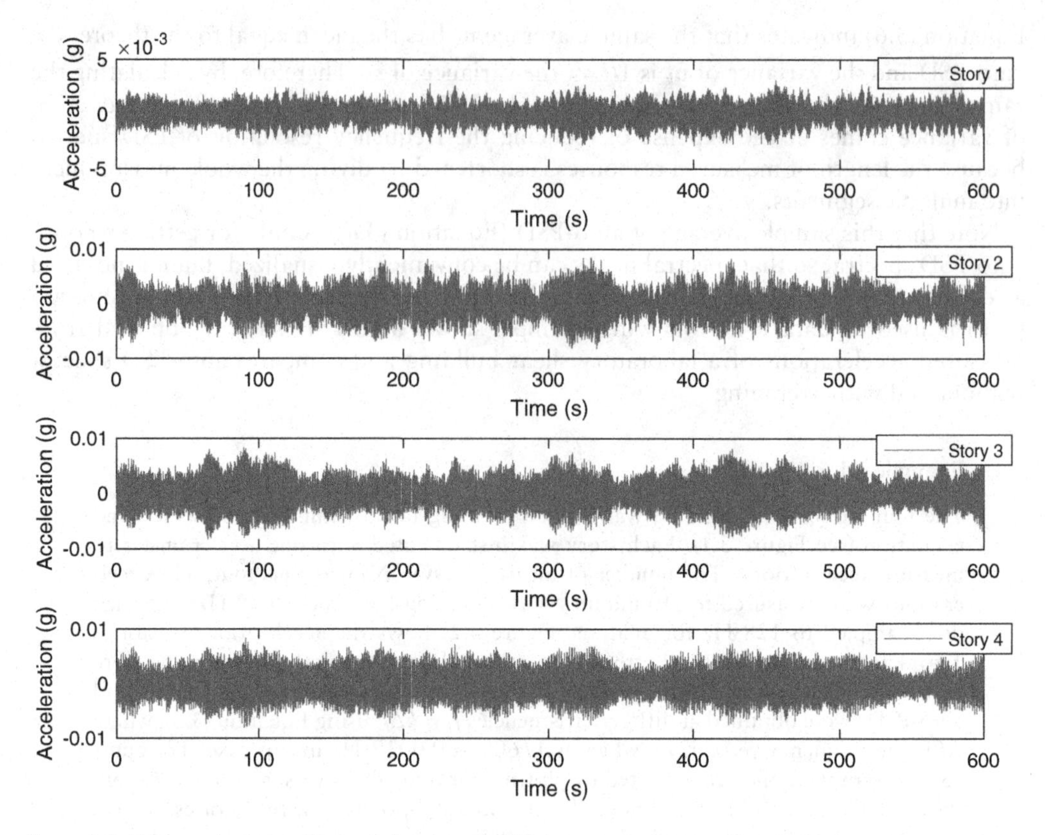

Figure 3.2 Measured accelerations of the shear building.

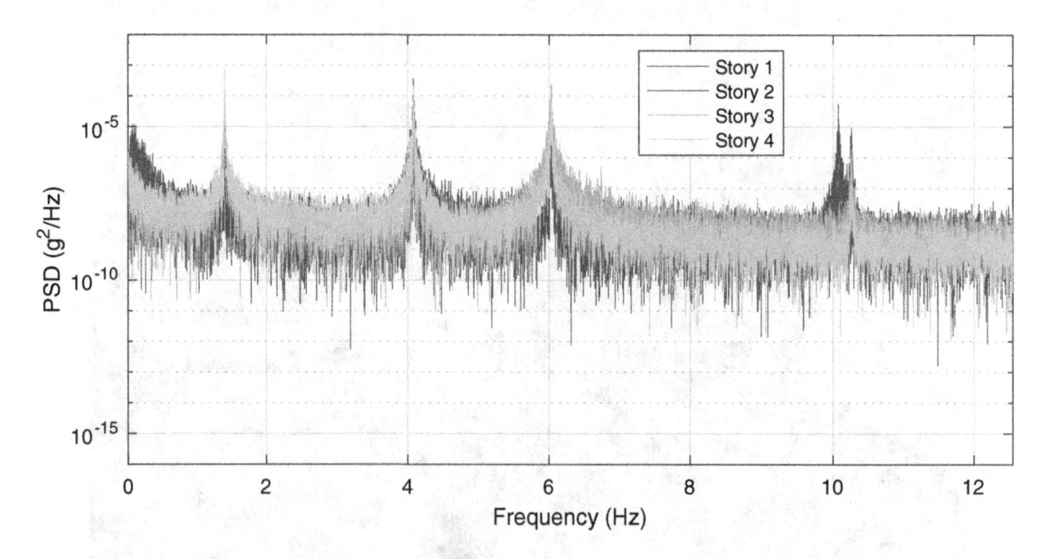

Figure 3.3 Auto-PSD spectra without averaging.

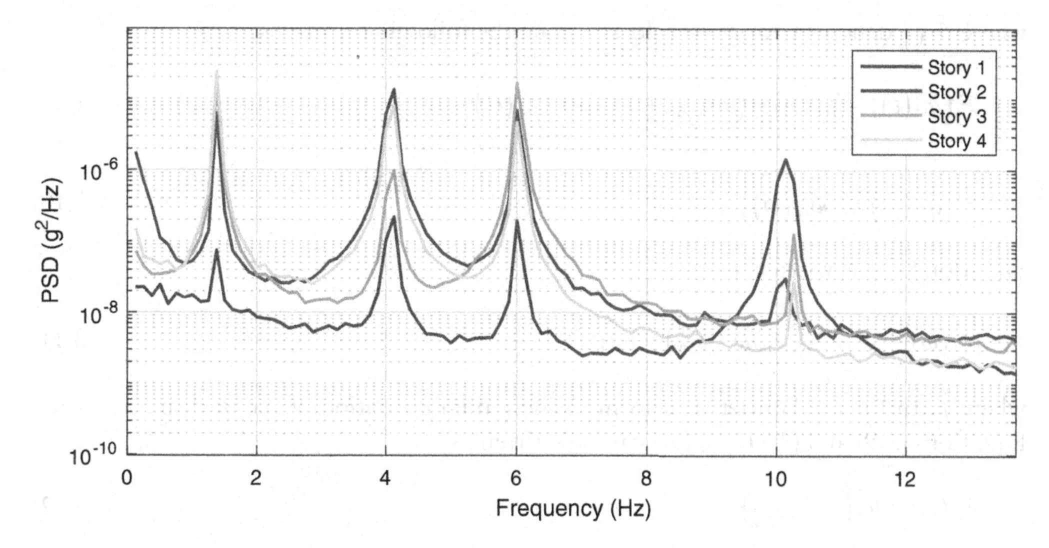

Figure 3.4 Auto-PSD spectra with averaging.

The averaged auto-PSD spectra were then constructed by dividing the original data into multiple segments, each of which has 1024 data points. Each segment of data was used to calculate one set of auto-PSD spectra, and then the averaged PSD spectra were obtained in Figure 3.4 (Equation (3.4)). It can be seen that after averaging the PSD spectra in Figure 3.4 are much smoother than the ones in Figure 3.3. The four spectral peaks can be clearly identified in Figure 3.4. Spectral peaks indicate that the vibration energy of a structure concentrates around these peaks, so they are related to structural vibration modes. The frequency resolution was reduced to 0.1251 Hz for the smoothed spectra.

3.2 MATHEMATICAL MODELING OF PSD

Having introduced experimental PSD, this section develops the theoretical model of PSD based on modal parameters, i.e., natural frequencies, damping ratios, mode shapes, and PSDs of modal excitations, in order to minimize the difference between experimental and theoretical PSDs for identifying modal parameters. According to modal superposition (see Chapter 2), the Fourier transform of acceleration can be written:

$$\ddot{\mathbf{x}}(\omega) = \sum_{m=1}^{N_m} \phi_m \ddot{q}_m(\omega) \tag{3.7}$$

where $\ddot{\mathbf{x}}(\omega)$ is the Fourier transform of acceleration; $\ddot{q}_m(\omega)$ is the Fourier transform of the acceleration in the m-th modal coordinate; ϕ_m is the mode shape of the m-th mode. It is assumed that mode shapes do not depend on time, so Fourier transform does not have effects on mode shapes. To get the expression of $\ddot{q}_m(\omega)$, consider the Fourier transform of the equation of motion of the m-th modal coordinate:

$$\mathcal{F}\left(\ddot{q}_m(t) + 2\omega_m \xi_m \dot{q}_m(t) + \omega_m^2 q_m(t)\right) = \mathcal{F}\left(p_m(t)\right) \tag{3.8}$$

where $\mathcal{F}(\cdot)$ denotes Fourier transform. Using the following relations,

$$\mathcal{F}(\ddot{q}_m(t)) = \ddot{q}_m(\omega) = (i\omega)^2 q_m(\omega) = -\omega^2 q_m(\omega) \tag{3.9}$$

$$\mathcal{F}(\dot{q}_m(t)) = i\omega q_m(\omega) \tag{3.10}$$

Equation (3.8) is written as

$$q_m(\omega)\left(-\omega^2 + 2\omega_m\xi_m i\omega + \omega_m^2\right) = p_m(\omega) \tag{3.11}$$

where $p_m(\omega)$ is the Fourier transform of the modal excitation of the m-th mode. Using Equations (3.9) and (3.11), $\ddot{q}_m(\omega)$ can be written as

$$\ddot{q}_m(\omega) = h_m(\omega)p_m(\omega) \tag{3.12}$$

where $h_m(\omega)$ is the transfer function of the m-th mode

$$h_m(\omega) = \frac{1}{1-\beta_m^2 - i2\xi_m\beta_m} \tag{3.13}$$

where $\beta_m = \frac{\omega_m}{\omega}$ is the ratio of the natural frequency to the frequency coordinate.

Moreover, in the neighborhood of one spectral peak, only one mode is dominant, so when considering the PSD data around one spectral peak, Equation (3.7) is rewritten to leave only the contribution of one mode (Au 2011b; Au 2018):

$$\ddot{\mathbf{x}}(\omega) = \boldsymbol{\phi}_m \ddot{q}_m(\omega) \tag{3.14}$$

Therefore, the theoretical PSD matrix is obtained based on Equations (3.13) and (3.14)

$$\begin{aligned} \mathbf{S}(\omega) &= \ddot{\mathbf{x}}(\omega)\ddot{\mathbf{x}}(\omega)^* \\ &= \boldsymbol{\phi}_m \ddot{q}_m(\omega)\boldsymbol{\phi}_m^T \ddot{q}_m^*(\omega) \\ &= S_m(\omega)h_m(\omega)h_m^*(\omega)\boldsymbol{\phi}_m\boldsymbol{\phi}_m^T \end{aligned} \tag{3.15}$$

where $S_m(\omega) = p_m(\omega)p_m^*(\omega)$ is the PSD of the modal excitation of the m-th mode. Because ambient excitations cannot be measured, a theoretical model of $S_m(\omega)$ needs to be built and identified together with other parameters. If the modal excitation is modeled as an arbitrary stochastic process, it is difficult to identify the value of $S_m(\omega)$ at each discrete frequency. Notice that a Gaussian process has the characteristic that its PSD is constant within a certain frequency range. It is thus assumed that for a structure under ambient excitations, the modal excitation is a Gaussian process, i.e., $S_m(\omega)$ is constant. This assumption only requires that the PSDs of excitations are constant around the spectral peaks corresponding to structural modes, but not constant over the whole frequency range, so it is not difficult to fulfill in practice. By using this assumption, only the PSD data around the spectral peaks of interest are included for identification of modal parameters. The PSD data at other frequency ranges are not used because they do not carry the information of the modes of interest.

By substituting a constant PSD S_m for $S_m(\omega)$ and using Equation (3.13), Equation (3.15) is written as

$$\mathbf{S}(\omega) = \frac{S_m}{\left(1-\beta_m^2\right)^2 + \left(2\xi_m\beta_m\right)^2}\boldsymbol{\phi}_m\boldsymbol{\phi}_m^T \tag{3.16}$$

The theoretical auto-PSD can then be obtained by taking the diagonal of Equation (3.16):

$$\mathbf{s}(\omega) = \frac{S_m}{\left(1-\beta_m^2\right)^2 + \left(2\xi_m\beta_m\right)^2}\mathbf{a}_m \tag{3.17}$$

where $\mathbf{a}_m$ is the diagonal of $\boldsymbol{\phi}_m\boldsymbol{\phi}_m^T$, which contains the squares of the components of the mode shape $\boldsymbol{\phi}_m$. It is more convenient to use root auto-PSD in practice. Taking the square root of Equation (3.17) gives

$$\mathbf{r}(\omega) = \sqrt{\mathbf{s}(\omega)} = \frac{\sqrt{S_m}}{\sqrt{\left(1-\beta_m^2\right)^2 + \left(2\xi_m\beta_m\right)^2}}\mathbf{b}_m \tag{3.18}$$

where $\mathbf{b}_m = \left[\left|\phi_{m,1}\right|,\ldots,\left|\phi_{m,d}\right|,\ldots,\left|\phi_{m,N_d}\right|\right]$ is defined as the mode shape amplitude vector of the m-th mode, and it contains the absolute values of the components of $\boldsymbol{\phi}_m$; $\left|\phi_{m,d}\right|$, denoting the absolute value (amplitude) of the mode shape at the d-th DOF.

3.3 THE OPTIMIZATION ALGORITHM

3.3.1 The objective function

Having introduced experimental and theoretical PSDs, it is now time to develop the algorithm for identifying modal parameters. Given experimental root auto-PSDs $\hat{\mathbf{r}}_k$, which are obtained by taking the square root of Equation (3.3) and theoretical root auto-PSDs $\mathbf{r}_k(\boldsymbol{\theta})$ at discrete frequencies ω_k around a spectral peak, the objective is to find the optimum of the uncertain parameters $\boldsymbol{\theta} = \{\omega_m, \xi_m, S_m, \boldsymbol{\phi}_m\}$, so that the difference between $\hat{\mathbf{r}}_k$ and $\mathbf{r}_k(\boldsymbol{\theta})$ is minimized. Note that only the modal parameters of one mode are included in $\boldsymbol{\theta}$, because only one spectral peak is considered. The modal parameters of different modes are identified by considering different spectral peaks, respectively. The objective function $J(\boldsymbol{\theta})$ is written for the m-th mode quantifying the errors between $\hat{\mathbf{r}}_k$ and $\mathbf{r}_k(\boldsymbol{\theta})$ at different frequencies:

$$J(\boldsymbol{\theta}) = \sum_{k=1}^{N_k}\left(\mathbf{r}_k(\boldsymbol{\theta}) - \hat{\mathbf{r}}_k\right)^T\left(\mathbf{r}_k(\boldsymbol{\theta}) - \hat{\mathbf{r}}_k\right) \tag{3.19}$$

Therefore, the problem is to find the optimal $\boldsymbol{\theta}$ so that $J(\boldsymbol{\theta})$ is minimized. In practice, the number of measured DOFs is usually large, so $\boldsymbol{\phi}_m$ (and thus, $\boldsymbol{\theta}$) is high-dimensional. Direct optimization of $J(\boldsymbol{\theta})$ is inefficient. This high-dimensional problem requires large computational power and, most of the time, cannot converge. An efficient algorithm is developed to tackle this high-dimensional problem. The idea is to divide the uncertain parameters into three groups, $\mathbf{b}_m$ (amplitude of mode shape), S_m, and $\left[\omega_m, \xi_m\right]$ and do the optimization for each group based on the other two groups. Iteration is conducted among the three groups until convergence is reached.

3.3.2 Optimization for $\mathbf{b}_m$

Because of mathematical convenience, the mode shape amplitude $\mathbf{b}_m$ is considered here instead of considering the mode shape $\boldsymbol{\phi}_m$ directly. $\boldsymbol{\phi}_m$ can be recovered from $\mathbf{b}_m$ by putting signs to the elements of $\mathbf{b}_m$, and the signs can be easily obtained by the experimental PSDs at the identified natural frequency. The procedures will be discussed later. The first derivative of $J(\boldsymbol{\theta})$ is taken with respect to the mode shape amplitude $\mathbf{b}_m$:

$$\frac{\partial J}{\partial \mathbf{b}_m} = \sum_{k=1}^{N_k} 2\left(\mathbf{r}_k(\theta) - \hat{\mathbf{r}}_k\right)^T \frac{\partial \mathbf{r}_k(\theta)}{\partial \mathbf{b}_m} \tag{3.20}$$

where $\frac{\partial \mathbf{r}_k(\theta)}{\partial \mathbf{b}_m}$ is obtained based on Equation (3.18):

$$\frac{\partial \mathbf{r}_k(\theta)}{\partial \mathbf{b}_m} = \sqrt{S_m B_k} \tag{3.21}$$

where it is defined that $B_k = \frac{1}{\left(1-\beta_{m,k}^2\right)^2 + \left(2\xi_m \beta_{m,k}\right)^2}$ for convenience; $\beta_{m,k} = \frac{\omega_k}{\omega_m}$ is the frequency ratio at the discrete frequency ω_k. The optimal $\mathbf{b}_m$, given other parameters, is obtained by setting Equation (3.20) to zero:

$$\begin{aligned} &\sum_{k=1}^{N_k} \mathbf{r}_k^T(\theta) \frac{\partial \mathbf{r}_k(\theta)}{\partial \mathbf{b}_m} = \sum_{k=1}^{N_k} \hat{\mathbf{r}}_k^T \frac{\partial \mathbf{r}_k(\theta)}{\partial \mathbf{b}_m} \Rightarrow \\ &\sum_{k=1}^{N_k} \sqrt{S_m B_k}\, \mathbf{b}_m \sqrt{S_m B_k} = \sum_{k=1}^{N_k} \hat{\mathbf{r}}_k \sqrt{S_m B_k} \Rightarrow \\ &\mathbf{b}_m = \frac{\sum_{k=1}^{N_k} \sqrt{B_k}\, \hat{\mathbf{r}}_k}{\sqrt{S_m} \sum_{k=1}^{N_k} B_k} \end{aligned} \tag{3.22}$$

3.3.3 Optimization for S_m

Using the same technique as in Section 3.3.2, the first derivative of $J(\boldsymbol{\theta})$ is taken with respect to the PSD of modal excitation S_m:

$$\frac{\partial J}{\partial S_m} = \sum_{k=1}^{N_k} 2\left(\mathbf{r}_k(\theta) - \hat{\mathbf{r}}_k\right)^T \frac{\partial \mathbf{r}_k(\theta)}{\partial S_m} \tag{3.23}$$

where $\frac{\partial \mathbf{r}_k(\theta)}{\partial S_m}$ is obtained based on Equation (3.18):

$$\frac{\partial \mathbf{r}_k(\theta)}{\partial S_m} = \frac{\sqrt{B_k}}{2\sqrt{S_m}} \mathbf{b}_m \tag{3.24}$$

Setting Equation (3.23) to zero and substituting Equations (3.18) and (3.24) into Equation (3.23) gives

$$\sum_{k=1}^{N_k} \left(B_k \mathbf{b}_m^T \mathbf{b}_m - \hat{\mathbf{r}}_k^T \frac{\sqrt{B_k}}{\sqrt{S_m}} \mathbf{b}_m \right) = 0 \tag{3.25}$$

Note that $\mathbf{b}_m$ is normalized such that its Euclidean norm equals unity, i.e., $\mathbf{b}_m = 1$. Therefore, the optimal S_m given other parameters is obtained from Equation (3.25)

$$S_m = \left(\frac{\sum_{k=1}^{N_k} \hat{\mathbf{r}}_k^T \mathbf{b}_m \sqrt{B_k}}{\sum_{k=1}^{N_k} B_k} \right)^2 \tag{3.26}$$

It is now left to find the optimal ω_m and ξ_m given other parameters. However, the optima of ω_m and ξ_m cannot be derived analytically like $\mathbf{b}_m$ and S_m. The optimization is thus done numerically by minimizing the objective function Equation (3.19) given the optima of $\mathbf{b}_m$ and S_m. It is very efficient because only two scalars are to be optimized.

3.3.4 Initialization for the iteration

To set the iteration in motion, one needs to input the initial values of modal parameters to the algorithm. The initial values of the natural frequency, damping ratio, and mode shape amplitude of the target mode, $\omega_{m,0}$, $\xi_{m,0}$, and $\mathbf{b}_{m,0}$, are required to be selected. Given $\omega_{m,0}$, $\xi_{m,0}$, and $\mathbf{b}_{m,0}$, the initial value of the modal excitation PSD $S_{m,0}$ can then be calculated using Equation (3.26). To obtain $\omega_{m,0}$, after constructing auto-PSD spectra using measured accelerations, $\omega_{m,0}$ is selected at a spectral peak, which corresponds to a target mode to be identified. $\xi_{m,0}$ can be initialized to be 1%. $\mathbf{b}_{m,0}$ needs some special treatment.

Using modal coordinates (Equation (3.7)), the PSD matrix is expressed as

$$\begin{aligned} \mathbf{S}(\omega) &= \ddot{\mathbf{x}}(\omega)\ddot{\mathbf{x}}(\omega)^* \\ &= \sum_{m=1}^{N_m} \phi_m \ddot{q}_m(\omega) \sum_{m=1}^{N_m} \phi_m^T \ddot{q}_m^*(\omega) \\ &= \Phi \ddot{\mathbf{q}}(\omega) \ddot{\mathbf{q}}^*(\omega) \Phi^* \\ &= \Phi \mathbf{Q}(\omega) \Phi^* \end{aligned} \tag{3.27}$$

where $\Phi = [\phi_1, \ldots, \phi_{N_m}]$; $\ddot{\mathbf{q}}(\omega) = [\ddot{q}_1(\omega), \ldots, \ddot{q}_{N_m}]^T$; $\mathbf{Q}(\omega) = \ddot{\mathbf{q}}(\omega)\ddot{\mathbf{q}}^*(\omega)$ are defined as the PSD matrix in modal coordinates at frequency ω; $\Phi^* = \Phi^T$ has been used in the last equality because the conjugate transpose is the same as the transpose for a real matrix.

On the other hand, notice that the singular value decomposition (SVD) of $\mathbf{S}(\omega)$ has a similar structure as Equation (3.27) (Yang et al. 2019):

$$\begin{aligned} \mathbf{S}(\omega) &= \mathbf{U}\mathbf{D}(\omega)\mathbf{V}^* \\ &= \mathbf{U}\mathbf{D}(\omega)\mathbf{U}^* \end{aligned} \tag{3.28}$$

where $\mathbf{D}(\omega) \in R^{N_d \times N_d}$ is a diagonal matrix that contains the non-negative real singular values of $\mathbf{S}(\omega)$ in the descending order on the diagonal. The columns of the two unitary matrices $\mathbf{U} \in R^{N_d \times N_d}$ and $\mathbf{V} \in R^{N_d \times N_d}$ are the left and right singular vectors, respectively, which are arranged corresponding to the order of the singular values; $\mathbf{V}^* = \mathbf{U}^*$ has been used in the second equality because $\mathbf{S}(\omega)$ is normal and positive definite. By comparing Equations (3.27) and (3.28), $\mathbf{D}(\omega)$ can be roughly considered to be the PSD matrix in the modal coordinates and $\mathbf{U}$ can be considered to be mode shapes.

Based on the similar structure expressed in Equations (3.27) and (3.28), $\mathbf{b}_{m,0}$ can be obtained by doing SVD for the PSD matrix at the target spectral peak, i.e., after picking $\omega_{m,0}$ at the spectral peak, SVD is done for the PSD matrix $\mathbf{S}(\omega_{m,0})$ at the frequency $\omega_{m,0}$. Note that the magnitude of a singular value quantifies how much the corresponding mode contributes at frequency ω, so the largest singular value at a spectral peak corresponds to the dominant mode at this peak. As a result, the singular vector corresponding to the largest singular value at $\omega_{m,0}$ is chosen as the initial mode shape $\boldsymbol{\phi}_{m,0}$. $\mathbf{b}_{m,0}$ is then taken as the amplitude of $\boldsymbol{\phi}_{m,0}$.

3.3.5 The structure of the optimization algorithm

The structure of the proposed algorithm is constituted by an outside optimization and an inside iteration. The outside optimization is to minimize the objective function Equation (3.19) with respect to ω_m and ξ_m, and this can be conveniently done using the MATLAB function **fminsearch**(). The inside iteration built inside the outside optimization is to iterate between $\mathbf{b}_m$ and S_m until convergence for each set of values of ω_m and ξ_m tried by **fminsearch**(). This is achieved by constructing the inside iteration in the objective function of **fminsearch**(). The details of the proposed algorithm are summarized in the following algorithm. Note that a line that begins with % is a comment for the algorithm.

The iterative algorithm for the optimization of modal parameters

Initialization

Provide the initial values of the natural frequency, damping ratio, and mode shape amplitude of the target mode, $\omega_{m,0}$, $\xi_{m,0}$ and $\mathbf{b}_{m,0}$.

Calculate $S_{m,0}$ using Equation (3.26) given $\omega_{m,0}$, $\xi_{m,0}$ and $\mathbf{b}_{m,0}$.

Set the maximum number of iterations N_{itr} (e.g., $N_{itr} = 2000$) and the iteration tolerance for mode shapes and PSDs of modal excitations ϵ (e.g., $\epsilon = 10^{-3}$).

Outside optimization

Input the initial values $\omega_{m,0}$, $\xi_{m,0}$, $S_{m,0}$, and $\mathbf{b}_{m,0}$ to **fminsearch**().

```
while fminsearch() is not converged do
fminsearch() generates the updated natural frequency ωm and damping
ratio ξm.
Inside iteration
for j = 1, ..., Nitr do
    % The following is to update the PSD of modal excitation.
```

```
    Given the updated ωm, ξm and the current bm,0, calculate the updated
    Sm using Equation (3.26).
    % The following is to update the mode shape amplitude.
    Given the updated ωm, ξm, and the current Sm,0, calculate the
    updated bm using Equation (3.22).
    % The following is to check the tolerance for the inside iteration.
    If the tolerance is small enough,
    % break out of the for loop.
    if ||bm − bm,0|| / ||bm,0|| < ε and |(Sm − Sm,0) / Sm,0| < ε do
    Sm,0 ← Sm
    bm,0 ← bm
    Break out of the for loop.
else do
    Sm,0 ← Sm
    bm,0 ← bm
    end if
end for
if j = Nitr do
    Display the warning message "WARNING: The maximum number of
    iterations has been reached for
    the mode shape and PSD of modal excitations!"
end if
Calculate the objective function value J(θ)/(NkNd) for fminsearch(), where
J(θ) is calculated using Equation (3.19).
end while
```

Get the identified optimal modal parameters $\hat{\omega}_m$, $\hat{\xi}_m$, $\hat{S}_m$, and $\hat{\mathbf{b}}_m$.

Note that after the algorithm is converged, the mode shape $\hat{\boldsymbol{\phi}}_m$ needs to be recovered from the identified $\hat{\mathbf{b}}_m$ by putting signs to the elements of $\hat{\mathbf{b}}_m$. It can be easily done using the technique in Section 3.4 for initializing mode shapes. SVD is done for the PSD matrix at the identified natural frequency $\hat{\omega}_m$. The signs of the singular vector corresponding to the largest singular value are then extracted and put to each element of $\hat{\mathbf{b}}_m$, respectively, and the optimal mode shape $\hat{\boldsymbol{\phi}}_m$ is obtained. The above algorithm can be repeatedly applied for different spectral peaks to identify modal parameters of different modes.

3.4 SIMULATED STUDY: A 12-STORY SHEAR BUILDING MODEL

In this section, a 12-story shear building model is used to illustrate the application details of the proposed method and verify its performance. This shear building is two-dimensional with a lumped mass $m_0 = 10^6$ kg and an inter-story stiffness $k_0 = 4 \times 10^6$ N/m uniform for each floor. With this information, the mass and stiffness matrices can be constructed, and natural frequencies and mode shapes can be obtained. The same damping ratios of 1% are assumed for all the modes. Accelerations were simulated from this shear building to be used as measured data. Gaussian excitations were added at each floor for simulating the accelerations. The same root PSD $S_0 = 98\,\text{N}/\sqrt{\text{Hz}}$ was assumed for the excitations at each floor. The root PSDs of the excitations were chosen so that the root PSD of the modal excitation of each mode (in terms of the acceleration unit) has the same value, $S_m = 0.1 \times 10^{-4}\,\text{g}/\sqrt{\text{Hz}}$, where m denotes the mode number. This root PSD of modal excitation is used as the exact value that can be checked against the identified value by the proposed method. Accelerations of the 12 degrees of freedom at the 12 floors were then simulated and added with Gaussian noise with root PSD $0.01 \times 10^{-4}\,\text{g}/\sqrt{\text{Hz}}$. Five minutes of accelerations were simulated with a sampling frequency of 256 Hz. These accelerations were used to construct PSDs to be used as "measured" data for modal analysis.

The five-minute simulated accelerations were divided into different segments, and the PSDs were calculated using each segment of data. The averaged root PSDs (the square roots of PSDs) of each degree of freedom were then constructed by averaging the PSDs from different segments of data (see Figure 3.5). The spectral peaks indicate the 12 structural modes of the shear building. Note that these smooth spectra are only for an easy visualization and picking the PSD data around each peak. The raw PSD data that contained the most information were used for modal analysis. According to the proposed method, the initial values of the natural frequencies were selected from Figure 3.5 at the 12 spectral peaks. The initial values of the mode shape amplitudes were obtained by the singular value decomposition of the PSD matrices at the initial natural frequencies. Based on the engineering experience, structures' damping ratios take values around 1%, so the initial values of damping ratios were chosen to be 1%. This choice of damping ratio will not affect the result because the final values of damping ratios are obtained by optimization based on measured data.

By using the proposed method, the modal parameters of the 12 modes $\theta = \{\omega_m, \xi_m, S_m, \phi_m : m = 1, 2, \cdots, 12\}$ were identified. The identified natural frequencies, damping ratios, and PSDs of modal excitations are compared with the corresponding exact values (see Table 3.1), the ones used to simulate the structural responses. The proposed method can accurately identify the modal parameters using the noisy responses. The accuracy of the identified mode shapes of the 12 modes are also checked. The mode shapes of all the modes can also be identified. It can be seen in Figure 3.6 that the identified

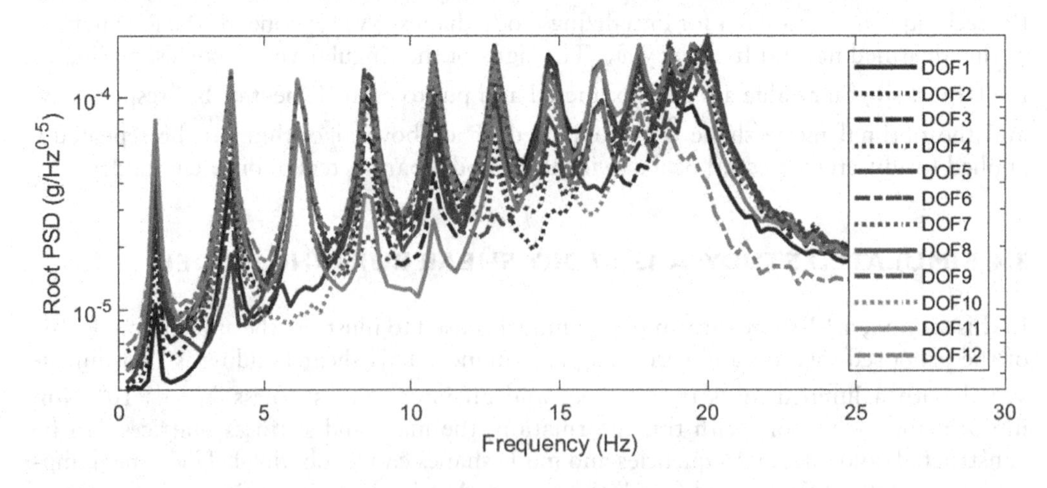

Figure 3.5 The root PSDs at the 12 degrees of freedom of the 12-story shear building.

Table 3.1 Comparison of the Modal Parameters of the 12-Story Shear Building

	f_m (Hz)		ξ_m		S_m ($\times 10^{-4}$ g/$\sqrt{\text{Hz}}$)	
	Identified	*Exact*	*Identified*	*Exact*	*Identified*	*Exact*
Mode 1	1.26	1.26	1.15%	1.00%	0.11	0.10
Mode 2	3.77	3.77	0.86%	1.00%	0.07	0.10
Mode 3	6.21	6.22	1.02%	1.00%	0.09	0.10
Mode 4	8.56	8.57	1.06%	1.00%	0.10	0.10
Mode 5	10.79	10.79	1.00%	1.00%	0.09	0.10
Mode 6	12.84	12.83	1.00%	1.00%	0.09	0.10
Mode 7	14.69	14.68	1.28%	1.00%	0.12	0.10
Mode 8	16.28	16.29	1.52%	1.00%	0.15	0.10
Mode 9	17.65	17.64	1.11%	1.00%	0.11	0.10
Mode 10	18.70	18.72	1.30%	1.00%	0.14	0.10
Mode 11	19.47	19.50	1.75%	1.00%	0.27	0.10
Mode 12	19.93	19.97	1.09%	1.00%	0.12	0.10

mode shapes almost overlap the exact ones, indicating the good performance of the proposed method. The PSDs of the accelerations were reconstructed based on the optimal modal parameters using Equation (3.18). Figure 3.7 shows the comparison between the reconstructed and experimental PSDs of the 12 DOFs around the 12 selected spectral peaks, where the dash and solid lines denote the reconstructed and experimental PSDs, respectively. It can be observed that the theoretical PSDs "smoothen" the stochastic PSDs, so the information consistent with the modeling assumptions is effectively extracted by the proposed method, and the stochastic averaging for the experimental PSDs is not required.

3.5 EXPERIMENTAL STUDY: MODAL ANALYSIS OF A COUPLED STRUCTURAL SYSTEM

In this section, the proposed PSD-based modal analysis method is applied for a full-scale coupled structural system. This structural system was also studied by Lam et al. (2019). It consists of the main building and the secondary building (see Figure 3.8). This structure is a construction training center. Activities inside the structure can cause violent vibration. A field test was conducted to study the coupled dynamics between the main and secondary buildings. Six triaxial accelerometers were available (see Figure 3.9), but many locations were to be measured to obtain detailed dynamic properties of the coupled system, so the measurement was divided into multiple setups, and the partial mode shapes identified in each setup were assembled into global mode shapes. Figure 3.10(a) shows the floor plan of the roof. The main building and the secondary building are connected, but a void exists between them. Coupled dynamics is expected for this structural system. The circles denote the measured locations; the squares with numbers denote the numbering of the measured locations; and S1 to S5 denote sensors 1 to 5. The measurement plan for covering the two buildings is as follows. For the main building, staircases 1 to 4 are measured independently, and then the four staircases are connected by the reference sensors on the roof, i.e., the sensors installed at the locations of the two columns of circles (with numbering 1, 2, 3, 4, 15, and 16).

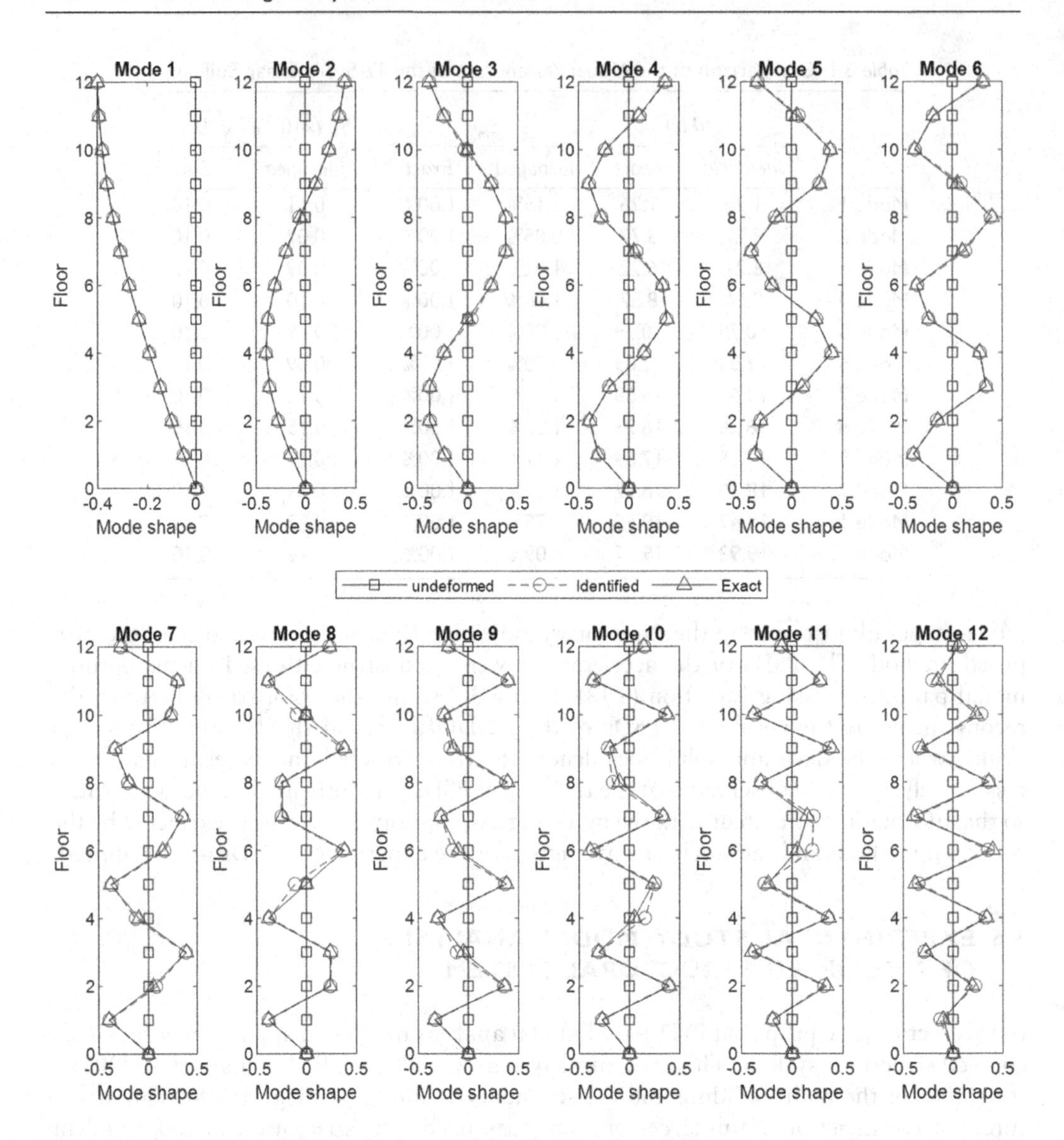

Figure 3.6 Comparison of the mode shapes of the 12-story shear building.

The measurement of staircase 1 is used to illustrate the measurement procedures. Figure 3.10(b) shows the sensor installation for staircase 1, where the underlined numbers denote the setup numbers; the triangle denotes the location of the console (the computer and signal processing system); and the circles with numbers again denote the sensors. In the first setup, sensors 1 to 4 (with underlined number "<u>1</u>") measured locations 1 to 4 on the roof (see also Figure 3.10(a)). Sensor 5 measured location 5 on the roof floor of staircase 1, and sensor 6 measured floor 7 inside staircase 1. The console was on the roof for the first setup. Sensors 5 and 6 were then fixed at the roof and floor 7 inside staircase 1 for setups 1 to 3 (see the underlined numbers "<u>1</u>," "<u>2</u>," and "<u>3</u>"). Sensors 5 and 6 were used as reference sensors for assembling mode shapes of staircase 1. In the second setup, sensors 1 to 4 were moved to be installed on floors 6, 5, 4, and 3 inside staircase 1, respectively. Only floors 1 and 2 were not measured after setup 2; so, in setup 3, sensors 3 and

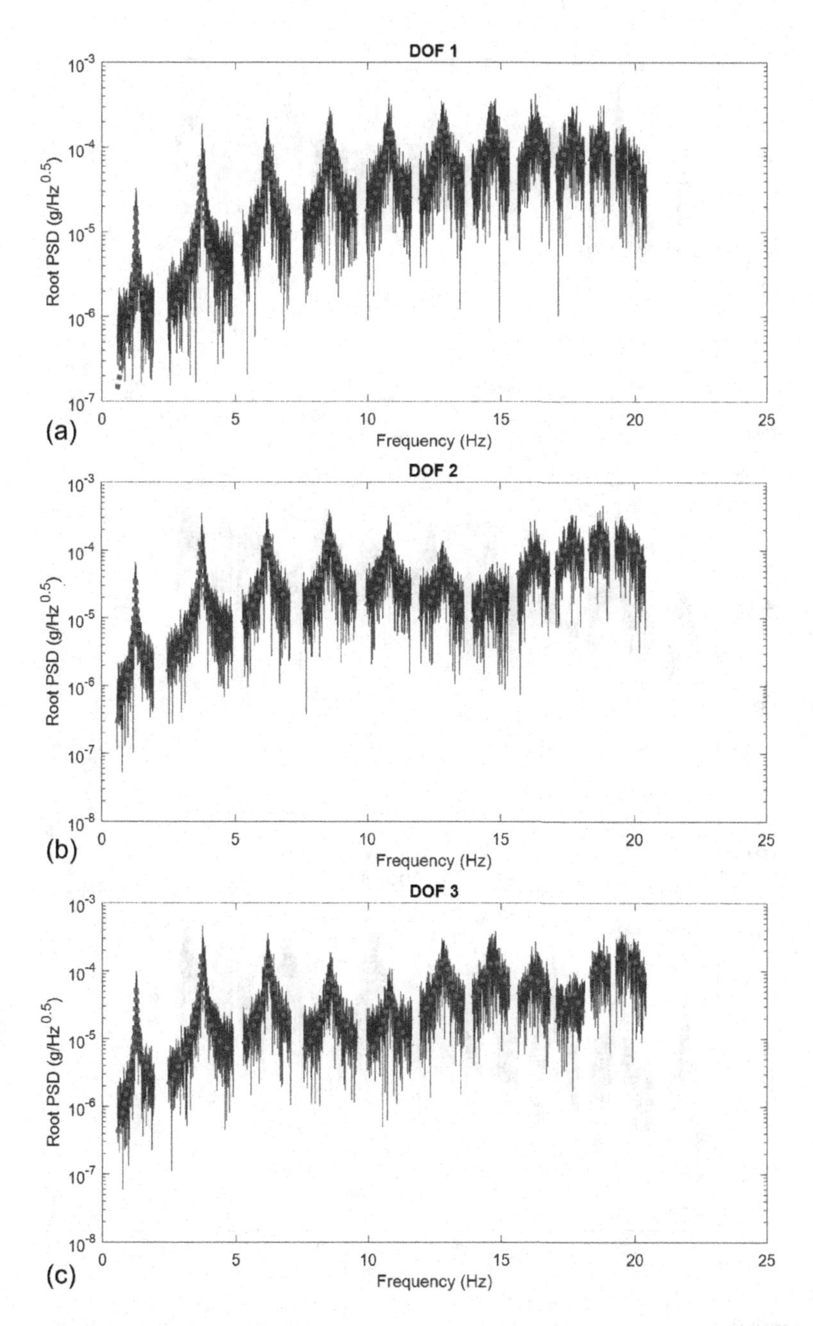

Figure 3.7 Comparison of measured and calculated PSDs: (a) DOF 1; (b) DOF 2; (c) DOF 3; (d) DOF 4; (e) DOF 5; (f) DOF 6; (g) DOF 7; (h) DOF 8; (i) DOF 9; (j) DOF 10; (k) DOF 11; (l) DOF 12.

4 were moved to be installed on floors 2 and 1, respectively, while sensors 1 and 2 were not moved. Note that due to the limited length of cables, the console was moved from the roof to floor 5 in setup 2. By using this plan, each two setups had common measured locations, so mode shapes in different setups could be assembled. After measuring staircase 1, the measurement team went back to the roof to measure staircase 2. In order to assemble

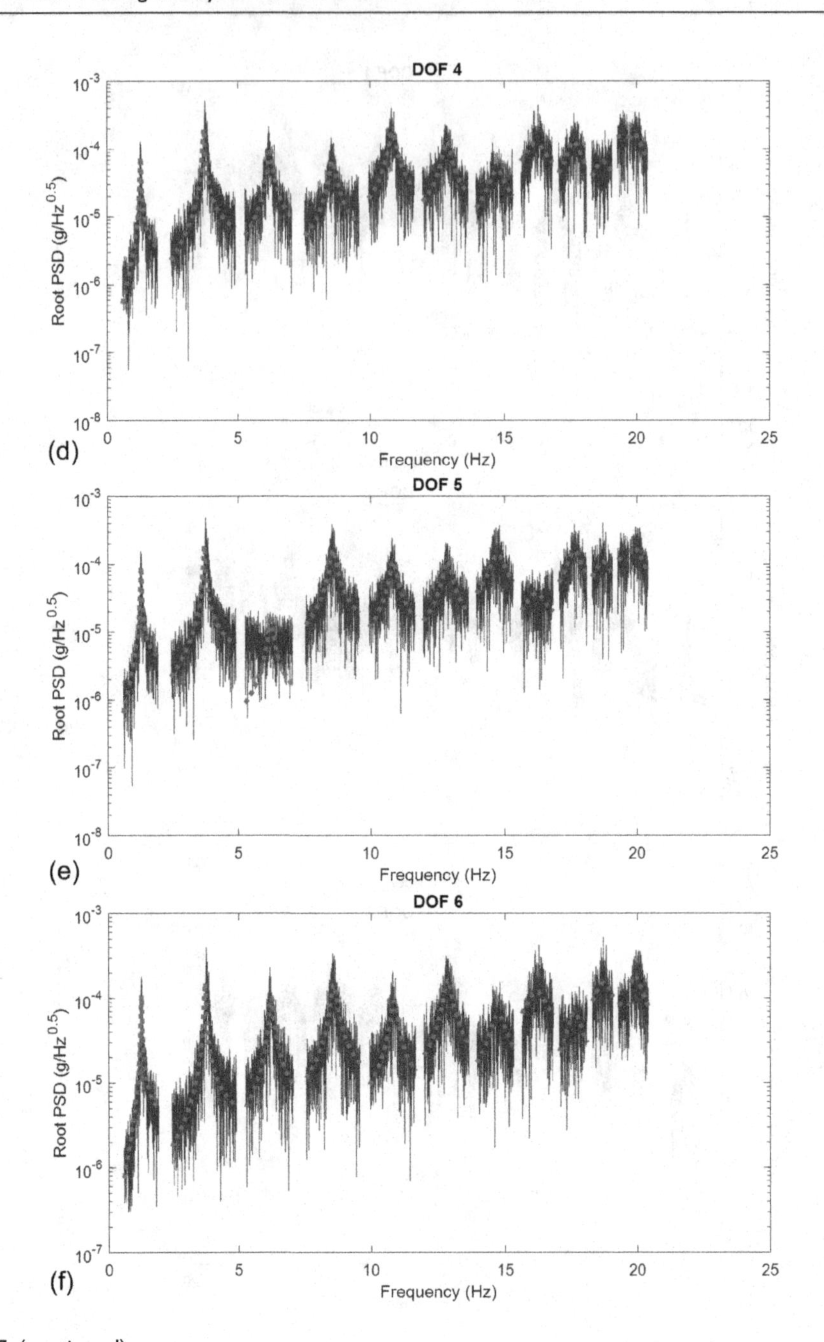

Figure 3.7 (continued)

mode shapes of staircases 1 and 2, sensors 1 to 4 measured locations 1 to 4 again to get reference locations. Sensors 5 and 6 measured the roof and floor 7 inside staircase 2. Using a similar plan as staircase 1, staircase 2 was measured.

When measuring staircases 3 and 4, due to the limited length of cables, locations 3 and 4 in the roof setup were changed to locations 15 and 16 because these two locations are close to staircases 3 and 4. Locations 1 and 2 were still measured for staircases 3 and 4,

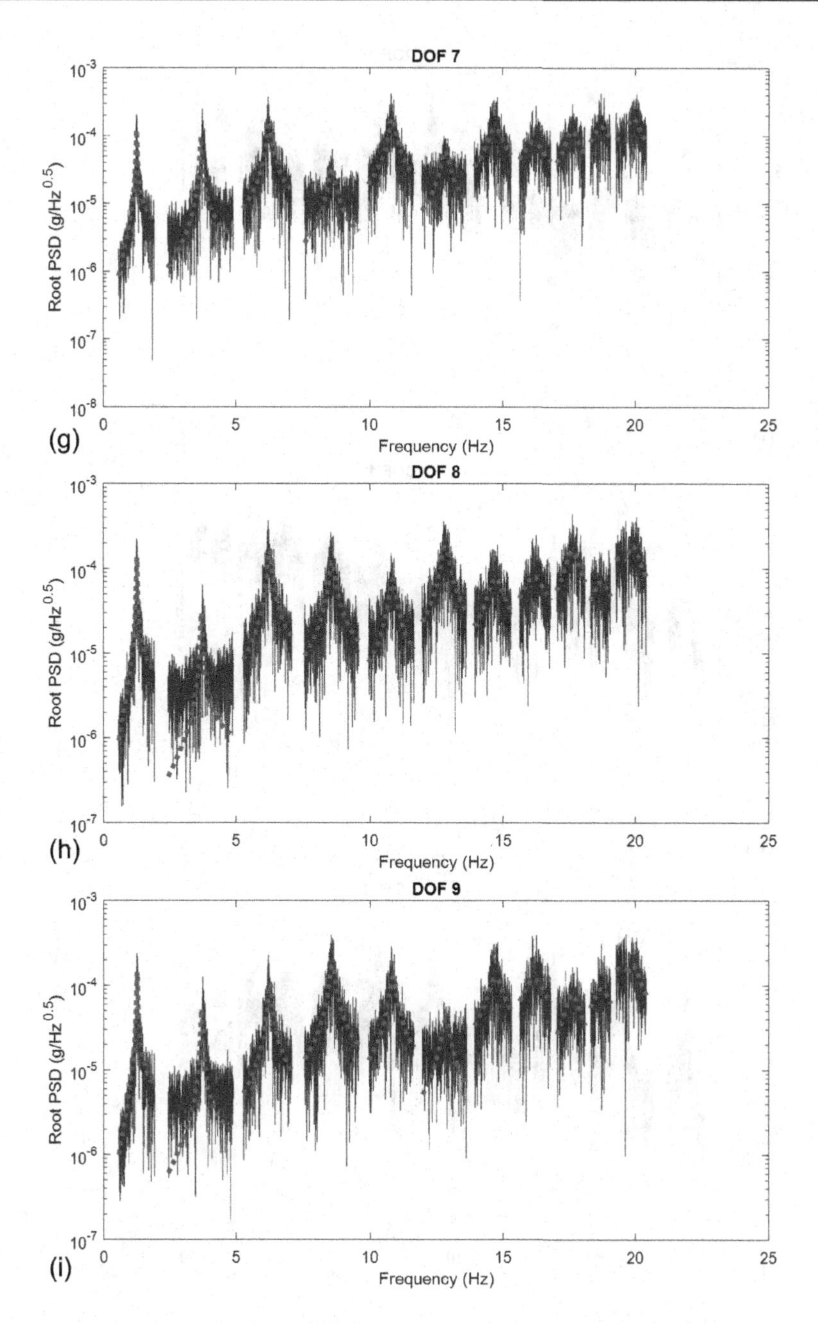

Figure 3.7 (continued)

so the measurements for the four staircases had common locations 1 and 2 for assembling their mode shapes. The rest of the measurement for staircases 3 and 4 is similar to that of staircases 1 and 2.

The secondary building was measured independently. Four corners at each floor of the secondary building were measured, and the measurements of different floors were connected by the reference sensors between two successive floors. The measurements of the

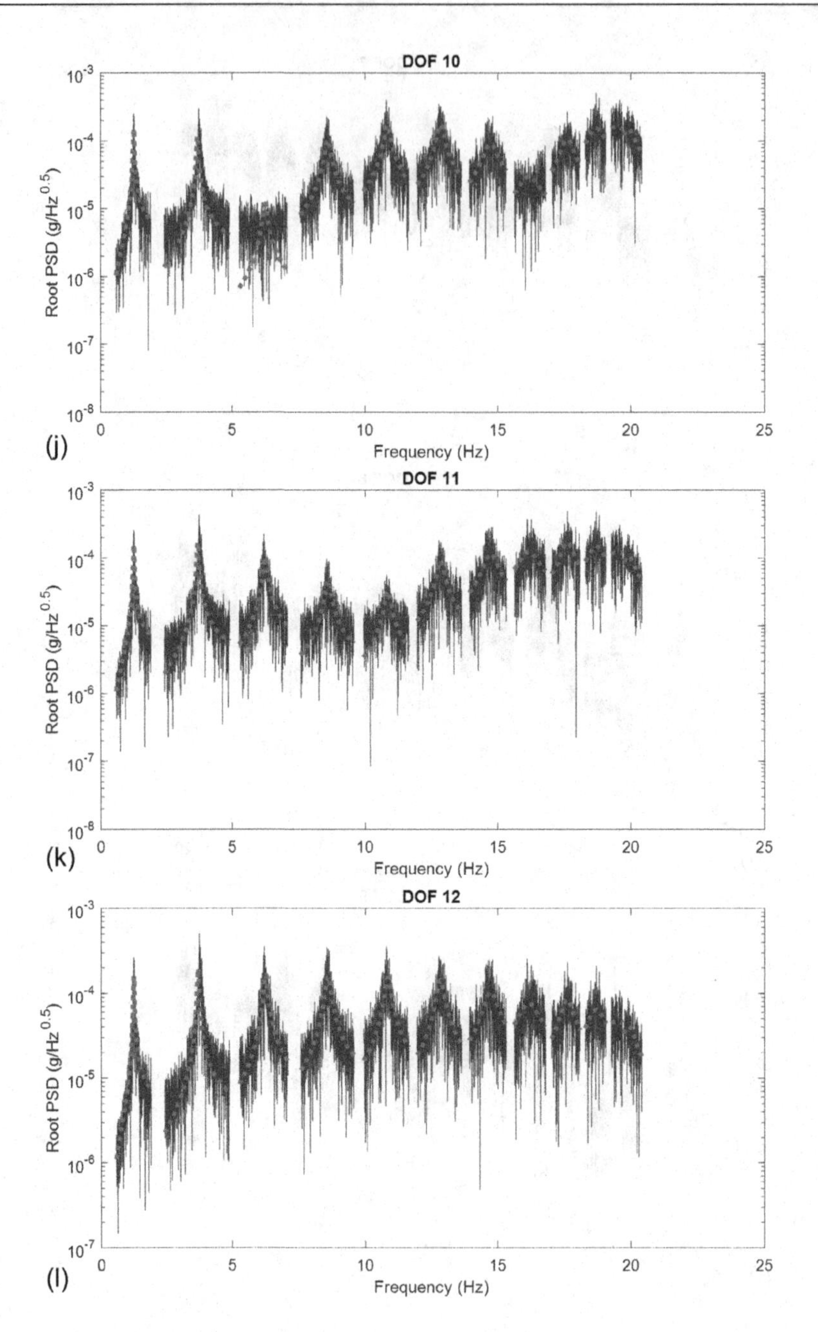

Figure 3.7 (continued)

main and secondary buildings were connected by the reference sensors at the line going through these two buildings, i.e., the line with locations 4, 3, 34, 33, 35, and 36 (see Figure 3.10(a)).

21 setups were conducted in total, and 10 minutes of acceleration data were measured for each setup with sampling frequency 2048 Hz. The data were then downsampled to

Figure 3.8 The coupled structural system (from Google Maps).

Figure 3.9 Equipment for the field test.

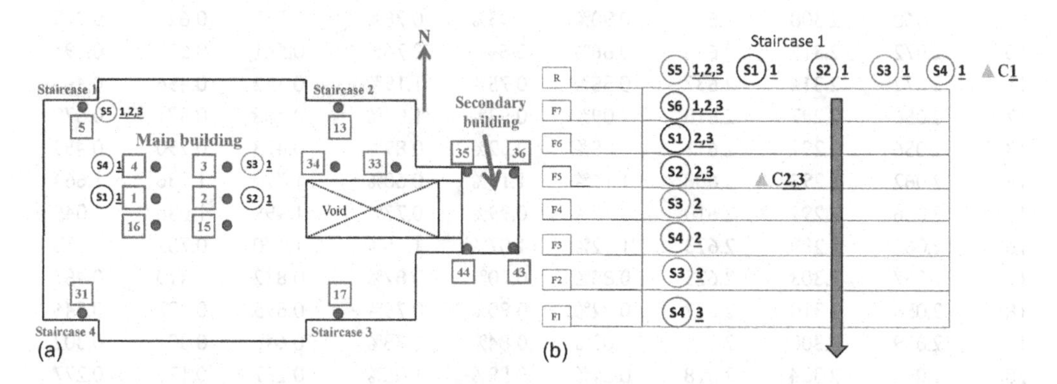

Figure 3.10 Measurement plan: (a) setups on the roof; (b) setups for staircase 1.

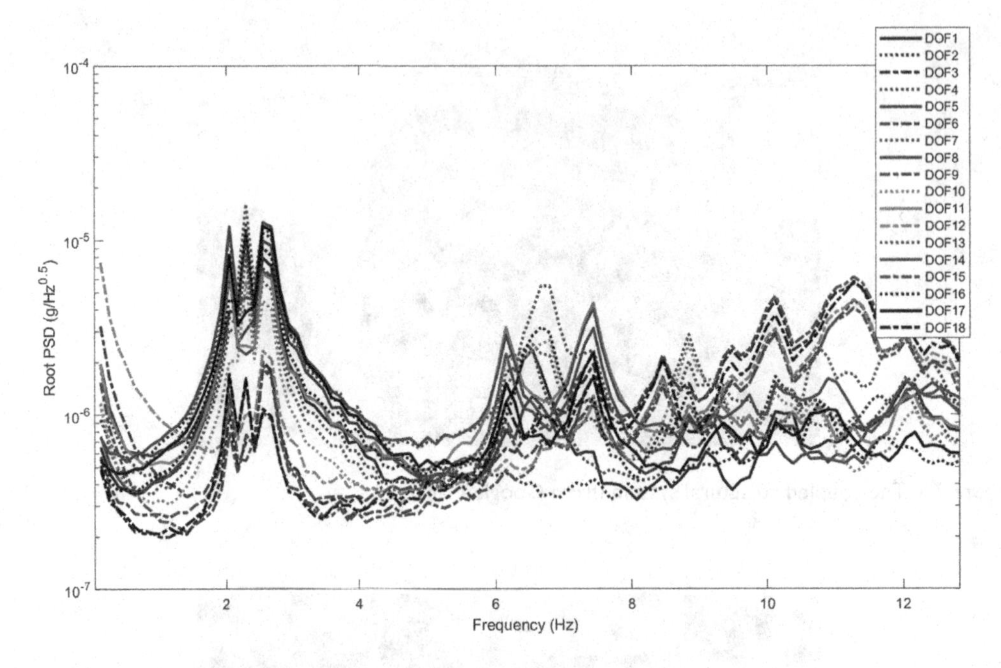

Figure 3.11 The PSDs of the coupled structural system.

Table 3.2 The Identified Frequencies, Damping Ratios, and Root PSD of the Coupled Structural System

Setup No.	Frequency (Hz)			Damping ratio			Root PSD ($\times 10^{-6}$ g / $\sqrt{\text{Hz}}$)		
	Mode 1	*Mode 2*	*Mode 3*	*Mode 1*	*Mode 2*	*Mode 3*	*Mode 1*	*Mode 2*	*Mode 3*
1	2.053	2.287	2.614	0.82%	0.99%	0.73%	1.045	0.990	0.656
2	2.058	2.291	2.621	0.98%	0.65%	0.68%	0.687	0.441	0.716
3	2.058	2.292	2.617	0.78%	1.27%	0.93%	0.510	0.731	0.778
4	2.069	2.298	2.620	0.98%	0.94%	0.83%	0.634	0.759	0.443
5	2.068	2.296	2.621	1.16%	0.83%	1.18%	0.988	0.672	0.477
6	2.064	2.300	2.625	0.97%	0.78%	0.71%	1.033	0.521	0.380
7	2.069	2.302	2.630	1.04%	0.70%	1.32%	0.828	0.440	0.760
8	2.074	2.303	2.630	0.97%	1.22%	0.68%	0.829	0.928	0.284
9	2.068	2.308	2.641	0.90%	0.93%	0.78%	0.760	0.628	0.285
10	2.072	2.315	2.646	0.66%	0.64%	0.76%	0.541	0.553	0.491
11	2.072	2.314	2.637	0.58%	0.75%	1.19%	0.352	0.456	0.460
12	2.056	2.297	2.618	1.09%	0.93%	1.09%	0.488	0.571	0.678
13	2.056	2.291	2.611	1.09%	0.52%	0.89%	0.423	0.290	0.453
14	2.062	2.293	2.608	1.16%	1.19%	0.68%	1.790	1.336	1.664
15	2.066	2.292	2.610	0.75%	0.99%	0.77%	1.499	1.036	1.045
16	2.062	2.289	2.623	1.02%	1.07%	1.15%	1.170	0.757	1.010
17	2.057	2.303	2.625	0.88%	0.70%	0.87%	0.832	0.410	0.487
18	2.056	2.310	2.632	0.82%	0.80%	0.75%	0.595	0.377	0.346
19	2.059	2.300	2.628	1.02%	0.84%	0.78%	0.640	0.301	0.302
20	2.062	2.304	2.628	0.82%	0.58%	1.05%	0.287	0.171	0.277
21	2.063	2.297	2.621	0.72%	1.22%	0.81%	0.309	0.335	0.208
Mean	2.063	2.299	2.624	0.92%	0.88%	0.89%	0.773	0.605	0.581

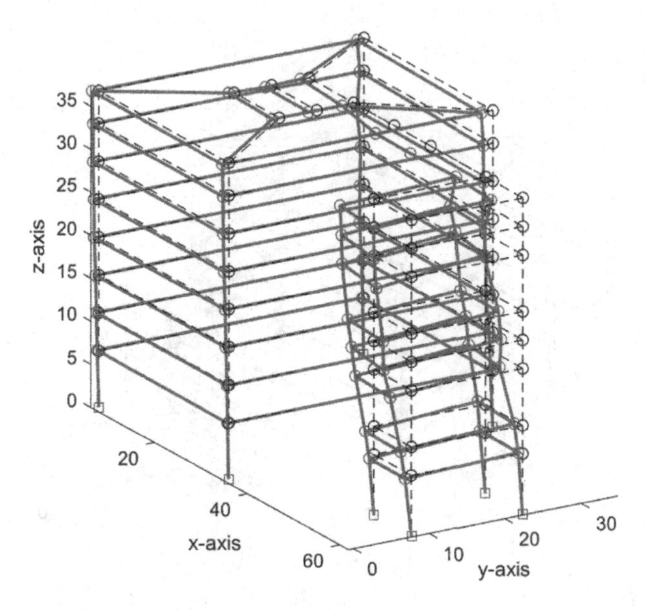

Figure 3.12 Mode shape of mode 1.

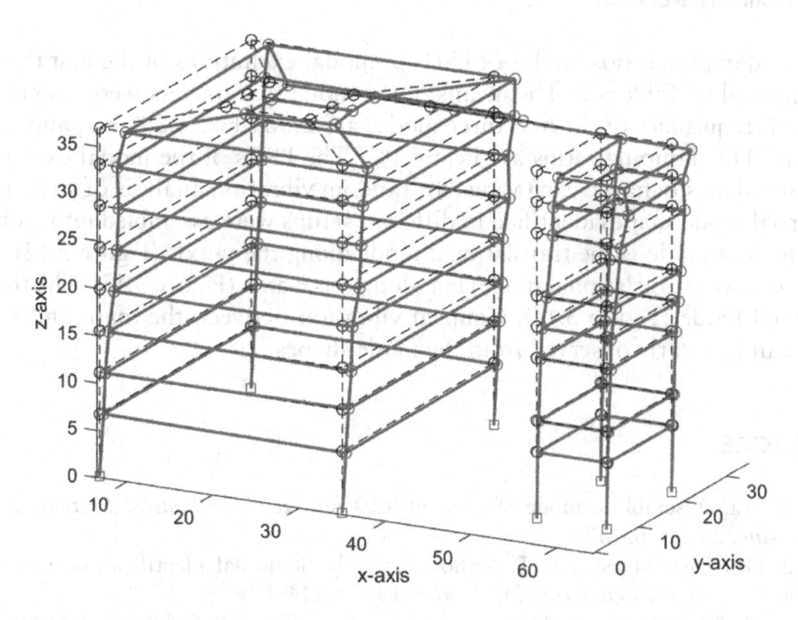

Figure 3.13 Mode shape of mode 2.

128 Hz for analysis. Figure 3.11 shows the PSDs of the 18 DOFs transformed from the accelerations in one setup. The peaks in this figure correspond to the structural modes. The PSDs of each setup were used as experimental data to identify the natural frequencies, damping ratios, root PSDs of modal excitations, and partial mode shapes using the proposed method in this chapter. The partial mode shapes of each setup were assembled into global mode shapes using the least-squares method developed by Au (2011a). In the present case, the first three modes (corresponding to the first three spectral peaks in the figure) were identified for illustration. After doing optimization for each setup, the natural

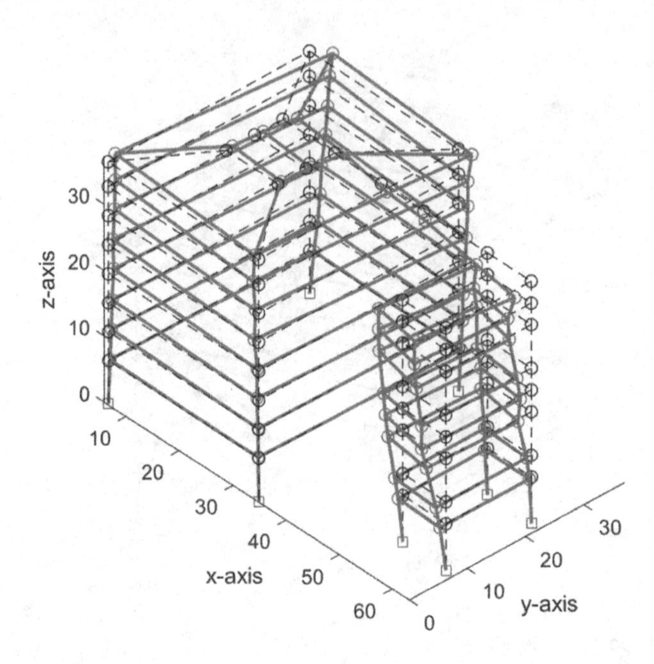

Figure 3.14 Mode shape of mode 3.

frequencies, damping ratios, and root PSDs of modal excitations of the first three modes are summarized in Table 3.2. The means of the modal parameters were also calculated. The natural frequencies of the first three modes are 2.063 Hz, 2.299 Hz, and 2.624 Hz, respectively. The damping ratios are below 1%. The PSDs of the modal excitations are useful to simulate stochastic excitations for random vibration analysis of structures.

The partial mode shapes identified in different setups were assembled into global mode shapes. The first mode is the translational mode along the y-axis (Figure 3.12). The second mode is also a translational mode, but along the x-axis (Figure 3.13). The third mode is a torsional mode (Figure 3.14). Coupled vibration between the main and secondary buildings can be clearly observed from the mode shapes.

REFERENCES

Au, S.K., 2011a. Assembling mode shapes by least squares. *Mechanical Systems and Signal Processing*, 25(1), 163–179.

Au, S.K., 2011b. Fast Bayesian FFT method for ambient modal identification with separated modes. *Journal of Engineering Mechanics*, 137(3), 214–226.

Au, S.K., 2018. *Operational modal analysis: Modeling, Bayesian inference, uncertainty laws*. Springer.

Lam, H.F., Yang, J.H. and Beck, J.L., 2019. Bayesian operational modal analysis and assessment of a full-scale coupled structural system using the Bayes-Mode-ID method. *Engineering Structures*, 186, 183–202.

Yang, J.H., Lam, H.F. and Beck, J.L., 2019. Bayes-Mode-ID: A Bayesian modal-component-sampling method for operational modal analysis. *Engineering Structures*, 189, 222–240.

Chapter 4

Modal analysis based on cross-correlation data

4.1 MATHEMATICAL MODEL OF A STRUCTURAL DYNAMIC SYSTEM

This section develops the mathematical model of a structural dynamic system for modal analysis. The model includes modal parameters (i.e., natural frequencies, damping ratios, and mode shapes) as its parameters, and given these parameters, the model predicts system responses to best fit the experimental data. Our objective is to do modal analysis of full-scale complex engineering structures, but excitations input to structures either cannot be measured or are very costly to measure. It is thus not possible to simulate accelerations, velocities, or displacements using the system model due to the absence of excitations. The measured data in practice are structural accelerations, so they are not suitable to be used as data for modal analysis of full-scale structures. It is proposed to use cross-correlation data transformed from measured accelerations. The cross-correlation data are just another form of measured data, and contain the same information of structural dynamic properties. The cross-correlation function of the responses at the i-th and j-th DOFs of a structure is defined as

$$R_{ij}(\tau) = E\left[x_i(t)x_j(t+\tau)\right] \tag{4.1}$$

where $R_{ij}(\tau)$ denotes the cross-correlation function with time lag τ related to DOFs i and j; $E[\cdot]$ is the expectation; $x_i(t)$ and $x_j(t+\tau)$ denotes the displacements at DOFs i and j, respectively. It is assumed in the above definition that displacements of the structure are stationary stochastic processes, so the cross-correlation function depends only on the time lag τ. The stationary assumption is easy to fulfill in practice for full-scale structures under ambient excitations. Given measured data in discrete time, the sample average estimation in discrete time of Equation (4.1) is

$$\hat{R}_{ij}(k) = \frac{1}{N_t}\sum_{h=0}^{N_t-1-k} x_i(h)x_j(h+k) \tag{4.2}$$

where the time lag index $k = 0,1,\cdots,N_t-1$; the discrete time index $h = 0,1,\cdots,N_t-1$.

The theoretical cross-correlation function will now be derived based on the dynamic system model to match the experimental counterpart in Equation (4.2). The model starts from the equation of motion of a linear structural system with N_d DOFs:

$$\mathbf{M}\ddot{\mathbf{u}}(t) + \mathbf{C}\dot{\mathbf{u}}(t) + \mathbf{K}\mathbf{u}(t) = \mathbf{F}(t) \tag{4.3}$$

DOI: 10.1201/9780429445866-4

where $\mathbf{M}, \mathbf{C}, \mathbf{K} \in R^{N_d \times N_d}$ are the system mass, damping, and stiffness matrices, respectively; $\ddot{\mathbf{u}}$, $\dot{\mathbf{u}}$, and $\mathbf{u} \in R^{N_d}$ are the acceleration, velocity, and displacement, respectively; and $\mathbf{F}(t) \in R^{N_d}$ is the stochastic excitation. Although $\mathbf{F}(t)$ cannot be measured in practice, its characteristics can be described using the PSDs. Considering that $\mathbf{F}(t)$ is stationary and only some modes of interest are focused on, it is assumed that the PSD of $\mathbf{F}(t)$ is flat (being constant) only around the frequency bands corresponding to the modes of interest. This condition is easy to fulfill in practice because it is not required that the PSD of the ambient excitation is flat in the whole frequency range. This assumption indicates that the modal excitations of the modes of interest can be modeled as Gaussian processes, which will provide nice analytical features for the theoretical cross-correlation functions. In practice, not all the N_d DOFs of the system can be measured. Let $N_o\left(N_o \le N_d\right)$ denote the number of DOFs chosen to model the actual structural system, i.e., sensors will be installed on these N_o DOFs, and $\ddot{\mathbf{x}}, \dot{\mathbf{x}}, \mathbf{x} \in R^{N_o}$ denote the output responses at the N_o DOFs. $\ddot{\mathbf{x}}, \dot{\mathbf{x}}$, and $\mathbf{x}$ are the sub-vectors of $\ddot{\mathbf{u}}$, $\dot{\mathbf{u}}$, and $\mathbf{u}$, respectively. Expressing $\mathbf{x}$ using modal superposition (see Chapter 2) gives

$$\mathbf{x}(t) = \sum_{m=1}^{N_m} \boldsymbol{\varphi}_m d_m(t) = \boldsymbol{\Phi}\mathbf{d}(t) \tag{4.4}$$

where only N_m modes are used considering that not all N_o modes can be identified in practice ($N_m \le N_o$); $\boldsymbol{\varphi}_m \in R^{N_o}$ is the mode shape of Mode m confined to the measured DOFs with unit Euclidean norm and d_m is the m-th modal coordinate; $\boldsymbol{\Phi} = \left[\boldsymbol{\varphi}_1, \boldsymbol{\varphi}_2, \cdots, \boldsymbol{\varphi}_{N_m}\right] \in R^{N_o \times N_m}$ includes all the mode shapes; $\mathbf{d} = \left[d_1, d_2, \cdots, d_{N_m}\right]^T \in R^{N_m}$. By assuming a classical viscous damping model, the single degree of freedom (SDOF) equation of motion for the m-th mode can be obtained from the decoupling of the multi-degree of freedom (MDOF) Equation (4.3):

$$\ddot{d}_m(t) + 2\xi_m \omega_m \dot{d}_m(t) + \omega_m^2 d_m(t) = \sum_{i=1}^{N_o} \varphi_{im} f_i(t) \tag{4.5}$$

where ξ_m and ω_m are the damping ratio and natural frequency of the m-th mode, respectively; $f_i(t)$ is the excitation function related to the i-th observed DOF; φ_{im} is the i-th DOF-component of the mode shape $\boldsymbol{\varphi}_m$.

The cross-correlation function is first derived in terms of modal parameters and then transformed into geometrical coordinates. Similar works have also been done by James et al. (1993) and Beck et al. (1995, 1994, 1998). It will be shown that the cross-correlation function in each modal coordinate has the same form as the free-decay response of an SDOF system under initial conditions. This analytical form of the modal cross-correlation function is helpful to develop an efficient algorithm of modal analysis. The displacement in the m-th modal coordinate can be expressed using Duhamel's integral:

$$d_m(t) = \int_{-\infty}^{t} h_m(t-\mu) \sum_{i=1}^{N_o} \varphi_{im} f_i(\mu)\, d\mu \tag{4.6}$$

where $h_m(t-\mu)$ is the unit-impulse response function derived in Chapter 2

$$h_m(t-\mu) = \frac{1}{\omega_m^D} \sin \omega_m^D (t-\mu) \exp\left(-\xi_m \omega_m (t-\mu)\right) \tag{4.7}$$

where $\omega_m^D = \omega_m\sqrt{1-\xi_m^2}$ is the damped natural frequency. Applying change of variable $\eta = t - \mu$ in the above equation gives

$$d_m(t) = \int_0^\infty h_m(\eta)\sum_{i=1}^{N_O}\varphi_{im}f_i(t-\eta)d\eta \tag{4.8}$$

Using the above equation, the cross-correlation function between the n-th and m-th modal coordinates is written according to the definition in Equation (4.1).

$$\begin{aligned} R_{d_nd_m}(\tau) &= E\left[d_n(t)d_m(t+\tau)\right] \\ &= \int_0^\infty\int_0^\infty h_n(\eta)h_m(\mu)\sum_{i=1}^{N_O}\sum_{j=1}^{N_O}\varphi_{in}\varphi_{jm}E\left[f_i(t-\eta)f_j(t+\tau-\mu)\right]d\eta d\mu \end{aligned} \tag{4.9}$$

By assuming that the excitation has a constant PSD in the frequency band of each mode of interest, the correlation function of the excitations at two DOFs is expressed as follows:

$$E\left[f_i(t-\eta)f_j(t+\tau-\mu)\right] = F_{ij}\delta(\tau-\mu+\eta) \tag{4.10}$$

where F_{ij} is constant and $\delta(\tau-\mu+\eta)$ is the Dirac delta function: $\delta(\tau-\mu+\eta)=1$ if $\mu=\tau+\eta$; otherwise $\delta(\tau-\mu+\eta)=0$. Equation (4.9) can be simplified by using Equation (4.10):

$$R_{d_nd_m}(\tau) = \left(\sum_{i=1}^{N_O}\sum_{j=1}^{N_O}\varphi_{in}\varphi_{jm}F_{ij}\right)\int_0^\infty h_n(\eta)h_m(\tau+\eta)d\eta \tag{4.11}$$

where only one integral variable η is left, so the integration is simplified. We then study the mathematical structure of the integration of unit-impulse response functions in the above equation. Substituting Equation (4.7) for the two unit-impulse response functions in the above integral gives

$$\begin{aligned} &\int_0^\infty h_n(\eta)h_m(\tau+\eta)d\eta \\ &= \int_0^\infty \frac{1}{\omega_n^D}\sin\omega_n^D\eta\exp(-\xi_n\omega_n\eta)\frac{1}{\omega_m^D}\sin\omega_m^D(\tau+\eta)\exp\left(-\xi_m\omega_m(\tau+\eta)\right)d\eta \\ &= \frac{1}{\omega_n^D\omega_m^D}\exp(-\xi_m\omega_m\tau)\int_0^\infty \exp\left((-\xi_n\omega_n-\xi_m\omega_m)\eta\right)\sin\omega_n^D\eta\left(\sin\omega_m^D\tau\cos\omega_m^D\eta\right. \\ &\quad \left.+\cos\omega_m^D\tau\sin\omega_m^D\eta\right)d\eta \\ &= \frac{1}{\omega_n^D\omega_m^D}\exp(-\xi_m\omega_m\tau)\left[\sin\omega_m^D\tau\int_0^\infty \exp\left((-\xi_n\omega_n-\xi_m\omega_m)\eta\right)\sin\omega_n^D\eta\cos\omega_m^D\eta d\eta\right. \\ &\quad \left.+\cos\omega_m^D\tau\int_0^\infty \exp\left((-\xi_n\omega_n-\xi_m\omega_m)\eta\right)\sin\omega_n^D\eta\sin\omega_m^D\eta d\eta\right] \end{aligned} \tag{4.12}$$

The first integral in the brackets of the above equation can be rewritten as

$$\begin{aligned}
&\int_0^\infty \exp\left(\left(-\xi_n\omega_n - \xi_m\omega_m\right)\eta\right)\sin\omega_n^D\eta\cos\omega_m^D\eta d\eta \\
&= \frac{1}{2}\int_0^\infty \exp\left(\left(-\xi_n\omega_n - \xi_m\omega_m\right)\eta\right)\sin\left(\omega_n^D + \omega_m^D\right)\eta d\eta \\
&+ \frac{1}{2}\int_0^\infty \exp\left(\left(-\xi_n\omega_n - \xi_m\omega_m\right)\eta\right)\sin\left(\omega_n^D - \omega_m^D\right)\eta d\eta
\end{aligned} \tag{4.13}$$

The second integral in the brackets of the above equation can be rewritten as

$$\begin{aligned}
&\int_0^\infty \exp\left(\left(-\xi_n\omega_n - \xi_m\omega_m\right)\eta\right)\sin\omega_n^D\eta\sin\omega_m^D\eta d\eta \\
&= \frac{1}{2}\int_0^\infty \exp\left(\left(-\xi_n\omega_n - \xi_m\omega_m\right)\eta\right)\cos\left(\omega_n^D - \omega_m^D\right)\eta d\eta \\
&- \frac{1}{2}\int_0^\infty \exp\left(\left(-\xi_n\omega_n - \xi_m\omega_m\right)\eta\right)\cos\left(\omega_n^D + \omega_m^D\right)\eta d\eta
\end{aligned} \tag{4.14}$$

The previous two equations have only two kinds of integrals, i.e., one about the product of the exponential function and the sine function, and the other about the product of the exponential function and the cosine function. These two kinds of integrals can be evaluated analytically, and their formulas are given as follows

$$\int \exp(av)\sin(bv)dv = \frac{\exp(av)}{a^2+b^2}\left(a\sin(bv) - b\cos(bv)\right) + D \tag{4.15}$$

$$\int \exp(av)\cos(bv)dv = \frac{\exp(av)}{a^2+b^2}\left(a\cos(bv) + b\sin(bv)\right) + D \tag{4.16}$$

By using Equations (4.15) and (4.16), Equations (4.13) and (4.14) can be evaluated to be two constants C_1 and C_2, respectively. These two constants will not affect the following derivation, so they are not evaluated explicitly. Substituting C_1 and C_2 for the two integrals in Equation (4.12) simplifies this equation to

$$\begin{aligned}
\int_0^\infty h_n(\eta)h_m(\tau+\eta)d\eta &= \frac{1}{\omega_n^D\omega_m^D}\exp\left(-\xi_m\omega_m\tau\right)\left[C_1\sin\omega_m^D\tau + C_2\cos\omega_m^D\tau\right] \\
&= C_{nm}\exp\left(-\xi_m\omega_m\tau\right)\sin\left(\omega_m^D\tau + \alpha_{nm}\right)
\end{aligned} \tag{4.17}$$

where C_{nm} and α_{nm} are constants. Using Equation (4.17) in Equation (4.11) gives the cross-correlation functions between the n-th and m-th modal coordinates:

$$R_{d_nd_m}(\tau) = C_{nm}\exp\left(-\xi_m\omega_m\tau\right)\sin\left(\omega_m^D\tau + \alpha_{nm}\right)\left(\sum_{i=1}^{N_o}\sum_{j=1}^{N_o}\varphi_{in}\varphi_{jm}F_{ij}\right) \tag{4.18}$$

To transform the cross-correlation function in modal coordinates to geometrical coordinates, the modal expansion (Equation (4.4)) is first applied for the displacements of the i-th DOF:

$$x_i(t) = \sum_{m=1}^{N_O} \varphi_{im} d_m(t) \tag{4.19}$$

Using Equation (4.19) in the definition of the cross-correlation function in Equation (4.1) gives

$$\begin{aligned} R_{ij}(\tau) &= E\left[x_i(t)x_j(t+\tau)\right] \\ &= E\left[\sum_{n=1}^{N_O} \varphi_{in} d_n(t) \sum_{m=1}^{N_O} \varphi_{jm} d_m(t+\tau)\right] \\ &= \sum_{n=1}^{N_O}\sum_{m=1}^{N_O} \varphi_{in}\varphi_{jm} E\left[d_n(t)d_m(t+\tau)\right] \\ &= \sum_{n=1}^{N_O}\sum_{m=1}^{N_O} \varphi_{in}\varphi_{jm} R_{d_n d_m}(\tau) \end{aligned} \tag{4.20}$$

Substituting Equation (4.18) into Equation (4.20) gives

$$R_{ij}(\tau) = \sum_{n=1}^{N_O}\sum_{m=1}^{N_O} \varphi_{in}\varphi_{jm} \left(\left(\sum_{k=1}^{N_O}\sum_{l=1}^{N_O} \varphi_{kn}\varphi_{lm} F_{kl}\right) C_{nm} \exp(-\xi_m\omega_m\tau)\sin\left(\omega_m^D\tau + \alpha_{nm}\right)\right) \tag{4.21}$$

By grouping the terms with index n, the above equation can be rewritten as

$$R_{ij}(\tau) = \sum_{m=1}^{N_O} \varphi_{jm} \exp(-\xi_m\omega_m\tau)\left(\sum_{n=1}^{N_O} K_{inm} \sin\left(\omega_m^D\tau + \alpha_{nm}\right)\right) \tag{4.22}$$

where K_{inm} is a constant that groups the relevant terms in Equation (4.21). Note that the summation of the sine functions in the above equation can be evaluated analytically:

$$\sum_{n=1}^{N_O} K_{inm} \sin\left(\omega_m^D\tau + \alpha_{nm}\right) = J_{im} \sin\left(\omega_m^D\tau + \beta_{im}\right) \tag{4.23}$$

where

$$J_{im}^2 = \sum_{n=1}^{N_O}\sum_{s=1}^{N_O} K_{inm}K_{ism} \cos(\alpha_{nm} - \alpha_{sm}) \tag{4.24}$$

$$\tan\beta_{im} = \frac{\sum_{n=1}^{N_O} K_{inm}\sin\alpha_{nm}}{\sum_{n=1}^{N_O} K_{inm}\cos\alpha_{nm}} \tag{4.25}$$

Substituting Equation (4.23) into Equation (4.22) simplifies the cross-correlation function between the i-th and the j-th DOF as follows:

$$R_{ij}(\tau) = \sum_{m=1}^{N_o} \varphi_{jm} J_{im} \exp(-\xi_m \omega_m \tau) \sin(\omega_m^D \tau + \beta_{im}) \tag{4.26}$$

The above equation indicates that the cross-correlation function of displacements can be written by superposing free-decay responses contributed by the modes of the original dynamic model in Equation (4.3), and the mode shapes at the j-th DOF of different modes can be obtained based on the cross-correlation function between the i-th and the j-th DOFs. If we fix the i-th DOF, i.e., choose a reference DOF S, and construct the cross-correlation function of each DOF with respect to the reference DOF S, we can assemble a correlation response vector similar to the displacement vector:

$$\mathbf{R}_x(\tau) = \left[R_{S1}(\tau), R_{S2}(\tau), \cdots, R_{SN_o}(\tau)\right]^T \tag{4.27}$$

The cross-correlation functions of the displacements are thus the solution of the reduced original dynamic model, i.e., the model confined to the measured DOFs, under initial conditions $\mathbf{R_x}(0)$ and $\dot{\mathbf{R}}_\mathbf{x}(0)$:

$$\bar{\mathbf{M}}\ddot{\mathbf{R}}_\mathbf{x}(\tau) + \bar{\mathbf{C}}\dot{\mathbf{R}}_\mathbf{x}(\tau) + \bar{\mathbf{K}}\mathbf{R}_\mathbf{x}(\tau) = 0 \tag{4.28}$$

where $\bar{\mathbf{M}}$, $\bar{\mathbf{C}}$, $\bar{\mathbf{K}} \in R^{N_o \times N_o}$ are the reduced system mass, damping, and stiffness matrices confined to the measured N_o DOFs, respectively; $\mathbf{R_x}(\tau)$ denotes the cross-correlation function vector of the displacements calculated at different DOFs; and the dots now denote derivatives of the correlation function with respect to time lag τ.

It has been shown above that the cross-correlation functions of displacements can be written as the superposition of free-decay responses of different modes. In practice, usually accelerations are measured, so we want to show that the cross-correlation functions constructed using accelerations also have the same mathematical form in Equation (4.26). Differentiating the equation of motion in modal coordinates, Equation (4.5), twice gives

$$\begin{gathered} \frac{d^2}{dt^2}\ddot{d}_m(t) + 2\xi_m\omega_m \frac{d^2}{dt^2}\dot{d}_m(t) + \omega_m^2 \frac{d^2}{dt^2} d_m(t) = \sum_{i=1}^{N_o} \varphi_{im} \frac{d^2}{dt^2} f_i(t) \Rightarrow \\ \frac{d^2}{dt^2}\ddot{d}_m(t) + 2\xi_m\omega_m \frac{d}{dt}\ddot{d}_m(t) + \omega_m^2 \ddot{d}_m(t) = \sum_{i=1}^{N_o} \varphi_{im} \frac{d^2}{dt^2} \ddot{f}_i(t) \end{gathered} \tag{4.29}$$

By comparing Equation (4.5) and Equation (4.29), it is known that accelerations and displacements follow the same equation of motion. Going through the same procedure above shows that the cross-correlation functions of accelerations can also be written as superposition of free-decay responses of different modes, i.e., Equation (4.26). Let $r_{ij}(\tau)$ denote the cross-correlation function of accelerations between the i-th and the j-th DOFs. The cross-correlation function vector of accelerations can be defined similarly with Equation (4.27):

$$\mathbf{r}(\tau) = \left[r_{S1}(\tau), r_{S2}(\tau), \cdots, r_{SN_o}(\tau)\right]^T \tag{4.30}$$

where S denotes the selected reference DOF number. According to Equation (4.28), $\mathbf{r}(\tau)$ can be expressed using different modal coordinates for convenient system identification

$$\mathbf{r}(\tau)=\sum_{m=1}^{N_m}\boldsymbol{\varphi}_m q_m(\tau) \tag{4.31}$$

where $q_m(\tau)$ is the m-th modal coordinate of $\mathbf{r}(\tau)$, and its free-decay response is written as

$$q_m(\tau)=q_m(0)u_m(\tau)+\dot{q}_m(0)v_m(\tau) \tag{4.32}$$

where $q_m(0)$ and $\dot{q}_m(0)$ are the modal initial conditions related to the m-th modal coordinate and its first derivative, respectively; the modal response functions $u_m(\tau)$ and $v_m(\tau)$ are given by

$$u_m(\tau)=e^{-\xi_m\omega_m\tau}\left(\cos\omega_m^D\tau+\frac{\xi_m}{\sqrt{1-\xi_m^2}}\sin\omega_m^D\tau\right) \tag{4.33}$$

$$v_m(\tau)=\frac{e^{-\xi_m\omega_m\tau}}{\omega_m^D}\sin\omega_m^D\tau \tag{4.34}$$

The cross-correlation function between the reference DOF S and the i-th DOF is thus expressed as

$$r_{Si}=\sum_{m=1}^{N_m}\varphi_{im}\left(q_m(0)u_m(\tau)+\dot{q}_m(0)v_m(\tau)\right) \tag{4.35}$$

The model for system identification is thus selected as Equation (4.35). This model indicates that the theoretical cross-correlation functions can be calculated given the modal parameters $\{\omega_m,\xi_m,\boldsymbol{\varphi}_m,q_m(0),\dot{q}_m(0):m=1,\cdots,N_m\}$, so they can fit to the experimental cross-correlation functions constructed by measured accelerations. Ambient excitations need not be measured (it is almost impossible to measure ambient excitations for full-scale structures).

4.2 IDENTIFYING MODAL PARAMETERS

The problem of identifying modal parameters is formulated as an optimization problem, where the difference between the model-predicted and experimental cross-correlation functions is to be minimized. We first formulate the objective function to be minimized. Let $\boldsymbol{\theta}$ contain the modal parameters to be identified, i.e., $\theta=\{\omega_m,\xi_m,\boldsymbol{\varphi}_m,q_m(0),\dot{q}_m(0):m=1,\cdots,N_m\}$. The model-predicted and experimental cross-correlation functions of the i-th DOF at the j-th instance of the time lag τ_j, where the discrete-time model has been considered and $\{\tau_j=j\Delta\tau:j=0,\cdots,N_\tau-1\}$, has a difference of b_{ij}:

$$\hat{r}_{Si}(\tau_j)=r_{Si}(\theta,\tau_j)+b_{ij} \tag{4.36}$$

where $\hat{r}_{Si}(\tau_j)$ denotes the experimental cross-correlation function of the i-th DOF at the j-th instance of the time lag τ_j corresponding to the reference DOF S; $r_{Si}(\boldsymbol{\theta},\tau_j)$ denotes the model-predicted one, whose dependence on the system parameters $\boldsymbol{\theta}$ is emphasized. The differences of different DOFs at different instances of the time lag need to be minimized to find optimal modal parameters, so the objective function is formulated as the averaged difference over different DOFs and time-lag instances:

$$J(\boldsymbol{\theta}) = \frac{1}{N_\tau N_o} \sum_{j=1}^{N_\tau} \sum_{i=1}^{N_o} \left(\hat{r}_{Si}(\tau_j) - r_{Si}(\boldsymbol{\theta},\tau_j) \right)^2 \tag{4.37}$$

4.2.1 Modal-component optimization

The minimization of the objective function in Equation (4.37) involves modal parameters of different modes, the number of which is large for full-scale structures. This is because the number of measured DOFs for full-scale structures is usually large (for getting detailed mode shapes to understand structural dynamic behaviors), and multiple modes are to be identified. If all the modal parameters are optimized simultaneously, the high-dimensional optimization will be very time-consuming and may not be convergent. To reduce the dimension of the optimization problem, a modal-component optimization method is proposed. This method is extended from Beck (Beck 1979; Beck & Jennings 1980; Werner et al. 1987). The basic idea is to divide $\boldsymbol{\theta}$ into different modal components, $\boldsymbol{\theta} = \{M_1,\cdots,M_m,\cdots,M_{N_m}\}$, where each component M_m contains modal parameters of one mode, $M_m = \{\omega_m,\xi_m,\varphi_m,q_m(0),\dot{q}_m(0)\}$, and optimizes in a "modal-sweep" fashion among different modes, i.e., optimizes for one modal component M_m at a time conditional on other modal components, and iterates among different modal components until convergence.

The framework of the modal-component optimization is summarized in Algorithm 4.1. For each iteration step, this algorithm will conduct N_m times of optimization. Each optimization is to update one modal component, i.e., the modal parameters of one mode, with other modal components fixed at their latest partial optima. Once one modal component is updated, its latest value will replace its previous value to be used in the optimization of the following modal components, as seen in the algorithm. After all the N_m modal components are updated, the current iteration step is finished, and the next iteration step is conducted. In the following, an efficient algorithm will be developed for optimizing each modal component.

4.2.1.1 Algorithm 4.1: Modal-component optimization

Given the modal parameters in the $(k-1)$-th iteration step $\boldsymbol{\theta}^{(k-1)}$, conduct optimization in the k-th iteration step as follows:

(1) Minimize the objective function with respect to the first modal component in the k-th iteration step $M_1^{(k)}$, given other modal components in the $(k-1)$-th iteration step $\left\{\tilde{M}_2^{(k-1)},\tilde{M}_3^{(k-1)},\cdots,\tilde{M}_m^{(k-1)},\cdots\tilde{M}_{N_m}^{(k-1)}\right\}$, where $\tilde{M}_m^{(k-1)}$ denotes the optimum of the m-th component obtained at the $(k-1)$-th iteration step:

$$\tilde{M}_1^{(k)} = \arg_{M_1^{(k)}} \min J\left(M_1^{(k)} \middle| \left\{\tilde{M}_2^{(k-1)}, \tilde{M}_3^{(k-1)}, \cdots, \tilde{M}_m^{(k-1)}, \cdots \tilde{M}_{N_m}^{(k-1)}\right\}\right)$$

(2) Minimize the objective function with respect to the second modal component in the k-th iteration step $M_2^{(k)}$, given the updated modal component $\tilde{M}_1^{(k)}$ in the k-th iteration step and other modal components in the $(k-1)$-th iteration step $\left\{\tilde{M}_1^{(k)}, \tilde{M}_3^{(k-1)}, \cdots, \tilde{M}_m^{(k-1)}, \cdots \tilde{M}_{N_m}^{(k-1)}\right\}$:

$$\tilde{M}_2^{(k)} = \arg_{M_2^{(k)}} \min J\left(M_2^{(k)} \middle| \left\{\tilde{M}_1^{(k)}, \tilde{M}_3^{(k-1)}, \cdots, \tilde{M}_m^{(k-1)}, \cdots \tilde{M}_{N_m}^{(k-1)}\right\}\right)$$

$$\vdots$$

(n) Minimize the objective function with respect to the m-th modal component in the k-th iteration step $M_m^{(k)}$, given the updated modal components $\left\{\tilde{M}_1^{(k)}, \tilde{M}_2^{(k)}, \cdots, \tilde{M}_{m-1}^{(k)}\right\}$ in the k-th iteration step and other modal components in the $(k-1)$-th iteration step $\left\{\tilde{M}_{m+1}^{(k-1)}, \cdots \tilde{M}_{N_m}^{(k-1)}\right\}$:

$$\tilde{M}_m^{(k)} = \arg_{M_m^{(k)}} \min J\left(M_m^{(k)} \middle| \left\{\tilde{M}_1^{(k)}, \tilde{M}_2^{(k)}, \cdots, \tilde{M}_{m-1}^{(k)}, \tilde{M}_{m+1}^{(k-1)}, \cdots \tilde{M}_{N_m}^{(k-1)}\right\}\right)$$

$$\vdots$$

(N_m) Minimize the objective function with respect to the N_m-th modal component (the final one) in the k-th iteration step $M_{N_m}^{(k)}$, given the updated modal components $\left\{\tilde{M}_1^{(k)}, \tilde{M}_2^{(k)}, \cdots, \tilde{M}_{N_m-1}^{(k)}\right\}$ in the k-th iteration step:

$$\tilde{M}_{N_m}^{(k)} = \arg_{M_{N_m}^{(k)}} \min J\left(M_n^{(k)} \middle| \left\{\tilde{M}_1^{(k)}, \tilde{M}_2^{(k)}, \cdots, \tilde{M}_{N_m-1}^{(k)}\right\}\right)$$

(N_m+1) Update the iteration step $k = k + 1$. Repeat Steps (1) to (N_m) until convergence.

4.2.2 Optimization of one modal component

Although decomposing the optimization into one modal component reduces the optimization dimension, direct optimization on one modal component can still be challenging mainly due to the high-dimensional nature of mode shape. To obtain useful dynamic properties of full-scale complex engineering structures, a moderate or large number of locations is required to be measured, rendering high-dimensional mode shape vectors. If the optimization is conducted with each element of a mode shape vector treated as an optimization parameter, the optimization will be very time-consuming and convergence cannot be guaranteed, because the number of optimization parameters is large. To apply the proposed method for full-scale structures, an efficient algorithm is developed for the optimization of each modal component. The main idea of this algorithm is to derive partial optima for the mode shape and modal initial conditions—conditional on the natural frequency and damping ratio—so that optimization is only conducted for the natural frequency and damping ratio (only two parameters), and the dimension of optimization can be further reduced after the reduction from multi-mode optimization to the single-mode one. The treatment of the mode shape, modal initial conditions, natural frequency, and damping ratio in each component is discussed, respectively.

4.2.2.1 Mode shape

According to the proposed system model, the modes of interest need to be selected. It is proposed to select the modes of interest from the resonant peaks in the singular-value spectra constructed using the measured accelerations. The singular-value spectra will also be used for bypassing the high-dimensional optimization of mode shapes. We first investigate the mathematical structure of mode shapes with respect to the data. Noticing that fast algorithms are available to decompose a matrix using an orthogonal basis (Strang 1993) and mode shapes constitute an orthogonal basis, it is proposed to decompose the PSD matrices (the counterpart of the cross-correlation data used in modal analysis) using mode shapes. With this decomposition method, it means that the PSD matrices provide information of mode shapes, so direct optimization of mode shape can be avoided.

Let $\mathcal{F}(\ddot{\mathbf{x}},\omega) \in R^{N_o}$ denote the Fourier transform of the acceleration $\ddot{\mathbf{x}} \in R^{N_o}$ of the measured DOFs at frequency ω. The PSD matrix at frequency ω, $\mathbf{P}(\omega) \in R^{N_o \times N_o}$, is expressed by the product of $\mathcal{F}(\ddot{\mathbf{x}},\omega)$ and its conjugate transpose $\mathcal{F}^*(\ddot{\mathbf{x}},\omega) \in R^{1 \times N_o}$, where * denotes the conjugate transpose operator for a matrix, and then expanded using mode shapes:

$$\begin{aligned} \mathbf{P}(\omega) &= \mathcal{F}(\ddot{\mathbf{x}},\omega)\mathcal{F}^*(\ddot{\mathbf{x}},\omega) \\ &= \mathcal{F}\left(\mathbf{\Phi}\ddot{\mathbf{d}},\omega\right)\mathcal{F}^*\left(\mathbf{\Phi}\ddot{\mathbf{d}},\omega\right) \\ &= \mathbf{\Phi}\left(\mathcal{F}\left(\ddot{\mathbf{d}},\omega\right)\mathcal{F}^*\left(\ddot{\mathbf{d}},\omega\right)\right)\mathbf{\Phi}^* \\ &= \mathbf{\Phi}\mathbf{Q}(\omega)\mathbf{\Phi}^* \end{aligned} \tag{4.38}$$

where $\mathbf{Q}(\omega) = \mathcal{F}\left(\ddot{\mathbf{d}},\omega\right)\mathcal{F}^*\left(\ddot{\mathbf{d}},\omega\right)$ is the PSD matrix in modal coordinates; $\ddot{\mathbf{x}} = \mathbf{\Phi}\ddot{\mathbf{d}}$ has been used; and mode shape matrix $\mathbf{\Phi}$ can be taken outside the Fourier transform operator because it is not frequency-dependent.

On the other hand, consider the singular value decomposition (SVD) of the PSD matrix $\mathbf{P}(\omega)$:

$$\begin{aligned} \mathbf{P}(\omega) &= \mathbf{U}\Sigma(\omega)\mathbf{V}^* \\ &= \mathbf{U}\Sigma(\omega)\mathbf{U}^* \end{aligned} \tag{4.39}$$

where $\Sigma(\omega) \in R^{N_o \times N_o}$ is the diagonal singular value matrix that contains the singular values on the diagonal in the descending order; $\mathbf{U} \in R^{N_o \times N_o}$ and $\mathbf{V} \in R^{N_o \times N_o}$ contain the left- and right- singular vectors of $\mathbf{P}(\omega)$, respectively, and they are unitary matrices; $\mathbf{U} = \mathbf{V}$ has been used because $\mathbf{P}(\omega)$ is symmetrical. In fact, because the PSD matrix is a normal and positive definite matrix, the SVD above is also the eigen-decomposition (Strang 1993).

The two expressions in Equations (4.38) and (4.39) show that the PSD matrix has the same structure of SVD, so the singular vector matrix can be interpreted as a mode shapes matrix. In practical applications, the PSD matrices can be obtained based on measured accelerations. SVD is then conducted for the PSD matrices at each frequency instance. The singular vector matrix will be the estimate of the mode shape matrix. However, when optimizing for one mode, only one column of the singular vector matrix corresponding to this mode is needed (but not the whole singular vector matrix). To select the singular vector corresponding to the mode under consideration, note that the singular value quantifies how much a mode contributes at frequency ω, i.e., the larger the singular value is, the higher the contribution of this mode at frequency ω. Moreover, it is proposed that optimization of one modal component is confined to a certain frequency band of which only this mode has significant contribution; i.e., during optimization, the natural frequency only varies within this frequency band. This frequency band is chosen as the band of the corresponding spectral peak in the singular-value spectra. As a result, only one mode has the dominant contribution for the optimization of one modal component. Given the frequency ω within the chosen frequency band, the mode shape of the dominant mode is thus obtained as the singular vector of $\mathbf{P}(\omega)$ corresponding to the largest singular value. Based on the previous analysis, the mode shape is not directly optimized in the optimization. The natural frequency is treated as an optimization parameter. When the value of the natural frequency varies within the chosen frequency band during the optimization, the mode shape will be obtained as the singular vector of the PSD matrix at each frequency value. The procedure of the treatment of mode shape in the optimization is summarized as follows:

(1) Calculate the Fourier transform at each discrete frequency ω_j.
(2) Construct the experimental PSD matrices at each discrete frequency ω_j using Equation (4.38).
(3) Perform SVD of the PSD matrices in Step (2) at each discrete frequency ω_j. Construct the figure of singular-value spectra by plotting the singular values against the discrete frequency instances. Note that this figure includes multiple lines because there are multiple singular values at each frequency instance. Each line corresponds to one singular value. The frequency band of each mode is chosen such that it includes the spectral peak corresponding to this mode of interest.
(4) Given an intermediate frequency ω_j during the optimization process, the mode shape is obtained as the singular vector corresponding to the largest singular value of the PSD matrix at ω_j. Once ω_j is changed in the optimization process, the associated singular vector is also changed at this frequency to update the associated mode shape. These updated modal parameters are used to calculate the objective function of the modal component until convergence.

4.2.2.2 Modal initial conditions

The modal initial displacement and velocity, $q_m(0)$ and $\dot{q}_m(0)$, are needed to calculate the objective function of each modal component $J\left(M_m^{(k)}\middle|\left\{\tilde{M}_1^{(k)},\tilde{M}_2^{(k)},\cdots,\tilde{M}_{m-1}^{(k)},\tilde{M}_{m+1}^{(k-1)},\cdots\tilde{M}_{N_m}^{(k-1)}\right\}\right)$. According to Equation (4.32), the objective function of $M_m^{(k)}$ is a linear function of $q_m(0)$ and $\dot{q}_m(0)$, with $\omega_m, \xi_m, \varphi_m$ and modal parameters of other modal components in the previous iteration step available. Moreover, the modal response functions $\{u_m(\tau), v_m(\tau)\}$ can be conveniently evaluated with the analytical formulas of Equations (4.33) and (4.34). Therefore, identifying modal initial conditions, given $\{\omega_m, \xi_m, \varphi_m\}$ (with their values at appropriate iteration steps), is a linear optimization problem that can be done analytically using the least-squares method (Au et al. 2005). The modal initial conditions will be identified together, so the modal initial conditions are put in $\mathbf{q}(0) \in R^{N_m \times 1}$ and $\dot{\mathbf{q}}(0) \in R^{N_m \times 1}$, respectively:

$$\mathbf{q}(0)=\left[q_1(0),q_2(0),\cdots,q_{N_m}(0)\right]^T \tag{4.40}$$

$$\dot{\mathbf{q}}(0)=\left[\dot{q}_1(0),\dot{q}_2(0),\cdots,\dot{q}_{N_m}(0)\right]^T \tag{4.41}$$

The cross-correlation functions of the i-th DOF at different time-lag instances are then written in a matrix form:

$$\mathbf{R}_{Si}=\mathbf{U}\Psi_i\mathbf{q}(0)+\mathbf{V}\Psi_i\dot{\mathbf{q}}(0) \tag{4.42}$$

where $\mathbf{R}_{Si}\in R^{N_\tau\times 1}$ is the cross-correlation vector of the i-th DOF

$$\mathbf{R}_{Si}=\left[r_{Si}(\tau_1),r_{Si}(\tau_2),\cdots,r_{Si}(\tau_{N_\tau})\right]^T \tag{4.43}$$

$\mathbf{U}\in R^{N_\tau\times N_m}$ contains the modal response functions of different modes $\{u_m(\tau_j)\}$, and the m-th column of $\mathbf{U}$, denoted by $\mathbf{U}(:,m)\in R^{N_\tau}$, contains the modal response functions of the m-th mode at different time lags:

$$\mathbf{U}(:,m)=\left[u_m(\tau_1),u_m(\tau_2),\cdots,u_m(\tau_{N_\tau})\right]^T \tag{4.44}$$

$\mathbf{V}\in R^{N_\tau\times N_m}$ is constructed in a similar way and its m-th column $\mathbf{V}(:,m)\in R^{N_\tau}$ contains $\{v_m(\tau_j)\}$:

$$\mathbf{V}(:,m)=\left[v_m(\tau_1),v_m(\tau_2),\cdots,v_m(\tau_{N_\tau})\right]^T \tag{4.45}$$

$\Psi_i\in R^{N_m\times N_m}$ is a diagonal matrix whose diagonal elements are the mode shapes at the i-th DOF of different modes:

$$\Psi_i=\begin{bmatrix}\varphi_{i1} & & & & \\ & \ddots & & & \\ & & \varphi_{im} & & \\ & & & \ddots & \\ & & & & \varphi_{iN_m}\end{bmatrix} \tag{4.46}$$

where φ_{im} is the mode shape at the i-th DOF of the m-th mode. Equation (4.42) is further transformed into a block-matrix form:

$$\begin{aligned}\mathbf{R}_{Si}&=\left[\mathbf{U}\Psi_i \quad \mathbf{V}\Psi_i\right]\begin{bmatrix}\mathbf{q}(0)\\ \dot{\mathbf{q}}(0)\end{bmatrix}\\ &=\mathbf{A}_i\begin{bmatrix}\mathbf{q}(0)\\ \dot{\mathbf{q}}(0)\end{bmatrix}\end{aligned} \tag{4.47}$$

where $\mathbf{A}_i=\left[\mathbf{U}\Psi_i \quad \mathbf{V}\Psi_i\right]$. Left-multiplying $\mathbf{A}_i^T$ on both sides of the above equation and considering all the measured DOFs gives the least-squares equation for the modal initial conditions:

$$\sum_{i=1}^{N_o} \mathbf{A}_i^T \hat{\mathbf{R}}_{Si} = \left(\sum_{i=1}^{N_o} \mathbf{A}_i^T \mathbf{A}_i \right) \begin{bmatrix} \mathrm{q}(0) \\ \dot{\mathrm{q}}(0) \end{bmatrix} \tag{4.48}$$

where the experimental cross-correlation functions $\hat{\mathbf{R}}_{Si}$ have been used. Solving the above equation for $\begin{bmatrix} \mathrm{q}(0) \\ \dot{\mathrm{q}}(0) \end{bmatrix}$ gives

$$\begin{bmatrix} \mathrm{q}(0) \\ \dot{\mathrm{q}}(0) \end{bmatrix} = \left(\sum_{i=1}^{N_o} \mathbf{A}_i^T \mathbf{A}_i \right)^{-1} \sum_{i=1}^{N_o} \mathbf{A}_i^T \hat{\mathbf{R}}_{Si} \tag{4.49}$$

The above equation indicates that the modal initial conditions depend on $\{\mathbf{A}_i : i = 1,2,\cdots,N_o\}$, which in turn need the system parameters of all the modal components. The remarks on how to choose appropriate values of the system parameters are given here. For example, if optimization is now conducted for the m-th modal component at the k-th iteration step $M_m^{(k)}$, then it means that partial optima of modal components 1 to m – 1 at the k-th iteration step, $\left\{\tilde{M}_1^{(k)}, \tilde{M}_2^{(k)}, \cdots, \tilde{M}_{m-1}^{(k)}\right\}$, are available, while optimization has not been conducted for other modal components, so only the partial optima at the (k – 1)-th iteration step can be used, i.e., $\left\{\tilde{M}_{m+1}^{(k-1)}, \cdots \tilde{M}_{N_m}^{(k-1)}\right\}$, the partial optima of modal components m + 1 to N_m at the (k – 1)-th iteration step. Moreover, the optimization algorithm will try different values of the m-th modal component $M_m^{(k)}$, which will be used together with $\left\{\tilde{M}_1^{(k)}, \tilde{M}_2^{(k)}, \cdots, \tilde{M}_{m-1}^{(k)}, \tilde{M}_{m+1}^{(k-1)}, \cdots \tilde{M}_{N_m}^{(k-1)}\right\}$ for calculating $\{\mathbf{A}_i : i = 1,2,\cdots,N_o\}$.

4.2.2.3 Natural frequency and damping ratio

Unlike the mode shape and modal initial conditions, the natural frequency and damping ratio is directly optimized when optimizing for one modal component. It is only a two-dimensional optimization problem and can be easily done by various optimization algorithms, e.g., the simplex method (Nelder & Mead 1965). During the optimization, once a set of intermediate values of the natural frequency and damping ratio is provided by the algorithm, the mode shape and modal initial conditions are obtained based on the intermediate natural frequency and damping ratio for evaluating the objective function of the modal component. The objective function is repeatedly evaluated until convergence is reached for the modal component. Having discussed the treatment of each system parameter, this section summarizes the optimization of one modal component in the following.

4.2.2.4 Algorithm 4.2: Optimization of one modal component $M_m^{(k)}$

(1) Initialization. If optimization is conducted for the first modal component in the first iteration step, $M_1^{(1)}$, initial modal parameters are required. The initial values' natural frequencies are picked at the spectral peaks of interest from the SVD spectra. The initial values of damping ratios are set to 1% according to engineering experience. The initial values of mode shapes are obtained from the singular vectors of the PSD matrices at the initial natural frequencies, based on the procedures introduced in Section 4.2.2.1. Given the initial values of natural frequencies, damping ratios, and mode shapes, the initial values of modal initial conditions are finally obtained by solving Equation (4.49).

(2) For a general modal component $M_m^{(k)}$, use the Nelder-Mead method (Nelder and Mead 1965) to minimize the objective function $J\left(M_m^{(k)}\middle|\left\{\tilde{M}_1^{(k)},\tilde{M}_2^{(k)},\cdots,\tilde{M}_{m-1}^{(k)},\tilde{M}_{m+1}^{(k-1)},\cdots\tilde{M}_{N_m}^{(k-1)}\right\}\right)$ (see Equation (4.37)) with respect to ω_m and ξ_m. Note that only these two parameters are the optimization parameters and the high-dimensional parameters are calculated based on the intermediate values of these two parameters. The optimization works as follows.

a. The Nelder-Mead method produces the intermediate $\bar{\omega}_m$ and $\bar{\xi}_m$.
b. The updated mode shape $\bar{\varphi}_m$ is obtained as the singular vector at the frequency $\bar{\omega}_m$. The mode shape matrix at the i-th DOF for computing the modal initial conditions is thus updated:

$$\Psi_i=\begin{bmatrix}\tilde{\varphi}_{i1}^{(k)} & & & & & \\ & \ddots & & & & \\ & & \tilde{\varphi}_{i(m-1)}^{(k)} & & & \\ & & & \bar{\varphi}_{i(m)} & & \\ & & & & \tilde{\varphi}_{i(m+1)}^{(k-1)} & \\ & & & & & \ddots \\ & & & & & & \tilde{\varphi}_{iN_m}^{(k-1)}\end{bmatrix},\quad i=1,2,\cdots,N_o \tag{4.50}$$

Notice that for modes 1 to m – 1 the optima in the k-th iteration step are used; for modes m + 1 to N_m the optima in the previous iteration step are used because optimization has not been conducted for these modal components; and the updated mode shape of mode m is used.

c. Based on Equations (4.43) and (4.44), calculate $\left\{u_n(\tau_j),v_n(\tau_j):n=1,\cdots,m-1;j=1,\cdots,N_\tau\right\}$ using the optima in the k-th iteration step $\left\{\tilde{\omega}_n^{(k)},\tilde{\xi}_n^{(k)}:n=1,\cdots,m-1\right\}$; calculate $\left\{u_n(\tau_j),v_n(\tau_j):n=m+1,\cdots,N_m;j=1,\cdots,N_\tau\right\}$ using the optima in the (k – 1)-th iteration step $\left\{\tilde{\omega}_n^{(k-1)},\tilde{\xi}_n^{(k-1)}:n=m+1,\cdots,N_m\right\}$; calculate $u_m(\tau_j)$ and $v_m(\tau_j)$ using $\bar{\omega}_m$ and $\bar{\xi}_m$ provided by the optimization method. Construct **U** and **V** using Equations (4.44) and (4.45).
d. Construct $\mathbf{A}_i$ in Equation (4.47) to obtain the updated modal initial conditions $\left[\bar{\mathrm{q}}(0)\quad \dot{\bar{\mathrm{q}}}(0)\right]^T$ by solving Equation (4.49).
e. Calculate the objective function value in Equation (4.37).
f. If the optimization does not converge, go back to Step a and repeat the above procedures; otherwise, the optimal modal component $\tilde{M}_m^{(k)}$ is obtained.

4.3 MODAL ANALYSIS OF A FOOTBRIDGE

A full-scale footbridge is used to illustrate the proposed method. This lightweight footbridge can be easily excited by pedestrians or traffic below to a considerable vibration. Field tests were conducted to obtain this footbridge's modal parameters for further dynamic analysis. Figure 4.1 shows the ongoing field test of the footbridge. In this case, six triaxial accelerometers were available but 18 locations on both sides of the footbridge were to be measured, so the measurement was divided into multiple setups, with each setup covering one part of the footbridge. Figure 4.2 shows the measurement plan. The circles with "S1" to "S6" denote sensors 1 to 6. These circles' different colors indicate different setups. The measured locations were numbered according to the order they are measured (see the numbers next to the circles).

Four setups were conducted to cover the 18 locations. 10 minutes of data were collected for each setup. In setup 1, sensors S1 to S6 were placed on locations 1 to 6. Note that S1 and S2 were fixed at locations 1 and 2 for the four setups to be used as reference sensors, such that the partial mode shapes could be assembled into global mode shapes. In setup 2, S3 to S6 were moved to install at locations 7 to 10 with S1 and S2 fixed at the original locations. In setup 3, S3 to S6 were again moved to locations 11 to 14, and finally in setup 4, S3 to S6 were moved locations to 15 to 18.

Figure 4.3 shows the accelerations of one accelerometer measured in a normal day. Figure 4.3(a) and (b) are the accelerations along the x and y directions, respectively, and Figure 4.3(c) is the acceleration perpendicular to the footbridge slab. The magnitude of Figure 4.3(c) is larger than those of Figure 4.3(a) and (b) because the excitation is mainly vertical.

When conducting modal analysis for each setup, the cross-correlation data were first prepared using the measured accelerations. The cross-correlation of each sensor was prepared against S1. Then, by using the proposed method, the modal parameters were

Figure 4.1 The footbridge.

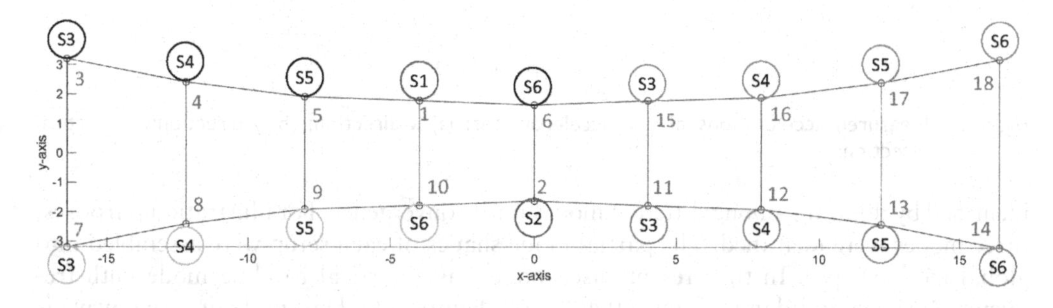

Figure 4.2 Measurement plan of the footbridge.

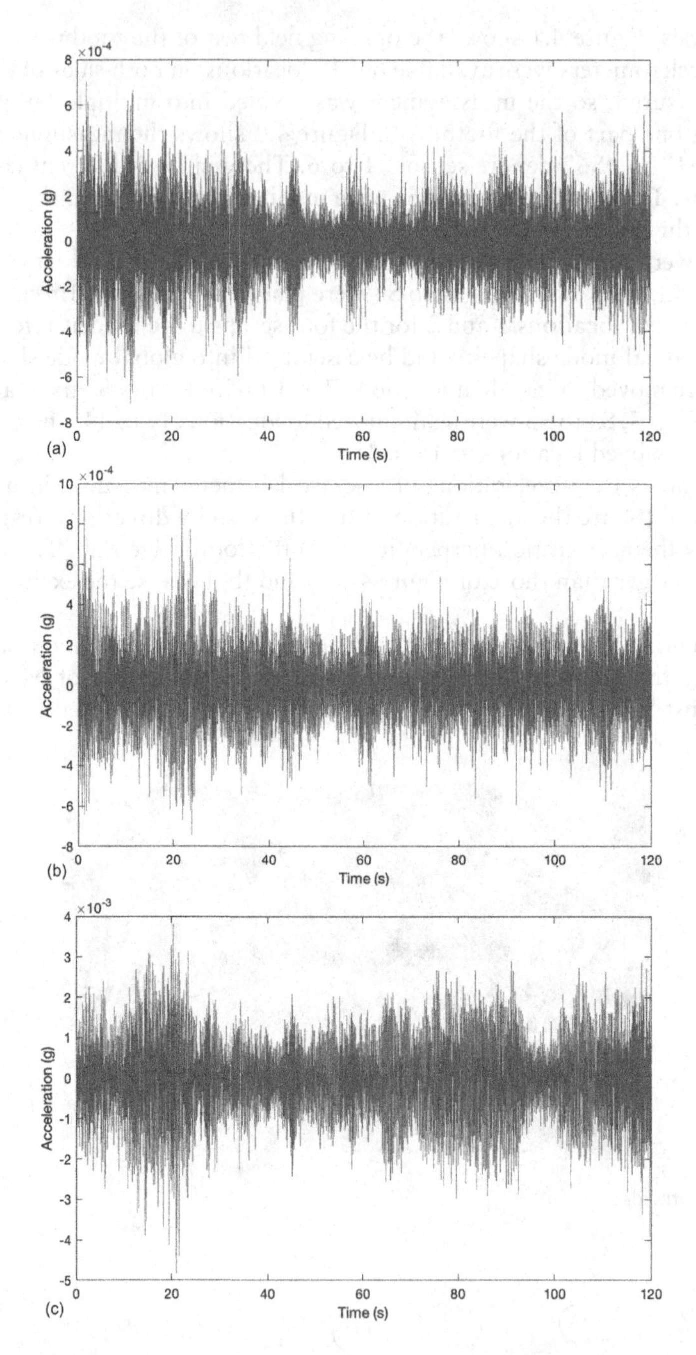

Figure 4.3 Measured accelerations of one accelerometer: (a) *x*-direction; (b) *y*-direction; (c) vertical direction.

identified by iterating among different modes until convergence. For illustration purposes, three modes were identified. The partial mode shapes of each setup were assembled into global mode shapes. In the present case, mode 1 is a vertical bending mode with frequency 4.45 Hz and damping ratio 0.40% (see Figure 4.4). Half cycle of a sine wave is observed for the mode shape. Mode 2 is a high-order mode with frequency 19.92 Hz and

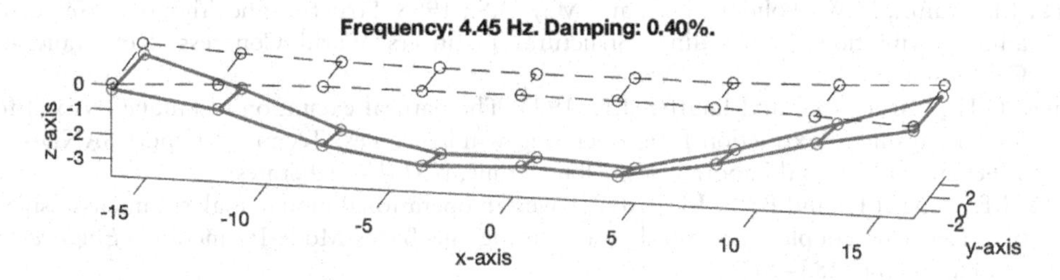

Figure 4.4 Modal parameters of mode 1.

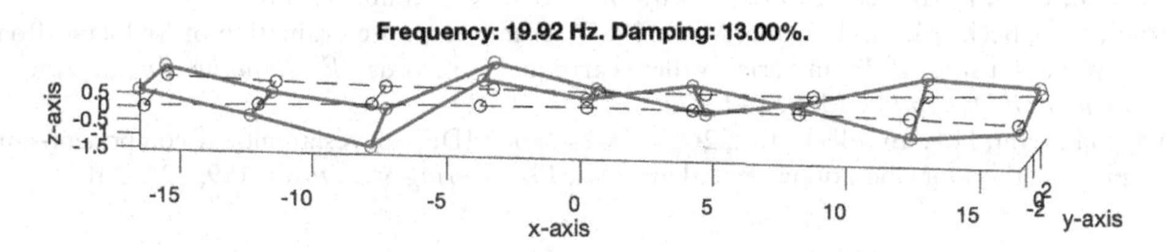

Figure 4.5 Modal parameters of mode 2.

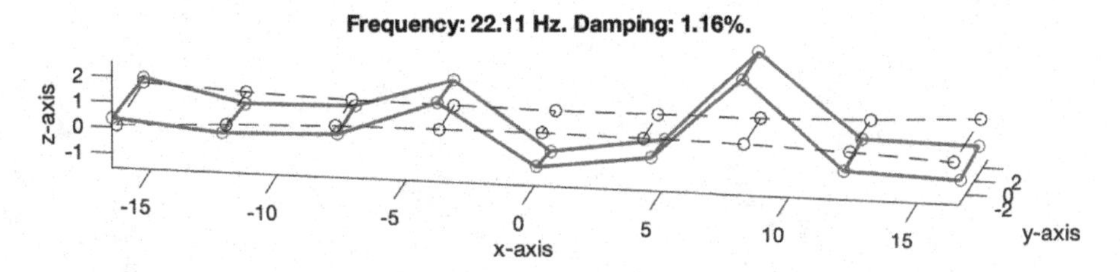

Figure 4.6 Modal parameters of mode 3.

damping ratio 13.00%. As shown in Figure 4.5, out-phase twisting is observed for the right part, and the two sides of the left part of the footbridge move in-phase vertically. Mode 3 is a high-order bending mode with frequency 22.11 Hz and damping ratio 1.16% (Figure 4.6). About two and a half cycles of the sine wave are observed.

REFERENCES

Au, S., Ng, C., Sien, H. and Chua, H., 2005. Modal identification of a suspension footbridge using free vibration signatures. *International Journal of Applied Mathematics and Mechanics*, 1(4), 55–73.

Beck, J.L., 1979. Determining models of structures from earthquake records (Doctoral dissertation). California Institute of Technology.

Beck, J.L. and Jennings, P.C., 1980. Structural identification using linear models and earthquake records. *Earthquake Engineering & Structural Dynamics*, 8(2), 145–160.

Beck, J.L., May, B.S. and Polidori, D.C., 1994. Determination of modal parameters from ambient vibration data for structural health monitoring. Proceedings First World Conference on Structural Control, Pasadena, California.

Beck, J.L., May, B.S., Polidori, D.C. and Vanik, M.W., 1995. Ambient vibration studies of three steel-frame buildings strongly shaken by the 1994 Northridge earthquake. Technical Report EERL95-06, California Institute of Technology.

Beck, J.L., Vanik, M.W., Polidori, D.C. and May, B.S., 1998. Structural health monitoring using ambient vibrations. Proceedings Structural Engineers World Congress, San Francisco, California.

James, G.H., Carne, T.G. and Lauffer, J.P., 1993. The natural excitation technique (NExT) for modal parameter extraction from operating wind turbines. Technical Report SAND--92-1666 Sandia National Laboratories, Albuquerque, NM (United States).

Lam, H.F., Yang, J.H. and Beck, J.L., 2019. Bayesian operational modal analysis and assessment of a full-scale coupled structural system using the Bayes-Mode-ID method. *Engineering Structures*, 186, 183–202.

Nelder, J.A. and Mead, R., 1965. A simplex method for function minimization. *The Computer Journal*, 7(4), 308–313.

Strang, G., 1993. *Introduction to Linear Algebra*. Wellesley-Cambridge Press.

Werner, S.D., Beck, J.L. and Levine, M.B., 1987. Seismic response evaluation of Meloland Road Overpass using 1979 imperial valley earthquake records. *Earthquake Engineering & Structural Dynamics*, 15(2), 249–274.

Yang, J.H., Lam, H.F. and Beck, J.L., 2019. Bayes-Mode-ID: A Bayesian modal-component-sampling method for operational modal analysis. *Engineering Structures*, 189, 222–240.

Chapter 5

System identification based on vector autoregressive moving average models

5.1 STATE-SPACE REPRESENTATION OF A DYNAMIC SYSTEM

For full-scale structures, it will be convenient and efficient to identify models in a simple mathematical form without constructing complicated finite element models. A vector autoregressive moving average model (VARMA) fits this purpose. It has a simple linear structure that provides us the advantage of working on measured data directly. This chapter exploits the advantages of VARMA models for identifying structural dynamic systems and extracting dynamic properties (e.g., natural frequencies, damping ratios, and mode shapes) from the identified VARMA models. To do this, the connection between a structural model and a VARMA model needs to be built. In this section, we first develop the state-space model for a structural model. The detailed study of state-space models can be found in the works by Meirovitch (2010, 1975) and Pi (1990). This state-space representation will serve as the foundation for connecting a structural model and a VARMA model.

The equation of motion of a multi–degree-of-freedom structural model can be written in the following form:

$$\begin{aligned}\dot{\mathbf{x}}(t) &= \dot{\mathbf{x}}(t)\\ \ddot{\mathbf{x}}(t) &= -\mathbf{M}^{-1}\mathbf{K}\mathbf{x}(t) - \mathbf{M}^{-1}\mathbf{C}\dot{\mathbf{x}}(t) + \mathbf{M}^{-1}\mathbf{F}(t)\end{aligned} \tag{5.1}$$

Combining the two equations in Equation (5.1) gives

$$\begin{bmatrix}\dot{\mathbf{x}}(t)\\ \ddot{\mathbf{x}}(t)\end{bmatrix} = \begin{bmatrix}\mathbf{0}_{N_d\times N_d} & \mathbf{I}_{N_d\times N_d}\\ -\mathbf{M}^{-1}\mathbf{K} & -\mathbf{M}^{-1}\mathbf{C}\end{bmatrix}\begin{bmatrix}\mathbf{x}(t)\\ \dot{\mathbf{x}}(t)\end{bmatrix} + \begin{bmatrix}0\\ \mathbf{M}^{-1}\end{bmatrix}\mathbf{F}(t) \tag{5.2}$$

By introducing the state vector $w(t) = \left[x(t)^T, \dot{x}(t)^T\right]^T \in R^{2N_d}$, Equation (5.2) can be written into a state-space form:

$$\dot{\mathbf{w}}(t) = \mathbf{A}\mathbf{w}(t) + \mathbf{B}\mathbf{F}(t) \tag{5.3}$$

where

$$\mathbf{A} = \begin{bmatrix}\mathbf{0}_{N_d\times N_d} & \mathbf{I}_{N_d\times N_d}\\ -\mathbf{M}^{-1}\mathbf{K} & -\mathbf{M}^{-1}\mathbf{C}\end{bmatrix}, \mathbf{B} = \begin{bmatrix}0\\ \mathbf{M}^{-1}\end{bmatrix} \tag{5.4}$$

Next, the solution of the state-space model is derived. Based on this solution, the transition matrix is obtained to facilitate the recursive formulation of dynamic responses.

DOI: 10.1201/9780429445866-5

Without excitations, the first-order homogeneous differential equation $\dot{\mathbf{w}}(t) = \mathbf{A}\mathbf{w}(t)$ has the solution

$$\mathbf{w}(t) = e^{\mathbf{A}t}\mathbf{w}(0) \tag{5.5}$$

where $\mathbf{w}(0)$ is the initial condition. The state transition matrix is defined using the matrix exponential function in the above equation:

$$\mathbf{\Phi}(t,\tau) = e^{\mathbf{A}(t-\tau)} \tag{5.6}$$

Equation (5.6) corresponds to the state transition from τ to t, so $\mathbf{\Phi}(t) = e^{\mathbf{A}t}$ in Equation (5.5) corresponds to the state transition from 0 to t. Using Taylor series expansion, the matrix exponential function $e^{\mathbf{A}t}$ can be expanded as follows:

$$e^{\mathbf{A}t} = \mathbf{I}_{2N_d \times 2N_d} + t\mathbf{A} + \frac{t^2}{2!}\mathbf{A}^2 + \frac{t^3}{3!}\mathbf{A}^3 + \cdots = \sum_{i=0}^{\infty} \frac{(\mathbf{A}t)^i}{i!} \tag{5.7}$$

Using Equation (5.7), we have

$$\dot{\mathbf{\Phi}}(t) = \frac{d\left(e^{\mathbf{A}t}\right)}{dt} = \sum_{i=0}^{\infty} \frac{i\mathbf{A}(\mathbf{A}t)^{i-1}}{i!} = \mathbf{A}\sum_{i=1}^{\infty} \frac{(\mathbf{A}t)^{i-1}}{(i-1)!} = \mathbf{A}e^{\mathbf{A}t} = e^{\mathbf{A}t}\mathbf{A} \tag{5.8}$$

Equation (5.8) is thus written as

$$\dot{\mathbf{\Phi}}(t) = \mathbf{A}\mathbf{\Phi}(t) = \mathbf{\Phi}(t)\mathbf{A} \tag{5.9}$$

The solution of the non-homogeneous equation Equation (5.3) will be obtained using the state transition matrix. Expressing $\mathbf{A}$ with Equation (5.9) gives

$$\mathbf{A} = \dot{\mathbf{\Phi}}(t)\mathbf{\Phi}^{-1}(t) = \dot{\mathbf{\Phi}}(t)\mathbf{\Phi}(-t) \tag{5.10}$$

where $\mathbf{\Phi}^{-1}(t) = \left(e^{\mathbf{A}t}\right)^{-1} = e^{\mathbf{A}(-t)} = \mathbf{\Phi}(-t)$ has been used. On the other hand,

$$\mathbf{\Phi}(t)\mathbf{\Phi}(-t) = \mathbf{I}_{2N_d \times 2N_d} \tag{5.11}$$

Differentiating Equation (5.11) gives

$$\dot{\mathbf{\Phi}}(t)\mathbf{\Phi}(-t) - \mathbf{\Phi}(t)\dot{\mathbf{\Phi}}(-t) = 0 \tag{5.12}$$

Substituting Equation (5.10) for the first term of Equation (5.12) gives another expression for $\mathbf{A}$

$$\mathbf{A} = \mathbf{\Phi}(t)\dot{\mathbf{\Phi}}(-t) \tag{5.13}$$

Substituting the system matrix $\mathbf{A}$ in Equation (5.13) into Equation (5.3) gives

$$\dot{\mathbf{w}}(t) = \mathbf{\Phi}(t)\dot{\mathbf{\Phi}}(-t)\mathbf{w}(t) + \mathbf{BF}(t) \tag{5.14}$$

Left-multiplying $\mathbf{\Phi}(-t)$ to Equation (5.14) gives

$$\begin{aligned} &\mathbf{\Phi}(-t)\dot{\mathbf{w}}(t) = \dot{\mathbf{\Phi}}(-t)\mathbf{w}(t) + \mathbf{\Phi}(-t)\mathbf{BF}(t) \\ &\Rightarrow \mathbf{\Phi}(-t)\dot{\mathbf{w}}(t) - \dot{\mathbf{\Phi}}(-t)\mathbf{w}(t) = \mathbf{\Phi}(-t)\mathbf{BF}(t) \\ &\Rightarrow \frac{d}{dt}\left(\mathbf{\Phi}(-t)w(t)\right) = \mathbf{\Phi}(-t)\mathbf{BF}(t) \end{aligned} \tag{5.15}$$

Integrating Equation (5.15) gives

$$\begin{aligned} &\mathbf{\Phi}(-t)\mathbf{w}(t) - \mathbf{\Phi}(0)\mathbf{w}(0) = \int_0^t \mathbf{\Phi}(-\tau)\mathbf{BF}(\tau)d\tau \\ &\Rightarrow \mathbf{\Phi}(-t)\mathbf{w}(t) = \mathbf{w}(0) + \int_0^t \mathbf{\Phi}(-\tau)\mathbf{BF}(\tau)d\tau \end{aligned} \tag{5.16}$$

where $\mathbf{\Phi}(0) = \mathbf{I}_{2N_d \times 2N_d}$ has been used. Left-multiplying $\mathbf{\Phi}(t)$ to Equation (5.16) finally gives the solution of the state-space model of a dynamic system

$$\mathbf{w}(t) = \mathbf{\Phi}(t)\mathbf{w}(0) + \int_0^t \mathbf{\Phi}(t-\tau)\mathbf{BF}(\tau)d\tau \tag{5.17}$$

where the first term on the right is due to the initial condition and the second term is due to excitations. Equation (5.17) is the continuous-time solution. In practice, responses are measured in discrete time, so the discrete-time solution based on Equation (5.17) is formulated to facilitate the recursive representation of responses. Suppose the continuous-time state is discretized with a constant time step Δt, so that states are available at discrete time instances $\{t_i = i\Delta t: i = 0,1,2,\cdots\}$, and the state at the i-th time instance is denoted by $\mathbf{w}_i = \mathbf{w}(i\Delta t)$. Consider the response at the i-th time instance $t_i = i\Delta t$

$$\begin{aligned} \mathbf{w}_i &= \mathbf{\Phi}(i\Delta t)\mathbf{w}_0 + \int_0^{i\Delta t} \mathbf{\Phi}(i\Delta t - \tau)\mathbf{BF}(\tau)d\tau \\ &= e^{\mathbf{A}i\Delta t}\mathbf{w}_0 + \int_0^{i\Delta t} e^{\mathbf{A}(i\Delta t - \tau)}\mathbf{BF}(\tau)d\tau \end{aligned} \tag{5.18}$$

Similarly, the response at the next time instance $t_{i+1} = (i+1)\Delta t$ can be written

$$\begin{aligned} \mathbf{w}_{i+1} &= e^{\mathbf{A}(i+1)\Delta t}\mathbf{w}_0 + \int_0^{(i+1)\Delta t} e^{\mathbf{A}((i+1)\Delta t - \tau)}\mathbf{BF}(\tau)d\tau \\ &= e^{\mathbf{A}(i+1)\Delta t}\mathbf{w}_0 + \underbrace{\int_0^{i\Delta t} e^{\mathbf{A}((i+1)\Delta t - \tau)}\mathbf{BF}(\tau)d\tau + \int_{i\Delta t}^{(i+1)\Delta t} e^{\mathbf{A}((i+1)\Delta t - \tau)}\mathbf{BF}(\tau)d\tau}_{\text{Divide the integral into two time segments.}} \\ &= e^{\mathbf{A}\Delta t}\underbrace{\left(e^{\mathbf{A}i\Delta t}\mathbf{w}_0 + \int_0^{i\Delta t} e^{\mathbf{A}(i\Delta t - \tau)}\mathbf{BF}(\tau)d\tau\right)}_{\mathbf{w}_i} + \int_{i\Delta t}^{(i+1)\Delta t} e^{\mathbf{A}((i+1)\Delta t - \tau)}\mathbf{BF}(\tau)d\tau \end{aligned} \tag{5.19}$$

By comparing Equations (5.18) and (5.19), $\mathbf{w}_i$ can be substituted in the first term of the last line in Equation (5.19). The second term of the last line in Equation (5.19) is the integral within one time step. In practice, excitations are usually measured with small time step Δt, so it is reasonable to assume that excitations are piecewise constant within each time step, and they are assigned with the values at the beginning of each time step, i.e., $\mathbf{F}(\tau) \approx \mathbf{F}(i\Delta t) = \mathbf{F}_i$. Changing the variable τ by $t = (i+1)\Delta t - \tau$ for the above integral then gives

$$\int_{i\Delta t}^{(i+1)\Delta t} e^{\mathbf{A}((i+1)\Delta t - \tau)} \mathbf{BF}(\tau) d\tau \approx \int_{\Delta t}^{0} e^{\mathbf{A}t} \mathbf{BF}_i d(-t) = \int_{0}^{\Delta t} e^{\mathbf{A}t} dt \mathbf{BF}_i \tag{5.20}$$

By defining two constants

$$\mathbf{\Phi} = e^{\mathbf{A}\Delta t} \tag{5.21}$$

and

$$\mathbf{J} = \int_{0}^{\Delta t} e^{\mathbf{A}t} dt\, \mathbf{B} = \mathbf{A}^{-1} e^{\mathbf{A}t}\Big|_0^{\Delta t} \mathbf{B} = \mathbf{A}^{-1}\left(e^{\mathbf{A}\Delta t} - \mathbf{I}_{2N_d \times 2N_d}\right)\mathbf{B} \tag{5.22}$$

the states at two consecutive time instances can be written recursively

$$\mathbf{w}_{i+1} = \mathbf{\Phi w}_i + \mathbf{JF}_i \tag{5.23}$$

where $\mathbf{\Phi}$ is interpreted as the state transition matrix within a time step. Based on Equation (5.23), we can obtain the general recursive relation between the state at the current time instance $\mathbf{w}_i$ and the state after j time steps $\mathbf{w}_{i+1}$. First, $\mathbf{w}_{i+2}$ can be obtained in terms of $\mathbf{w}_{i+1}$ by following Equation (5.23)

$$\begin{aligned} \mathbf{w}_{i+2} &= \mathbf{\Phi w}_{i+1} + \mathbf{JF}_{i+1} \\ &= \mathbf{\Phi}\left(\mathbf{\Phi w}_i + \mathbf{JF}_i\right) + \mathbf{JF}_{i+1} \\ &= \mathbf{\Phi}^2 \mathbf{w}_i + \mathbf{\Phi JF}_i + \mathbf{JF}_{i+1} \end{aligned} \tag{5.24}$$

where $\mathbf{w}_{i+1}$ has been substituted with Equation (5.23). Similarly, using Equation (5.23) again and substituting $\mathbf{w}_{i+2}$ with Equation (5.24) gives $\mathbf{w}_{i+3}$ in terms of $\mathbf{w}_i$:

$$\begin{aligned} \mathbf{w}_{i+3} &= \mathbf{\Phi w}_{i+2} + \mathbf{JF}_{i+2} \\ &= \mathbf{\Phi}\left(\mathbf{\Phi}^2 \mathbf{w}_i + \mathbf{\Phi JF}_i + \mathbf{JF}_{i+1}\right) + \mathbf{JF}_{i+2} \\ &= \mathbf{\Phi}^3 \mathbf{w}_i + \mathbf{\Phi}^2 \mathbf{JF}_i + \mathbf{\Phi JF}_{i+1} + \mathbf{JF}_{i+2} \end{aligned} \tag{5.25}$$

To find the pattern, we write $\mathbf{w}_{i+4}$ in terms of $\mathbf{w}_i$, using the same manner as follows:

$$\begin{aligned} \mathbf{w}_{i+4} &= \mathbf{\Phi w}_{i+3} + \mathbf{JF}_{i+3} \\ &= \mathbf{\Phi}\left(\mathbf{\Phi}^3 \mathbf{w}_i + \mathbf{\Phi}^2 \mathbf{JF}_i + \mathbf{\Phi JF}_{i+1} + \mathbf{JF}_{i+2}\right) + \mathbf{JF}_{i+3} \\ &= \mathbf{\Phi}^4 \mathbf{w}_i + \mathbf{\Phi}^3 \mathbf{JF}_i + \mathbf{\Phi}^2 \mathbf{JF}_{i+1} + \mathbf{\Phi JF}_{i+2} + \mathbf{JF}_{i+3} \end{aligned} \tag{5.26}$$

The general recursive relation between $\mathbf{w}_{i+j}$ and $\mathbf{w}_i$ can then be obtained from the pattern of Equations (5.23) to (5.26):

$$\mathbf{w}_{i+j} = \mathbf{\Phi}^j \mathbf{w}_i + \sum_{k=0}^{j-1} \mathbf{\Phi}^{j-1-k} \mathbf{J}\mathbf{F}_{i+k} \tag{5.27}$$

5.2 TRANSFORMING STRUCTURAL MODELS TO VAR MODELS FOR FREE VIBRATION

In this section, we show how to transform a structural state-space model to a vector autoregressive (VAR) model for free vibration. The techniques used for free vibration will then be applied for constructing a VARMA model from a structural state-space model under forced vibration in the next section. These methods are due to the pioneering works of Pi (1990) and Pi and Mickleborough (1989). In practice, measurements are not taken for all responses at all degrees of freedom (DOFs). A selection matrix $\mathbf{L} \in R^{N_o \times 2N_d}$ is thus introduced to select the responses corresponding to the measured DOFs:

$$\mathbf{y}_i = \mathbf{L}\mathbf{w}_i \tag{5.28}$$

where $\mathbf{y}_i \in R^{N_o}$ is the response at the N_o measured degrees of freedom. The selection matrix $\mathbf{L}$ could have the form $\mathbf{L} = \begin{bmatrix} \mathbf{I}_{N_o \times N_o} & \mathbf{0}_{N_o \times (2N_d - N_o)} \end{bmatrix}$, which means that the first N_o components of $\mathbf{w}_i$ are measured. A VAR model is defined here:

$$\mathbf{y}_i = \begin{bmatrix} \mathbf{P}_1 & \mathbf{P}_2 & \cdots & \mathbf{P}_n \end{bmatrix} \begin{bmatrix} \mathbf{y}_{i-1} \\ \mathbf{y}_{i-2} \\ \vdots \\ \mathbf{y}_{i-n} \end{bmatrix} \tag{5.29}$$

where $\{\mathbf{P}_j \in R^{N_o \times N_o} : j = 1, 2, \cdots, n\}$ are parameter matrices and n is the order of the VAR model. Our objective is to show that a structural state-space model under free vibration is equivalent to this VAR model. When a structure is under free vibration, Equation (5.27) is used to construct an augmented response vector $\mathbf{u}_n$ without considering the excitation terms:

$$\mathbf{u}_n = \begin{bmatrix} \mathbf{y}_{i+1} \\ \mathbf{y}_{i+2} \\ \vdots \\ \mathbf{y}_{i+j} \\ \vdots \\ \mathbf{y}_{i+n} \end{bmatrix} = \begin{bmatrix} \mathbf{L}\mathbf{\Phi}\mathbf{w}_i \\ \mathbf{L}\mathbf{\Phi}^2\mathbf{w}_i \\ \vdots \\ \mathbf{L}\mathbf{\Phi}^j\mathbf{w}_i \\ \vdots \\ \mathbf{L}\mathbf{\Phi}^n\mathbf{w}_i \end{bmatrix} = \begin{bmatrix} \mathbf{L} \\ \mathbf{L}\mathbf{\Phi} \\ \vdots \\ \mathbf{L}\mathbf{\Phi}^{j-1} \\ \vdots \\ \mathbf{L}\mathbf{\Phi}^{n-1} \end{bmatrix} \mathbf{\Phi}\mathbf{w}_i = \mathbf{G}\mathbf{\Phi}\mathbf{w}_i \tag{5.30}$$

where

$$\mathbf{G} = \begin{bmatrix} \mathbf{L} \\ \mathbf{L}\mathbf{\Phi} \\ \vdots \\ \mathbf{L}\mathbf{\Phi}^{j-1} \\ \vdots \\ \mathbf{L}\mathbf{\Phi}^{n-1} \end{bmatrix} \tag{5.31}$$

is known to be the observability matrix, which is assumed to be non-singular. Following Equation (5.30), $\mathbf{u}_{n+1}$ can be written as

$$\mathbf{u}_{n+1} = \begin{bmatrix} \mathbf{y}_{i+2} \\ \mathbf{y}_{i+3} \\ \vdots \\ \mathbf{y}_{i+j} \\ \vdots \\ \mathbf{y}_{i+n+1} \end{bmatrix} = \begin{bmatrix} \mathbf{L\Phi}^2\mathbf{w}_i \\ \mathbf{L\Phi}^3\mathbf{w}_i \\ \vdots \\ \mathbf{L\Phi}^j\mathbf{w}_i \\ \vdots \\ \mathbf{L\Phi}^{n+1}\mathbf{w}_i \end{bmatrix} = \begin{bmatrix} \mathbf{L} \\ \mathbf{L\Phi} \\ \vdots \\ \mathbf{L\Phi}^{j-1} \\ \vdots \\ \mathbf{L\Phi}^{n-1} \end{bmatrix} \mathbf{\Phi}^2\mathbf{w}_i = \mathbf{G\Phi}^2\mathbf{w}_i \tag{5.32}$$

Now we want to construct $\mathbf{u}_n$ in Equation (5.32) in order to build the recursive formulation that occurs in a VAR model. With the assumption that G is non-singular, the identity matrix is represented by $\mathbf{I} = \mathbf{G}^{-1}\mathbf{G}$ and inserted into Equation (5.32)

$$\begin{aligned} \mathbf{u}_{n+1} &= \mathbf{G\Phi I\Phi w}_i \\ &= \mathbf{G\Phi G}^{-1}\underbrace{\mathbf{G\Phi w}_i}_{\mathbf{u}_n} \\ &= \mathbf{Hu}_n \end{aligned} \tag{5.33}$$

where it is defined that:

$$\mathbf{H} = \mathbf{G\Phi G}^{-1} \tag{5.34}$$

and Equation (5.30) has been substituted. The structure of **H** is investigated to produce the required VAR model from Equation (5.33). Putting $\mathbf{\Phi}$ into the inverse $\mathbf{G}^{-1}$ gives

$$\mathbf{H} = \begin{bmatrix} \mathbf{L} \\ \mathbf{L\Phi} \\ \vdots \\ \mathbf{L\Phi}^{j-1} \\ \vdots \\ \mathbf{L\Phi}^{n-1} \end{bmatrix} \begin{bmatrix} \mathbf{L\Phi}^{-1} \\ \mathbf{L} \\ \vdots \\ \mathbf{L\Phi}^{j-2} \\ \vdots \\ \mathbf{L\Phi}^{n-2} \end{bmatrix}^{-1} \tag{5.35}$$

Next, a technique proposed by Pi (Pi 1990; Pi & Mickleborough 1989) is applied for revealing an elegant structure of H. Observing that many common terms occur in Equation (5.35), we define some matrices that will facilitate the derivation:

$$\mathbf{O} = \begin{bmatrix} \mathbf{L} \\ \mathbf{L\Phi} \\ \vdots \\ \mathbf{L\Phi}^{n-2} \end{bmatrix} \tag{5.36}$$

$$\mathbf{Q} = \mathbf{L\Phi}^{n-1} \tag{5.37}$$

$$\mathbf{R} = \mathbf{L\Phi}^{-1} \tag{5.38}$$

Using Equations (5.36) to (5.38), Equation (5.35) can be written as

$$\mathbf{H} = \begin{bmatrix} \mathbf{O} \\ \mathbf{Q} \end{bmatrix} \begin{bmatrix} \mathbf{R} \\ \mathbf{O} \end{bmatrix}^{-1} \tag{5.39}$$

Note that the matrix inverse in the above equation can also be written in a block matrix form

$$\begin{bmatrix} \mathbf{R} \\ \mathbf{O} \end{bmatrix}^{-1} = \begin{bmatrix} \mathbf{R}^{\{-1\}} & \mathbf{O}^{\{-1\}} \end{bmatrix} \tag{5.40}$$

where $\mathbf{R}^{\{-1\}}$ and $\mathbf{O}^{\{-1\}}$ are not necessary the ordinary inverse matrices $\mathbf{R}^{-1}$ and $\mathbf{O}^{-1}$. Note also that left-multiplying $\begin{bmatrix} \mathbf{R} \\ \mathbf{O} \end{bmatrix}$ to both sides of Equation (5.40) gives the identity matrix:

$$\begin{bmatrix} \mathbf{R} \\ \mathbf{O} \end{bmatrix} \begin{bmatrix} \mathbf{R} \\ \mathbf{O} \end{bmatrix}^{-1} = \begin{bmatrix} \mathbf{R} \\ \mathbf{O} \end{bmatrix} \begin{bmatrix} \mathbf{R}^{\{-1\}} & \mathbf{O}^{\{-1\}} \end{bmatrix} = \begin{bmatrix} \mathbf{R}\mathbf{R}^{\{-1\}} & \mathbf{R}\mathbf{O}^{\{-1\}} \\ \mathbf{O}\mathbf{R}^{\{-1\}} & \mathbf{O}\mathbf{O}^{\{-1\}} \end{bmatrix} = \mathbf{I} \tag{5.41}$$

So we have the following properties

$$\mathbf{R}\mathbf{R}^{\{-1\}} = \mathbf{I} \tag{5.42}$$

$$\mathbf{O}\mathbf{O}^{\{-1\}} = \mathbf{I} \tag{5.43}$$

$$\mathbf{R}\mathbf{O}^{\{-1\}} = \mathbf{O}\mathbf{R}^{\{-1\}} = 0 \tag{5.44}$$

Substituting Equations (5.40) and (5.42) to (5.44) into Equation (5.39) gives

$$\begin{aligned} \mathbf{H} &= \begin{bmatrix} \mathbf{O} \\ \mathbf{Q} \end{bmatrix} \begin{bmatrix} \mathbf{R}^{\{-1\}} & \mathbf{O}^{\{-1\}} \end{bmatrix} \\ &= \begin{bmatrix} \mathbf{O}\mathbf{R}^{\{-1\}} & \mathbf{O}\mathbf{O}^{\{-1\}} \\ \mathbf{Q}\mathbf{R}^{\{-1\}} & \mathbf{Q}\mathbf{O}^{\{-1\}} \end{bmatrix} \\ &= \begin{bmatrix} 0 & \mathbf{I} \\ \mathbf{Q}\mathbf{R}^{\{-1\}} & \mathbf{Q}\mathbf{O}^{\{-1\}} \end{bmatrix} \end{aligned} \tag{5.45}$$

Equation (5.45) can be re-formed into the following form:

$$\mathbf{H} = \begin{bmatrix} \mathbf{0}_{N_O \times N_O} & \mathbf{I}_{N_O \times N_O} & \cdots & \mathbf{0}_{N_O \times N_O} & \mathbf{0}_{N_O \times N_O} \\ \mathbf{0}_{N_O \times N_O} & \mathbf{0}_{N_O \times N_O} & \mathbf{I}_{N_O \times N_O} & \cdots & \mathbf{0}_{N_O \times N_O} \\ & & \vdots & & \\ \mathbf{0}_{N_O \times N_O} & \mathbf{0}_{N_O \times N_O} & \mathbf{0}_{N_O \times N_O} & \cdots & \mathbf{I}_{N_O \times N_O} \\ \mathbf{P}_n & \mathbf{P}_{n-1} & \mathbf{P}_{n-2} & \cdots & \mathbf{P}_1 \end{bmatrix} \tag{5.46}$$

where the zero and identity matrices in Equation (5.45) are re-grouped into matrices with smaller dimensions according to the dimension of state vector, and $\mathbf{QR}^{\{-1\}}$ and $\mathbf{QO}^{\{-1\}}$ are re-grouped into matrices $\mathbf{P}_1$ to $\mathbf{P}_n$ with proper dimensions. Substituting Equation (5.46) into Equation (5.33) gives

$$\begin{bmatrix} \mathbf{y}_{i+2} \\ \mathbf{y}_{i+3} \\ \vdots \\ \mathbf{y}_{i+j} \\ \vdots \\ \mathbf{y}_{i+n+1} \end{bmatrix} = \begin{bmatrix} \mathbf{0}_{N_O \times N_O} & \mathbf{I}_{N_O \times N_O} & \cdots & \mathbf{0}_{N_O \times N_O} & \mathbf{0}_{N_O \times N_O} \\ \mathbf{0}_{N_O \times N_O} & \mathbf{0}_{N_O \times N_O} & \mathbf{I}_{N_O \times N_O} & \cdots & \mathbf{0}_{N_O \times N_O} \\ & & \vdots & & \\ \mathbf{0}_{N_O \times N_O} & \mathbf{0}_{N_O \times N_O} & \mathbf{0}_{N_O \times N_O} & \cdots & \mathbf{I}_{N_O \times N_O} \\ \mathbf{P}_n & \mathbf{P}_{n-1} & \mathbf{P}_{n-2} & \cdots & \mathbf{P}_1 \end{bmatrix} \begin{bmatrix} \mathbf{y}_{i+1} \\ \mathbf{y}_{i+2} \\ \vdots \\ \mathbf{y}_{i+j} \\ \vdots \\ \mathbf{y}_{i+n} \end{bmatrix} \tag{5.47}$$

The last row of Equation (5.47) gives

$$\mathbf{y}_{i+n+1} = \begin{bmatrix} \mathbf{P}_n & \mathbf{P}_{n-1} & \cdots & \mathbf{P}_1 \end{bmatrix} \begin{bmatrix} \mathbf{y}_{i+1} \\ \mathbf{y}_{i+2} \\ \vdots \\ \mathbf{y}_{i+n} \end{bmatrix} \tag{5.48}$$

Changing the variable $j = i + 1 + n$ for the subscripts of the states in Equation (5.48) gives

$$\mathbf{y}_j = \begin{bmatrix} \mathbf{P}_n & \mathbf{P}_{n-1} & \cdots & \mathbf{P}_1 \end{bmatrix} \begin{bmatrix} y_{j-n} \\ y_{j-n+1} \\ \vdots \\ y_{j-1} \end{bmatrix} \tag{5.49}$$

This equation is the same VAR model in Equation (5.29). We have shown that a structural state-space model Equation (5.3) is equivalent to a VAR model Equation (5.29). This indicates that if given responses $\{\hat{\mathbf{y}}_j : j = 1, 2, \cdots, N_t\}$ measured at N_t time instances, the parameter matrices $\{\mathbf{P}_i : i = 1, 2, \cdots, n\}$ can be identified and modal parameters can then be extracted from the identified parameter matrices. The detailed derivation will be postponed after we show that a structural state-space model under forced vibration is equivalent to a VARMA model.

5.3 TRANSFORMING STRUCTURAL MODELS TO VARMA MODELS FOR FORCED VIBRATION

Similar to the free-vibration case, an augmented response vector is constructed for the forced-vibration case using Equation (5.27)

$$\mathbf{u}_n = \begin{bmatrix} \mathbf{y}_{i+1} \\ \mathbf{y}_{i+2} \\ \vdots \\ \mathbf{y}_{i+j} \\ \vdots \\ \mathbf{y}_{i+n} \end{bmatrix} = \begin{bmatrix} \mathbf{L\Phi w}_i \\ \mathbf{L\Phi}^2\mathbf{w}_i \\ \vdots \\ \mathbf{L\Phi}^j\mathbf{w}_i \\ \vdots \\ \mathbf{L\Phi}^{n-1}\mathbf{w}_i \\ \mathbf{L\Phi}^n\mathbf{w}_i \end{bmatrix} + \begin{bmatrix} \mathbf{LJ} & \mathbf{0} & \cdots & \mathbf{0} & \mathbf{0} & \mathbf{0} \\ \mathbf{L\Phi J} & \mathbf{LJ} & \cdots & \mathbf{0} & \mathbf{0} & \mathbf{0} \\ & & \vdots & & & \\ \mathbf{L\Phi}^{j-1}\mathbf{J} & \mathbf{L\Phi}^{j-2}\mathbf{J} & \cdots & \mathbf{0} & \mathbf{0} & \mathbf{0} \\ & & \vdots & & & \\ \mathbf{L\Phi}^{n-2}\mathbf{J} & \mathbf{L\Phi}^{n-3}\mathbf{J} & \cdots & \mathbf{L\Phi J} & \mathbf{LJ} & \mathbf{0} \\ \mathbf{L\Phi}^{n-1}\mathbf{J} & \mathbf{L\Phi}^{n-2}\mathbf{J} & \cdots & \mathbf{L\Phi}^2\mathbf{J} & \mathbf{L\Phi J} & \mathbf{LJ} \end{bmatrix} \begin{bmatrix} \mathbf{F}_i \\ \mathbf{F}_{i+1} \\ \vdots \\ \mathbf{F}_{i+j-1} \\ \vdots \\ \mathbf{F}_{i+n-2} \\ \mathbf{F}_{i+n-1} \end{bmatrix}$$

$$= \underbrace{\begin{bmatrix} \mathbf{L} \\ \mathbf{L\Phi} \\ \vdots \\ \mathbf{L\Phi}^{j-1} \\ \vdots \\ \mathbf{L\Phi}^{n-2} \\ \mathbf{L\Phi}^{n-1} \end{bmatrix}}_{\mathbf{G}} \mathbf{\Phi w}_i + \begin{bmatrix} \mathbf{LJ} & \mathbf{0} & \cdots & \mathbf{0} & \mathbf{0} & \mathbf{0} \\ \mathbf{L\Phi J} & \mathbf{LJ} & \cdots & \mathbf{0} & \mathbf{0} & \mathbf{0} \\ & & \vdots & & & \\ \mathbf{L\Phi}^{j-1}\mathbf{J} & \mathbf{L\Phi}^{j-2}\mathbf{J} & \cdots & \mathbf{0} & \mathbf{0} & \mathbf{0} \\ & & \vdots & & & \\ \mathbf{L\Phi}^{n-2}\mathbf{J} & \mathbf{L\Phi}^{n-3}\mathbf{J} & \cdots & \mathbf{L\Phi J} & \mathbf{LJ} & \mathbf{0} \\ \mathbf{L\Phi}^{n-1}\mathbf{J} & \mathbf{L\Phi}^{n-2}\mathbf{J} & \cdots & \mathbf{L\Phi}^2\mathbf{J} & \mathbf{L\Phi J} & \mathbf{LJ} \end{bmatrix} \begin{bmatrix} \mathbf{F}_i \\ \mathbf{F}_{i+1} \\ \vdots \\ \mathbf{F}_{i+j-1} \\ \vdots \\ \mathbf{F}_{i+n-2} \\ \mathbf{F}_{i+n-1} \end{bmatrix} \tag{5.50}$$

By defining two matrices

$$\mathbf{E} = \begin{bmatrix} \mathbf{LJ} & \mathbf{0} & \cdots & \mathbf{0} & \mathbf{0} & \mathbf{0} \\ \mathbf{L\Phi J} & \mathbf{LJ} & \cdots & \mathbf{0} & \mathbf{0} & \mathbf{0} \\ & & \vdots & & & \\ \mathbf{L\Phi}^{j-1}\mathbf{J} & \mathbf{L\Phi}^{j-2}\mathbf{J} & \cdots & \mathbf{0} & \mathbf{0} & \mathbf{0} \\ & & \vdots & & & \\ \mathbf{L\Phi}^{n-2}\mathbf{J} & \mathbf{L\Phi}^{n-3}\mathbf{J} & \cdots & \mathbf{L\Phi J} & \mathbf{LJ} & \mathbf{0} \\ \mathbf{L\Phi}^{n-1}\mathbf{J} & \mathbf{L\Phi}^{n-2}\mathbf{J} & \cdots & \mathbf{L\Phi}^2\mathbf{J} & \mathbf{L\Phi J} & \mathbf{LJ} \end{bmatrix} \tag{5.51}$$

$$\tilde{\mathbf{F}}_{n-1} = \begin{bmatrix} \mathbf{F}_i \\ \mathbf{F}_{i+1} \\ \vdots \\ \mathbf{F}_{i+j-1} \\ \vdots \\ \mathbf{F}_{i+n-2} \\ \mathbf{F}_{i+n-1} \end{bmatrix} \tag{5.52}$$

and using $\mathbf{G}$ defined in Equation (5.30), Equation (5.50) is written as

$$\mathbf{u}_n = \mathbf{G\Phi w}_i + \mathbf{E}\tilde{\mathbf{F}}_{n-1} \tag{5.53}$$

Similarly, shifting one time step forward gives $\mathbf{u}_{n+1}$

$$\mathbf{u}_{n+1} = \begin{bmatrix} \mathbf{y}_{i+2} \\ \mathbf{y}_{i+3} \\ \vdots \\ \mathbf{y}_{i+j+1} \\ \vdots \\ \mathbf{y}_{i+n+1} \end{bmatrix}$$

$$= \begin{bmatrix} \mathbf{L\Phi}^2\mathbf{w}_i \\ \mathbf{L\Phi}^3\mathbf{w}_i \\ \vdots \\ \mathbf{L\Phi}^{j+1}\mathbf{w}_i \\ \vdots \\ \mathbf{L\Phi}^n\mathbf{w}_i \\ \mathbf{L\Phi}^{n+1}\mathbf{w}_i \end{bmatrix} + \begin{bmatrix} \mathbf{L\Phi J} & \mathbf{LJ} & 0 & \cdots & 0 & 0 & 0 \\ \mathbf{L\Phi}^2\mathbf{J} & \mathbf{L\Phi J} & \mathbf{LJ} & \cdots & 0 & 0 & 0 \\ \vdots & & & \vdots & & & \\ \mathbf{L\Phi}^j\mathbf{J} & \mathbf{L\Phi}^{j-1}\mathbf{J} & \mathbf{L\Phi}^{j-2}\mathbf{J} & \cdots & 0 & 0 & 0 \\ \vdots & & & \vdots & & & \\ \mathbf{L\Phi}^{n-1}\mathbf{J} & \mathbf{L\Phi}^{n-2}\mathbf{J} & \mathbf{L\Phi}^{n-3}\mathbf{J} & \cdots & \mathbf{L\Phi J} & \mathbf{LJ} & 0 \\ \mathbf{L\Phi}^n\mathbf{J} & \mathbf{L\Phi}^{n-1}\mathbf{J} & \mathbf{L\Phi}^{n-2}\mathbf{J} & \cdots & \mathbf{L\Phi}^2\mathbf{J} & \mathbf{L\Phi J} & \mathbf{LJ} \end{bmatrix} \begin{bmatrix} \mathbf{F}_i \\ \mathbf{F}_{i+1} \\ \vdots \\ \mathbf{F}_{i+j-1} \\ \vdots \\ \mathbf{F}_{i+n-2} \\ \mathbf{F}_{i+n-1} \\ \mathbf{F}_{i+n} \end{bmatrix}$$

$$= \underbrace{\begin{bmatrix} \mathbf{L} \\ \mathbf{L\Phi} \\ \vdots \\ \mathbf{L\Phi}^{j-1} \\ \vdots \\ \mathbf{L\Phi}^{n-2} \\ \mathbf{L\Phi}^{n-1} \end{bmatrix}}_{\mathbf{G}} \mathbf{\Phi}^2\mathbf{w}_i + \left[\underbrace{\begin{bmatrix} \mathbf{L} \\ \mathbf{L\Phi} \\ \vdots \\ \mathbf{L\Phi}^{j-1} \\ \vdots \\ \mathbf{L\Phi}^{n-2} \\ \mathbf{L\Phi}^{n-1} \end{bmatrix}}_{\mathbf{G}} \mathbf{\Phi J} \;\middle|\; \underbrace{\begin{matrix} \mathbf{LJ} & 0 & \cdots & 0 & 0 & 0 \\ \mathbf{L\Phi J} & \mathbf{LJ} & \cdots & 0 & 0 & 0 \\ & & \vdots & & & \\ \mathbf{L\Phi}^{j-1}\mathbf{J} & \mathbf{L\Phi}^{j-2}\mathbf{J} & \cdots & 0 & 0 & 0 \\ & & \vdots & & & \\ \mathbf{L\Phi}^{n-2}\mathbf{J} & \mathbf{L\Phi}^{n-3}\mathbf{J} & \cdots & \mathbf{L\Phi J} & \mathbf{LJ} & 0 \\ \mathbf{L\Phi}^{n-1}\mathbf{J} & \mathbf{L\Phi}^{n-2}\mathbf{J} & \cdots & \mathbf{L\Phi}^2\mathbf{J} & \mathbf{L\Phi J} & \mathbf{LJ} \end{matrix}}_{\mathbf{E}} \right] \begin{bmatrix} \mathbf{F}_i \\ \mathbf{F}_{i+1} \\ \vdots \\ \mathbf{F}_{i+j-1} \\ \vdots \\ \mathbf{F}_{i+n-2} \\ \mathbf{F}_{i+n-1} \\ \mathbf{F}_{i+n} \end{bmatrix} \tag{5.54}$$

By comparing Equations (5.50) and (5.54), we can identify matrices **G** and **E**. Moreover, according to the definition of Equation (5.52), we can also define the excitation matrix by shifting one time step forward:

$$\tilde{\mathbf{F}}_n = \begin{bmatrix} \mathbf{F}_{i+1} \\ \mathbf{F}_{i+2} \\ \vdots \\ \mathbf{F}_{i+j-1} \\ \vdots \\ \mathbf{F}_{i+n-1} \\ \mathbf{F}_{i+n} \end{bmatrix} \tag{5.55}$$

Re-grouping Equation (5.54) into block matrices for using **G**, **E**, and $\tilde{\mathbf{F}}_n$ gives

$$\begin{aligned}\mathbf{u}_{n+1} &= \mathbf{G\Phi}^2\mathbf{w}_i + \begin{bmatrix}\mathbf{G\Phi J} & \mathbf{E}\end{bmatrix}\begin{bmatrix}\mathbf{F}_i \\ \tilde{\mathbf{F}}_n\end{bmatrix} \\ &= \mathbf{G\Phi I\Phi w}_i + \mathbf{G\Phi JF}_i + \mathbf{E}\tilde{\mathbf{F}}_n \\ &= \mathbf{G\Phi G}^{-1}\mathbf{G\Phi w}_i + \mathbf{G\Phi JF}_i + \mathbf{E}\tilde{\mathbf{F}}_n\end{aligned} \tag{5.56}$$

Substituting Equation (5.34) for $\mathbf{G\Phi G}^{-1}$ and Equation (5.53) for $\mathbf{G\Phi w}_i$ into Equation (5.56) gives

$$\begin{aligned}\mathbf{u}_{n+1} &= \mathbf{H}\left(\mathbf{u}_n - \mathbf{E}\tilde{\mathbf{F}}_{n-1}\right) + \mathbf{G\Phi JF}_i + \mathbf{E}\tilde{\mathbf{F}}_n \\ &= \mathbf{Hu}_n - \mathbf{HE}\tilde{\mathbf{F}}_{n-1} + \mathbf{HGJF}_i + \mathbf{E}\tilde{\mathbf{F}}_n \\ &= \mathbf{Hu}_n - \mathbf{H}\left(\mathbf{E}\tilde{\mathbf{F}}_{n-1} - \mathbf{GJF}_i\right) + \mathbf{E}\tilde{\mathbf{F}}_n\end{aligned} \tag{5.57}$$

where $\mathbf{G\Phi} = \mathbf{HG}$ by Equation (5.34) has been used to reach the second equality. In order to calculate $\mathbf{E}\tilde{\mathbf{F}}_{n-1} - \mathbf{GJF}_i$ in the above equation, we want to construct $\tilde{\mathbf{F}}_{n-1}$ in $\mathbf{GJF}_i$. To do this, augmenting $\mathbf{F}_i$ with the terms from $\mathbf{F}_{i+1}$ to $\mathbf{F}_{i+n-1}$ transforms $\mathbf{GJF}_i$ to

$$\mathbf{GJF}_i = \begin{bmatrix}\mathbf{L} \\ \mathbf{L\Phi} \\ \vdots \\ \mathbf{L\Phi}^{j-1} \\ \vdots \\ \mathbf{L\Phi}^{n-2} \\ \mathbf{L\Phi}^{n-1}\end{bmatrix}\mathbf{JF}_i = \begin{bmatrix}\mathbf{LJ} & 0 & \cdots & 0 \\ \mathbf{L\Phi J} & 0 & \cdots & 0 \\ & \vdots & & \\ \mathbf{L\Phi}^{j-1}\mathbf{J} & 0 & \cdots & 0 \\ & \vdots & & \\ \mathbf{L\Phi}^{n-2}\mathbf{J} & 0 & \cdots & 0 \\ \mathbf{L\Phi}^{n-1}\mathbf{J} & 0 & \cdots & 0\end{bmatrix}\begin{bmatrix}\mathbf{F}_i \\ \mathbf{F}_{i+1} \\ \vdots \\ \mathbf{F}_{i+j-1} \\ \vdots \\ \mathbf{F}_{i+n-2} \\ \mathbf{F}_{i+n-1}\end{bmatrix} \tag{5.58}$$

With Equation (5.58), we have

$$\mathbf{E}\tilde{\mathbf{F}}_{n-1} - \mathbf{GJF}_i = \begin{bmatrix}\mathbf{LJ} & 0 & \cdots & 0 & 0 & 0 \\ \mathbf{L\Phi J} & \mathbf{LJ} & \cdots & 0 & 0 & 0 \\ & & \vdots & & & \\ \mathbf{L\Phi}^{j-1}\mathbf{J} & \mathbf{L\Phi}^{j-2}\mathbf{J} & \cdots & 0 & 0 & 0 \\ & & \vdots & & & \\ \mathbf{L\Phi}^{n-2}\mathbf{J} & \mathbf{L\Phi}^{n-3}\mathbf{J} & \cdots & \mathbf{L\Phi J} & \mathbf{LJ} & 0 \\ \mathbf{L\Phi}^{n-1}\mathbf{J} & \mathbf{L\Phi}^{n-2}\mathbf{J} & \cdots & \mathbf{L\Phi}^2\mathbf{J} & \mathbf{L\Phi J} & \mathbf{LJ}\end{bmatrix}\begin{bmatrix}\mathbf{F}_i \\ \mathbf{F}_{i+1} \\ \vdots \\ \mathbf{F}_{i+j-1} \\ \vdots \\ \mathbf{F}_{i+n-2} \\ \mathbf{F}_{j+n-1}\end{bmatrix} - \begin{bmatrix}\mathbf{LJ} & 0 & \cdots & 0 \\ \mathbf{L\Phi J} & 0 & \cdots & 0 \\ & \vdots & & \\ \mathbf{L\Phi}^{j-1}\mathbf{J} & 0 & \cdots & 0 \\ & \vdots & & \\ \mathbf{L\Phi}^{n-2}\mathbf{J} & 0 & \cdots & 0 \\ \mathbf{L\Phi}^{n-1}\mathbf{J} & 0 & \cdots & 0\end{bmatrix}\begin{bmatrix}\mathbf{F}_i \\ \mathbf{F}_{i+1} \\ \vdots \\ \mathbf{F}_{i+j-1} \\ \vdots \\ \mathbf{F}_{i+n-2} \\ \mathbf{F}_{i+n-1}\end{bmatrix}$$

$$
= \begin{bmatrix} 0 & 0 & \cdots & 0 & 0 & 0 \\ 0 & \mathbf{LJ} & \cdots & 0 & 0 & 0 \\ & & \vdots & & & \\ 0 & \mathbf{L\Phi}^{j-2}\mathbf{J} & \cdots & 0 & 0 & 0 \\ & & \vdots & & & \\ 0 & \mathbf{L\Phi}^{n-3}\mathbf{J} & \cdots & \mathbf{L\Phi J} & \mathbf{LJ} & 0 \\ 0 & \mathbf{L\Phi}^{n-2}\mathbf{J} & \cdots & \mathbf{L\Phi}^2\mathbf{J} & \mathbf{L\Phi J} & \mathbf{LJ} \end{bmatrix} \begin{bmatrix} \mathbf{F}_i \\ \mathbf{F}_{i+1} \\ \vdots \\ \mathbf{F}_{i+j-1} \\ \vdots \\ \mathbf{F}_{i+n-2} \\ \mathbf{F}_{i+n-1} \end{bmatrix} \tag{5.59}
$$

By deleting the first zero column of the left matrix and $\mathbf{F}_i$ and by adding a zero column at the end of the left matrix and $\mathbf{F}_{i+n}$ at the bottom of the right matrix, Equation (5.59) becomes

$$
\mathbf{E}\tilde{\mathbf{F}}_{n-1} - \mathbf{GJF}_i = \begin{bmatrix} 0 & 0 & \cdots & 0 & 0 & 0 \\ \mathbf{LJ} & 0 & \cdots & 0 & 0 & 0 \\ & & \vdots & & & \\ \mathbf{L\Phi}^{j-2}\mathbf{J} & \mathbf{L\Phi}^{j-3}\mathbf{J} & \cdots & 0 & 0 & 0 \\ & & \vdots & & & \\ \mathbf{L\Phi}^{n-3}\mathbf{J} & \mathbf{L\Phi}^{n-4}\mathbf{J} & \cdots & \mathbf{LJ} & 0 & 0 \\ \mathbf{L\Phi}^{n-2}\mathbf{J} & \mathbf{L\Phi}^{n-3}\mathbf{J} & \cdots & \mathbf{L\Phi J} & \mathbf{LJ} & 0 \end{bmatrix} \begin{bmatrix} \mathbf{F}_{i+1} \\ \mathbf{F}_{i+2} \\ \vdots \\ \mathbf{F}_{i+j-1} \\ \vdots \\ \mathbf{F}_{i+n-1} \\ \mathbf{F}_{i+n} \end{bmatrix} \tag{5.60}
$$

Therefore, the term $\mathbf{H}\left(\mathbf{E}\tilde{\mathbf{F}}_{n-1} - \mathbf{GJF}_i\right)$ in Equation (5.57) is obtained

$$
\mathbf{H}\left(\mathbf{E}\tilde{\mathbf{F}}_{n-1} - \mathbf{GJF}_i\right)
$$

$$
= \begin{bmatrix} 0 & \mathbf{I} & \cdots & 0 & 0 \\ 0 & 0 & \mathbf{I} & \cdots & 0 \\ & & \vdots & & \\ 0 & 0 & 0 & \cdots & \mathbf{I} \\ \mathbf{P}_n & \mathbf{P}_{n-1} & \mathbf{P}_{n-2} & \cdots & \mathbf{P}_1 \end{bmatrix} \begin{bmatrix} 0 & 0 & \cdots & 0 & 0 & 0 \\ \mathbf{LJ} & 0 & \cdots & 0 & 0 & 0 \\ & & \vdots & & & \\ \mathbf{L\Phi}^{j-2}\mathbf{J} & \mathbf{L\Phi}^{j-3}\mathbf{J} & \cdots & 0 & 0 & 0 \\ & & \vdots & & & \\ \mathbf{L\Phi}^{n-3}\mathbf{J} & \mathbf{L\Phi}^{n-4}\mathbf{J} & \cdots & \mathbf{LJ} & 0 & 0 \\ \mathbf{L\Phi}^{n-2}\mathbf{J} & \mathbf{L\Phi}^{n-3}\mathbf{J} & \cdots & \mathbf{L\Phi J} & \mathbf{LJ} & 0 \end{bmatrix}
$$

$$
= \begin{bmatrix} \mathbf{LJ} & 0 & \cdots & 0 & 0 & 0 \\ \mathbf{L\Phi J} & \mathbf{LJ} & \cdots & 0 & 0 & 0 \\ & & \vdots & & & \\ \mathbf{L\Phi}^{j-2}\mathbf{J} & \mathbf{L\Phi}^{j-3}\mathbf{J} & \cdots & 0 & 0 & 0 \\ & & \vdots & & & \\ \mathbf{L\Phi}^{n-2}\mathbf{J} & \mathbf{L\Phi}^{n-3}\mathbf{J} & \cdots & \mathbf{L\Phi J} & \mathbf{LJ} & 0 \\ \sum_{j=1}^{n-1} \mathbf{P}_j\mathbf{L\Phi}^{n-j-1}\mathbf{J} & \sum_{j=1}^{n-2} \mathbf{P}_j\mathbf{L\Phi}^{n-j-2}\mathbf{J} & \cdots & \sum_{j=1}^{2} \mathbf{P}_j\mathbf{L\Phi}^{2-j}\mathbf{J} & \mathbf{P}_1\mathbf{LJ} & 0 \end{bmatrix} \underbrace{\begin{bmatrix} \mathbf{F}_{i+1} \\ \mathbf{F}_{i+2} \\ \vdots \\ \mathbf{F}_{i+j-1} \\ \vdots \\ \mathbf{F}_{i+n-1} \\ \mathbf{F}_{i+n} \end{bmatrix}}_{\tilde{\mathbf{F}}_n} \tag{5.61}
$$

With Equation (5.61) and the definition of $\mathbf{E}$ in Equation (5.51), the last two terms of Equation (5.57) are combined:

$$-\mathbf{H}\left(\mathbf{E}\tilde{\mathbf{F}}_{n-1}-\mathbf{FJF}_i\right)+\mathbf{E}\tilde{\mathbf{F}}_n$$

$$=\left(\begin{bmatrix} \mathbf{LJ} & \mathbf{0} & \cdots & \mathbf{0} & \mathbf{0} & \mathbf{0} \\ \mathbf{L\Phi J} & \mathbf{LJ} & \cdots & \mathbf{0} & \mathbf{0} & \mathbf{0} \\ \vdots & \vdots & \vdots & \vdots & \vdots & \vdots \\ \mathbf{L\Phi}^{j-1}\mathbf{J} & \mathbf{L\Phi}^{j-2}\mathbf{J} & \cdots & \mathbf{0} & \mathbf{0} & \mathbf{0} \\ \vdots & \vdots & \vdots & \vdots & \vdots & \vdots \\ \mathbf{L\Phi}^{n-2}\mathbf{J} & \mathbf{L\Phi}^{n-3}\mathbf{J} & \cdots & \mathbf{L\Phi J} & \mathbf{LJ} & \mathbf{0} \\ \mathbf{L\Phi}^{n-1}\mathbf{J} & \mathbf{L\Phi}^{n-2}\mathbf{J} & \cdots & \mathbf{L\Phi}^2\mathbf{J} & \mathbf{L\Phi J} & \mathbf{LJ} \end{bmatrix}\right.$$

$$\left.-\begin{bmatrix} \mathbf{LJ} & \mathbf{0} & \cdots & \mathbf{0} & \mathbf{0} & \mathbf{0} \\ \mathbf{L\Phi J} & \mathbf{LJ} & \cdots & \mathbf{0} & \mathbf{0} & \mathbf{0} \\ \vdots & \vdots & \vdots & \vdots & \vdots & \vdots \\ \mathbf{L\Phi}^{j-1}\mathbf{J} & \mathbf{L\Phi}^{j-2}\mathbf{J} & \cdots & \mathbf{0} & \mathbf{0} & \mathbf{0} \\ \vdots & \vdots & \vdots & \vdots & \vdots & \vdots \\ \mathbf{L\Phi}^{n-2}\mathbf{J} & \mathbf{L\Phi}^{n-3}\mathbf{J} & \cdots & \mathbf{L\Phi J} & \mathbf{LJ} & \mathbf{0} \\ \sum_{j=1}^{n-1}\mathbf{P}_j\mathbf{L\Phi}^{n-j-1}\mathbf{J} & \sum_{j=1}^{n-2}\mathbf{P}_j\mathbf{L\Phi}^{n-j-2}\mathbf{J} & \cdots & \sum_{j=1}^{2}\mathbf{P}_j\mathbf{L\Phi}^{2-j}\mathbf{J} & \mathbf{P}_1\mathbf{LJ} & \mathbf{0} \end{bmatrix}\right)\begin{bmatrix} \mathbf{F}_{i+1} \\ \mathbf{F}_{i+2} \\ \vdots \\ \mathbf{F}_{i+j-1} \\ \vdots \\ \mathbf{F}_{i+n-1} \\ \mathbf{F}_{i+n} \end{bmatrix}$$

$$=\begin{bmatrix} \mathbf{0} & \mathbf{0} & \cdots & \mathbf{0} & \mathbf{0} & \mathbf{0} \\ \mathbf{0} & \mathbf{0} & \cdots & \mathbf{0} & \mathbf{0} & \mathbf{0} \\ \vdots & \vdots & \vdots & \vdots & \vdots & \vdots \\ \mathbf{0} & \mathbf{0} & \cdots & \mathbf{0} & \mathbf{0} & \mathbf{0} \\ \vdots & \vdots & \vdots & \vdots & \vdots & \vdots \\ \mathbf{0} & \mathbf{0} & \cdots & \mathbf{0} & \mathbf{0} & \mathbf{0} \\ \mathbf{L\Phi}^{n-1}\mathbf{J}-\sum_{j=1}^{n-1}\mathbf{P}_j\mathbf{L\Phi}^{n-j-1}\mathbf{J} & \mathbf{L\Phi}^{n-2}\mathbf{J}-\sum_{j=1}^{n-2}\mathbf{P}_j\mathbf{L\Phi}^{n-j-2}\mathbf{J} & \cdots & \mathbf{L\Phi}^2\mathbf{J}-\sum_{j=1}^{2}\mathbf{P}_j\mathbf{L\Phi}^{2-j}\mathbf{J} & \mathbf{L\Phi J}-\mathbf{P}_1\mathbf{LJ} & \mathbf{LJ} \end{bmatrix} \tag{5.62}$$

It can be seen that Equation (5.62) is obtained as the product of two matrices. Only the last block row of the first matrix is non-zero; other blocks are zero. Using matrices $\{\mathbf{D}_i : i = 1,2,\cdots,n\}$ to represent each of the non-zero blocks and substituting Equation (5.62) in (5.57) gives

$$\begin{bmatrix} \mathbf{y}_{i+2} \\ \mathbf{y}_{i+3} \\ \vdots \\ \mathbf{y}_{i+j+1} \\ \vdots \\ \mathbf{y}_{i+n} \\ \mathbf{y}_{i+n+1} \end{bmatrix} = \begin{bmatrix} \mathbf{0} & \mathbf{I} & \cdots & \mathbf{0} & \mathbf{0} & \mathbf{0} \\ \mathbf{0} & \mathbf{0} & \mathbf{I} \cdots & \mathbf{0} & \mathbf{0} & \mathbf{0} \\ \vdots & \vdots & \vdots & \vdots & \vdots & \vdots \\ \mathbf{0} & \mathbf{0} & \cdots & \mathbf{0} & \mathbf{0} & \mathbf{0} \\ \vdots & \vdots & \vdots & \vdots & \vdots & \vdots \\ \mathbf{0} & \mathbf{0} & \cdots & \mathbf{0} & \mathbf{0} & \mathbf{I} \\ \mathbf{P}_n & \mathbf{P}_{n-1} & \cdots & \mathbf{P}_3 & \mathbf{P}_2 & \mathbf{P}_1 \end{bmatrix}\begin{bmatrix} \mathbf{y}_{i+1} \\ \mathbf{y}_{i+2} \\ \vdots \\ \mathbf{y}_{i+j} \\ \vdots \\ \mathbf{y}_{i+n-1} \\ \mathbf{y}_{i+n} \end{bmatrix}$$

$$+\begin{bmatrix} 0 & 0 & \cdots & 0 & 0 & 0 \\ 0 & 0 & \cdots & 0 & 0 & 0 \\ \vdots & \vdots & \vdots & \vdots & \vdots & \vdots \\ 0 & 0 & \cdots & 0 & 0 & 0 \\ \vdots & \vdots & \vdots & \vdots & \vdots & \vdots \\ 0 & 0 & \cdots & 0 & 0 & 0 \\ \mathbf{D}_n & \mathbf{D}_{n-1} & \cdots & \mathbf{D}_3 & \mathbf{D}_2 & \mathbf{D}_1 \end{bmatrix} \begin{bmatrix} \mathbf{F}_{i+1} \\ \mathbf{F}_{i+2} \\ \vdots \\ \mathbf{F}_{i+j} \\ \vdots \\ \mathbf{F}_{i+n-1} \\ \mathbf{F}_{i+n} \end{bmatrix} \tag{5.63}$$

Expanding the last block row of the above equation gives

$$\mathbf{y}_{i+n+1} = \begin{bmatrix} \mathbf{P}_n & \mathbf{P}_{n-1} & \cdots & \mathbf{P}_1 \end{bmatrix} \begin{bmatrix} \mathbf{y}_{i+1} \\ \mathbf{y}_{i+2} \\ \vdots \\ \mathbf{y}_{i+n} \end{bmatrix} + \begin{bmatrix} \mathbf{D}_n & \mathbf{D}_{n-1} & \cdots & \mathbf{D}_1 \end{bmatrix} \begin{bmatrix} \mathbf{F}_{i+1} \\ \mathbf{F}_{i+2} \\ \vdots \\ \mathbf{F}_{i+n} \end{bmatrix} \tag{5.64}$$

Changing the variable $j = i + 1 + n$ for the subscripts of the states in the above equation gives the VARMA model for forced vibration:

$$y_j = \begin{bmatrix} \mathbf{P}_n & \mathbf{P}_{n-1} & \cdots & \mathbf{P}_1 \end{bmatrix} \begin{bmatrix} y_{j-n} \\ y_{j-n+1} \\ \vdots \\ y_{j-1} \end{bmatrix} + \begin{bmatrix} \mathbf{D}_n & \mathbf{D}_{n-1} & \cdots & \mathbf{D}_1 \end{bmatrix} \begin{bmatrix} \mathbf{F}_{j-n} \\ \mathbf{F}_{j-n+1} \\ \vdots \\ \mathbf{F}_{j-1} \end{bmatrix} \tag{5.65}$$

For ambient vibration, excitations can be assumed to be Gaussian, so the excitation terms of Equation (5.65) may be lumped into one Gaussian term ε_j. A structural system under ambient excitation can be also described by a VAR model:

$$y_j = \begin{bmatrix} \mathbf{P}_n & \mathbf{P}_{n-1} & \cdots & \mathbf{P}_1 \end{bmatrix} \begin{bmatrix} y_{j-n} \\ y_{j-n+1} \\ \vdots \\ y_{j-1} \end{bmatrix} + \varepsilon_j \tag{5.66}$$

5.4 EXTRACTING MODAL PARAMETERS FROM A VARMA MODEL

Modal parameters are obtained from the eigenvalue problem of the second-order differential equation of a structural model, so we need to solve the eigenvalue problem of a VARMA model and see how the eigenvalues and eigenvectors are related to the modal parameters of a structural model. First, we study the eigenvalue problem, or modal analysis, of a state-space model Equation (5.3). To do this, consider the free-vibration problem and set the excitation term to zero:

$$\dot{\mathbf{w}}(t) = \mathbf{A}\mathbf{w}(t) \tag{5.67}$$

The solution of this homogeneous equation takes the form that separates the time and spatial effects:

$$\mathbf{w}(t) = e^{\mu t}\boldsymbol{\psi} \tag{5.68}$$

where both μ and $\boldsymbol{\psi}$ are constant. Substituting Equation (5.68) into Equation (5.67) gives

$$\begin{aligned} e^{\mu t}\mu\boldsymbol{\psi} &= e^{\mu t}\mathbf{A}\boldsymbol{\psi} \Rightarrow \\ \mathbf{A}\boldsymbol{\psi} &= \mu\boldsymbol{\psi} \end{aligned} \tag{5.69}$$

which is the eigenvalue problem for the state-space model with μ and $\boldsymbol{\psi} \in R^{2N_d}$ being the eigenvalue and eigenvector, respectively. Because $\mathbf{A}$ is not symmetric, the eigenvectors do not have the orthogonal property with respect to $\mathbf{A}$, and μ and $\boldsymbol{\psi}$ can be real or complex. There will be $2N_d$ eigenvalues or eigenvectors. Because $\mathbf{A}$ is a real matrix, the eigenvalues or eigenvectors will come in complex conjugate pairs, i.e., if μ_m and $\boldsymbol{\psi}_m$ are the eigenvalue and eigenvector, $\bar{\mu}_m$ and $\bar{\boldsymbol{\psi}}_m$ will also be the eigenvalue and eigenvector, where the over bar $\bar{\square}$ denotes complex conjugate. Therefore, there are N_d pairs of eigenvalues or eigenvectors for a state-space model, $\{\mu_m, \bar{\mu}_m; \boldsymbol{\psi}_m, \bar{\boldsymbol{\psi}}_m : m = 1,2,\cdots,N_d\}$.

Next, consider the classically damped system $\mathbf{M}\ddot{\mathbf{x}}(t) + \mathbf{C}\dot{\mathbf{x}}(t) + \mathbf{K}\mathbf{x}(t) = 0$ corresponding to the state-space model. Its solution has the same form as Equation (5.68):

$$\mathbf{x}(t) = e^{\mu t}\boldsymbol{\phi} \tag{5.70}$$

Substituting Equation (5.70) into the differential equation gives

$$\begin{aligned} \mathbf{M}\mu^2 e^{\mu t}\boldsymbol{\phi} + \mathbf{C}\mu e^{\mu t}\boldsymbol{\phi} + \mathbf{K}e^{\mu t}\boldsymbol{\phi} &= 0 \Rightarrow \\ \left(\mathbf{M}\mu^2 + \mathbf{C}\mu + \mathbf{K}\right)\boldsymbol{\phi} &= 0 \end{aligned} \tag{5.71}$$

Because the system has a classical damping matrix, the eigenvalue problem Equation (5.71) gives the mode shapes $\boldsymbol{\phi} \in R^{N_d}$ that are independent of $\mathbf{C}$ and depend only on $\mathbf{M}$ and $\mathbf{K}$; i.e., the eigenvalue problem Equation (5.71) depends only on $\mathbf{M}$ and $\mathbf{K}$. Note that both the second-order differential equation and its corresponding state-space model represent the same dynamic system, so Equations (5.68) and (5.70) have the same eigenvalues that contain natural frequencies and damping ratios:

$$\mu_m = -\xi_m\omega_m \pm i\omega_m\sqrt{1-\xi_m^2}, m = 1,2,\cdots,N_d \tag{5.72}$$

but the eigenvectors $\boldsymbol{\psi}$ and $\boldsymbol{\phi}$ are different. In the following, it is shown how the eigenvectors of a state-space model are connected to mode shapes. By dividing $\boldsymbol{\psi}$ into two vectors $\boldsymbol{\psi} = \begin{bmatrix} \boldsymbol{\psi}^t \\ \boldsymbol{\psi}^b \end{bmatrix}$ according to the block form of $\mathbf{A}$, Equation (5.69) is expanded as

$$\begin{bmatrix} \mathbf{0}_{N_d \times N_d} & \mathbf{I}_{N_d \times N_d} \\ -\mathbf{M}^{-1}\mathbf{K} & -\mathbf{M}^{-1}\mathbf{C} \end{bmatrix} \begin{bmatrix} \boldsymbol{\psi}^t \\ \boldsymbol{\psi}^b \end{bmatrix} = \mu \begin{bmatrix} \boldsymbol{\psi}^t \\ \boldsymbol{\psi}^b \end{bmatrix} \tag{5.73}$$

The first row gives

$$\boldsymbol{\psi}^b = \mu\boldsymbol{\psi}^t \tag{5.74}$$

The second row gives

$$\mu \mathbf{M}\psi^b + \mathbf{C}\psi^b + \mathbf{K}\psi^t = 0 \tag{5.75}$$

Substituting Equation (5.74) into Equation (5.75) gives

$$\left(\mathbf{M}\mu^2 + \mathbf{C}\mu + \mathbf{K}\right)\psi^t = 0 \tag{5.76}$$

By comparing Equations (5.71) and (5.76), it is known that

$$\psi^t = \phi \tag{5.77}$$

Therefore, the eigenvectors of a state-space model can be written in terms of the mode shapes of the same dynamic system:

$$\psi = \begin{bmatrix} \phi \\ \hline \mu\phi \end{bmatrix} \tag{5.78}$$

According to the derivation of VAR and VARMA models for free and forced vibration cases, it is known that the dynamics of a structural system is contained in the AR matrix **H** (see Equation (5.46)). The structure of the eigenvalue problem of **H** is thus investigated based on Equation (5.34)

$$\mathbf{HG} = \mathbf{G\Phi} \tag{5.79}$$

It is seen that the above equation has a form similar to the eigenvalue problem. A diagonal matrix will be constructed to transform this equation to the eigenvalue problem. Substituting Equation (5.31) into the above equation gives

$$\mathbf{H}\begin{bmatrix} \mathbf{L} \\ \mathbf{L\Phi} \\ \vdots \\ \mathbf{L\Phi}^{j-1} \\ \vdots \\ \mathbf{L\Phi}^{n-1} \end{bmatrix} = \begin{bmatrix} \mathbf{L} \\ \mathbf{L\Phi} \\ \vdots \\ \mathbf{L\Phi}^{j-1} \\ \vdots \\ \mathbf{L\Phi}^{n-1} \end{bmatrix}\mathbf{\Phi} \tag{5.80}$$

Equation (5.80) contains the powers of the transition matrix, $\mathbf{\Phi}$, so we study the eigenvalue problem of $\mathbf{\Phi}$ first. By expressing **A** with its eigenvalues and eigenvectors

$$\mathbf{A} = \Psi\Lambda\Psi^{-1} \tag{5.81}$$

where

$$\Lambda = \begin{bmatrix} \mu_1 & & & & \\ & \ddots & & & \\ & & \mu_m & & \\ & & & \ddots & \\ & & & & \mu_{2N_d} \end{bmatrix} \tag{5.82}$$

and substituting Equation (5.81) into the expansion of $\mathbf{\Phi}$ in Equation (5.7), we have

$$\begin{aligned}
\mathbf{\Phi} &= \mathbf{I}_{2N_d\times 2N_d} + t\mathbf{\Psi\Lambda\Psi}^{-1} + \frac{t^2}{2!}\mathbf{\Psi\Lambda}\underbrace{\mathbf{\Psi}^{-1}\mathbf{\Psi}}_{\mathbf{I}}\mathbf{\Lambda\Psi}^{-1} + \frac{t^3}{3!}\mathbf{\Psi\Lambda}\underbrace{\mathbf{\Psi}^{-1}\mathbf{\Psi}}_{\mathbf{I}}\mathbf{\Lambda}\underbrace{\mathbf{\Psi}^{-1}\mathbf{\Psi}}_{\mathbf{I}}\mathbf{\Lambda\Psi}^{-1} + \cdots \\
&= \mathbf{I}_{2N_d\times 2N_d} + t\mathbf{\Psi\Lambda\Psi}^{-1} + \frac{t^2}{2!}\mathbf{\Psi\Lambda}^2\mathbf{\Psi}^{-1} + \frac{t^3}{3!}\mathbf{\Psi\Lambda}^3\mathbf{\Psi}^{-1} + \cdots \\
&= \mathbf{\Psi}\left(\mathbf{I}_{2N_d\times 2N_d} + t\mathbf{\Lambda} + \frac{t^2}{2!}\mathbf{\Lambda}^2 + \frac{t^3}{3!}\mathbf{\Lambda}^3 + \cdots\right)\mathbf{\Psi}^{-1} \\
&= \mathbf{\Psi}\begin{bmatrix} \sum_{i=1}^{\infty}\frac{t^{i-1}\mu_1^{i-1}}{(i-1)!} & & & & \\ & \ddots & & & \\ & & \sum_{i=1}^{\infty}\frac{t^{i-1}\mu_m^{i-1}}{(i-1)!} & & \\ & & & \ddots & \\ & & & & \sum_{i=1}^{\infty}\frac{t^{i-1}\mu_{2N_d}^{i-1}}{(i-1)!} \end{bmatrix}\mathbf{\Psi}^{-1}
\end{aligned} \tag{5.83}$$

Note that the series sums in the above equation are

$$\lambda_m = \sum_{i=1}^{\infty}\frac{t^{i-1}\mu_m^{i-1}}{(i-1)!} = e^{\mu_m t} \tag{5.84}$$

where λ_m is known as the eigenvalue of the discrete-time system, which is obtained with t equal to the sampling time Δt in practice; and μ_m is known as the eigenvalue of the continuous-time system. As a result, Equation (5.83) can be expressed as

$$\begin{aligned}
&\mathbf{\Phi} = e^{\mathbf{A}t} = \mathbf{\Psi}e^{\mathbf{\Lambda}t}\mathbf{\Psi}^{-1} \Rightarrow \\
&\mathbf{\Phi\Psi} = \mathbf{\Psi}e^{\mathbf{\Lambda}t}
\end{aligned} \tag{5.85}$$

where $e^{\mathbf{\Lambda}t}$ is a diagonal matrix

$$\mathbf{\Upsilon} = e^{\mathbf{\Lambda}t} = \begin{bmatrix} \lambda_1 & & & & \\ & \ddots & & & \\ & & \lambda_m & & \\ & & & \ddots & \\ & & & & \lambda_{2N_d} \end{bmatrix} \tag{5.86}$$

Equation (5.85) indicates that $\mathbf{\Phi}$ has the same eigenvectors as $\mathbf{A}$, and the eigenvalues of $\mathbf{\Phi}$ are the exponential functions based on the eigenvalues of $\mathbf{A}$, i.e., $\left\{\lambda_m = e^{\mu_m t} : m = 1,2,\cdots,2N_d\right\}$.

By using a similar derivation, the j-th power of the transition matrix can be expressed as

$$
\begin{aligned}
\Phi^j = e^{j\mathbf{A}t} &= \mathbf{I} + t\Psi(j\Lambda)\Psi^{-1} + \frac{t^2}{2!}\Psi(j\Lambda)^2\Psi^{-1} + \frac{t^3}{3!}\Psi(j\Lambda)^3\Psi^{-1} + \cdots \\
&= \Psi e^{j\Lambda t}\Psi^{-1} \\
&= \Psi\Upsilon^j\Psi^{-1}
\end{aligned}
\tag{5.87}
$$

Equation (5.87) can be rewritten as

$$
\Phi^j\Psi = \Psi\Upsilon^j \tag{5.88}
$$

Because Υ^j is diagonal, Equation (5.88) is the eigenvalue problem of the j-th power of the transition matrix. This equation indicates that $\mathbf{\Phi}^j$ has the same eigenvectors as $\mathbf{A}$, and its eigenvalues are $\left\{\lambda_m^j : m = 1,2,\cdots,2N_d\right\}$.

Having obtained the structure of the eigenvalue problem of the transition matrix, we can construct the eigenvalue problem for Equation (5.80). Post-multiplying the eigenvectors $\mathbf{\Psi}$ to both sides of Equation (5.80) gives

$$
\mathbf{H}\begin{bmatrix} \mathbf{L}\Psi \\ \mathbf{L}\Phi\Psi \\ \vdots \\ \mathbf{L}\Phi^{j-1}\Psi \\ \vdots \\ \mathbf{L}\Phi^{n-1}\Psi \end{bmatrix} = \begin{bmatrix} \mathbf{L} \\ \mathbf{L}\Phi \\ \vdots \\ \mathbf{L}\Phi^{j-1} \\ \vdots \\ \mathbf{L}\Phi^{n-1} \end{bmatrix}\Phi\Psi \tag{5.89}
$$

Applying Equation (5.88) in the above equation gives

$$
\mathbf{H}\begin{bmatrix} \mathbf{L}\Psi \\ \mathbf{L}\Psi\Upsilon \\ \vdots \\ \mathbf{L}\Psi\Upsilon^{j-1} \\ \vdots \\ \mathbf{L}\Psi\Upsilon^{n-1} \end{bmatrix} = \begin{bmatrix} \mathbf{L} \\ \mathbf{L}\Phi \\ \vdots \\ \mathbf{L}\Phi^{j-1} \\ \vdots \\ \mathbf{L}\Phi^{n-1} \end{bmatrix}\Psi\Upsilon \Rightarrow \mathbf{H}\begin{bmatrix} \mathbf{L}\Psi \\ \mathbf{L}\Psi\Upsilon \\ \vdots \\ \mathbf{L}\Psi\Upsilon^{j-1} \\ \vdots \\ \mathbf{L}\Psi\Upsilon^{n-1} \end{bmatrix} = \begin{bmatrix} \mathbf{L}\Psi \\ \mathbf{L}\Psi\Upsilon \\ \vdots \\ \mathbf{L}\Psi\Upsilon^{j-1} \\ \vdots \\ \mathbf{L}\Psi\Upsilon^{n-1} \end{bmatrix}\Upsilon \tag{5.90}
$$

The eigenvalue problem of $\mathbf{H}$ has been established by Equation (5.90). The m-th eigenvalue and eigenvector of $\mathbf{H}$ are

$$
\lambda_m \text{ and } \varphi_m = \begin{bmatrix} \mathbf{L}\psi_m \\ \lambda_m\mathbf{L}\psi_m \\ \vdots \\ \lambda_m^{j-1}\mathbf{L}\psi_m \\ \vdots \\ \lambda_m^{n-1}\mathbf{L}\psi_m \end{bmatrix} \tag{5.91}
$$

With Equations (5.72), (5.84), and (5.91), natural frequencies, damping ratios, and mode shapes can be extracted from a VAR model. Note that although the previous equations are derived for velocities and displacements, these equations also apply for accelerations, because accelerations follow the same equations of motion as velocities and displacements. In practice, given measured accelerations, the procedures for identifying modal parameters from a VAR model are as follows:

(1) The AR matrices $\{\mathbf{P}_j\}$ are first identified based on Equation (5.66).
(2) $\mathbf{H}$ is then constructed using the identified $\{\mathbf{P}_j\}$.
(3) The eigenvalue problem of $\mathbf{H}$ is solved to get the eigenvalues of the discrete-time system λ_m and the eigenvectors $\boldsymbol{\varphi}_m$.
(4) To obtain natural frequencies and damping ratios, the eigenvalue of the continuous-time system is calculated using Equation (5.84)

$$\mu_m = \frac{\ln \lambda_m}{\Delta t} \tag{5.92}$$

where the sampling time Δt has been used instead of t. Based on Equation (5.72) of μ_m, natural frequencies and damping ratios are then obtained

$$\omega_m = |\mu_m|, \xi_m = -\frac{\mu_m^R}{|\mu_m|} \tag{5.93}$$

where $|\mu_m| = \sqrt{\left(\mu_m^R\right)^2 + \left(\mu_m^I\right)^2}$ is the norm of the complex number μ_m; μ_m^R and μ_m^I are the real and imaginary parts of μ_m, respectively.
(5) According to Equation (5.78), the first half of $\boldsymbol{\psi}_m$ is the mode shape $\boldsymbol{\phi}_m$. If the first N_o degrees of freedom are measured, then the first N_o components of $\boldsymbol{\varphi}_m$ constitute the mode shape $\boldsymbol{\phi}_m$.

5.5 IDENTIFICATION OF VAR MODELS

Having established the connection between a VAR model and a structural model, the next step is to derive the parameter matrices of a VAR model given measured data. The derivation will give the optimal AR parameter matrices in step (1) of the above procedures. The method is based on Yang and Lam (2019). The response at the j-th time step can be predicted by a VAR model with order n (Equation (5.66))

$$\mathbf{y}_j = \mathbf{T}\hat{\mathbf{z}}_j + \boldsymbol{\varepsilon}_j \tag{5.94}$$

where $\mathbf{T}$ consists of the n VAR matrices to be identified

$$\mathbf{T} = \begin{bmatrix} \mathbf{P}_1 & \mathbf{P}_2 & \cdots & \mathbf{P}_n \end{bmatrix} \tag{5.95}$$

$\hat{\mathbf{z}}_j \in R^{nN_o}$ is treated as the data vector at time step j that contains responses of n previous time steps before time step j

$$\hat{\mathbf{z}}_j = \begin{bmatrix} \hat{\mathbf{y}}_{j-1} \\ \hat{\mathbf{y}}_{j-2} \\ \vdots \\ \hat{\mathbf{y}}_{j-n} \end{bmatrix} \tag{5.96}$$

where "$\hat{}$" denotes that the considered quantities are measured data; ε_j is treated as the difference between the model-predicted and measured response at time step j. By defining the response matrix $\mathbf{Y} = [\mathbf{y}_1, \mathbf{y}_2, \cdots, \mathbf{y}_{N_t}] \in R^{N_o \times N_t}$ that contains the model-predicted responses at all time steps, the data matrix $\hat{\mathbf{Z}} = [\hat{\mathbf{z}}_1, \hat{\mathbf{z}}_2, \cdots, \hat{\mathbf{z}}_{N_t}] \in R^{(nN_o) \times N_t}$ that contains the data vectors at all time steps, and the difference matrix $\mathbf{E} = [\varepsilon_1, \varepsilon_2, \cdots, \varepsilon_{N_t}] \in R^{N_o \times N_t}$ that contains the differences at all time steps, the model-predicted responses at different time steps can be written compactly based on Equation (5.94)

$$\mathbf{Y} = \mathbf{T}\hat{\mathbf{Z}} + \mathbf{E} \tag{5.97}$$

To obtain the optimal value of **T**, the sum of square differences at all time steps needs to be minimized. Instead of considering the difference vectors in the matrix **E**, it will be mathematically convenient if they are arranged in a long vector. This can be done by using the *vec* operator, which stacks the columns of a matrix on top of each other. The details can be found in the Appendix. Applying the *vec* operator for Equation (5.97) transforms the matrix equation into a vector equation

$$\begin{aligned} vec(\mathbf{Y}) &= vec\left(\mathbf{T}\hat{\mathbf{Z}}\right) + vec(\mathbf{E}) \Rightarrow \\ \mathbf{y} &= \left(\hat{\mathbf{Z}}^T \otimes \mathbf{I}_{N_o \times N_o}\right)\boldsymbol{\theta} + \mathbf{e} \end{aligned} \tag{5.98}$$

where the following definitions have been used

$$\mathbf{y} = vec(\mathbf{Y}) \tag{5.99}$$

$$\boldsymbol{\theta} = vec(\mathbf{T}) \tag{5.100}$$

$$\mathbf{e} = vec(\mathbf{E}) \tag{5.101}$$

and Equation (A.4) in the Appendix has been used for $vec\left(\mathbf{T}\hat{\mathbf{Z}}\right)$. Measured responses can be represented using Equation (5.98)

$$\hat{\mathbf{y}} = \left(\hat{\mathbf{Z}}^T \otimes \mathbf{I}_{N_o \times N_o}\right)\boldsymbol{\theta} + \mathbf{e} \tag{5.102}$$

where $\hat{\mathbf{y}}$ has the same form as Equation (5.99) with measured responses substituted. In order to obtain the optimal $\boldsymbol{\theta}$, an objective function that denotes the sum of squared differences is formulated

$$J = \mathbf{e}^T\mathbf{e} = \left(\hat{\mathbf{y}} - \left(\hat{\mathbf{Z}}^T \otimes \mathbf{I}_{N_o \times N_o}\right)\boldsymbol{\theta}\right)^T \left(\hat{\mathbf{y}} - \left(\hat{\mathbf{Z}}^T \otimes \mathbf{I}_{N_o \times N_o}\right)\boldsymbol{\theta}\right) \tag{5.103}$$

The optimal value of the uncertain parameter vector is then obtained by minimizing the objective function, or the sum of squared differences between the measured and model-predicted responses. This is done by calculating the first derivative of the objective function with respect to $\boldsymbol{\theta}$

$$\begin{aligned}\frac{\partial J}{\partial \boldsymbol{\theta}} &= 2\mathbf{e}^T \frac{\partial \mathbf{e}}{\partial \boldsymbol{\theta}} \\ &= -2\left(\hat{\mathbf{y}} - \left(\hat{\mathbf{Z}}^T \otimes \mathbf{I}_{N_o \times N_o}\right)\boldsymbol{\theta}\right)^T \left(\hat{\mathbf{Z}}^T \otimes \mathbf{I}_{N_o \times N_o}\right)\end{aligned} \tag{5.104}$$

Setting the first derivative to zero gives

$$\begin{aligned}\boldsymbol{\theta}^T \left(\hat{\mathbf{Z}}^T \otimes \mathbf{I}_{N_o \times N_o}\right)^T \left(\hat{\mathbf{Z}}^T \otimes \mathbf{I}_{N_o \times N_o}\right) &= \hat{\mathbf{y}}^T \left(\hat{\mathbf{Z}}^T \otimes \mathbf{I}_{N_o \times N_o}\right) \Rightarrow \\ \left(\left(\hat{\mathbf{Z}}^T \otimes \mathbf{I}_{N_o \times N_o}\right)^T \left(\hat{\mathbf{Z}}^T \otimes \mathbf{I}_{N_o \times N_o}\right)\right)^T \boldsymbol{\theta} &= \left(\hat{\mathbf{Z}}^T \otimes \mathbf{I}_{N_o \times N_o}\right)^T \hat{\mathbf{y}} \Rightarrow \\ \left(\hat{\mathbf{Z}} \otimes \mathbf{I}_{N_o \times N_o}\right)\left(\hat{\mathbf{Z}}^T \otimes \mathbf{I}_{N_o \times N_o}\right)\boldsymbol{\theta} &= \left(\hat{\mathbf{Z}} \otimes \mathbf{I}_{N_o \times N_o}\right)\hat{\mathbf{y}} \Rightarrow \\ \left(\hat{\mathbf{Z}}\hat{\mathbf{Z}}^T \otimes \mathbf{I}_{N_o \times N_o}\right)\boldsymbol{\theta} &= \left(\hat{\mathbf{Z}} \otimes \mathbf{I}_{N_o \times N_o}\right)\hat{\mathbf{y}}\end{aligned} \tag{5.105}$$

where Equation (A.6) has been used for the transpose of the Kronecker product to reach the third line and Equation (A.7) has been used for the left-hand side of the third line, i.e., the product of two Kronecker products. Based on Equation (5.105), the optimal uncertain parameter vector $\hat{\boldsymbol{\theta}}$ is obtained

$$\begin{aligned}\hat{\boldsymbol{\theta}} &= \left(\hat{\mathbf{Z}}\hat{\mathbf{Z}}^T \otimes \mathbf{I}_{N_o \times N_o}\right)^{-1} \left(\hat{\mathbf{Z}} \otimes \mathbf{I}_{N_o \times N_o}\right)\hat{\mathbf{y}} \Rightarrow \\ &= \left(\left(\hat{\mathbf{Z}}\hat{\mathbf{Z}}^T\right)^{-1} \otimes \mathbf{I}_{N_o \times N_o}\right)\left(\hat{\mathbf{Z}} \otimes \mathbf{I}_{N_o \times N_o}\right)\hat{\mathbf{y}} \Rightarrow \\ &= \left(\left(\left(\hat{\mathbf{Z}}\hat{\mathbf{Z}}^T\right)^{-1} \hat{\mathbf{Z}}\right) \otimes \mathbf{I}_{N_o \times N_o}\right)\hat{\mathbf{y}}\end{aligned} \tag{5.106}$$

where Equation (A.8) has been used for the inverse of the Kronecker product to reach the second line.

Although transforming the matrix form to the vector form makes the derivation convenient, the intermediate calculation of the Kronecker product for obtaining $\hat{\boldsymbol{\theta}}$ usually exhausts computer memory quickly, in practice, because many degrees of freedom (N_o) and a huge number of data points (N_t) are usually involved for full-scale structures, and the dimension of the matrix from the Kronecker product is huge. To see this, note that the number of elements of the Kronecker product of $\hat{\boldsymbol{\theta}}$ in Equation (5.106) is $n \times N_o \times N_t \times N_o^2 = nN_tN_o^3$. For a typical field test, if the order of the VAR model is $n = 20$, $N_o = 20$ degrees of freedom are measured, and 10 minutes of data with a sampling frequency of 128 Hz are used for analysis so that $N_t = 10 \times 60 \times 128 = 76800$, the number of elements will be 1.2288×10^{10}. The computer memory required for storing these elements is large. For an efficient implementation, $\hat{\theta}$ is transformed back to its matrix form to obtain the optimal AR parameter matrix $\hat{\mathbf{T}}$. First, introducing the vectorization operator *vec* in Equation (5.106) with the definitions of $\hat{\mathbf{y}}$ (Equation (5.99)) and $\hat{\theta}$ (Equation (5.100)) gives

$$vec\left(\hat{\mathbf{T}}\right) = \left(\left(\left(\hat{\mathbf{Z}}\hat{\mathbf{Z}}^T\right)^{-1}\hat{\mathbf{Z}}\right) \otimes \mathbf{I}_{N_o \times N_o}\right) vec\left(\hat{\mathbf{Y}}\right) \tag{5.107}$$

If the Kronecker product of the above equation can be put into the *vec* operator, the transformation will be done. This can be achieved by noticing that the right-hand side of the above equation has the same form as that of Equation (A.4), so applying Equation (A.4) for Equation (5.107) gives

$$vec\left(\hat{\mathbf{T}}\right) = vec\left(\hat{\mathbf{Y}}\left(\left(\hat{\mathbf{Z}}\hat{\mathbf{Z}}^T\right)^{-1}\hat{\mathbf{Z}}\right)^T\right) = vec\left(\hat{\mathbf{Y}}\hat{\mathbf{Z}}^T\left(\hat{\mathbf{Z}}\hat{\mathbf{Z}}^T\right)^{-1}\right) \tag{5.108}$$

The optimal AR parameter matrix is thus obtained

$$\begin{aligned}
\hat{\mathbf{T}} &= \hat{\mathbf{Y}}\hat{\mathbf{Z}}^T\left(\hat{\mathbf{Z}}\hat{\mathbf{Z}}^T\right)^{-1} \\
&= \left[\hat{\mathbf{y}}_1, \hat{\mathbf{y}}_2, \cdots, \hat{\mathbf{y}}_{N_t}\right]\begin{bmatrix}\hat{\mathbf{z}}_1^T \\ \hat{\mathbf{z}}_2^T \\ \vdots \\ \hat{\mathbf{z}}_{N_t}^T\end{bmatrix}\left(\left[\hat{\mathbf{z}}_1, \hat{\mathbf{z}}_2, \cdots, \hat{\mathbf{z}}_{N_t}\right]\begin{bmatrix}\hat{\mathbf{z}}_1^T \\ \hat{\mathbf{z}}_2^T \\ \vdots \\ \hat{\mathbf{z}}_{N_t}^T\end{bmatrix}\right)^{-1} \\
&= \sum_j \hat{\mathbf{y}}_j\hat{\mathbf{z}}_j^T\left(\sum_j \hat{\mathbf{z}}_j\hat{\mathbf{z}}_j^T\right)^{-1}
\end{aligned} \tag{5.109}$$

5.6 ILLUSTRATIVE EXAMPLE

The four-story shear building in Chapter 3 was used to illustrate the system identification method based on VAR models. Four sensors were installed on the building to measure accelerations of each floor, respectively (see Chapter 3). The accelerations were measured with a sampling frequency of 2048 Hz, and downsampled to 64 Hz for analysis. Five minutes of data were used in this case. Following the procedures at the end of Section 5.4, the first step is to identify the optimal AR parameter matrices $\left\{\hat{\mathbf{P}}_j \in R^{4\times4}\right\}$. The model order needs to be selected first. A practical criterion for selecting an appropriate VAR model order is that the modes produced by a VAR model can match with the spectral peaks of interest in PSD spectra (Yang & Lam 2019). The model order was chosen to be n = 20 for the shear building. It indicates that there are 20 AR parameter matrices for this case. $\left\{\hat{\mathbf{z}}_j \in R^{80} : j = 1,2,\cdots,N_t\right\}$ were then constructed using Equation (5.96) given measured data $\left\{\hat{\mathbf{y}}_j \in R^4 : j = 1,2,\cdots,N_t\right\}$. The optimal AR matrix $\hat{\mathbf{T}}$ was obtained

Table 5.1 The First Six Optimal AR Parameter Matrices

$\hat{\mathbf{P}}_1$				$\hat{\mathbf{P}}_2$			
−0.6664	−0.0127	0.0057	−0.0190	0.1065	−0.0168	0.0114	−0.0073
−0.0460	−0.8417	−0.0012	−0.0002	−0.0088	0.2310	−0.0501	−0.0442
−0.0118	0.0338	−0.7505	0.0824	−0.0112	−0.0409	0.1241	−0.0841
0.0205	−0.0117	−0.0010	−0.7980	−0.0020	−0.0107	−0.0866	−0.0216
$\hat{\mathbf{P}}_3$				$\hat{\mathbf{P}}_4$			
−0.0265	0.0084	−0.0041	0.0211	0.0482	0.0151	−0.0129	0.0136
0.0156	−0.2261	−0.0068	0.0594	0.0190	0.0777	−0.0303	−0.0316
0.0037	−0.0081	−0.0528	−0.0271	0.0229	−0.0656	0.1060	−0.0824
−0.0023	0.0192	−0.0455	−0.0941	−0.0106	−0.0301	−0.0595	0.0826
$\hat{\mathbf{P}}_5$				$\hat{\mathbf{P}}_6$			
−0.1739	0.0223	−0.0106	0.0143	−0.0584	0.0052	−0.0043	−0.0187
0.0257	0.0041	0.0018	0.0181	0.0267	0.0712	−0.0166	−0.0118
−0.0124	−0.0007	−0.0533	−0.0051	0.0064	−0.0350	−0.0046	−0.0397
0.0029	0.0182	0.0079	−0.0236	−0.0005	−0.0365	0.0038	0.0698

using Equation (5.109). $\left\{\hat{\mathbf{P}}_i \in R^{4\times4}\right\}$ were then obtained according to the partition of $\hat{\mathbf{T}}$ in Equation (5.95). The first six optimal AR parameter matrices are shown in Table 5.1 for illustration purposes. To check whether the identified VAR model can predict the measured accelerations, the predicted accelerations were calculated using Equation (5.94) without considering the differences, and compared with the measured ones in Figure 5.1. Figure 5.1(a) shows the comparison for the whole set of measured data and Figure 5.1(b) shows the detailed view for the segments between 100 and 105. It can be seen that the accelerations predicted by the identified VAR model fit the measured acceleration well.

Next, the optimal AR parameter matrices were used to construct **H** according to Equation (5.46). After solving the eigenvalue problem of **H**, the natural frequencies and damping ratios of the shear building were calculated based on the eigenvalues of **H** using Equations (5.92) and (5.93). Each mode shape was obtained by the first four components of each eigenvector of **H**. The natural frequencies, damping ratios, and mode shapes of four modes of the shear building are summarized in Figure 5.2. The natural frequencies of the four modes are 1.39 Hz, 4.08 Hz, 6.03 Hz, and 10.11 Hz, respectively. The damping ratios of this shear building are all smaller than 1%. The mode shapes look like sine waves with increasing wave numbers from the lowest mode to the highest mode.

5.7 SYSTEM IDENTIFICATION OF AN OFFICE BUILDING

The VAR–model-based method was then demonstrated on an office building (see Figure 5.3). This building is approximately 97.00 m high and its floor slab measures about 55.00 m×38.84 m. The building vibration is significant due to the daily excitation from building users and the yearly typhoon excitation; therefore, it is important to identify the dynamic properties of this building for a better understanding of the structural behaviors and further assessing structural performance. The vibration test was conducted for the 19th floor to measure accelerations of the building for identifying the VAR model. Modal

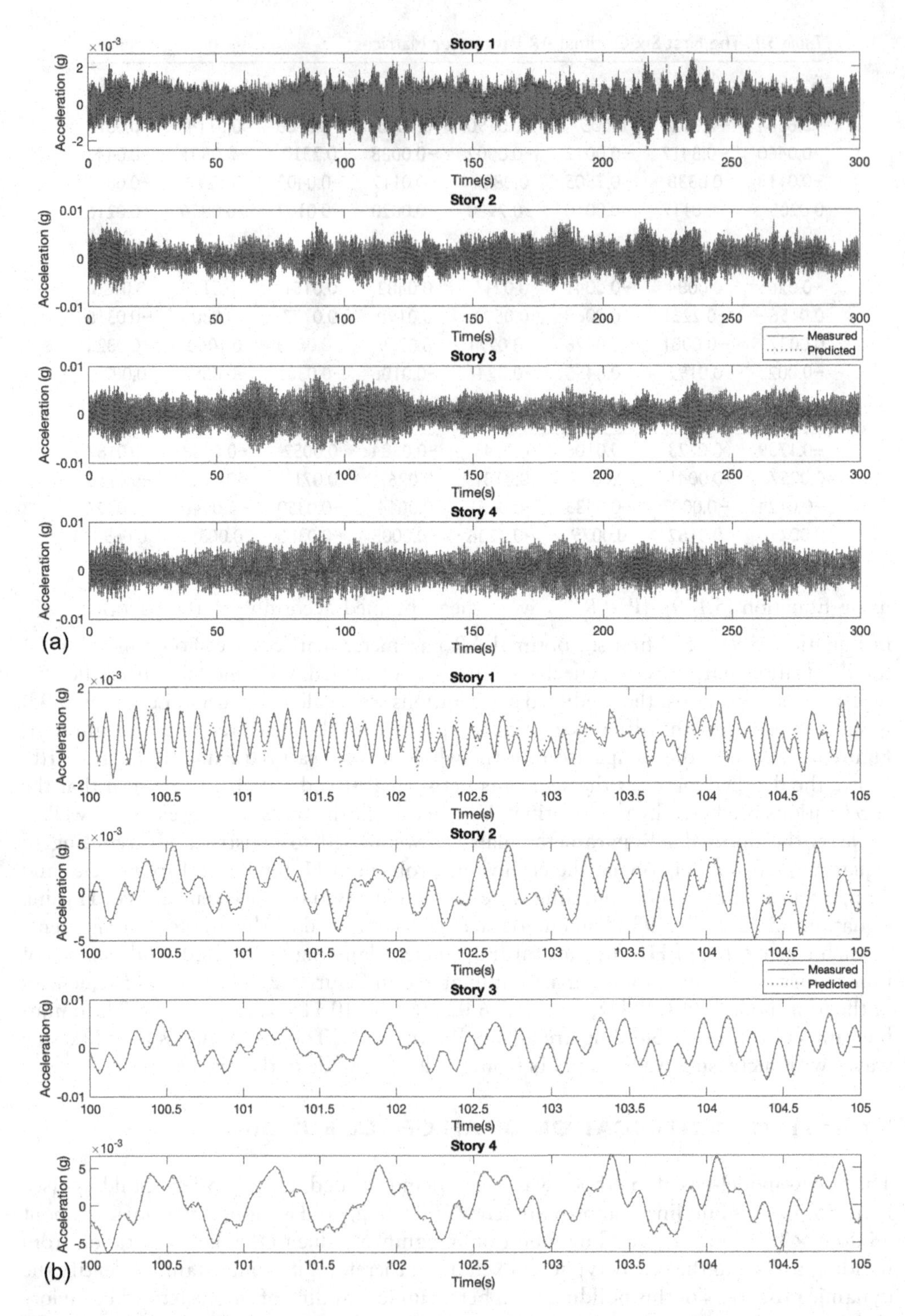

Figure 5.1 Comparison of measured and model-predicted accelerations: (a) the whole set of data; (b) detailed view.

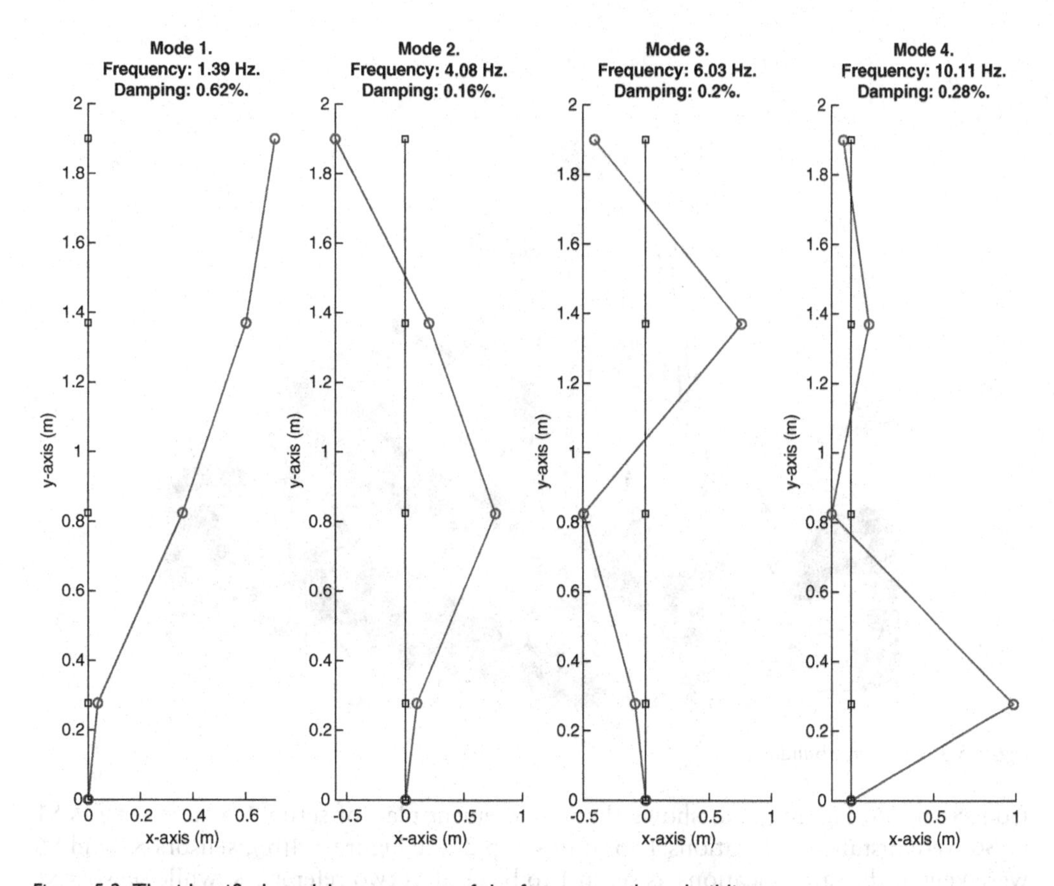

Figure 5.2 The identified modal parameters of the four-story shear building.

parameters of the building were extracted from the identified VAR model. Note that the proposed method does not require excitation information for system identification, and only responses are needed. This is an important advantage for practical applications. This is because usually excitations of full-scale structures are impossible to measure, e.g., excitations of building users, wind, and traffic. Artificially generated excitations can be measured, but the amplitude of artificial excitation must be very large to produce responses large enough for system identification. Nevertheless, these large excitations are not desirable because they may damage structures. System identification methods requiring excitation information are thus not suitable for full-scale structures.

5.7.1 Ambient vibration test

Figure 5.4 shows the measurement plan of the 19th floor. In this field test, the sensors were installed in the middle and around the perimeter of the floor slab. Ten locations were to be measured in total, but only six sensors were available, so the measurement was divided into two setups. In each setup, only part of ten locations were measured. Between the two setups, common locations were measured so that mode shapes identified from the two setups could be assembled into global mode shapes. In Figure 5.4, the measured locations were denoted by triangles and numbered from 1 to 10. The sensors were numbered

Figure 5.3 The office building.

from S1 to S6. Figure 5.4(a) shows the measurement plan of setup 1, where sensors S1 to S6 were installed at locations 1 to 6. In setup 2 (see Figure 5.4(b)), sensors S5 and S6 were kept at the same locations as setup 1 to be used as two references, while sensors S1 to S4 were moved to install at locations 7 to 10, respectively. With the mode shapes at locations 5 and 6 identified at both setups, the partial mode shapes of the two setups can be assembled into global mode shapes.

Figure 5.5 shows the equipment used in the vibration test. The six triaxial accelerometers can measure accelerations along EW, NS, and vertical directions at the same time. Measured data are transferred through cables into the portable measurement box, which contains A/D cards and a power supply. The laptop is also connected to the measurement box for monitoring and storing measured data. In each setup, measurement was taken for 5 minutes with the sampling frequency 2048 Hz. The data were then downsampled to 64 Hz for system identification.

5.7.2 Identifying the VAR model and modal parameters

Based on the measured accelerations, the VAR model of this floor system was identified using Equation (5.109). Only one identified VAR parameter matrix is shown in Table 5.2 for illustration purposes. The building accelerations were reconstructed using the identified VAR model and compared with the measured ones. Figure 5.6 compares the measured and model-predicted accelerations along EW, NS, and vertical directions at location 1. The predicted accelerations by the identified VAR model match the measured data well. The identified matrix $\hat{\mathbf{T}}$ was used to construct $\mathbf{H}$ (Equation (5.46)). By solving

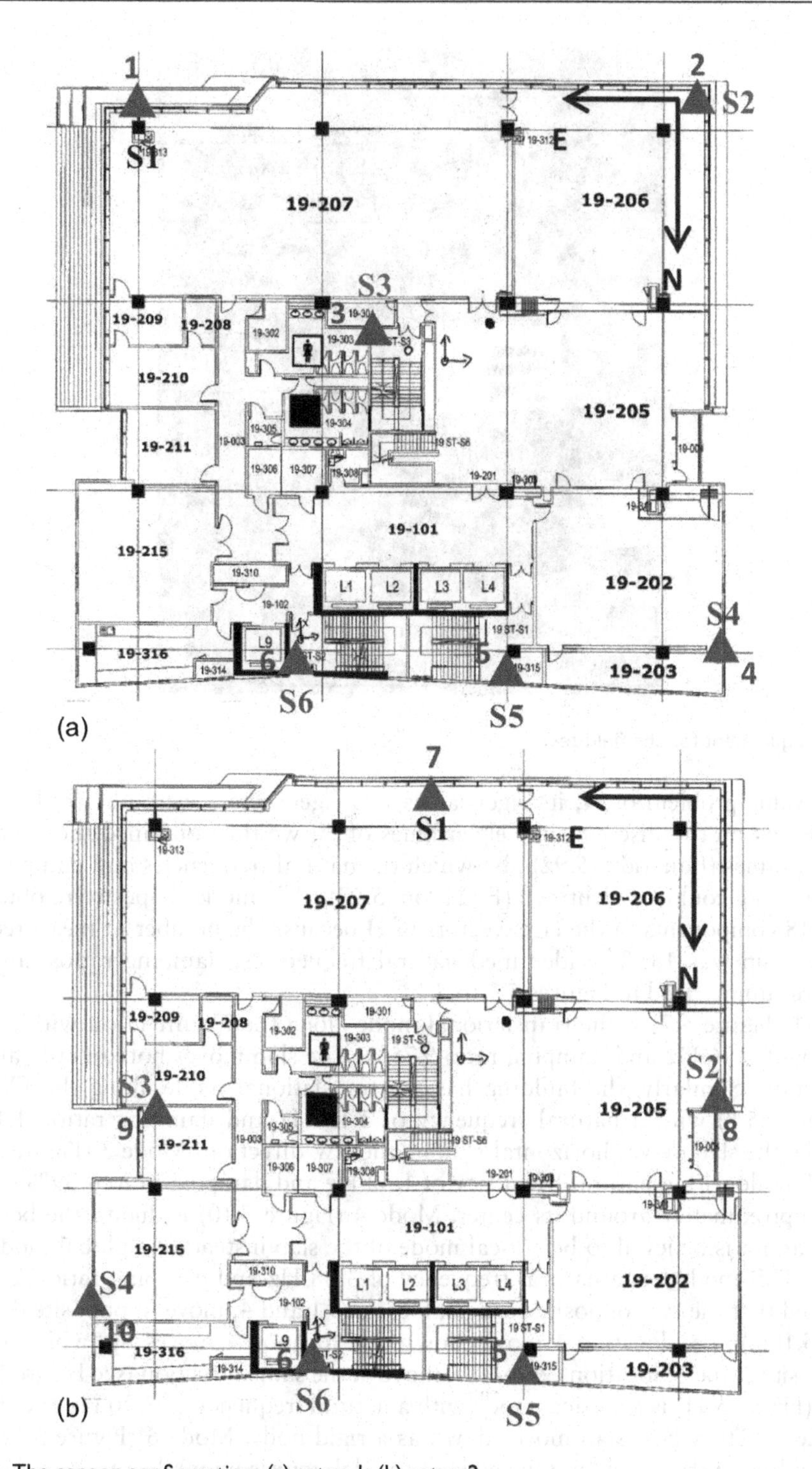

Figure 5.4 The sensor configuration: (a) setup 1; (b) setup 2.

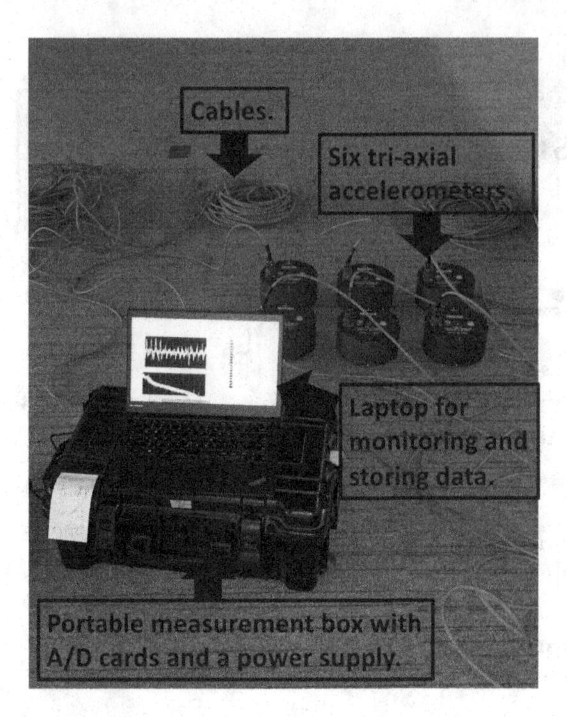

Figure 5.5 Equipment for the field test.

the eigenvalue problem of H, its eigenvalues and eigenvectors were obtained (Equation (5.91)). Based on the discrete-time eigenvalues of H, we then obtained the continuous-time eigenvalues (Equation (5.92)), by which the natural frequencies and damping ratios of this building could be identified (Equation (5.93)). The mode shapes were obtained as the first 18 components of the eigenvectors of **H** because the number of measured DOFs for each setup was 18. The identified natural frequencies, damping ratios, and mode shapes are summarized in Figures 5.7 to 5.12.

Mode 1 (Figure 5.7) is the translational mode along the NS direction with a natural frequency of 0.96 Hz and damping ratio 9.75%. The slab moves horizontally along the NS direction. Similarly, the building has the translational mode along the EW direction (Figure 5.8) with a natural frequency of 1.12 Hz and damping ratio 11.11%. In this mode, the slab moves horizontally along the EW direction. Mode 3 (Figure 5.9) is a torsional mode with a natural frequency of 1.52 Hz and damping ratio 6.39%. The slab rotates approximately around its center. Mode 4 (Figure 5.10) is due to the bending of the slab, and it is believed to be a local mode of the slab instead of a global mode of the building. This mode has a natural frequency of 3.38 Hz and damping ratio 12.66%. It is observed that the two opposite corners, locations 1 and 4, move in opposite directions (Figure 5.10(a)); i.e., location 1 moves down and location 4 moves up, while the other two opposite corners, locations 2 and 10, move in the same direction (see Figure 5.10(b)). Mode 5 (Figure 5.11) is a vertical mode with a natural frequency of 5.00 Hz and damping ratio 1.26%. The whole slab moves down as a rigid body. Mode 6 (Figure 5.12) is due to the bending of the building. One edge of the slab moves up while another edge moves down. It has a natural frequency of 6.37 Hz and damping ratio 3.68%.

Table 5.2 The Second Optimal VAR Parameter Matrix $\hat{\boldsymbol{P}}_2$

Columns 1 to 9								
1.9136	−0.0145	−0.0192	0.2072	−0.0332	−0.0142	−1.4348	−0.0529	0.0747
0.2871	1.7442	−0.0080	0.0140	0.0867	−0.0073	−0.9130	−0.2676	0.0641
−0.1093	−0.0120	1.4733	0.2930	0.2609	−0.0402	−0.3379	0.3592	−0.2392
0.4099	−0.0292	−0.0092	1.6607	−0.1440	−0.0107	−1.4235	−0.1294	0.0586
−0.0709	0.0008	0.0123	−0.2862	1.4887	0.0248	1.2089	0.1875	0.0168
−0.5593	−0.2807	−0.0691	2.3295	2.2893	1.2257	−2.0768	−0.6021	−0.2178
0.5617	−0.1368	−0.0143	0.2094	0.0548	−0.0091	−0.0247	0.0786	0.0498
−0.0732	0.2757	0.0094	0.0378	0.1752	−0.0117	0.0033	0.5278	0.0605
−0.0499	−0.1352	−0.0037	0.1371	0.0713	−0.0282	−0.2830	0.1951	0.9175
0.1478	0.1798	0.0189	−0.0984	−0.1083	0.0127	−0.0917	0.3405	0.0574
0.0234	0.0955	0.0171	−0.1433	0.0707	0.0096	0.8192	0.1582	0.0336
−0.2175	−0.0531	−0.0057	0.1174	−0.0954	−0.0008	0.0917	−0.2362	−0.0749
0.1426	0.1907	0.0208	−0.0893	−0.1292	0.0128	−0.1165	0.4192	0.0700
−0.1275	0.0099	0.0210	0.0176	0.2234	−0.0181	0.7442	−0.4230	0.0219
−0.2006	0.1981	0.0285	0.0315	−0.0473	0.0081	0.0680	0.1620	−0.1011
0.0226	−0.0461	0.0349	0.0175	−0.0505	−0.0033	−0.0537	0.0897	0.0179
0.0383	0.2747	−0.0049	−0.0092	0.0901	−0.0087	−0.3603	−0.0676	0.0214
0.0435	−0.0161	0.0132	0.1433	0.1025	−0.0068	−0.1428	0.2295	−0.0481
Columns 10 to 18								
−0.4929	0.1367	0.0205	0.4878	0.7866	0.0102	0.3846	−1.1765	−0.0108
−0.4546	−0.0663	0.0930	0.7347	0.1967	−0.0134	0.5696	−0.9507	−0.0285
−0.3754	−0.0041	0.0745	0.4237	−0.1627	0.3985	0.0475	−0.1327	−0.3216
−0.5371	0.1036	0.0272	0.5245	0.9937	−0.0045	0.4464	−1.2207	0.0090
0.2593	0.1197	−0.1453	−0.5313	−0.5329	−0.0567	−0.5350	−0.0849	−0.0314
−0.7991	−0.7191	−0.0217	0.5522	−0.9917	0.1116	0.4646	0.1644	0.2919
−0.1432	0.0150	−0.1185	0.1396	0.7232	−0.0122	−0.0246	−0.9467	0.0211
−0.0653	−0.1220	0.0124	0.1103	0.0808	−0.0822	0.1267	−0.1841	0.0249
0.0001	−0.2701	−0.0710	0.0712	0.0269	0.0302	0.1298	0.0207	−0.0327
0.5999	0.1606	−0.1613	0.5912	−0.2316	0.0646	−0.5619	−0.2020	−0.0207
0.2200	1.5506	−0.0791	−0.4407	−0.7601	0.0424	−0.5302	−0.1640	−0.1200
0.7927	0.4432	1.2445	−1.4305	−0.3404	−0.9340	0.3038	0.0876	0.4786
0.2549	0.2733	−0.1537	0.8337	−0.3872	0.0491	−0.5351	−0.1870	−0.0091
0.0812	−0.0102	−0.0212	−0.2812	1.1087	−0.0392	−0.3766	−0.3360	0.0065
0.3533	0.2036	−0.1403	−0.3635	−0.1867	1.2352	−0.2003	−0.1961	−0.2668
0.0403	−0.0141	−0.1227	0.0202	0.0033	0.0993	0.5851	0.1512	−0.1139
−0.0866	−0.1894	−0.0044	0.1219	−0.0636	−0.0758	0.4086	0.5217	0.0583
0.2101	0.0430	−0.0387	0.1080	−0.2704	−0.3012	−0.1552	0.0683	1.3094

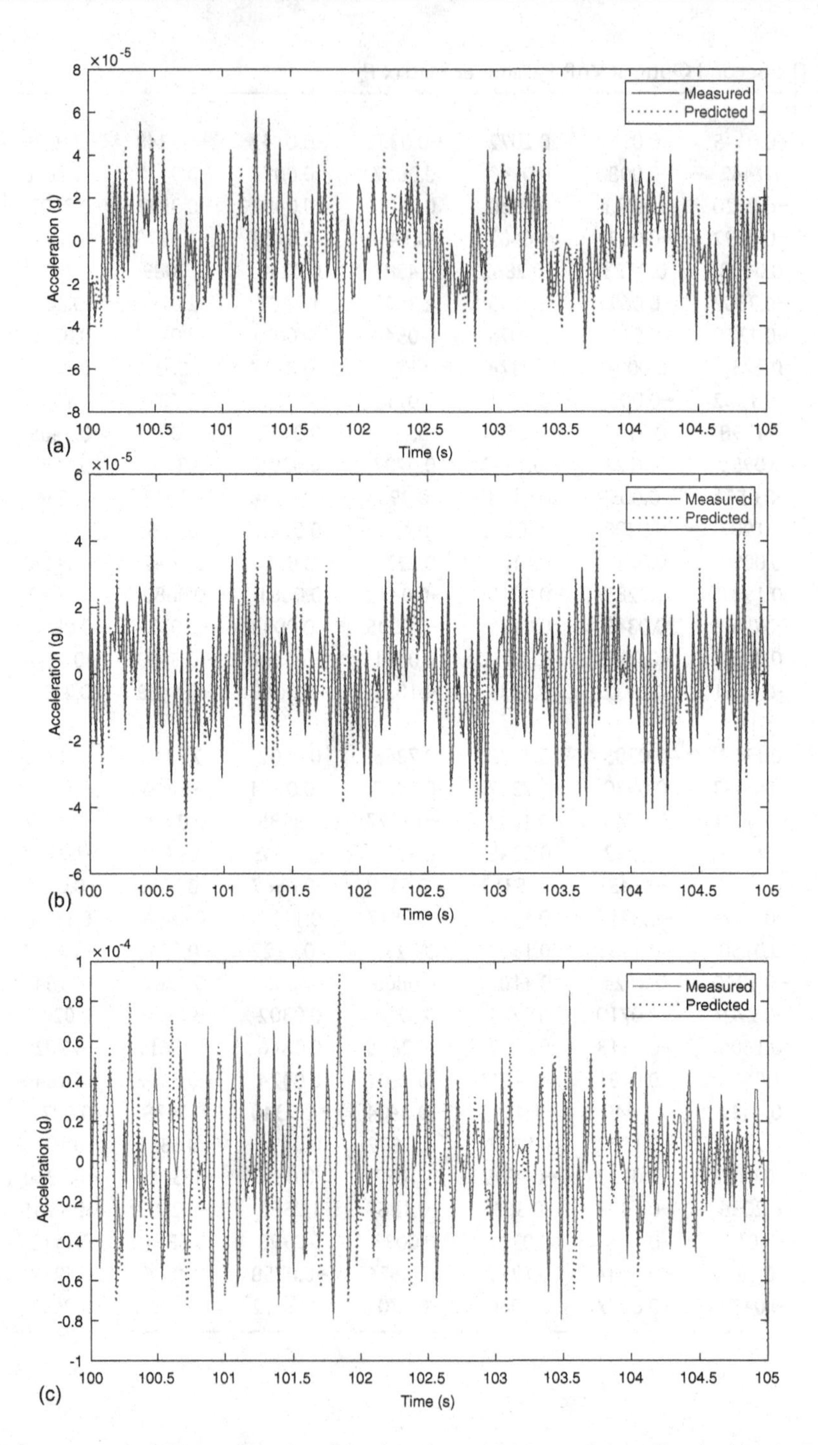

Figure 5.6 Comparison of the measured and predicted accelerations of the office building at location 1: (a) EW direction; (b) NS direction; (c) vertical direction.

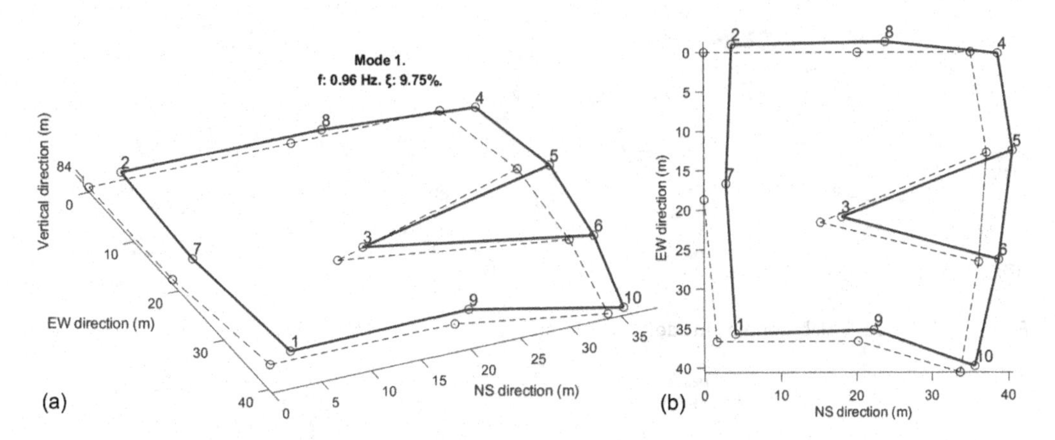

Figure 5.7 Mode 1: translational mode along the NS direction.

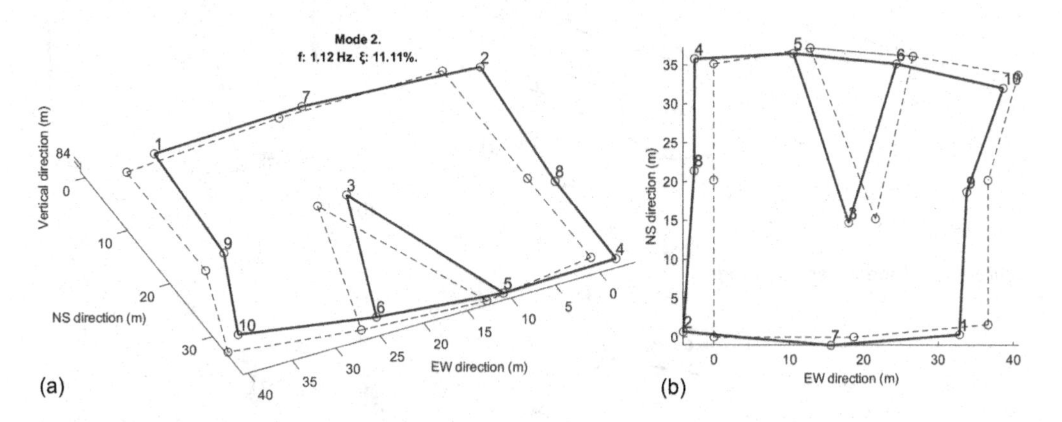

Figure 5.8 Mode 2: translational mode along the EW direction.

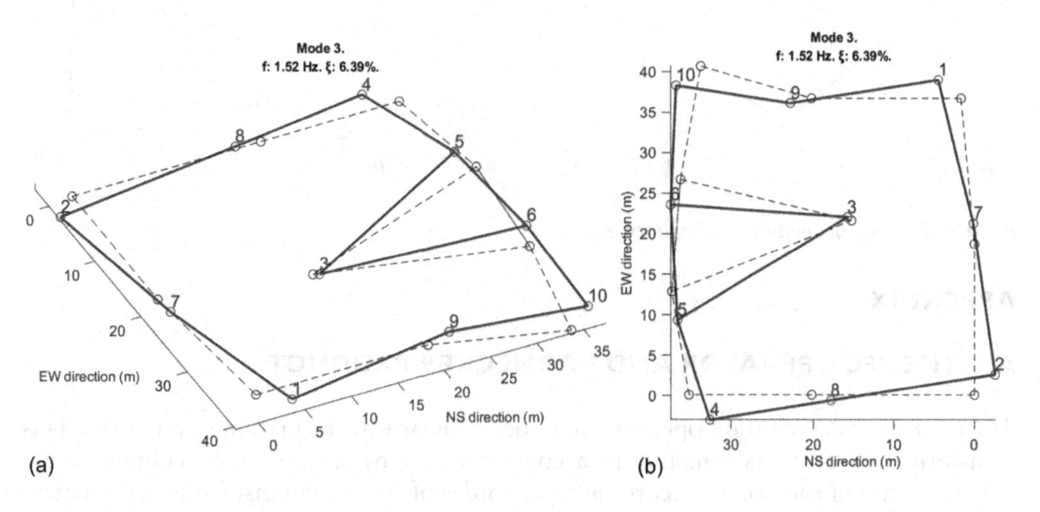

Figure 5.9 Mode 3: torsional mode.

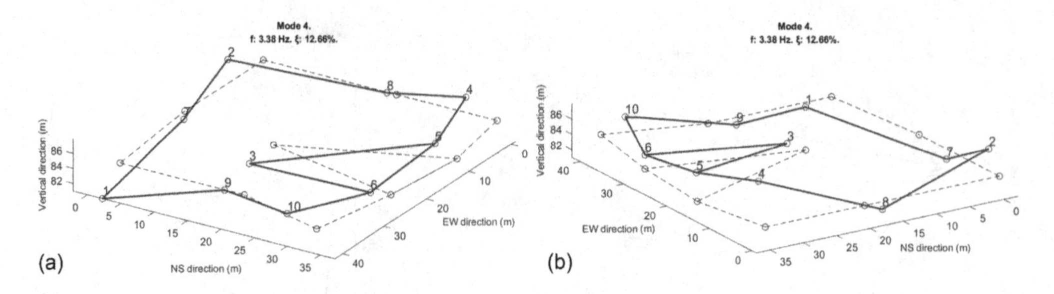

Figure 5.10 Mode 4: slab bending mode.

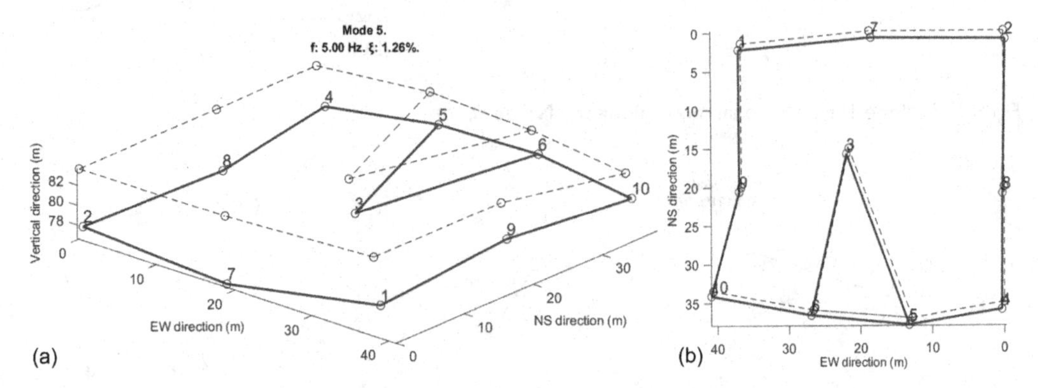

Figure 5.11 Mode 5: vertical mode.

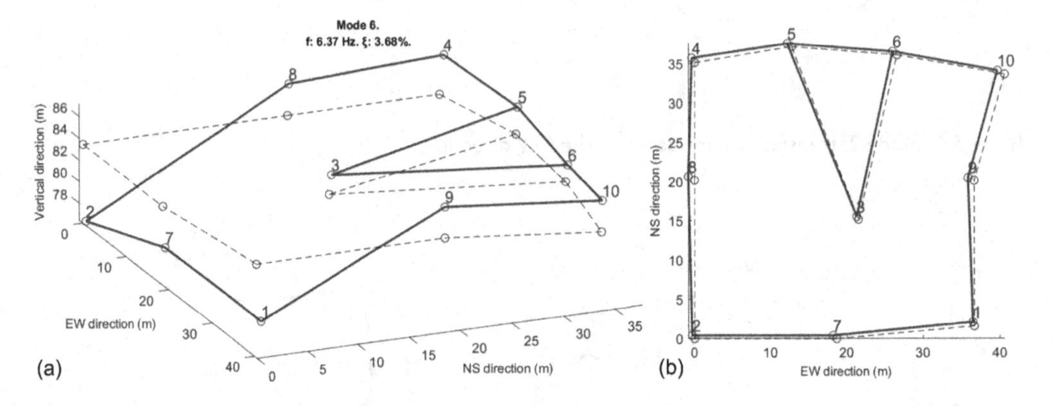

Figure 5.12 Mode 6: building bending mode.

APPENDIX

A.1 THE *VEC* OPERATOR AND KRONECKER PRODUCT

The *vec* is the vectorization operator that does a linear transformation to a matrix. This transformation converts a matrix to a column vector by stacking each column of the matrix on top of each other according to the order of these columns. Consider a matrix

$\mathbf{B} = [\mathbf{b}_1 \quad \mathbf{b}_2 \quad \cdots \quad \mathbf{b}_N] \in R^{M\times N}$, where $\{\mathbf{b}_i \in R^M : i = 1,2,\cdots,N\}$ are the columns of **B**. The *vec* operator transforms **B** into an $MN \times 1$ column vector:

$$vec(\mathbf{B}) = \begin{bmatrix} \mathbf{b}_1 \\ \mathbf{b}_2 \\ \vdots \\ \mathbf{b}_N \end{bmatrix} \tag{A.1}$$

The Kronecker product operator is denoted by $\otimes$. Consider the matrix $G \in R^{K\times L} = \{g_{kl} : k = 1,2,\cdots,K; l = 1,2,\cdots,L\}$, where g_{kl} is the element of G with k and l being the row and column indices, respectively, and the matrix $H \in R^{M\times N} = \{h_{mn} : m = 1,2,\cdots,M; n = 1,2,\cdots,N\}$, where h_{mn} is the element of H with m and n being the row and column indices, respectively. The Kronecker product of G and H generates a $KM \times LN$ matrix:

$$\mathbf{G} \otimes \mathbf{H} = \begin{bmatrix} g_{11}\mathbf{H} & \cdots & g_{1L}\mathbf{H} \\ \vdots & \ddots & \vdots \\ g_{K1}\mathbf{H} & \cdots & g_{KL}\mathbf{H} \end{bmatrix} \tag{A.2}$$

$$\mathbf{H} \otimes \mathbf{G} = \begin{bmatrix} h_{11}\mathbf{G} & \cdots & h_{1N}\mathbf{G} \\ \vdots & \ddots & \vdots \\ h_{M1}\mathbf{G} & \cdots & h_{MN}\mathbf{G} \end{bmatrix} \tag{A.3}$$

Some useful properties related to the vectorization and the Kronecker product are provided in the following. Note that it is assumed that the matrices have appropriate dimensions in the following equations.

$$vec(\mathbf{GH}) = (\mathbf{H}^T \otimes \mathbf{I})vec(\mathbf{G}) \tag{A.4}$$

$$vec(\mathbf{BGH}) = (\mathbf{H}^T \otimes \mathbf{B})vec(\mathbf{G}) \tag{A.5}$$

$$(\mathbf{H} \otimes \mathbf{G})^T = \mathbf{H}^T \otimes \mathbf{G}^T \tag{A.6}$$

$$(\mathbf{H} \otimes \mathbf{G})(\mathbf{J} \otimes \mathbf{B}) = \mathbf{HJ} \otimes \mathbf{GB} \tag{A.7}$$

$$(\mathbf{H} \otimes \mathbf{G})^{-1} = \mathbf{H}^{-1} \otimes \mathbf{G}^{-1} \tag{A.8}$$

If G and H are $K \times K$ and $M \times M$ square matrices, then

$$|\mathbf{G} \otimes \mathbf{H}| = |\mathbf{G}|^M |\mathbf{H}|^N \tag{A.9}$$

REFERENCES

Meirovitch, L., 1975. *Elements of vibration analysis*. McGraw-Hill Science, Engineering & Mathematics.

Meirovitch, L., 2010. *Fundamentals of vibrations*. Waveland Press.

Pi, Y., 1990. Parameter identification of vibrating structures (PhD thesis). University of New South Wales, Australia.

Pi, Y.L. and Mickleborough, N.C., 1989. Modal identification of vibrating structures using ARMA model. *Journal of Engineering Mechanics*, 115(10), 2232–2250.

Yang, J.H. and Lam, H.F., 2019. An innovative Bayesian system identification method using autoregressive model. *Mechanical Systems and Signal Processing*, 133, 106289.

Chapter 6

Model updating by minimizing errors in modal parameters

6.1 BASIC FORMULATION OF MODEL UPDATING

One of the very important advantages of carrying out the model updating in the modal domain is that it is not necessary to apply artificial external forces onto the target structure. With the advances in modal identification technologies, such as those introduced in Chapters 3, 4, and 5, it is possible to extract a set of reliable modal parameters from the measured dynamic responses of a structural system, such as a building or a bridge, under ambient vibration. In general, forced vibration tests with measured force are believed to yield more accurate natural frequencies and mode shapes of the target structure owing to (1) a higher signal-to-noise ratio and (2) additional information from the applied force. However, the magnitude of the external force must be high enough to induce detectable vibration from the ambient level. This certainly will affect the normal operation of the target structure (e.g., users must be moved out from the building during the forced vibration test). Furthermore, the applied force may cause damage to the non-structural or even structural components of the structure. In fact, if the purpose of model updating is for structural damage detection (or structural health monitoring), the owners of the structures will not allow the application of unnecessary dynamic loading on their structures. As a result, model updating based on ambient data is usually the only available option.

6.1.1 Parameterization of the model class

Consider the modeling of a structural dynamics system by the finite element method, the structural system can be uniquely defined by a set of uncertain model parameters once the class of models, $\mathcal{M}$, which defines the parameterization of the model, is confirmed. By assigning numerical values to this set of uncertain model parameters, a model can be obtained from this class of models. Assuming that the model class consists of N_θ uncertain model parameters, which is represented as a column vector, $\boldsymbol{\theta}$:

$$\theta = \{\theta_1, \theta_2, \cdots, \theta_{N_\theta}\}^T \tag{6.1}$$

where the superscript T means "transpose." To ensure the efficiency of the numerical optimization process, the elements in $\boldsymbol{\theta}$ are not directly the physical parameters but the corresponding non-dimensional scaling factor. Consider a shear building model as an example. The uncertain physical parameter can be the inter-story stiffness at the i-th story, k_i, and θ_i is the non-dimensional factor that scales its nominal value, say k_0, through the following relationship:

$$k_i = \theta_i k_0 \quad \text{for} \quad i = 1, \ldots, N_{\text{story}} \tag{6.2}$$

DOI: 10.1201/9780429445866-6

where N_{story} is the number of stories of the shear building. By doing this, all uncertain model parameters in $\boldsymbol{\theta}$ will have a value reasonably close to unity, or at least with the same order of magnitude. Thus, the numerical optimization process will not be biased towards parameters with much larger values than others. To consider a more general situation of finite element modeling, the target structure can be considered as an assembly of a given number of sub-structures. Assuming $\mathbf{K}_i$ to be the stiffness matrix of the i-th sub-structure for $i = 1,\ldots,N_\theta$, the system stiffness matrix $\mathbf{K}(\boldsymbol{\theta})$ can be obtained as:

$$\mathbf{K}(\boldsymbol{\theta}) = \mathbf{K}_0 + \sum_{i=1}^{N_\theta} \theta_i \mathbf{K}_i \tag{6.3}$$

where $\mathbf{K}_0$ is the assembly of the stiffness matrices for all sub-structures that are not uncertain (i.e., not necessary to be considered in model updating). Equation (6.3) is particularly suitable for structural damage detection by model updating. For all undamaged sub-structures, the corresponding θ value should be very close to unity. If a sub-structure is damaged, its θ value should be very small compared to the undamaged ones.

6.1.2 Objective functions

By following the finite element method, the system stiffness matrix $\mathbf{K}(\boldsymbol{\theta})$ and the system mass matrix $\mathbf{M}(\boldsymbol{\theta})$ can be obtained as functions of $\boldsymbol{\theta}$. It must be pointed out that most existing methods from the literature assume that the mass distribution of the structure is not uncertain, and the system mass matrix is independent of $\boldsymbol{\theta}$. The natural frequencies and mode shapes of a structural system can be obtained by solving the eigenvalue problem:

$$\left[\mathbf{K}(\boldsymbol{\theta}) - \omega_r^2 \mathbf{M}(\boldsymbol{\theta})\right] \boldsymbol{\phi}_r = \mathbf{0} \tag{6.4}$$

where $\omega_r = 2\pi f_r$ is the natural circular frequency in rad/s, and f_r is the natural frequency in Hz; $\boldsymbol{\phi}_r$ is the mode shape of the r-th mode. It is very clear from Equation (6.4) that both the natural frequencies and mode shapes are functions of the uncertain model parameters, $\boldsymbol{\theta}$. Thus, the model-predicted natural frequencies and mode shapes can be expressed as $f_r(\boldsymbol{\theta})$ and $\boldsymbol{\phi}_r(\boldsymbol{\theta})$, for $r = 1,\ldots,N_m$, where N_m is the number of modes to be considered in model updating.

It is always good to carry out the model updating in the easiest way, such that only natural frequencies are considered, as a preliminary study. Assuming that the first N_m modes were experimentally measured, the objective function for model updating can be formulated as:

$$J_f(\boldsymbol{\theta}) = \sum_{r=1}^{N_m} \left(\frac{\hat{f}_r - f_r(\boldsymbol{\theta})}{\hat{f}_r} \right)^2 \tag{6.5}$$

where $J_f(\boldsymbol{\theta})$ is the objective function, and the subscript f represents that only natural frequencies are considered; $\hat{f}_r$ is the measured natural frequency of the r-th mode. It is important to normalize the discrepancy between the measured and model-predicted natural frequencies, $\hat{f}_r - f_r(\boldsymbol{\theta})$, by the corresponding measured natural frequency, $\hat{f}_r$, so that the optimization process will not be biased towards higher modes with larger natural frequencies. The identified model can be obtained if one can get the global minimum, $\hat{\theta}$, through the numerical optimization of Equation (6.5). Since the number of measured modes (i.e., N_m) is usually a small number, the amount of information that can be extracted from the measured natural frequencies is very limited. Depending on the complexity of the physical structure and the assumed model class, the numerical minimization problem may have multiple solutions or some local minima. It is difficult, if not impossible, to ensure that the solution from numerical minimization is the global minimum. Thus, most researchers/engineers will repeat the numerical minimization many times with different initial trials. The initial trials are usually distributed evenly in the parameter space of interest. Consider a two-dimensional problem as an example (i.e., $\theta = \{\theta_1, \theta_2\}^T$). One may generate a set of initial trials as grid points on the (θ_1, θ_2) space for θ_1 and θ_2 to go from, say, 0.2 to 1.8 with a step size of 0.2. This can increase the confidence for obtaining a reasonable solution in the parameter space of interest, but the solution obtained may not necessarily be the global minimum. Please refer to the numerical case study in Section 6.1.3 for a detailed illustration by a simple two-story shear building model.

Another well-known problem in model updating (or damage detection) using Equation (6.5) is that it cannot distinguish model parameters at symmetrical locations (if the target structure has one or more symmetrical axes). Please refer to the numerical case study in Section 6.1.4 for a simple illustration on this problem.

Another formulation of the objective function different from Equation (6.5) is to consider only the mode shapes, in order to extract more information from the measured data. By following the idea of minimizing the discrepancy between the measured and model-predicted mode shapes, one may formulate the objective function as:

$$J_1(\theta) = \sum_{r=1}^{N_m} \sum_{i=1}^{N_n} \left(\hat{\psi}_{i,r} - \psi_{i,r}(\theta) \right)^2 \tag{6.6}$$

where N_n is the number of measured degrees-of-freedom (DOFs), which is smaller than the total number of DOFs of the model, N_d. In real situations, $N_d \gg N_n$; $\hat{\psi}_{i,r}$ is the measured model shape value at the i-th measured DOF and r-th mode. The matrix of measured mode shape, $\hat{\Psi}_{N_n \times N_m}$, is defined as:

$$\hat{\Psi} = \left[\hat{\psi}_1, \hat{\psi}_2, \cdots, \hat{\psi}_{N_m} \right] = \begin{bmatrix} \hat{\psi}_{1,1} & \hat{\psi}_{1,2} & \cdots & \hat{\psi}_{1,N_m} \\ \hat{\psi}_{2,1} & \hat{\psi}_{2,2} & \cdots & \hat{\psi}_{2,N_m} \\ \vdots & \vdots & \ddots & \vdots \\ \hat{\psi}_{N_n,1} & \hat{\psi}_{N_n,2} & \cdots & \hat{\psi}_{N_n,N_m} \end{bmatrix} \tag{6.7}$$

where $\hat{\psi}_r$ is the measured mode shape of the r-th mode; $\psi_{i,r}(\theta)$ represents the model-predicted mode shape at the i-th measured DOF and r-th mode. To match with the measured mode shape, the measured DOFs are extracted from the original model-predicted mode shape, $\phi_r(\theta)$, with a length of N_d as:

$$\psi_r(\theta) = \Gamma \phi_r(\theta) = \Gamma \begin{Bmatrix} \phi_{1,r}(\theta) \\ \phi_{2,r}(\theta) \\ \vdots \\ \phi_{N_d,r}(\theta) \end{Bmatrix} = \begin{Bmatrix} \psi_{1,r}(\theta) \\ \psi_{2,r}(\theta) \\ \vdots \\ \psi_{N_n,r}(\theta) \end{Bmatrix} \quad \text{for} \quad r = 1, \ldots, N_m \tag{6.8}$$

where $\mathbf{\Gamma}$ is a selection matrix with dimensions $N_n \times N_d$. The i-th row of $\mathbf{\Gamma}$ represents the i-th measured DOF, and the j-th column of $\mathbf{\Gamma}$ represents the j-th DOF of the model. Consider a structural system with 10 DOFs, where if DOFs 1, 4, 7, and 10 of the model are measured, then the selection matrix is:

$$\Gamma = \begin{bmatrix} 1 & 0 & 0 & 0 & 0 & 0 & 0 & 0 & 0 & 0 \\ 0 & 0 & 0 & 1 & 0 & 0 & 0 & 0 & 0 & 0 \\ 0 & 0 & 0 & 0 & 0 & 0 & 1 & 0 & 0 & 0 \\ 0 & 0 & 0 & 0 & 0 & 0 & 0 & 0 & 0 & 1 \end{bmatrix} \tag{6.9}$$

In this example, $N_d = 10$ and $N_n = 4$.

The objective function $J_1(\theta)$ in Equation (6.6) is not popular, as mode shape values are not exact. When a mode shape is multiplied by a constant (no matter whether it is positive or negative), it is also a mode shape that satisfies the eigenvalue problem, as shown in Equation (6.4). For the objective function in Equation (6.6) to be meaningful, both the measured mode shape, $\hat{\psi}_r$, and the corresponding model-predicted mode shape, $\psi_r(\theta)$, must be in the same normalization and the same sign. These processes introduce additional steps and uncertainties in the model updating process.

The most popular method to determine the discrepancy between two mode shapes is the modal assurance criterion (MAC). The MAC of two column vectors **a** and **b** with the same length is defined as (Moller & Friberg 1998):

$$\text{MAC}(\mathbf{a},\mathbf{b}) = \frac{\left(\mathbf{a}^T\mathbf{b}\right)^2}{\left(\mathbf{a}^T\mathbf{a}\right)\left(\mathbf{b}^T\mathbf{b}\right)} \tag{6.10}$$

where $\mathbf{a}^T\mathbf{b}$ represents the dot product between the column vectors **a** and **b**. An MAC value of 0 means that the two vectors are orthogonal to each other (i.e., they are completely different), while an MAC value of 1 means that the two vectors are parallel to each other (i.e., they are the same). By using MAC, the objective function for minimizing the discrepancy in mode shapes can be defined as:

$$J_\phi(\theta) = \sum_{r=1}^{N_m} \left(1 - \text{MAC}\left(\hat{\psi}_r, \Gamma\phi_r(\theta)\right)\right) \tag{6.11}$$

When the discrepancy between the measured and model-predicted mode shapes is large (small), the MAC value will be very close to zero (unity) and the objective function will be large (small). By minimizing the objective function in Equation (6.11), the updated model of the target structure may be obtained. Similar to the objective function in Equation (6.5), different initial trials need to be considered. Even with the MAC, it is better to have the measured and model-predicted mode shape in the same normalization (but not necessarily the same sign). Before model updating, it is not easy to have an accurate system mass matrix. Thus, the mass normalization method (commonly used for model-predicted full mode shapes) is not employed here. In this chapter, both mode shapes are normalized such that their lengths are equal to unity, $\|\hat{\psi}_r\| = \|\psi_r(\theta)\| = 1$, for $r = 1,\ldots,N_m$, where $\|\cdot\|$ represents the second norm of the vector. Thus, the MAC value can be simplified to:

$$\text{MAC}\left(\hat{\psi}_r, \Gamma\phi_r(\theta)\right) = \text{MAC}\left(\hat{\psi}_r, \psi_r(\theta)\right) = \left(\hat{\psi}_r^{\,T}\psi_r(\theta)\right)^2 \tag{6.12}$$

To fully utilize the measured data, most researchers will consider both natural frequencies and mode shapes in formulating the objective function, resulting in:

$$J_2(\theta) = \sum_{r=1}^{N_m}\left[\left(\frac{\hat{f}_r - f_r(\theta)}{\hat{f}_r}\right)^2 + \left(1 - \left(\hat{\psi}_r^{\,T}\psi_r(\theta)\right)^2\right)\right] \tag{6.13}$$

In general, it is believed that natural frequencies can be measured (or identified) with much higher accuracy when compared to mode shapes. It is also believed that the measurement noises associated with higher modes are usually higher than those associated with lower modes. If Equation (6.13) is directly used as the objective function for model updating, the identified model will be contaminated by measured quantities with high measurement noise. To overcome this limitation, Equation (6.13) can be extended to:

$$J_3(\theta) = \sum_{r=1}^{N_m}\left[w_{f,r}\left(\frac{\hat{f}_r - f_r(\theta)}{\hat{f}_r}\right)^2 + w_{\phi,r}\left(1 - \left(\hat{\psi}_r^{\,T}\psi_r(\theta)\right)^2\right)\right] \tag{6.14}$$

where $w_{f,r}$ and $w_{\phi,r}$ are the weighting factors for natural frequency and mode shape, respectively, of the r-th mode. These factors show the relative importance of natural frequency and mode shape in model updating. They can also reflect the relative importance of measured modal parameters from different modes. Intuitively, the weighting factors for natural frequency should take higher values than those for mode shape. Furthermore, the weighting factors for lower modes should take higher values than those for higher modes. However, assigning values to these factors based on the subjective judgment of users is inappropriate. A theoretical method for calculating these weighting factors following the Bayesian approach will be presented in the next section.

In vibration tests, the most time-consuming procedure is the installation of sensors for measuring the target DOFs and connecting cables from sensors to the console (if wired sensors are employed). Once the setup is completed, it is preferred to continuously measure the ambient vibration of the structure for a long period of time. The measured time-domain data can then be divided into different segments, and each segment can be used to identify a set of modal parameters. For example, if the total duration of the measurement is one hour, it can be divided into six 10-minute segments without overlap resulting in six data sets. Thus, multiple sets of measured modal parameters can be obtained. Equation (6.14) can be easily extended to consider this situation.

$$J(\theta) = \sum_{s=1}^{N_s}\sum_{r=1}^{N_m}\left[w_{f,r,s}\left(\frac{\hat{f}_{r,s} - f_r(\theta)}{\hat{f}_r}\right)^2 + w_{\phi,r,s}\left(1 - \left(\hat{\psi}_{r,s}^{\,T}\psi_r(\theta)\right)^2\right)\right] \tag{6.15}$$

where s is the counter for different data sets, and N_s is the total number of data sets.

6.1.3 Numerical case study of a two-story shear building model: non-uniqueness problem

The main purpose of this numerical case study is to demonstrate the effect of reducing the amount of measured data in model updating, so as to illustrate the concept of non-uniqueness in model updating problems.

Consider a two-story shear building model as shown in Figure 6.1, the columns are made of aluminum (with the modulus of elasticity, $E = 70$ GPa), given the cross-sectional dimensions of all columns are the same and are equal to $b = 30\,\text{mm}$ and $t = 5\,\text{mm}$ (the column is bending about the minor axis). The two floor masses are the same and are equal to $m_1 = m_2 = m = 20\,\text{kg}$.

a) Determine the nominal values of the inter-story stiffness for the first and second stories (i.e., $k_{0,1}$ and $k_{0,2}$).
b) To simulate the effect of modeling error in a real situation, the real inter-story stiffnesses are expressed as $k_1 = \theta_1 k_{0,1}$ and $k_2 = \theta_2 k_{0,2}$, where θ_1 and θ_2 are the non-dimensional scaling factors to be identified in the model updating process. Simulate the measured modal parameters with $\theta = (\theta_1, \theta_2) = (1.1, 0.9)$.
c) Use the measured (simulated) modal parameters in part (b) to identify the inter-story stiffness (i.e., (θ_1, θ_2)) by minimizing the discrepancy between the measured and model-predicted modal parameters under the following cases:
   i. Using all measured data (i.e., both natural frequencies and mode shapes in both modes);
   ii. Using only the measured natural frequencies for both modes without using the mode shapes;
   iii. Using only the measured natural frequency of the first mode; and
   iv. Using only the measured natural frequency of the second mode;
d) Discuss the results in part (c).

In part (a), the second moment of area for a column is needed, and it is equal to $I = bt^3/12 = 3.1250 \times 10^{-10}\,\text{m}^4$. The nominal values of the inter-story stiffness for both stories are the same and are given by:

$$k_{0,1} = k_{0,2} = 4\frac{12EI}{L^3} = 4\frac{12\left(70\times10^9\right)\left(3.1250\times10^{-10}\right)}{0.3^3} = 38888.89\,\text{N/m} \tag{6.16}$$

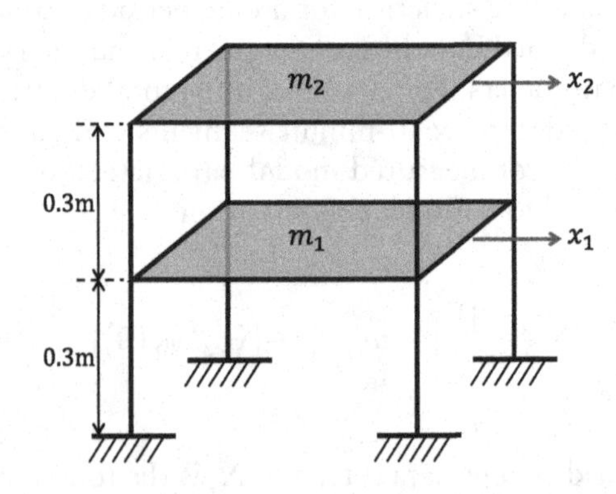

Figure 6.1 Numerical case study of a two-story shear building model.

In part (b), it is assumed that the inter-story stiffnesses of the real structure are equal to:

$$k_1 = 1.1k_{0,1} = 42777.78\,\text{N/m} \quad \text{and} \quad k_2 = 0.9k_{0,2} = 35000.00\,\text{N/m} \tag{6.17}$$

The system stiffness matrix becomes:

$$\mathbf{K} = \begin{bmatrix} k_1 + k_2 & -k_2 \\ -k_2 & k_2 \end{bmatrix} = \begin{bmatrix} 77777.78 & -35000.00 \\ -35000.00 & 35000.00 \end{bmatrix} \text{N/m} \tag{6.18}$$

The system mass matrix is:

$$\mathbf{M} = \begin{bmatrix} m_1 & 0 \\ 0 & m_2 \end{bmatrix} = \begin{bmatrix} 20 & 0 \\ 0 & 20 \end{bmatrix} \text{kg} \tag{6.19}$$

After solving the eigenvalue problem, the natural frequencies and mode shapes are presented in Figure 6.2 (readers are redirected to Section 2.2.1 for solving the eigenvalue problem of a multi-DOF system). Note that the mode shapes are normalized to have unit length.

In part (c), model updating is carried out with different amounts of measured information.

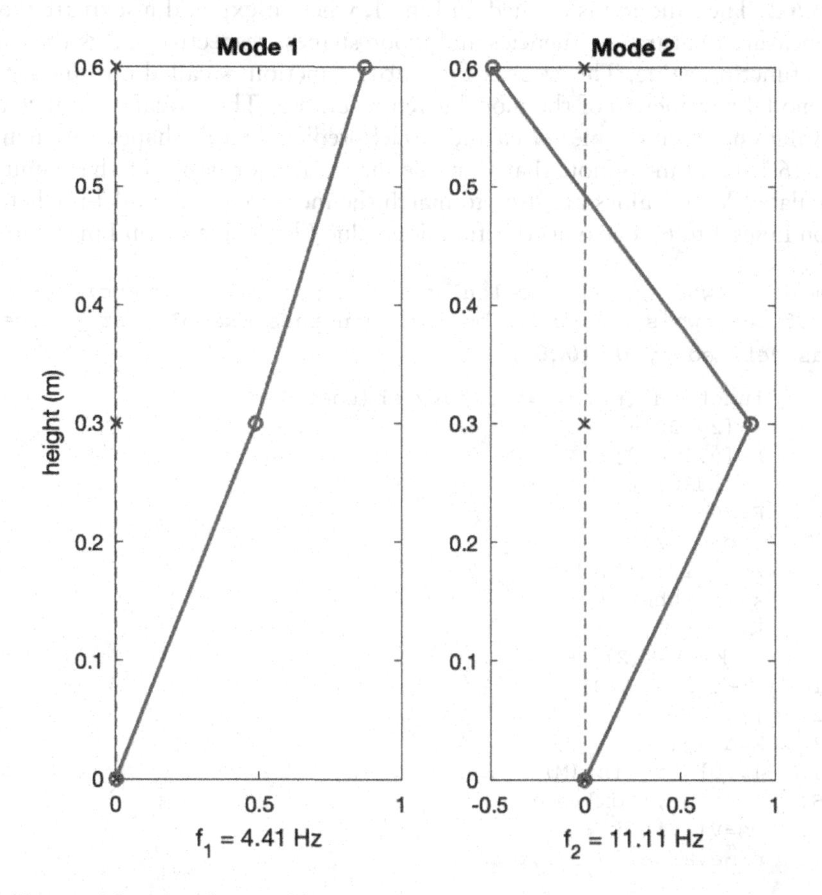

Figure 6.2 Measured (simulated) natural frequencies and mode shapes of the shear building model.

In case (c-i), the natural frequencies and mode shapes of both modes are considered. The MATLAB function `twostory _ sb()` is developed for calculating the natural frequencies and mode shapes of the two-story shear building model in this example (see Program 6.1). The function is defined on Line 1, where `nf` and `ms` are the calculated natural frequencies and mode shapes, respectively; and `theta` is the model defined by $\boldsymbol{\theta} = (\theta_1, \theta_2)$. Lines 2, 3, 4, and 5 define the two floor masses, two story heights, the column width and thickness, and the modulus of elasticity, respectively. The second moment of area of columns is calculated on Line 6, while the nominal value of the inter-story stiffness is calculated on Line 7. The inter-story stiffnesses are calculated on Line 8 based on the model defined by `theta`. The system stiffness matrix is determined on Lines 9 to 12, and the system mass matrix is calculated on Line 13. The eigenvalue problem of the shear building model is solved on Line 14, where `D` is a square matrix with eigenvalues at the diagonal, and V is the eigenvector matrix (all eigenvectors are column vectors). The eigenvalues are sorted in ascending order and stored in the variable `evalue` on Line 15. The order of the eigenvectors is arranged to be the same as that of the eigenvalues on Line 16. The sorted eigenvectors are stored in `ms` (that is, the calculated mode shapes). The natural frequencies in Hz are calculated from the eigenvalues on Line 17. Finally, the mode shapes in `ms` are normalized to have unit length on Lines 19 to 21, and the MATLAB function ends on Line 22.

By using the same weighting factors for natural frequencies and mode shapes, Equation (6.13) is the objective function which is implemented in MATLAB as `jcalnfms()` in Program 6.2. The function is defined on Line 1, where nfexp and msexp are the experimental measured natural frequencies and mode shapes, respectively; `J` is the calculated objective function value. The `twostory _ sb()` function is called on Line 2 to calculate the modal parameters of the model given in `theta`. The modal assurance criterion (MAC) values between the measured and model-predicted mode shapes are calculated by Equation (6.10) on Line 3 (note that all mode shapes are normalized to have unit length). The calculated MAC values are used to match the measured and model-predicted mode shapes on Lines 4 to 6. The objective function value J is calculated on Lines 8 to 10.

Program 6.1: twostory_sb() calculates the natural frequencies and mode shapes of the two-story shear building in this example for a given model as defined by $\boldsymbol{\theta} = (\theta_1, \theta_2)$.

```
Line 1.    function [nf,ms]=twostory_sb(theta)
Line 2.    m=[20 20];
Line 3.    L=[0.3 0.3];
Line 4.    b=0.030; t=0.005;
Line 5.    E=70e9;
Line 6.    I=b*t^3/12;
Line 7.    k0=4*12*E*I./L.^3;
Line 8.    k=k0.*theta;
Line 9.    K=[
Line 10.      k(1)+k(2)  -k(2)
Line 11.     -k(2)    k(2)
Line 12.   ];
Line 13.   M=diag(m);
Line 14.   [V,D]=eig(inv(M)*K);
Line 15.   [evalue,indx]=sort(diag(D));
Line 16.   ms=V(:,indx);
Line 17.   nf=evalue.^0.5/2/pi
```

```
Line 18.    mm=ms'*ms;
Line 19.    for ii=1:size(mm,2)
Line 20.      ms(:,ii)=ms(:,ii)./mm(ii,ii)^0.5;
Line 21.    end
Line 22.    end
```

Program 6.2: jcalnfms() calculates the objective function value using both natural frequencies and mode shapes.

```
Line 1.     function J=jcalnfms(theta,nfexp,msexp)
Line 2.     [nf,ms]=twostory_sb(theta);
Line 3.     MAC=(msexp'*ms).^2;
Line 4.     [maxMAC,indx]=max(MAC');
Line 5.     nf=nf(indx);
Line 6.     ms=ms(:,indx);
Line 7.     J=0;
Line 8.     for ii=1:length(nfexp)
Line 9.       V1=msexp(:,ii);
Line 10.      V2=ms(:,ii);
Line 11.      J=J+((nfexp(ii)-nf(ii))/nfexp(ii))^2+(1-(V1'*V2).^2);
Line 12.    end
Line 13.    end
```

The MATLAB optimization toolbox is used to minimize the objective function as given by `jcalnfms()`. After loading the measured modal parameters into the workspace (i.e., `nfexp` and `msexp`), the MATLAB function `fmincon()` is employed to minimize the objective function as:

```
>> x0=[1 1]; xL=x0./5; xU=x0*5;
>> [theta,j]=fmincon('jcalnfms',x0,[],[],[],[],xL,xU,[],[],nfexp,msexp);
   0.0000     1.1000      0.9000
```

where `x0` is the initial trial, `xL` and `xU` are the lower and upper bounds. The name of the objective function to be minimized and the initial trial are given as the first and second input parameters of `fmincon()`, respectively. The lower and upper bounds for the minimization are given as the seventh and eighth input parameters, respectively. The inputs `[]` (i.e., empty in MATLAB) instruct `fmincon()` to use the default values for the corresponding input parameters (i.e., the third, fourth, fifth, sixth, ninth, and tenth). Referring to Program 6.2, the objective function `jcalnfms()` requires three input parameters. They are the vector of model parameters to be updated (i.e., `theta`), the experimental measured natural frequencies (i.e., `nfexp`) and mode shapes (i.e., `msexp`). When `jcalnfms()` is called by `fmincon()`, the value of the first input parameter (i.e., `theta`) will be provided by `fmincon()`. Other input parametrers required by the objective function can be passed through `fmincon()` after its eleventh input parameters. In this example, the values of `nfexp` and `msexp` are provided by the user through the eleventh and the twelfth input parameters of `fmincon()`. The optimal solution and the corresponding objective function value are stored in `theta` and `j`, respectively. They are also shown on the last line, where the value of `j` is 0.0000, and the optimal solution in `theta` is $\hat{\theta}=(1.1,0.9)$.

The model updating (i.e., the minimization) is repeated with different initial trials to ensure it is reliable. For both scaling factors (i.e., (θ_1,θ_2)), the initial trials are from 0.4 to 2.0 with a step size of 0.2 (i.e., 9 trials in each dimension, and the total number of initial trials is 81). Figure 6.3 shows the model updating results from all initial trials, where the cross (x) markers represent the initial trials, and the circle (o) markers show the optimal solution. It is very clear from the figure that all initial trials go to the same optimal solution $\hat{\theta} = (1.1, 0.9)$ when all measured information is considered in the model updating process. It can be concluded that the model is successfully identified in case (c-i).

In case (c-ii), only the measured natural frequencies of both modes are considered, and therefore, Equation (6.5) is employed as the objective function. The same set of 81 initial trials are employed, and the model updating results are presented in Figure 6.4. The results in case (c-ii) are different from that of case (c-i). When only the natural frequency information is provided, two different solutions are obtained from the minimization processes with different initial trials. They are denoted as $\hat{\theta}_1 = (1.1, 0.9)$ and $\hat{\theta}_2 = (1.8, 0.55)$. It must be pointed out that the objective function values for both solutions are the same and are equal to zero. Without the information from the measured mode shape, it is impossible for one to conclude which one of these two identified solutions is the "correct" one. This case clearly shows the problem of non-uniqueness. Under the umbrella of the deterministic approach, it is difficult to explain the existence of two different models for the same structure. The most suitable method to handle this non-uniqueness problem is to follow the probabilistic approach (as illustrated in Chapter 7). It must be pointed out that the situation in this case is not the worst (i.e., a discrete number of models are identified in the parameter space of interest). The non-uniqueness problem can be more serious if the amount of information that can be extracted from the measured data is further reduced.

In case (c-iii), only the mode 1 natural frequency (i.e., $f_1 = \omega_1 / (2\pi) = 4.41\text{Hz}$) is employed in the model updating process. With the same set of 81 initial trials, the model updating results are presented in Figure 6.5. When only mode 1 natural frequency is considered in model updating, each initial trial goes to a different optimal solution. It must be pointed out that all solutions have zero objective function value. From the figure, the solutions all converge to a "curve." It is believed that all models on this curve return a

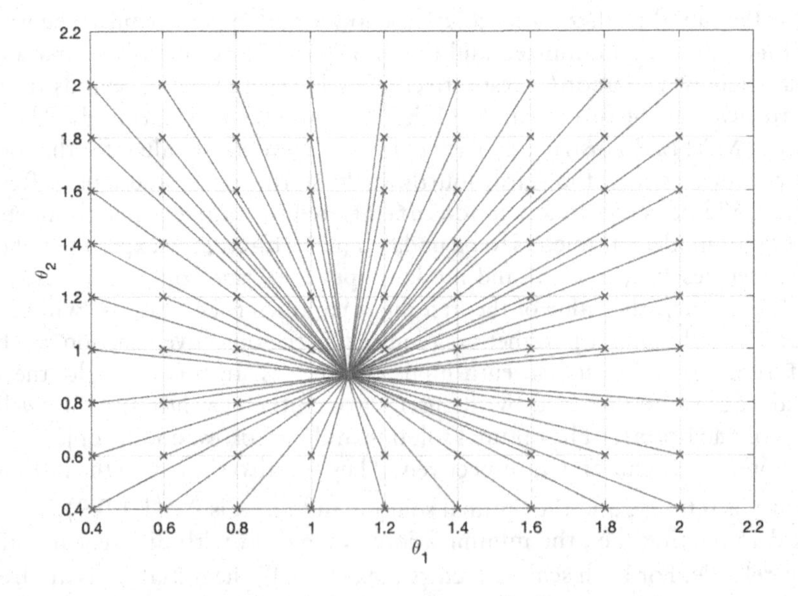

Figure 6.3 Model updating results in case (c-i) with all measured information.

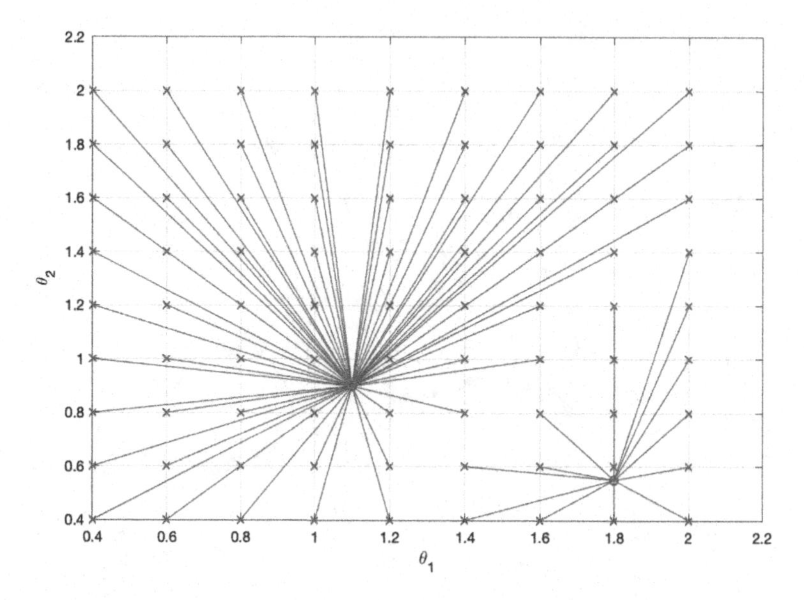

Figure 6.4 Model updating results in case (c-ii) with only natural frequencies for both modes.

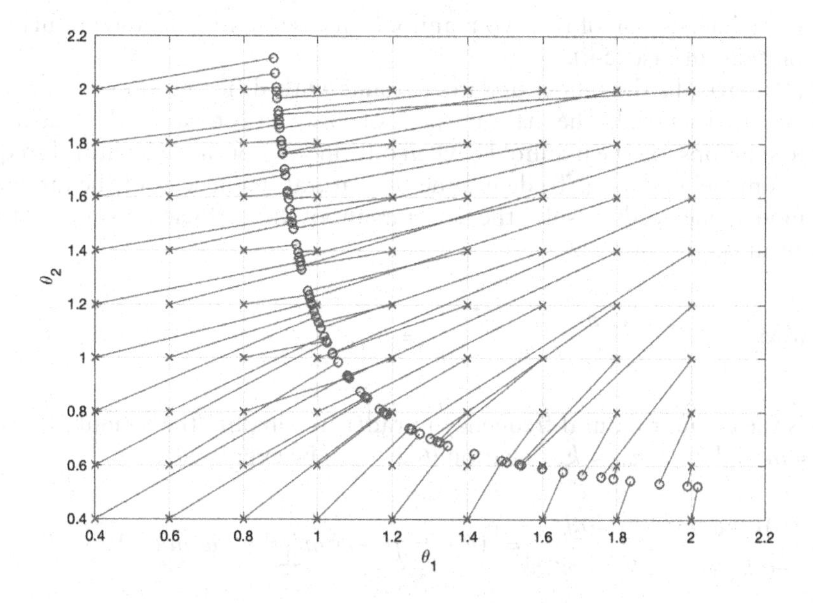

Figure 6.5 Model updating results in case (c-iii) with only natural frequencies for mode 1.

zero objective function value (to be proved later), and therefore, there are infinite optimal solutions in this case (unlike the previous case that has a discrete number of solutions). This curve of optimal solutions is called the manifold (usually has a lower dimension than the parameter space). This case clearly shows a very serious non-uniqueness problem.

Case (c-iv) uses only the mode 2 natural frequency (i.e., $f_2 = \omega_2 / (2\pi) = 11.11\text{Hz}$) in model updating. Figure 6.6 shows the model updating results. It is clear that the optimal solutions are on another manifold. By comparing Figure 6.4, Figure 6.5, and Figure 6.6, it

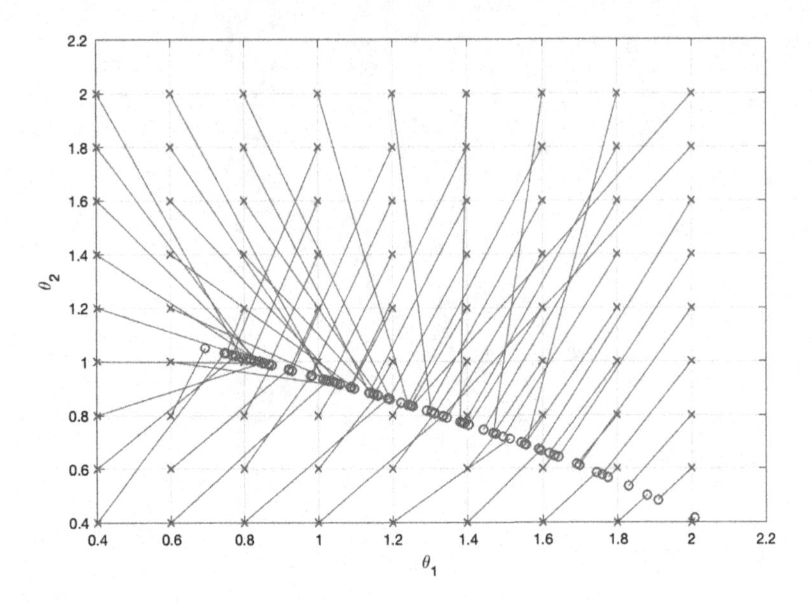

Figure 6.6 Model updating results in case (c-iv) with only natural frequencies for mode 2.

seems that the intersection of the two manifolds in cases (c-iii) and (c-iv) returns the two discrete solutions in case (c-ii).

Finally, in part (d), the model updating results with different amounts of measured information are discussed. The main purpose of this part is to study the reason for having infinite solutions (on the manifolds) when only mode 1 or mode 2 natural frequency is employed in model updating. To determine the natural frequencies of the targeted shear building model, one needs to solve the determinant of the coefficient matrix of the eigenvalue problem (see Section 2.2.1):

$$\left|\mathbf{K}-\omega^2\mathbf{M}\right| = \begin{vmatrix} k_1+k_2-\omega^2 m_1 & -k_2 \\ -k_2 & k_2-\omega^2 m_2 \end{vmatrix} = 0 \tag{6.20}$$

where ω is the circular natural frequency in rad/s. To simplify the formulation, consider $m_1 = m_2 = m$ and $k_{0,1} = k_{0,2} = k$, Equation (6.20) can be expressed as:

$$\begin{vmatrix} (\theta_1+\theta_2)k-\omega^2 m & -\theta_2 k \\ -\theta_2 k & \theta_2 k-\omega^2 m \end{vmatrix} = \left[(\theta_1+\theta_2)k-\omega^2 m\right]\left[\theta_2 k-\omega^2 m\right]-\theta_2^2 k^2 = 0 \tag{6.21}$$

where θ_1 and θ_2 are the uncertain model parameters to scale the inter-story stiffness of the first and second stories, respectively. Equation (6.21) can be rearranged to:

$$\left(\omega^2\right)^2-(\theta_1+2\theta_2)\frac{k}{m}\omega^2+\theta_1\theta_2\left(\frac{k}{m}\right)^2 = 0 \tag{6.22}$$

Let $\lambda = k/m$. The first and second roots of this quadradic equation (i.e., λ_1 and λ_2) can be expressed as:

$$\lambda_1 = \frac{\lambda}{2}\left(\theta_1 + 2\theta_2 - \sqrt{\theta_1^2 + (2\theta_2)^2}\right) \quad \text{and} \quad \lambda_2 = \frac{\lambda}{2}\left(\theta_1 + 2\theta_2 + \sqrt{\theta_1^2 + (2\theta_2)^2}\right) \tag{6.23}$$

where $\lambda_1 = \omega_1^2 \leq \lambda_2 = \omega_2^2$. If only the first mode natural frequency is given (i.e., f_1 is a constant), the first part of Equation (6.23) can be expressed as:

$$\frac{2\lambda_1}{\lambda} = c_1 = \theta_1 + 2\theta_2 - \sqrt{\theta_1^2 + (2\theta_2)^2} \tag{6.24}$$

This can be rearranged to express θ_2 as a function of θ_1 as:

$$\theta_2 = \frac{\theta_1^2 - (\theta_1 - c_1)^2}{4(\theta_1 - c_1)} \tag{6.25}$$

This curve is the manifold when only mode 1 frequency is employed in model updating. In this case study, $c_1 = 2\lambda_1 / \lambda = 0.7905$. This manifold is shown in Figure 6.7 as the one with the label "f_1 is a constant." The analytical manifold by Equation (6.25) in Figure 6.7 is the same as the numerical one in Figure 6.5.

If only the second mode natural frequency is given (i.e., f_2 is a constant), the second part of Equation (6.23) can be expressed as:

$$\frac{2\lambda_2}{\lambda} = c_2 = \theta_1 + 2\theta_2 + \sqrt{\theta_1^2 + (2\theta_2)^2} \tag{6.26}$$

This can be rearranged as:

$$\theta_2 = \frac{(c_2 - \theta_1)^2 - \theta_1^2}{4(c_2 - \theta_1)} \tag{6.27}$$

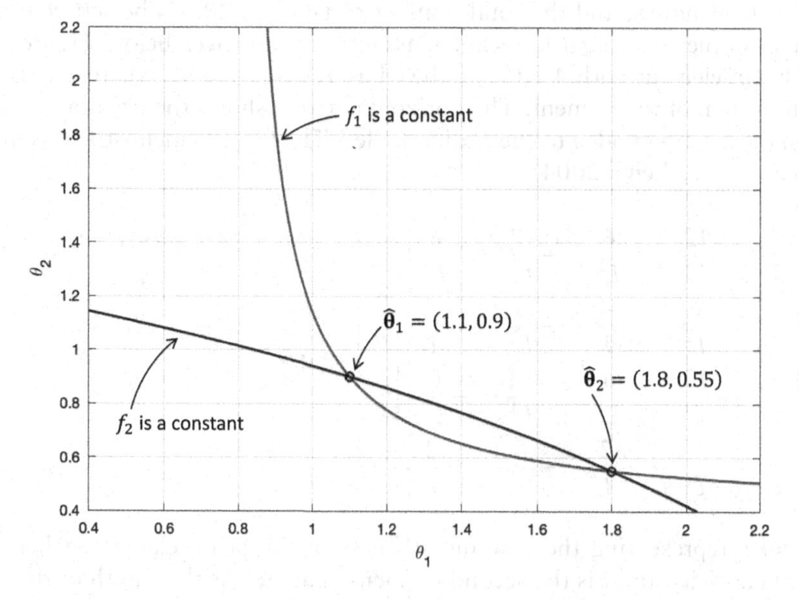

Figure 6.7 The analytical manifolds for given natural frequency information for mode 1 or 2.

where $c_2 = 2\lambda_2 / \lambda = 5.0095$ in this case study. Equation (6.27) is the manifold when only mode 2 frequency is considered in model updating. It is also shown in Figure 6.7 (the curve with the label "f_2 is a constant"). This manifold is the same as the numerical one in Figure 6.6. The two intersection points can be easily obtained by solving Equations (6.25) and (6.27). They are indicated in Figure 6.7 as $\hat{\theta}_1 = (1.1, 0.9)$ and $\hat{\theta}_2 = (1.8, 0.55)$. These two solutions are the same as those numerically obtained in case (ii) as shown in Figure 6.4 when the natural frequencies of both modes are considered in mode updating.

It must be pointed out that the existence of manifolds is common in model updating of real structures when the amount of available measured information is limited. The manifolds in this example are very simple, so they can be analytically formulated (see Equations (6.25) and (6.27)). For general model updating problems, it is extremely difficult, if not impossible, to derive the manifolds analytically. The approximation of manifolds is an interesting research topic in the field of model updating (especially when the complexity of model class is high, and the amount of available measured information is low), and this type of research problem is usually addressed through a probabilistic approach. Interested readers are directed to Chapter 7 for model updating following the Bayesian statistical system identification framework.

6.1.4 Numerical case study of a simple beam: symmetrical problem

Non-uniqueness is a difficult problem in structural model updating and damage detection, especially when the number of measured DOFs is limited and the number of uncertain model parameters to be identified is large. If only the measured natural frequencies are considered in the model updating process, the symmetrical property may cause this problem even if the targeted structure is a very simple one. This numerical case study uses a simple beam to illustrate this problem. The beam structure in this case study is shown in Figure 6.9. Both the left and right supports are pinned. The beam is divided into 10 elements with 11 nodes, and the total number of DOFs is 20. To be self-contained, the formulations of the element stiffness and mass matrices are given below. Figure 6.8 shows a typical beam element with 4 DOFs (labeled as 1, 2, 3, and 4), where $x'y'$ is the local coordinate system of the element. The horizontal arrow shows the orientation of the element from the starting node i to the ending node j. The local element stiffness matrix, $\mathbf{k}'$, is defined as (Paz & Leigh 2004):

$$\begin{Bmatrix} V_i \\ M_i \\ V_j \\ M_j \end{Bmatrix} = k_e \begin{bmatrix} \frac{12}{L^3} & \frac{6}{L^2} & -\frac{12}{L^3} & \frac{6}{L^2} \\ \frac{6}{L^2} & \frac{4}{L} & -\frac{6}{L^2} & \frac{2}{L} \\ -\frac{12}{L^3} & -\frac{6}{L^2} & \frac{12}{L^3} & -\frac{6}{L^2} \\ \frac{6}{L^2} & \frac{2}{L} & -\frac{6}{L^2} & \frac{4}{L} \end{bmatrix} \begin{Bmatrix} \Delta_i \\ \theta_i \\ \Delta_j \\ \theta_j \end{Bmatrix} \Rightarrow \mathbf{q} = \mathbf{k}'\mathbf{d} \tag{6.28}$$

where $k_e = EI$ representing the bending stiffness of the beam element, where E is the modulus of elasticity and I is the second moment of area; L is the length of the beam element; $\mathbf{q}$ is the force vector, where V_i and M_i are the shear force and bending moment at

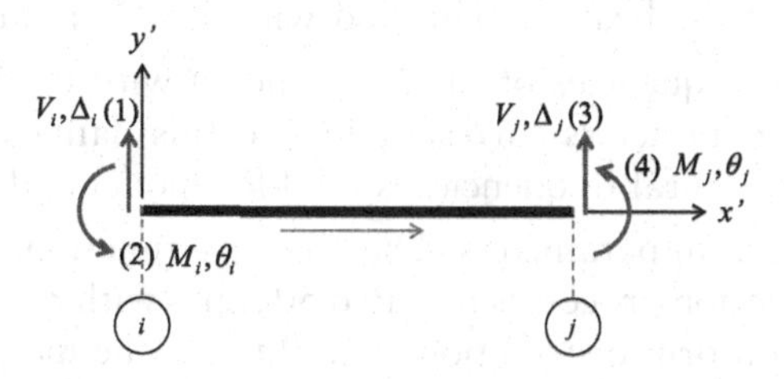

Figure 6.8 A beam element with the definition of various DOFs.

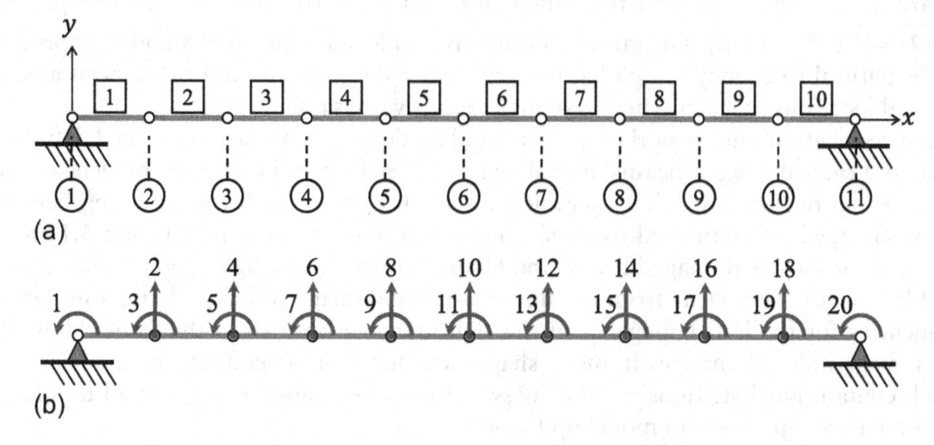

Figure 6.9 The model of the simple beam. (a) Numbering of nodes and elements. (b) Numbering of degrees-of-freedom.

node i; **d** is the displacement vector, where Δ_i and θ_i are the displacement and rotation at node i. The consistent mass matrix for the local beam element can be expressed as (Paz & Leigh 2004):

$$\mathrm{m} = \frac{mL}{420}\begin{bmatrix} 156 & 22L & 54 & -13L \\ 22L & 4L^2 & 13L & -3L^2 \\ 54 & 13L & 156 & -22L \\ -13L & -3L^2 & -22L & 4L^2 \end{bmatrix} \tag{6.29}$$

where m is the mass per unit length of the beam element (in kg/m). For the detailed derivation, readers are redirected to reference (Paz & Leigh 2004). For beam structures, the global coordinate system (see Figure 6.9(a)) can be easily selected to be the same as the local coordinate system (see Figure 6.8). Thus, it is not necessary to transform the element stiffness and mass matrices from the local to the global coordinate system. The element stiffness and mass matrices can be directly assembled to form the system stiffness and mass matrices.

In this case study, only the natural frequencies of the first 3 modes were measured. To demonstrate the symmetrical problem, the beam is divided into two parts (i.e., the left and right parts). The bending stiffnesses of the left and right parts are calculated by multiplying the nominal value, k_0, to the scaling factors θ_L and θ_R, respectively. That is:

$$k_e = \theta_L k_0 \quad \text{for} \quad e = 1,2,\ldots,5 \quad \text{and} \quad k_e = \theta_R k_0 \quad \text{for} \quad e = 6,7,\ldots,10 \tag{6.30}$$

Consider the left part of the beam is damaged with a 50% reduction in bending stiffness, where the natural frequencies of the beam model with $\theta = \{\theta_L, \theta_R\}^T = \{0.5, 1\}^T$ are calculated. The modal parameters corresponding to this damaged beam are shown in Figure 6.10. This set of natural frequencies is used for model updating of the beam with $\theta = \{\theta_L, \theta_R\}^T$ as the uncertain parameters. The objective function given in Equation (6.5) is used, and the minimization process is repeated 50 times with randomly generated initial trials, which follow a uniform distribution from 0 to 2. The model updating results are presented in Figure 6.11, where the cross markers are the initial trials and the circle markers are the minima. It is clear from the figure that there are two solutions $\hat{\theta}_1 = \{0.5, 1\}^T$ and $\hat{\theta}_2 = \{1, 0.5\}^T$ if only natural frequencies are employed in the model updating process. As the natural frequency is a global property, the changes in natural frequencies at symmetrical locations of a structural system are exactly the same.

As a verification, the natural frequencies and mode shapes of the undamaged, left-damaged, and right-damaged beams are calculated. The changes in natural frequencies (i.e., **df** in the figure title) are calculated and presented together with the matching between the undamaged and damaged mode shapes in Figure 6.12. Note that Figure 6.12(a) on the left is for the left-damaged beam, and Figure 6.12(b) on the right is for the right-damaged beam. It is very clear from the figure that the damage-induced changes in natural frequencies for the left-damaged and the right-damage beams are the same. However, the damage-induced changes in mode shapes are different. Therefore, the mode shapes, which contain local stiffness (and/or mass) information, must be considered to address the symmetrical problem in model updating.

6.1.5 Numerical case study of a truss: structural damage detection by model updating

This example is about the structural damage detection of a planar truss, as shown in Figure 6.13, based on model updating utilizing simulated noisy vibration data. The main purpose here is not only to illustrate the procedures for model updating by minimizing the errors in modal parameters but also to demonstrate the steps for generating measured vibration data by computer simulation.

6.1.5.1 Modeling of the truss and cases considered

The target truss (as shown in Figure 6.13) consists of eleven steel members made of UKB $127 \times 76 \times 13$ section with length L = 4 m. Thus, the cross-sectional area, mass density, and modulus of elasticity for all members are $A = 16.5\,\text{cm}^2$, $\rho = 13\,\text{kg/m}$ and $E = 200\,\text{GPa}$, respectively. In Figure 6.13(a), the circled numbers are node IDs, and the squared numbers are member IDs. The finite element model of the truss has 7 nodes (2 are supported and 5 movable) and 11 members (or truss elements). The total number of degrees-of-freedom (DOFs) is 14, with 10 movable DOFs and 4 supported DOFs. The numbering of DOFs is presented in Figure 6.13(b). The finite element model of a truss structure is simple, and it is briefly covered here. Interested readers are redirected to (Papadrakakis & Sapountzakis 2018) for the detailed formulations. Figure 6.14 shows the truss element employed in this case study, where i and j represent the starting and ending nodes of the truss element; e is the element number; and ϕ is the angle between the global x-axis (horizontal) and the

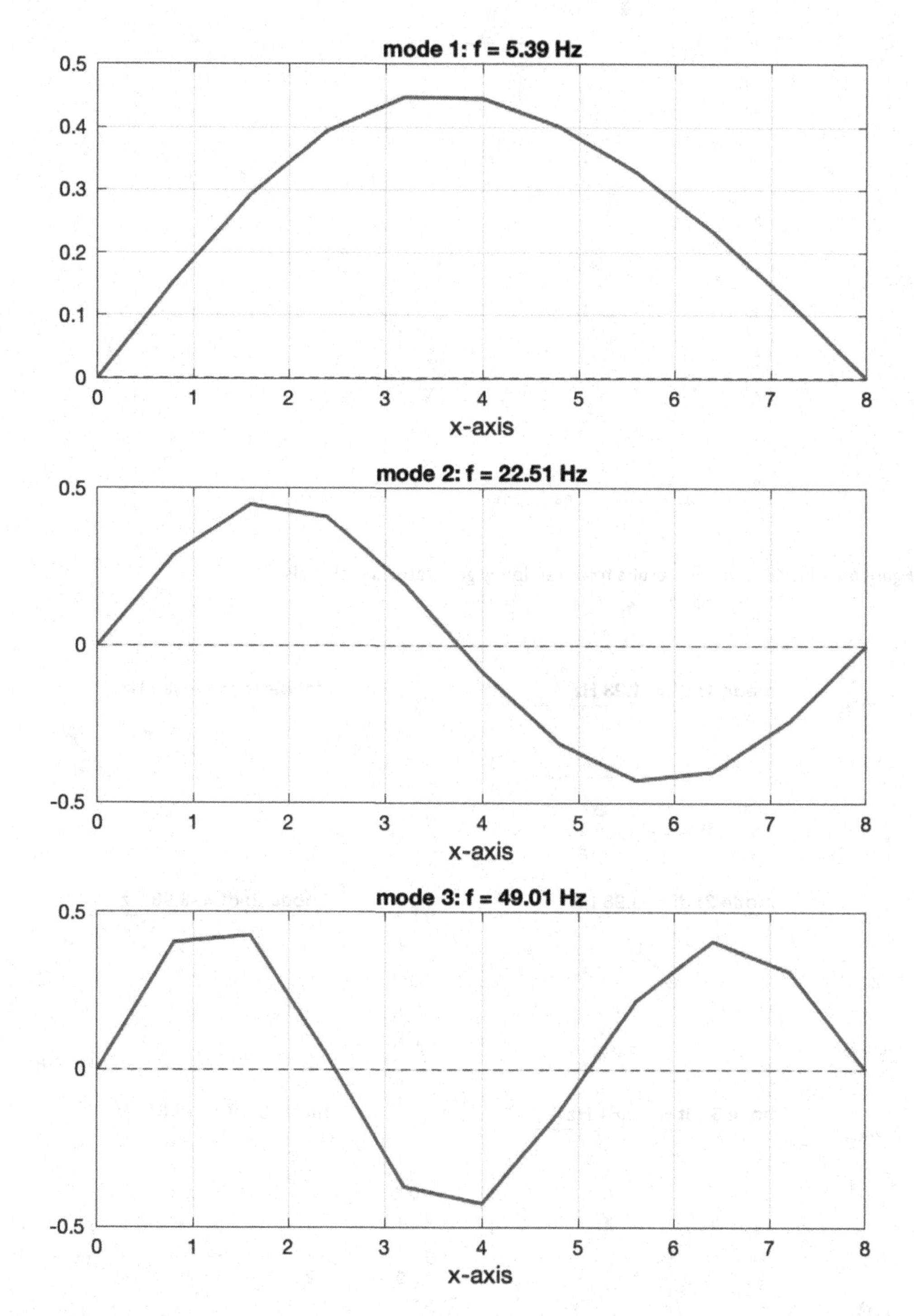

Figure 6.10 The modal parameters of the first three modes for the beam damaged on the left part.

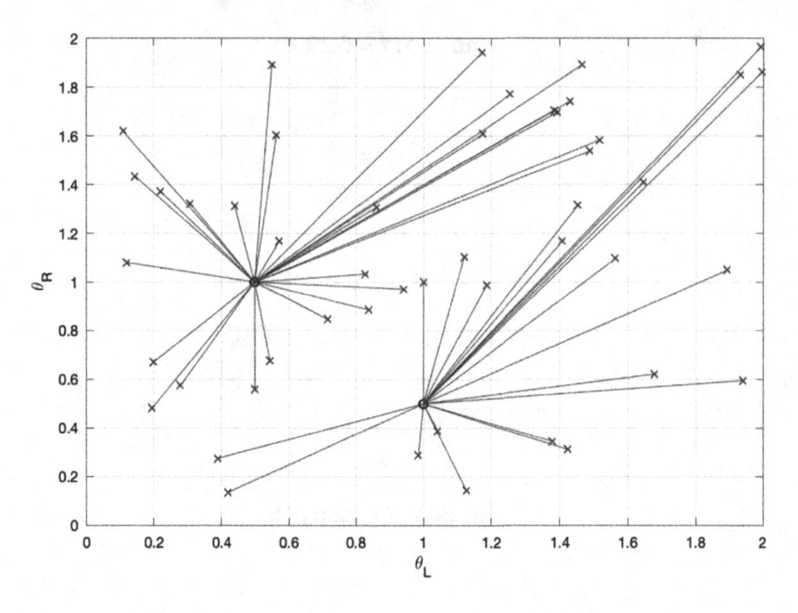

Figure 6.11 Model updating results from randomly generated initial trials.

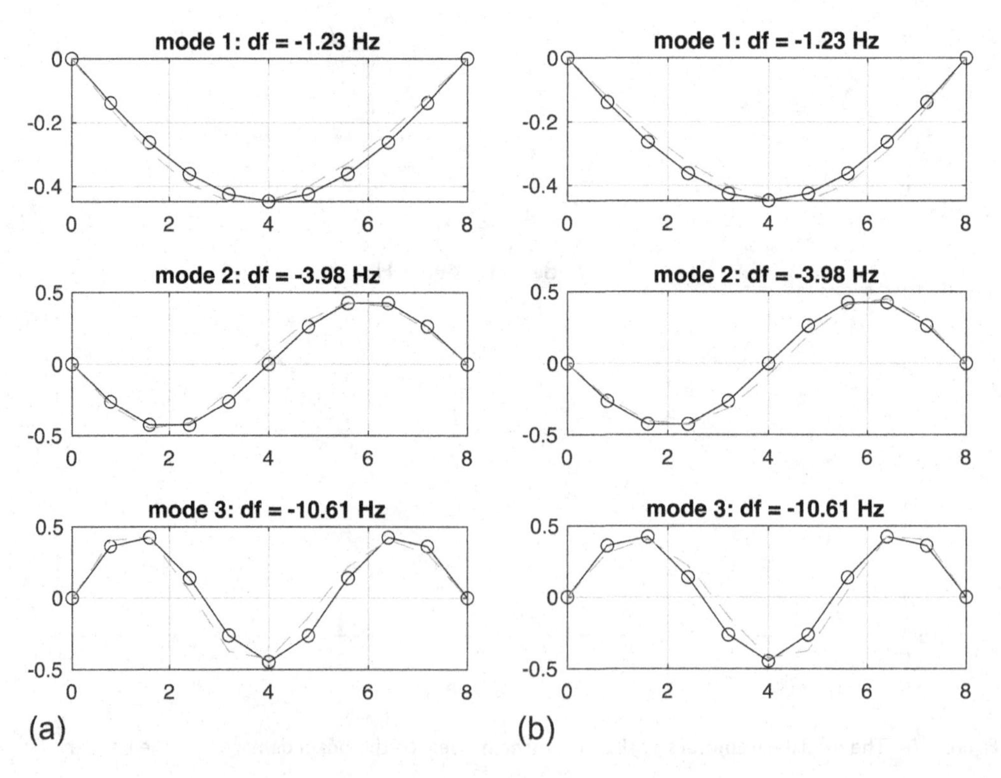

Figure 6.12 Changes in natural frequency and matching between undamaged (circle markers) and damaged beam (solid line). (a) Left-damaged. (b) Right-damaged.

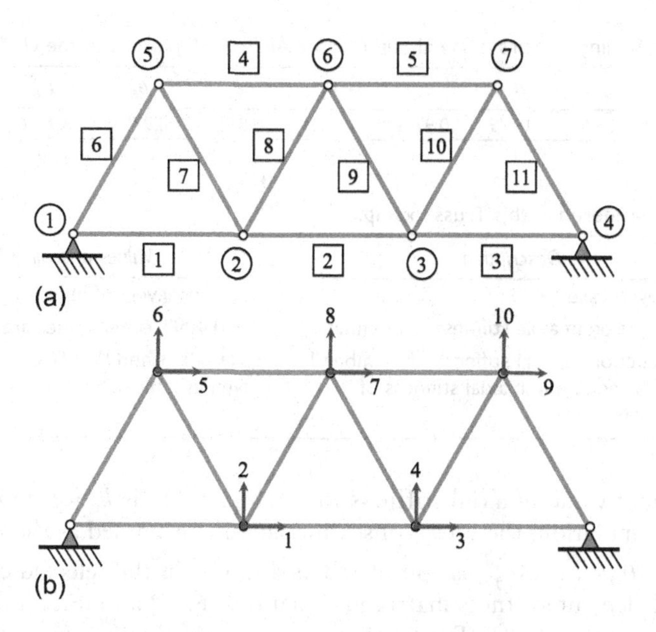

Figure 6.13 The truss example. (a) Numbering of nodes (in circles) and elements (in squares). (b) Numbering of movable DOFs.

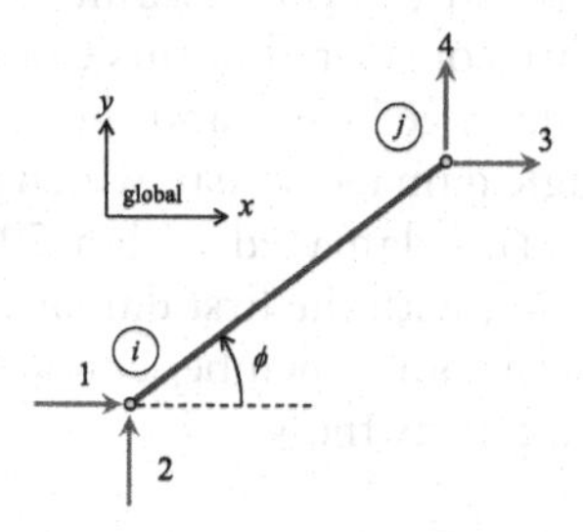

Figure 6.14 The truss element employed in this case study.

local axis of the element (i.e., along the member), which defines the orientation of the member; the four arrows show the DOFs (i.e., DOFs 1, 2, 3, and 4) of the truss element. The global element stiffness matrix of the e-th member is given by:

$$\mathbf{k}_e = k_e \begin{bmatrix} c^2 & cs & -c^2 & -cs \\ cs & s^2 & -cs & -s^2 \\ -c^2 & -cs & c^2 & cs \\ -cs & -s^2 & cs & s^2 \end{bmatrix}, \quad \text{for} \quad e = 1,\ldots,N_e \tag{6.31}$$

where $c = \cos\phi$ and $s = \sin\phi$; k_e is the axial stiffness of the e-th member; N_e is the total number of members (it is equal to 11 in this example). To simulate the variation of member stiffness due to imperfection, the real axial stiffness of each member is calculated by multiplying the nominal value, k_0, with the non-dimensional scaling factor as:

$$k_e = \theta_e k_0 \quad \text{for} \quad e = 1,2,\ldots,N_e \tag{6.32}$$

Table 6.1 Values of Scaling Factors of Axial Stiffness for All Truss Members in the Undamaged Case

θ_1	θ_2	θ_3	θ_4	θ_5	θ_6	θ_7	θ_8	θ_9	θ_{10}	θ_{11}
0.8185	0.8389	1.1294	1.0779	0.9268	1.1801	0.8138	0.9755	0.9526	1.1062	1.1181

Table 6.2 Cases Considered in this Truss Example

Case ID	*Description*	*Values of scaling factors*
UD	Undamaged case	All values given in Table 6.1.
D1	50% reduction in axial stiffness of member 1.	$\theta_1 = 0.4093$, other values are given in Table 6.1.
D2	50% reduction in axial stiffness of member 1 and 30% reduction in axial stiffness of member 6.	$\theta_1 = 0.4093$, and $\theta_6 = 0.8261$, other values are given in Table 6.1.

where the nominal value of axial stiffness for each member is $k_0 = EA / L = 82.5\,\text{MN/m}$. Based on this formulation, the vector of scaling factors (or named as the vector of model parameters) $\boldsymbol{\theta} = \{\theta_1, \theta_2, \ldots, \theta_{N_e}\}^T$ is considered as a model in the selected class of models. With the global element stiffness matrix in Equations (6.31) and (6.32), the system stiffness matrix can be assembled. For truss structures, it is convenient to use the lumped mass matrix as given in reference (Paz & Leigh 2004).

The undamaged truss is simulated using the scaling factors (randomly generated) as given in Table 6.1. Three cases are considered in this example, and they are summarized in Table 6.2. Case UD is the undamaged case, and the scaling factor values in Table 6.1 are employed. Case D1 is the single damage case in which the bottom chord member next to the left support (i.e., member 1) is damaged with a 50% reduction in axial stiffness. Case D2 is the double-damage case, with the first damage being the same as that in Case D1, and the second damage is on the brace member connecting the left support (i.e., member 6) with a 30% reduction in axial stiffness.

6.1.5.2 Simulation of measured time-domain vibrations

Once the system stiffness and mass matrices are generated, the time-domain responses of the truss structure under a given external applied force can be determined by the method of modal superposition as introduced in Section 2.2.3 (with the MATLAB function provided).

Consider a series of moving vehicles on the bridge deck of the truss bridge, where impulses will be applied to nodes 2 and 3 when the wheels go over them. Thus, the vehicle moving load is simplified to a series of impulses at nodes 2 and 3. All impulses are assumed to be identical with the same magnitude (=10 kN) and the duration of each impulse is assumed to be 0.0103s. For a sampling frequency of 2048 Hz, the impulse duration is equal to 20 time steps (i.e., the impulse takes ten time steps to go linearly from zero to 10 kN, and takes another ten time steps to go from 10 kN back to zero). The time between the front and back wheels of a vehicle is 0.5 s, and the time between two vehicles is 1.5 s. Figure 6.15 shows the first 3 s of simulated time-domain responses of the truss structure. The total duration of simulated responses is 120 s. It must be pointed out that a 5% root-mean-square (RMS) white noise is added to the model-predicted time-domain responses to capture the effect of measurement noise in the real situation. Program 6.3 shows the

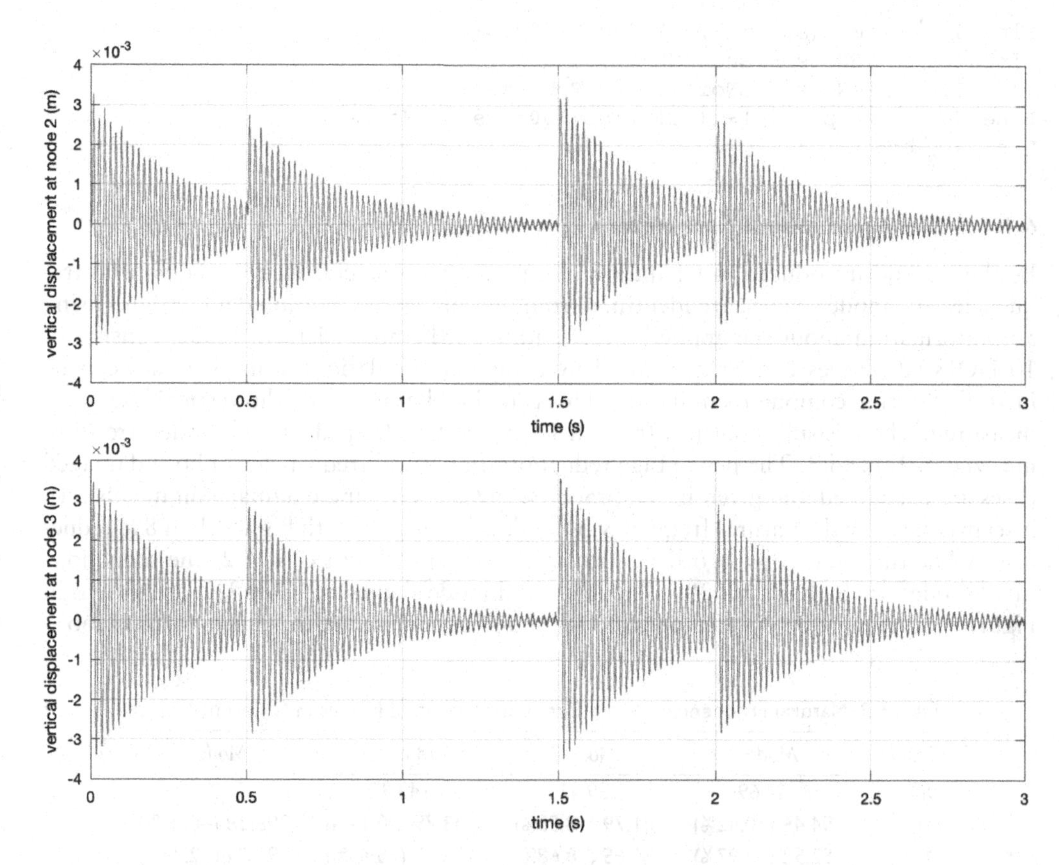

Figure 6.15 Simulated time-domain responses of the truss structure (with noise).

MATLAB script that can be used to add a certain percentage of RMS noise to the model-predicted time-domain responses (x). The dimensions of x are obtained on Line 1, where `Nt` is the number of time steps, and `Nmdof` is the number of measured DOFs. As the RMS values of responses at different measured DOFs are different, the noises must be added separately by the for-loop, as shown in Line 2. The RMS value of the response at a given measured DOF is calculated on Line 3. The noise vector (`RdNoise`) is generated by a normal distribution with zero mean and unit standard deviation on Line 4. The generated noise vector is divided by its own standard deviation on Line 5 (ensuring that the standard deviation is equal to unity). Finally, the measured responses (`xexp`) are simulated by adding the scaled noise vector to the calculated responses (x). The noise level is specified by the variable `NL`. In this case, `NL` is equal to 5.

Note that the number of sensors is usually limited in real measurement especially in field tests. In this case study, it is assumed that only vertical vibrations are measured. Thus, the measured DOFs are 2, 4, 6, 8, and 10. That is, the dimension of the measured mode shapes is 5×1.

Program 6.3: MATLAB script for adding a certain percentage of RMS noise to the model-predicted responses.

```
Line 1.   [Nt,Nmdof]=size(x);
Line 2.   for ii=1:Nmdof
```

```
Line 3.     rms=(sum(x(:,ii).^2)/Nt)^0.5;
Line 4.     RdNoise=randn(Nt,1);
Line 5.     RdNoise=RdNoise./std(RdNoise);
Line 6.     xexp(:,ii)=x(:,ii)+NL/100*rms.*RdNoise;
Line 7.  end
```

6.1.5.3 Identified modal parameters

By the method introduced in Chapter 3, the modal parameters, such as the natural frequencies and mode shapes, are identified from the simulated vibration data without using any information about the applied force. Although the finite element model consists of 10 DOFs (10 modes can be calculated by computer simulation), only 4 modes can be identified during computer simulation. This is mainly because only the vertical DOFs are measured. The measured natural frequencies of the first four identified modes are summarized in Table 6.3. The percentage reductions in natural frequencies in both damaged cases are calculated and given in the brackets. In Case D1 (single damage on member 1), it seems that only the natural frequency of mode 2 is sensitive with higher than 8% reduction, while the reductions in other modes are less than 1%. In Case D2, the reductions in all modes are higher than 2%. The measured mode shapes for Case UD are shown in Figure 6.16. It must be pointed out that the mode 2 mode shape (in Figure 6.16(b)) looks

Table 6.3 Natural Frequencies of the First Four Identified Modes in Different Cases

Case ID	*Mode 1*	*Mode 2*	*Mode 3*	*Mode 4*
UD	54.69	89.41	143.97	298.54
D1	54.46 (–0.42%)	81.79 (–8.53%)	143.79 (–0.12%)	298.38 (–0.05%)
D2	52.52 (–3.97%)	81.65 (–8.68%)	136.15 (–5.43%)	291.89 (–2.23%)

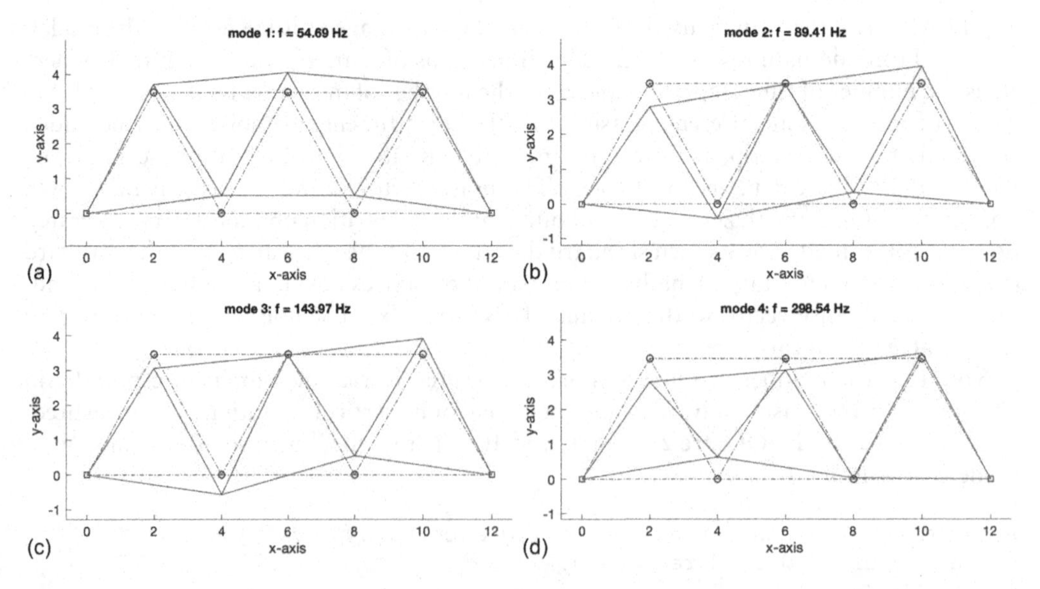

Figure 6.16 The four measured modes identified from the simulated responses (vertical DOFs only). (a) Mode 1. (b) Mode 2. (c) Mode 3. (d) Mode 4.

similar to the mode 3 mode shape (in Figure 6.16(c)), as only the vertical DOFs are measured. As the mode shapes for other cases are similar, they are not shown. The set of measured natural frequencies and mode shapes for Case UD is employed for model updating to identify the model of the damaged truss. The sets of measured modal parameters for Cases D1 and D2 are used for structural damage detection following the model updating approach.

6.1.5.4 Model updating: Case UD

The objective function as given in Equation (6.13) is employed for model updating in this case study. The scaling factors $\boldsymbol{\theta} = \{\theta_1, \theta_2, \ldots, \theta_{14}\}^T$ as defined in Equation (6.32) are treated as the design variables. To ensure the reliability of the model updating result, the minimization of the objective function is repeated 100 times with different initial trials, which are generated following a uniform distribution from 0 to 2. After generating the initial trial, the lower and upper bounds for the corresponding minimization are set to be 0.2 and 5 multiplied with the initial trial. For some initial trials, it is unavoidable that the optimization process gets trapped by a local minimum (or the numerical optimization process is ended because the prescribed maximum number of steps is achieved). When the identified 100 solutions are sorted in the ascending order of the objective function value, the first 20 solutions have almost the same objective function value of 0.0007, and the corresponding solutions are the same or very similar. The optimal solution with the smallest objective function value is presented in Table 6.4 (Case UD). Owing to the adding of 5% RMS noise in simulating the measured data, the identified values in Table 6.4 are not the same (but similar) as the simulated values in Table 6.1.

The matchings between the measured and model-predicted modal parameters are shown in Figure 6.17. The updated natural frequencies are very close to the corresponding measured natural frequencies in all four modes. The mode shapes are almost overlapping with each other in all four modes. Note that only vertical DOFs are considered in this case study. It must be pointed out that when the measured mode shapes are matched to the model-predicted mode shapes by the modal assurance criterion (MAC), the measured modes 1, 2, 3, and 4 are matched to the model-predicted modes 1, 2, 3, and 7. The model-predicted modes 4, 5, and 6 are missing in the measured modes. This case study shows the importance of mode matching between the measured and model-predicted modal parameters. If wrong modes are matched in the calculation of the objective function value, the minimization result becomes meaningless.

The identified undamaged model in Case UD is used as the baseline for structural damage detection in Cases D1 and D2 in the next section.

6.1.5.5 Structural damage detection by model updating: Cases D1 and D2

The model updating processes in Cases D1 and D2 are very similar to that in Case UD. In each of them, the minimization is repeated 100 times with different initial trials. In both damaged cases, the optimal solution in the undamaged case (UD) is treated as the first initial trial. In both Cases D1 and D2, the minima corresponding to the smallest 20 objective function values are almost the same. The model updating results are presented in Table 6.4 (see rows for Cases D1 and D2). The changes in axial stiffness for Cases D1 and D2 with respect to Case UD are calculated and presented in rows "%1" and "%2," respectively.

Table 6.4 Values of Scaling Factors of Axial Stiffness for All Truss Members in the Undamaged Case

Case	θ_1	θ_2	θ_3	θ_4	θ_5	θ_6	θ_7	θ_8	θ_9	θ_{10}	θ_{11}
UD	0.7863	0.6455	1.1927	1.0746	1.0545	1.1399	0.8168	0.9799	0.8986	1.0572	1.0756
D1	0.4174	0.6415	1.1634	1.0788	1.0848	1.1575	0.8100	0.9823	0.8870	1.0251	1.0745
%1	**–46.92**	–0.61	–2.46	0.40	2.87	1.54	–0.83	0.24	–1.29	–3.03	–0.1
D2	0.3665	0.6624	1.2727	1.1489	1.1498	0.8078	0.8080	0.9625	0.8328	0.9899	0.9899
%2	**–53.38**	2.62	6.70	6.92	9.03	**–29.14**	–1.07	–1.78	–7.31	–6.37	–0.81

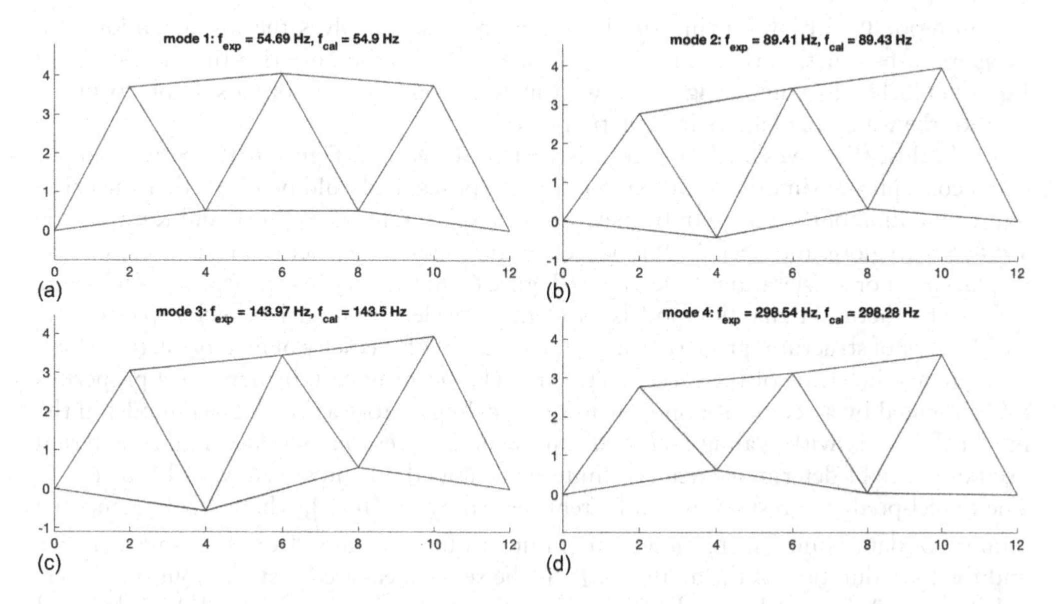

Figure 6.17 The matchings between the measured and updating modal parameters in Case UD. (a) Mode 1. (b) Mode 2. (c) Mode 3. (d) Mode 4.

Considering the percentage changes for Case D1, except for member 1, the percentage changes in other members are very small (with the largest being –3.03 for member 10). It is clear that the percentage reduction of axial stiffness for member 1 is about 47%, which is very close to the true value of 50%, as shown in Table 6.2. It can be concluded that the model updating method successfully identified the simulated damage in Case D1.

Next, the percentage changes for Case D2 are considered. It is clear that the percentage reductions for members 1 and 6 are about 53% and 29%, respectively. These values are very close to the true values of 50% and 30%, as shown in Table 6.2. However, it is also observed that the errors in this case are relatively high. For example, the percentage change for member 5 is about 9%. Nevertheless, the simulated damages in Case D2 are also successfully identified.

It must be pointed out that the measured modal parameters in this case study are obtained from modal identification (by the method in Chapter 3) of the simulated time-domain responses of the target structure, but not from solving the eigenvalue problem of the truss model. This process makes the model updating (or damage detection) processes closer to reality (even though it is a numerical case study). The limitation in this case study is that the problem of error in model class is not considered. That is, the class of models for simulating the measured data is the class of models employed in model updating. This limitation can be eliminated by using experimental case studies (or field tests) as done at the end of this chapter.

6.1.6 Determination of the weighting factors following the Bayesian approach

The deterministic model updating methods aim at pinpointing a model in the parameter space. However, the probabilistic methods focus on calculating the posterior (updated) probability density function (PDF) of uncertain model parameters. The calculation of the posterior PDF by following the Bayesian approach involves the minimization of a measure-of-fit function that is in a very similar form to the objective function given in Equation (6.15). By comparing these two functions, one can find out a suitable formulation for the weighting factors in Equation (6.15).

To do this, the Bayesian framework is first reviewed (see Figure 6.18). Some complicated concepts are simplified for explanation purposes. It should be noted that the original formulation of the Bayesian framework in classical references (Beck and Katafygiotis 1998; Katafygiotis and Beck 1998) focused on time-domain structural responses, such as displacement or acceleration. Referring to Figure 6.18, the very first procedure is to define a class of structural dynamics models $\mathcal{M}$ (e.g., finite element model), which is parameterized by a set of structural properties (e.g., the modulus of elasticity of members), to capture the dynamic behavior of the target structure. The set of uncertain structural properties is represented by a vector of non-dimensional scaling factors **a**. For a given model in the model class (i.e., with scaling factors $\boldsymbol{\theta}$), the structural responses under a given external excitation can be determined (e.g., by finite element method and the Newmark β method). The model-predicted responses can be represented by $Q_N\left(\boldsymbol{\theta},\hat{\mathbf{I}}_N\right)$, where N represents the number of data points in the time-domain (it can be calculated from the time step size and the total duration of the analysis); $\hat{\mathbf{I}}_N$ is the set of measured system input (the external excitation). It must be emphasized that different sets of responses will be calculated for different models $\boldsymbol{\theta}$ from the model class $\mathcal{M}$. On top of this deterministic model class, a class of probabilistic model class $\mathcal{P}$ for the output error E_N must also be defined. With the principle of maximum entropy, the output error can be assumed to follow Gaussian distributions with uncertainty $\boldsymbol{\sigma}$ (the parameters of the probabilistic model). Similarly, different values of $\boldsymbol{\sigma}$ will result in different PDFs, $p\left(E_N|\sigma\right)$. The two classes of models can be combined and represented by $\mathcal{M}_P$. With a set of measured responses, $\hat{\mathbf{Y}}_N$, the PDF of calculating the set of measured responses conditional on the model (i.e., $\boldsymbol{\theta}$ and $\boldsymbol{\sigma}$) and the measured system input can be calculated. With Bayes's theorem, the prior PDF, $p(\theta,\sigma)$, (from the subjective judgment of the users) can be incorporated, and the posterior PDF of uncertain parameters conditional on the set of measured data (including both the measured system input and output), $p\left(\theta,\sigma|\hat{\mathbf{I}}_N,\hat{\mathbf{Y}}_N\right)$, can be determined.

Under the original formulation in the time-domain, the measured system input, $\hat{\mathbf{I}}_N$, is essential. However, it is usually difficult or very expensive to measure. Consider

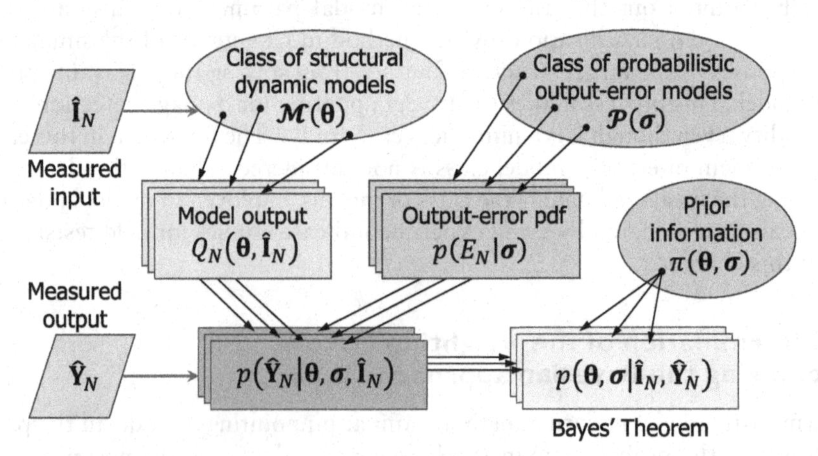

Figure 6.18 The Bayesian system identification framework (simplified).

a vibration test of a box-girder bridge as an example; one may install an actuator at the mid-span of the bridge and generate forces with a sine sweep signal in the time-domain. A load cell must be installed on the actuator to measure the system input to the structure. From the vibration test viewpoint, the magnitude of the applied force should be large in order to obtain a reasonable signal-to-noise ratio. However, a large force may cause damage to the bridge, and it is difficult to get approval from the bridge owner (usually the highways department of the country) to carry out the field test. Furthermore, the bridge must be closed during the vibration test as vehicle loads, which are extremely difficult to measure with acceptable accuracy, will certainly induce significant vibration on the bridge. Thus, the normal operation of the structure will be seriously affected by the vibration test. Furthermore, the measured responses may not necessarily be induced solely by the applied excitation (from the actuator). Other sources of possible excitations, such as wind action and traffic actions from the surrounding region, will contribute to the measured structural responses. As a result, it is more convenient to carry out model updating in the modal-domain. In 2000, the time-domain formulation was extended to modal-domain by Vanik et al. (2000). By following Bayes's theorem, the posterior PDF of uncertain model parameters $\boldsymbol{\theta}$ conditional on the set of measured natural frequencies and mode shapes, $\mathcal{D}$, and the assumed class of models, $\mathcal{M}_P$, is expressed as:

$$p(\theta|\mathcal{D},\mathcal{M}_P) = c_0 p(\theta|\mathcal{M}_P) p(\mathcal{D}|\theta,\mathcal{M}_P) \tag{6.33}$$

where c_0 is a normalizing constant to ensure the integration of the posterior PDF in the parameter space is equal to unity; $p(\theta|\mathcal{M}_P)$ is the prior PDF of uncertain model parameters conditional on the model class; $p(\mathcal{D}|\theta,\mathcal{M}_P)$ is the likelihood which is the most important term as it is where the measured data come into the picture. If one wants the posterior PDF to completely rely on the set of measured data, a non-informative prior PDF may be employed. The model parameters that maximize the posterior PDF are the most probable values (MPVs) of model parameters. The formulation of the posterior PDF of model parameters will be given below.

According to Lam et al. (2014), the error between the measured and model-predicted natural frequency (i.e., the prediction error of natural frequency) of the r-th mode and the s-th set of data can be expressed as:

$$\varepsilon_{\hat{f}_{r,s}} = \hat{f}_{r,s} - f_r(\theta) \tag{6.34}$$

Assuming that the prediction error for natural frequency follows a zero-mean multi-variate Gaussian distribution, the PDFs for the measured natural frequency of the r-th mode and the s-th set of data can be expressed as:

$$p(\hat{f}_{r,s}|\theta,\mathcal{M}_P) = c_1 \exp\left(-\frac{1}{2\sigma^2_{\hat{f}_{r,s}}}\left(\hat{f}_{r,s} - f_r(\theta)\right)^2\right) \tag{6.35}$$

where $\sigma^2_{\hat{f}_{r,s}}$ is the variance of the measured natural frequency of the r-th mode and the s-th set of data. The prediction error for mode shape of the r-th mode and the s-th set of data can be expressed as:

$$\varepsilon_{\hat{\psi}_{r,s}} = \hat{\psi}_{r,s} - \Gamma_s \phi_r(\theta) = \hat{\psi}_{r,s} - \psi_r(\theta) \tag{6.36}$$

where $\mathbf{\Gamma}_s$ is the selection matrix for the s-th set of data, which is defined in the same way as the one in Equation (6.9). Assuming that the prediction error for mode shape follows a zero-mean multi-variate Gaussian distribution, the PDFs for the measured mode shape of the r-th mode and the s-th set of data can be expressed as:

$$\begin{aligned} p\left(\hat{\psi}_{r,s} \middle| \theta, \mathcal{M}_P\right) &= c_2 \exp\left(-\frac{1}{2\sigma_{\hat{\psi}_{r,s}}^2} \left(\hat{\psi}_{r,s} - \psi_r(\theta)\right)^T \left(\hat{\psi}_{r,s} - \psi_r(\theta)\right)\right) \\ &= c_2 \exp\left(-\frac{1}{2\sigma_{\hat{\psi}_{r,s}}^2}\left(\hat{\psi}_{r,s}^T\hat{\psi}_{r,s} - \hat{\psi}_{r,s}^T\psi_r(\theta) - \psi_r^T(\theta)\hat{\psi}_{r,s} + \psi_r^T(\theta)\psi_r(\theta)\right)\right) \\ &= c_2 \exp\left(-\frac{1}{2\sigma_{\hat{\psi}_{r,s}}^2}\left(\left\|\hat{\psi}_{r,s}\right\|^2 - 2\hat{\psi}_{r,s}^T\psi_r(\theta) + \left\|\psi_r(\theta)\right\|^2\right)\right) \end{aligned} \tag{6.37}$$

Note that $\mathbf{a}^T\mathbf{b}$ $(= \mathbf{b}^T\mathbf{a})$ represents the dot product between the column vectors $\mathbf{a}$ and $\mathbf{b}$, and $\mathbf{a}^T\mathbf{a}$ is the square of the length of the vector $\mathbf{a}$ (and is equal to $\|\mathbf{a}\|^2$). It should be remembered that both $\hat{\psi}_{r,s}$ and $\psi_r(\theta)$ are normalized to have unit length. Therefore:

$$p\left(\hat{\psi}_{r,s} \middle| \theta, \mathcal{M}_P\right) = c_2 \exp\left(-\frac{1}{2\sigma_{\hat{\psi}_{r,s}}^2} 2\left(1 - \hat{\psi}_{r,s}^T\psi_r(\theta)\right)\right) \tag{6.38}$$

where c_1 in Equation (6.35) and c_2 in Equation (6.38) are two normalizing constants. Assuming that the measured natural frequency and mode shape are independent, the PDF of measured modal parameters of the r-th mode and the s-th set of data can be expressed as:

$$p\left(\hat{f}_{r,s}, \hat{\psi}_{r,s} \middle| \theta, \mathcal{M}_P\right) = c_1 c_2 \exp\left(-\frac{1}{2}\left(\frac{\left(\hat{f}_{r,s} - f_r(\theta)\right)^2}{\sigma_{\hat{f}_{r,s}}^2} + \frac{2\left(1 - \hat{\psi}_{r,s}^T\psi_r(\theta)\right)}{\sigma_{\hat{\psi}_{r,s}}^2}\right)\right) \tag{6.39}$$

Assuming that the measured modal parameters from different modes and different data sets are independent, the likelihood $p(\mathcal{D}|\theta, \mathcal{M}_P)$ can be expressed as:

$$\begin{aligned} p\left(\mathcal{D} \middle| \theta, \mathcal{M}_P\right) &= \prod_{r=1}^{N_m}\prod_{s=1}^{N_s} p\left(\hat{f}_{r,s}, \hat{\psi}_{r,s} \middle| \theta, \mathcal{M}_P\right) \\ &= c_3 \exp\left(-\frac{1}{2}\sum_{r=1}^{N_m}\sum_{s=1}^{N_s}\left(\frac{\left(\hat{f}_{r,s} - f_r(\theta)\right)^2}{\sigma_{\hat{f}_{r,s}}^2} + \frac{2\left(1 - \hat{\psi}_{r,s}^T\psi_r(\theta)\right)}{\sigma_{\hat{\psi}_{r,s}}^2}\right)\right) \end{aligned} \tag{6.40}$$

where c_3 is another normalizing constant. By substituting the likelihood in Equation (6.40) into Equation (6.33), one can get the posterior PDF of model parameters as:

$$p\left(\theta \middle| \mathcal{D}, \mathcal{M}_P\right) = c p\left(\theta \middle| \mathcal{M}_P\right) \exp\left(-\frac{1}{2} J(\theta)\right) \tag{6.41}$$

where $c = c_3c_0$ is also a normalizing constant. $J(\boldsymbol{\theta})$ is the measure-of-fit function, and it is given by:

$$J(\boldsymbol{\theta}) = \sum_{r=1}^{N_m}\sum_{s=1}^{N_s}\left(\frac{\left(\hat{f}_{r,s} - f_r(\boldsymbol{\theta})\right)^2}{\sigma^2_{\hat{f}_{r,s}}} + \frac{2\left(1 - \hat{\psi}^T_{r,s}\psi_r(\boldsymbol{\theta})\right)}{\sigma^2_{\hat{\psi}_{r,s}}}\right) \tag{6.42}$$

Finally, one can compare the measure-of-fit function in Equation (6.42) to the objective function in Equation (6.15). The weighting factors of natural frequencies and mode shapes of the r-th mode and the s-th set of data in Equation (6.15) are:

$$w_{f,r,s} = \frac{1}{\text{COV}^2\left(\hat{f}_{r,s}\right)} \quad \text{and} \quad w_{\phi,r,s} = \frac{2}{\text{COV}^2\left(\hat{\psi}_{r,s}\right)} \tag{6.43}$$

where $\text{COV}(\cdot)$ represents the coefficient of variation. This result shows that the weighting factor of a modal parameter is inversely proportional to the square of its COV. It is very clear that a larger weighting should be applied to the measured data with higher accuracy and vice versa.

6.2 NUMERICAL OPTIMIZATION ALGORITHMS

After formulating the objective function (e.g., the discrepancy between the measured and model-predicted modal parameters as shown in Equation (6.15)), the model updating problem is converted to an optimization problem. As this kind of objective function usually involves the formation of system stiffness and mass matrices as well as solving of eigenvalue problems, it is extremely difficult if not impossible to express the objective function explicitly in terms of the design variables (i.e., the uncertain model parameters—the non-dimensional scaling factors of different physical member properties, such as the modulus of elasticity of various structural members). Thus, analytically solving the optimization problem becomes impossible, and numerical methods have to be adopted. One may carry out the minimization using the MATLAB optimization toolbox as demonstrated in the case studies, in which the optimization algorithm is considered as a black box. In fact, researchers may develop their own numerical optimization algorithms instead of relying on commercial applications in minimizing the errors in modal parameters for the purpose of model updating. The main objective of this section is to introduce several commonly used numerical optimization algorithms for structural model updating. It must be pointed out that this chapter is not for general optimization theories or the usages of optimization in other fields, such as the optimization design of structural systems.

Depending on the complexity of the model class of the target structure, the amount of information that can be extracted from the measured data, and the number of design variables, the result of optimization can be classified into several categories:

1. The most desirable type of result is a single unique solution. This may happen when the quality and quantity of measured data are high and the number of design variables is small.

2. The second type of result is the multiple discrete solutions in the parameter space of interest. This is a typical non-uniqueness problem. If one carries out the numerical optimization many times with different initial trials, the results of different trials will go to several solutions with the same (or very similar) objective function value.
3. The third type of result is the infinite number of solutions in the parameter space of interest. There are "regions" at which the objective function values are the same or very close. These regions usually have dimensions lower than the dimensions of the parameter space. For example, if there are 2 design variables (i.e., the parameter space is 2-dimensional (2D)), the regions may be in 1D. Therefore, the "regions" are some lines or curves on the 2D area of the parameter space. This is a very serious non-uniqueness problem, and it may happen when the quality and quantity of measured data are low while the number of design variables is large. Therefore, the available information is far too small to identify the model (or the design variables).

Most numerical optimization algorithms should be able to handle the type 1 optimization results. For the type 2 results, one may have to use probabilistic optimization algorithms. If deterministic algorithms have to be used, the optimization process may need to be repeated many times with a different initial trial each time. It is preferable to have the initial trials distributed throughout the parameter space of interest to have a higher chance of obtaining all the solutions with the same or similar objective function value.

In general, the computational time for solving the optimization problem, increases significantly as the number of design variables increases. Furthermore, the computational cost will be higher if the size of the structural model to be updated is larger (i.e., the number of DOFs is large). In this section, MATLAB will be used to code and implement the introduced numerical optimization algorithm.

6.2.1 General formulation

The optimization problem must be formulated before it can be solved by different optimization techniques. The standard formulation of an optimization problem can be expressed as:

Minimize:

$$J(\theta) \tag{6.44}$$

Subject to:

$$g_i(\theta) \le 0 \quad \text{for} \quad i = 1, \ldots, N_g \tag{6.45}$$

$$h_j(\theta) = 0 \quad \text{for} \quad j = 1, \ldots, N_h \tag{6.46}$$

$$\theta_k^L \le \theta_k \le \theta_k^U \quad \text{for} \quad k = 1, \ldots, N_\theta \tag{6.47}$$

with design variables:

$$\theta = \{\theta_1, \theta_2, \ldots, \theta_{N_\theta}\}^T \tag{6.48}$$

In the formulation of an optimization problem (especially for the purpose of structural model updating), one needs to identify:

1. **Design variables:** A vector of design variables $\boldsymbol{\theta}$ as shown in Equation (6.48) must be defined at the very beginning (this is the same as the one in Equation (6.1)), where N_θ is the total number of design variables. For the model updating problem as defined in Section 6.1, this is the vector of scaling factors of the uncertain model properties. For example, $\boldsymbol{\theta}$ can be the non-dimensional scaling factor of the inter-story stiffness for a shear building as defined in Equation (6.2). For a structure that can be divided into a given number of sub-systems, the design variables $\boldsymbol{\theta}$ can be defined by Equation (6.3). The values of design variables will be updated in each iteration during the optimization process.
2. **Objective function:** For the purpose of model updating, an objective function, $J(\boldsymbol{\theta})$, as given in Equation (6.44) is a measure of how "good" or "bad" the model can be used to represent the real structure at the measured DOFs. Its value certainly depends on the values of the selected design variables. In the general formulation, the objective function is always minimized. In case one needs to maximize the objective function, the maximization problem can be transformed into a minimization problem in some simple ways. For example, by multiplying the objective function with –1 (i.e., $-J(\boldsymbol{\theta})$), or taking the inverse of the objective function (i.e., $1/J(\boldsymbol{\theta})$). Optimization with more than one objective function is called multi-objective optimization. This kind of optimization problem is also called Pareto optimization, which can be used to address the optimal sensor configuration problems as shown in the literature (Papadimitriou 2005).
3. **Constraints:** Constraints are series of restrictions (or limitations) that must be satisfied in order to produce a physical and stable structural model. Three kinds of constraints are considered here:
   a) **Upper and lower bounds of design variables:** It is very important to define the upper and lower bounds of uncertain model parameters for the purpose of structural model updating. Almost all mechanical properties of structural members or structural components must have non-zero positive values. Consider the modulus of elasticity of a structural member as an example, if a negative value is assigned to its design variable (by the optimization algorithm), the structural system may become unstable. If modal parameters are employed in formulating the objective function, the analysis may be stopped with an error. Not only negative values, a value equal to zero or very close to zero for the design variables may also result in numerical difficulties. Equation (6.47) in the standard formulation shows the upper bound (i.e., θ_k^U) and lower bound (i.e., θ_k^L) of the k-th design valuable (i.e., θ_k), for $k = 1,\ldots,N_\theta$. In fact, these bounds define the parameter space of interest. The computational time required will be increased if the assigned bounds cover a very wide range of space. An appropriate set of upper/lower bounds is important for the success of the optimization and the model updating problem.
   b) **Inequality constraints:** In the general formulation, only the "smaller than or equal to" inequality constraints are considered as shown in Equation (6.45). If the "larger than or equal to" constraints are necessary, one may consider multiplying the corresponding constraints with –1 to convert them to the "smaller than or equal to" constraints. The total number of inequality constraints is N_g.

The inequality constraints can also be used to define the upper/lower bounds of design variables. However, this is not recommended. In most existing numerical optimization applications (e.g., the optimization toolbox in MATLAB), values of design variables that violate the inequality constraints may still be tried in evaluating the objective function during the optimization process (they will not be considered in the solution). If such values result in a numerical problem, the optimization process will be forced to stop. For model updating problems in the modal-domain, one may use inequality constraints to enforce the natural frequencies of the structure to be larger or smaller than some pre-defined values. One should avoid using unnecessary constraints, as it will certainly increase the required computational time. Furthermore, it may result in some unexpected effects or problems with the optimization result.

c) **Equality constraints:** In the general formulation, equality constraints are represented by Equation (6.46). The total number of equality constraints is N_h. Equality constraints are very "strong," and it may result in the "no solution" problem. In fact, equality constraints are not common in optimization problems for model updating.

From the numerical optimization viewpoint, the use of non-dimensional scaling factors (instead of the physical model parameters) for optimization is very important and common. As discussed at the beginning of Section 6.1, the order of magnitude may differ from one uncertain model parameter to another. Consider the model updating of a simply supported steel beam using measured acceleration responses as an example, one may consider the modulus of elasticity of the beam and the damping ratio of the system (assuming that the system is classically damped) as uncertain parameters. The value of the modulus of elasticity is about 200×10^9 Pa, while that of the damping ratio is about 0.01. The difference in values is so large that it will affect the efficiency of the optimization process.

Since both the inequality and equality constraints are not common in optimization processes for solving model updating problems, both of them will be discarded in the following discussions and formulations.

6.2.2 Uniqueness of optimization solution

In the application of optimization techniques to model updating problems of civil engineering structures, such as buildings and bridges, it is seldom possible to ensure that a unique global optimum solution will be found. The reason can either be that the algorithm cannot identify the unique global solution or such a unique solution does not exist.

From a practical standpoint, a simple (but time-consuming) approach is to start the optimization process from several different initial trials, and if the optimization results are essentially the same, the chance of obtaining the true optimum is high. It will be useful if one can mathematically check if a result (from the numerical optimization algorithm) is really a minimum (at least a local minimum).

Let $\boldsymbol{\theta}^*$ be the result from the optimization algorithm, the necessary condition for it to be a minimum is:

$$\nabla J\left(\boldsymbol{\theta}^*\right) = 0 \tag{6.49}$$

where ∇ is the gradient operator, and it returns the gradient vector of the scalar function $J(\boldsymbol{\theta})$ (i.e., the vector of the first partial derivative of the objective function with respect to the design variables) as:

$$\nabla J(\boldsymbol{\theta}) = \left\{ \frac{\partial J(\boldsymbol{\theta})}{\partial \theta_1}, \frac{\partial J(\boldsymbol{\theta})}{\partial \theta_2}, \ldots, \frac{\partial J(\boldsymbol{\theta})}{\partial \theta_{N_\theta}} \right\}^T \tag{6.50}$$

Note that:

$$\nabla J(\boldsymbol{\theta}^*) = \nabla J(\boldsymbol{\theta})\Big|_{\boldsymbol{\theta}=\boldsymbol{\theta}^*} \quad \text{and} \quad \frac{\partial J(\boldsymbol{\theta}^*)}{\partial \theta_i} = \frac{\partial J(\boldsymbol{\theta})}{\partial \theta_i}\bigg|_{\boldsymbol{\theta}=\boldsymbol{\theta}^*} \quad \text{for} \quad i = 1, \ldots, N_\theta \tag{6.51}$$

If $\boldsymbol{\theta}^*$ can vanish all "slopes" in the gradient vector, it satisfies the necessary condition to be a minimum. Certainly, this is only a necessary but not sufficient condition. It is clear that points (i.e., different models $\boldsymbol{\theta}$ in the parameter space) that can satisfy the condition in Equation (6.49) can be a minimum, a maximum, or a point of inflection.

To obtain the sufficient condition, the Taylor series expansion of the objective function, $J(\boldsymbol{\theta})$, at the point $\boldsymbol{\theta}^*$, is considered:

$$J(\boldsymbol{\theta}) = J(\boldsymbol{\theta}^*) + d\boldsymbol{\theta}^T \nabla J(\boldsymbol{\theta}^*) + \frac{1}{2} d\boldsymbol{\theta}^T \nabla^2 J(\boldsymbol{\theta}^*) d\boldsymbol{\theta} + O(d\boldsymbol{\theta}^3) \tag{6.52}$$

where $d\boldsymbol{\theta} = \boldsymbol{\theta} - \boldsymbol{\theta}^*$ and $O(d\boldsymbol{\theta}^3)$ represent all the terms with third order or higher. These terms will be sufficiently small if $d\boldsymbol{\theta}$ is small, and $\nabla^2 J(\boldsymbol{\theta}^*)$ is the Hessian matrix, $\mathbf{H}(\boldsymbol{\theta}^*)$, of the objective function (i.e., the matrix of second partial derivatives of the objective function with respect to the design variables) evaluated at $\boldsymbol{\theta}^*$. The Hessian matrix $\mathbf{H}(\boldsymbol{\theta})$ can be expressed as:

$$\nabla^2 J(\boldsymbol{\theta}) = \mathbf{H}(\boldsymbol{\theta}) = \begin{bmatrix} \frac{\partial^2 J(\boldsymbol{\theta})}{\partial \theta_1^2} & \frac{\partial^2 J(\boldsymbol{\theta})}{\partial \theta_1 \partial \theta_2} & \cdots & \frac{\partial^2 J(\boldsymbol{\theta})}{\partial \theta_1 \partial \theta_{N_\theta}} \\ \frac{\partial^2 J(\boldsymbol{\theta})}{\partial \theta_2 \partial \theta_1} & \frac{\partial^2 J(\boldsymbol{\theta})}{\partial \theta_2^2} & \cdots & \frac{\partial^2 J(\boldsymbol{\theta})}{\partial \theta_2 \partial \theta_{N_\theta}} \\ \vdots & \vdots & \ddots & \vdots \\ \frac{\partial^2 J(\boldsymbol{\theta})}{\partial \theta_{N_\theta} \partial \theta_1} & \frac{\partial^2 J(\boldsymbol{\theta})}{\partial \theta_{N_\theta} \partial \theta_2} & \cdots & \frac{\partial^2 J(\boldsymbol{\theta})}{\partial \theta_{N_\theta}^2} \end{bmatrix} \tag{6.53}$$

If $\boldsymbol{\theta}^*$ is a minimum (or at least a local minimum) of the objective function, the objective function value at $\boldsymbol{\theta}^*$ (i.e., $J(\boldsymbol{\theta}^*)$) must be smaller than all other objective function values in its close neighborhood. Therefore:

$$J(\boldsymbol{\theta}) - J(\boldsymbol{\theta}^*) > 0 \tag{6.54}$$

By using Equation (6.52) and neglecting the terms with order higher than 3, the condition in Equation (6.54) can be expressed as:

$$\nabla J(\boldsymbol{\theta}^*)^T d\boldsymbol{\theta} + \frac{1}{2} d\boldsymbol{\theta}^T \nabla^2 J(\boldsymbol{\theta}^*) d\boldsymbol{\theta} > 0 \tag{6.55}$$

Referring to the necessary condition in Equation (6.49), Equation (6.55) can be simplified to:

$$d\theta^T \mathbf{H}\left(\theta^*\right) d\theta > 0 \tag{6.56}$$

Thus, the sufficient condition is that the Hessian matrix of the objective function evaluated at θ^* must be a positive definite matrix. In other words, all eigenvalues of $\mathbf{H}\left(\theta^*\right)$ must be positive. An alternative to solving the eigenvalue problem is to calculate the determinants of all the principal minors of $\mathbf{H}\left(\theta^*\right)$. If all determinants are positive, the sufficient condition in Equation (6.56) will also be satisfied.

Example 6.1

Determine if the matrix **A** as defined below is positive definite by (a) checking the determinants of all principal minors; (b) checking the eigenvalues.

$$\mathbf{A} = \begin{bmatrix} 9 & -4 & -1 & 9 \\ 12 & 30 & 17 & 8 \\ 12 & 16 & 24 & -15 \\ 29 & 3 & 19 & 31 \end{bmatrix}$$

SOLUTION

(a) The determinant of the first principal minor is:

$$|9| = 9$$

The determinant of the second principal minor is:

$$\begin{vmatrix} 9 & -4 \\ 12 & 30 \end{vmatrix} = 318$$

The determinant of the third principal minor is:

$$\begin{vmatrix} 9 & -4 & -1 \\ 12 & 30 & 17 \\ 12 & 16 & 24 \end{vmatrix} = 4536$$

The determinant of the fourth principal minor is:

$$\begin{vmatrix} 9 & -4 & -1 & 9 \\ 12 & 30 & 17 & 8 \\ 12 & 16 & 24 & -15 \\ 29 & 3 & 19 & 31 \end{vmatrix} = 126511$$

As all determinants are positive, matrix **A** is positive definite.

(b) MATLAB is employed to solve the eigenvalue problem of **A** and return the eigenvalues.

```
>> A= [9     -4      -1      9

      12     30      17      8
      12     16      24      -15
      29     3       19      31];
>> eig(A)
ans =
      6.3289
      16.1812
      42.2531
      29.2368
```

As all eigenvalues (i.e., 6.3289, 16.1812, 42.2531, and 29.2368) are positive, matrix **A** is positive definite, as expected.

It is concluded that if the gradient vector $\nabla J\left(\boldsymbol{\theta}^*\right)$ is equal to a zero vector and the Hessian matrix $\mathbf{H}\left(\boldsymbol{\theta}^*\right)$ is positive definite, the numerical optimization result $\boldsymbol{\theta}^*$ is a minimum (at least a local minimum). However, there is no guarantee that it is a global minimum.

6.2.3 Single-variable unconstrained optimization

Although many applications of optimization algorithms involve multiple design variables with constraints, single-variable unconstrained optimization techniques provide the fundamental solution method for the development of constrained optimization techniques for multiple design variables. Moreover, most model updating optimization problems are unconstrained (the upper/lower bounds of design variables are not considered through inequality or equality constraints). Therefore, this section focuses on numerical algorithms for addressing single-variable unconstrained optimization problems as background knowledge. In this section, several fundamental algorithms that can help the basic understanding of numerical optimization algorithms are introduced.

6.2.3.1 Golden-section method

The golden-section method is believed to be one of the most famous single-variable minimization methods. It is a zero-order method, and therefore, higher-order information, such as the gradient, is not required in the calculation. As the method is simple, it is easy to code and implement. This method is designed for objective functions that are unimodal.

One of the most important elements in the golden-section method is the golden ratio, for which the definition is illustrated in Figure 6.19. Consider a straight-line AB with length L, where there exists a point C that cuts the line into two segments with lengths a and b, where $a > b$, such that the ratio between the length of the large segment to that of the short segment (a/b) is equal to the ratio between the length of the line to that of the large segment ($L/a = (a+b)/a$). This ratio is called the golden ratio, G, and it can be expressed as:

$$G = \frac{a}{b} = \frac{a+b}{a} \quad \text{for} \quad a > b \tag{6.57}$$

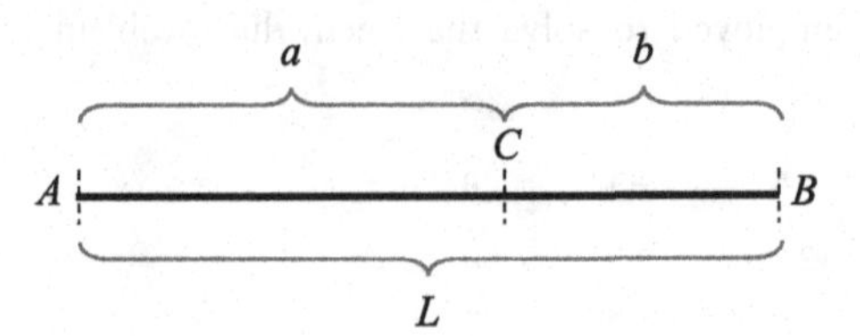

Figure 6.19 Definition of the golden ratio.

Express everything in terms of G, one obtains:

$$G = 1 + \frac{1}{G} \tag{6.58}$$

After rearranging:

$$G^2 - G - 1 = 0 \tag{6.59}$$

This is a quadratic equation (i.e., $ax^2 + bx + c = 0$, with $a = 1$, and $b = c = -1$). The roots for G are:

$$G = \frac{-b \pm \sqrt{b^2 - 4ac}}{2a} = \frac{1 \pm \sqrt{1+4}}{2} = \frac{1 \pm \sqrt{5}}{2} \tag{6.60}$$

By rejecting the negative root (as the ratio between two lengths cannot be negative), the golden ratio can be calculated by:

$$G = \frac{1 + \sqrt{5}}{2} = 1.618033988749895\ldots \tag{6.61}$$

The golden ratio is closely related to the Fibonacci numbers, F_n, that forms the Fibonacci sequence, in which the n-th number is equal to the sum of the previous two numbers (i.e., the $(n-1)$-th and $(n-2)$-th numbers) for $n > 1$. That is:

$$F_0 = 0,\ F_1 = 1, \quad \text{and} \quad F_n = F_{n-1} + F_{n-2} \quad \text{for} \quad n > 1 \tag{6.62}$$

The Fibonacci sequence (and the golden ratio) is famous, as it describes many regular patterns in nature (e.g., the number of petals of flowers, the pattern of seeds within a sunflower, the nautilus shell). The Fibonacci spiral is shown in Figure 6.20, where the first 10 Fibonacci numbers are 0, 1, 1, 2, 3, 5, 8, 13, 21, 34. Does it look like a nautilus shell?

To show the relationship between the Fibonacci sequence and the golden ratio, the ratio between two successive Fibonacci numbers is assumed to converge to a constant—say x, when n is approaching infinity. That is:

$$\frac{F_n}{F_{n-1}} = \frac{F_{n+1}}{F_n} = x \quad \text{for} \quad n \to \infty \tag{6.63}$$

As $F_{n+1} = F_n + F_{n-1}$ by definition, Equation (6.63) becomes:

$$\frac{F_n}{F_{n-1}} = \frac{F_n + F_{n-1}}{F_n} = x \tag{6.64}$$

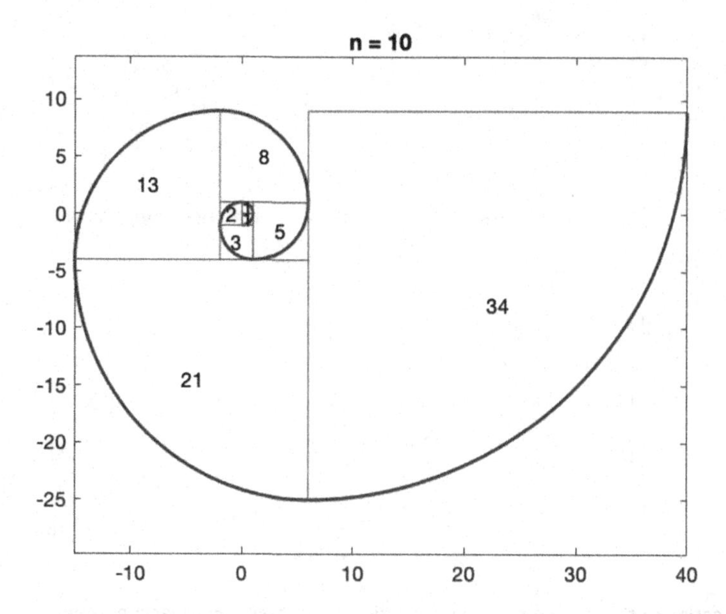

Figure 6.20 Fibonacci spiral.

Expressing everything in terms of x, it becomes:

$$1+\frac{1}{x}=x \Rightarrow x^2-x-1=0 \tag{6.65}$$

This quadratic equation of x is the same as the one for the golden ratio in Equation (6.59). Thus, it can be concluded that if the ratio of successive Fibonacci numbers converges, it will converge to the golden ratio. Figure 6.21 shows the ratio of successive Fibonacci numbers for increasing n. It is clear from the figure that the ratio converges very fast to the golden ratio.

The algorithm of the golden-section method for minimizing the objective function, $J(\theta)$, can be divided into two phases. The interval that contains the optimum is first identified in phase 1. Then, this interval will be reduced by moving the upper and lower bounds towards the optimum until the interval is sufficiently small in phase 2. The detailed procedure of the algorithm is presented below.

Phase 1:

1. Define a small step size, $\Delta(>0)$, for searching the interval (starting from the origin).
2. Set the iteration counter $i = 0$. Evaluate $J(0)$ and $J(\Delta)$. If $J(0)>J(\Delta)$, the minimum is on the right-hand side of the origin, and the search will be in the positive direction. If $J(0)<J(\Delta)$, the minimum is on the left-hand side of the origin. Thus, the search needs to be carried out in the negative direction, and $\Delta = -\Delta$ is assigned.
3. Increase the iteration counter $i=i+1$, and start the iteration.
4. In a general i-th iteration, calculate the value of θ_i by:

$$\theta_i=\sum_{j=1}^{i} G^j\Delta \tag{6.66}$$

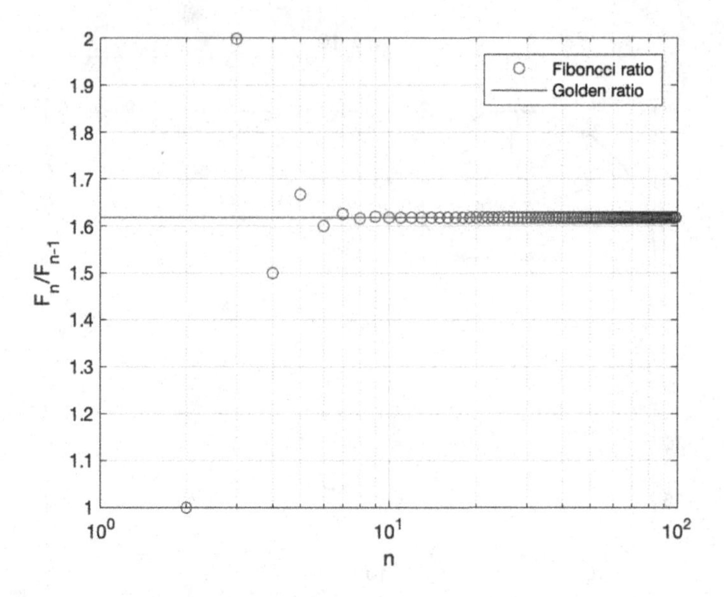

Figure 6.21 To estimate the golden ratio by the ratio of successive Fibonacci numbers.

where $G = \left(1+\sqrt{5}\right)/2$ is the golden ratio.

5. Evaluate $J(\theta_i)$.
6. Repeat steps 4 and 5 until $J(\theta_{i-1}) < J(\theta_i)$. Phase 1 completed.

As shown in Figure 6.22, the minimum is bounded by $\theta_{i-2}(=\theta_L)$ and $\theta_i(=\theta_U)$, where θ_L and θ_U are the lower and upper bounds, respectively. By Equation (6.66), the values of the last three points in phase 1 can be determined as:

$$\theta_{i-2} = \left(1+G+G^2+\cdots+G^{i-3}+G^{i-2}\right)\Delta \tag{6.67}$$

$$\theta_{i-1} = \left(1+G+G^2+\cdots+G^{i-3}+G^{i-2}+G^{i-1}\right)\Delta \tag{6.68}$$

$$\theta_{i} = \left(1+G+G^2+\cdots+G^{i-3}+G^{i-2}+G^{i-1}+G^{i}\right)\Delta \tag{6.69}$$

Therefore, the distance from θ_{i-2} to θ_{i-1} is $G^{i-1}\Delta$, and the distance between θ_{i-1} and θ_i is $G^i\Delta$. Furthermore, the interval between the lower and upper bounds is $I = \left(G^{i-1}+G^i\right)\Delta$ as shown in Figure 6.22. It must be pointed out that the lower and upper bounds (i.e., θ_{i-2} and θ_i, respectively) will move towards the minimum in phase 2 in a similar way as θ_{i-2} moves to θ_{i-1}. Therefore, it is necessary to find out the ratio between the distance from θ_{i-2} to θ_{i-1} and the interval, *I*. Assuming this ratio is *g*, the distance from θ_{i-2} to θ_{i-1} can be expressed as (see Figure 6.22):

$$G^{i-1}\Delta = gI = g\left(G^{i-1}+G^i\right)\Delta \tag{6.70}$$

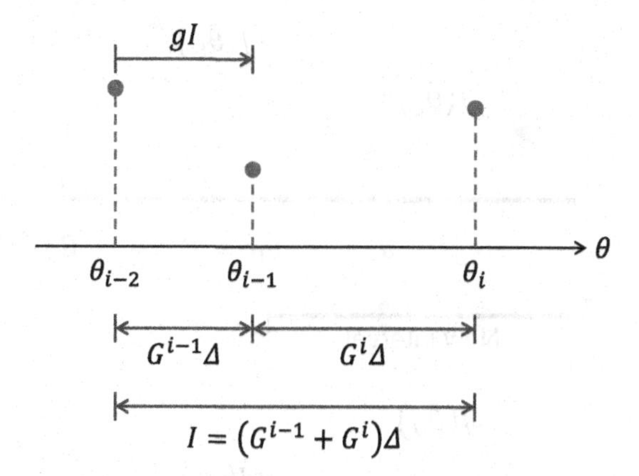

Figure 6.22 The geometry of the last three points in phase 1.

Now, g can be expressed as:

$$g = \frac{G^{i-1}}{G^{i-1} + G^i} = \frac{1}{1+G} \tag{6.71}$$

Substitute $G = \left(1+\sqrt{5}\right)/2$:

$$g = \frac{1}{1+\frac{1+\sqrt{5}}{2}} = \frac{2}{3+\sqrt{5}} = 0.381966\ldots \tag{6.72}$$

Phase 2:

This phase involves a series of iterations. In each iteration, two new points will be generated by moving the lower and upper bounds towards the minimum with a distance gI. Depending on the objective function value at these two new points, the new interval will be defined. From phase 1, the lower and upper bounds are $\theta_L = \theta_{i-2}$ and $\theta_U = \theta_i$.

1. Calculate the interval $I = \theta_U - \theta_L$, and generate the two new points:

$$\tilde{\theta}_L = \theta_L + gI \quad \text{and} \quad \tilde{\theta}_U = \theta_U - gI \tag{6.73}$$

   where $\tilde{\theta}_L$ and $\tilde{\theta}_U$ are the potential new lower and upper bounds, respectively.
2. Evaluate $J\left(\tilde{\theta}_L\right)$ and $J\left(\tilde{\theta}_U\right)$.
3. Update the interval and so as the new points of the next iteration based on the values of $J\left(\tilde{\theta}_L\right)$ and $J\left(\tilde{\theta}_U\right)$.
   a. If $J\left(\tilde{\theta}_L\right) < J\left(\tilde{\theta}_U\right)$, the minimum is bounded by θ_L and $\tilde{\theta}_U$ as shown in Figure 6.23(a).
      i. No change in the lower bound, and update the upper bound to $\theta_U = \tilde{\theta}_U$.
      ii. Update the interval $I = \theta_U - \theta_L$ based on the updated θ_U.
      iii. Update $\tilde{\theta}_U = \tilde{\theta}_L$, and update $\tilde{\theta}_L = \theta_L + gI$.
      iv. Evaluate $J\left(\tilde{\theta}_L\right)$, and update $J\left(\tilde{\theta}_U\right) = J\left(\tilde{\theta}_L\right)$.

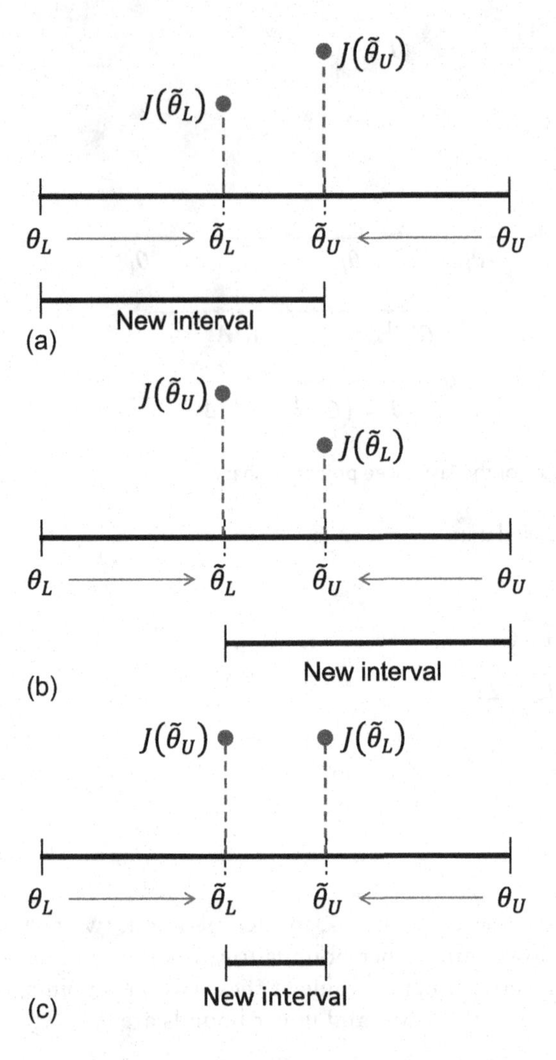

Figure 6.23 Defining new interval in phase 2 of the golden-section method.

b. If $J(\tilde{\theta}_L) > J(\tilde{\theta}_U)$, the minimum is bounded by $\tilde{\theta}_L$ and θ_U as shown in Figure 6.23(b).
   i. Update the lower bound to $\theta_L = \tilde{\theta}_L$, and no change in the upper bound.
   ii. Update the interval $I = \theta_U - \theta_L$ based on the updated θ_L.
   iii. Update $\tilde{\theta}_L = \tilde{\theta}_U$, and update $\tilde{\theta}_U = \theta_U - gI$.
   iv. Evaluate $J(\tilde{\theta}_U)$, and update $J(\tilde{\theta}_L) = J(\tilde{\theta}_U)$.
c. If $J(\tilde{\theta}_L) = J(\tilde{\theta}_U)$, the minimum is bounded by $\tilde{\theta}_L$ and $\tilde{\theta}_U$ as shown in Figure 6.23(c).
   i. Update both bounds to $\theta_L = \tilde{\theta}_L$ and $\theta_U = \tilde{\theta}_U$.
   ii. Update the interval $I = \theta_U - \theta_L$ based on the updated θ_L and θ_U.
   iii. Update $\tilde{\theta}_L = \theta_L + gI$ and $\tilde{\theta}_U = \theta_U - gI$.
   iv. Evaluate both $J(\tilde{\theta}_L)$ and $J(\tilde{\theta}_U)$.

4. Repeat step 3 until $I < \delta$, where δ is the pre-defined tolerance of iteration.

5. Calculate the minimum

$$\theta^* = \frac{\theta_L + \theta_U}{2} \tag{6.74}$$

6. Phase 2 is completed. Return θ^* to the user.

Note that the chance to have $J(\tilde{\theta}_L) = J(\tilde{\theta}_U)$ as in step 3c is very low. Thus, there is usually only one evaluation of the objective function in an iteration similar to the situation in steps 3a and 3b. With the help of the golden ratio, the number of function evaluations can be reduced.

The algorithm is simple and easy to code. A self-developed MATLAB function gold() is shown in Program 6.4 to implement both phases of the golden-section method for single-variable unconstrained minimization. To run gold(), the objective function must be provided. This can be done by developing another MATLAB function to implement the objective function. The file name of this MATLAB file will be used as one of the input parameters of gold().

Program 6.4: gold() for single-variable unconstrained minimization by golden-section method.

```
function [optx,infor]=gold(fname,stx,deta,tol,op)
if nargin < 1, fname=[]; end
if nargin < 2, stx=[]; end
if nargin < 3, deta=[]; end
if nargin < 4, tol=[]; end
if nargin < 5, op=[]; end
if isempty(fname)==1, error('fname is missing ...'); end
if isempty(stx)==1, stx=0; end
if isempty(deta)==1, deta=1.e-6; end
if isempty(tol)==1, tol=1.e-6; end
if isempty(op)==1, op=0; end
G=1.6180; g=0.3820;
fun=[fname '(x)'];
chk1=0; chk2=0;
x=stx; xL=x;
y=eval(fun); yL=y;
x=x+deta; xa=x;
y=eval(fun); ya=y;
if ya > yL
  deta=-deta;
  x=x+2*deta; xa=x;
  y=eval(fun); ya=y;
  if op(1)==1, disp('minimum at another direction ...'); end
end
if op(1)==1
  disp('locating upper and lower bound ...');
  disp('iteration     x     y');
end
count1=1;
while chk1==0
  x=x+deta*G^count1;
```

```
  y=eval(fun);
  if (count1==1)
   if (y >= ya), chk1=1; end
   yU=y; xU=x;
  else
   if (y >= yU), chk1=1; end
   yL=ya; ya=yU; yU=y;
   xL=xa; xa=xU; xU=x;
  end
  count1=count1+1;
  if op(1) == 1, disp([count1 x y]); end
end
I=xU-xL;
xb=xL+(1-g)*I;
if op(1) == 1
  disp('locating optimal point ...');
  disp('iteration     xLow      xUp');
end
count2=0;
while chk2==0
  count2=count2+1;
  x=xa; ya=eval(fun);
  x=xb; yb=eval(fun);
  if ya < yb
   xU=xb; xb=xa;
   I=xU-xL; xa=xL+g*I;
  end
  if ya > yb
   xL=xa; xa=xb;
   I=xU-xL; xb=xL+(1-g)*I;
  end
  if ya == yb
   xL=xa; xU=xb;
   I=xU-xL; xa=xL+g*I; xb=xL+(1-g)*I;
  end
  if abs(I) < tol
   optx=(xL+xU)/2;
   x=optx; opty=eval(fun);
   chk2=1;
  end
  if op(1) == 1, disp([count2 xL xU]); end
end
infor=[opty count1 count2];
end
```

Example 6.2

The developed function `gold()` is employed for model updating of a simple beam example as given in Section 6.1.4 (see Figure 6.9 for the beam model). The beam is divided into ten elements, and the bending stiffness of all ten elements are assumed to be different in simulating the measured natural frequencies and mode shapes. The model for simulating the measurement is generated from a uniform distribution from 0.8 to 1.2 as:

$$\theta = \{1.04 \quad 1.09 \quad 0.86 \quad 0.93 \quad 1.20 \quad 0.85 \quad 0.87 \quad 1.10 \quad 1.06 \quad 1.11\}^T$$

To consider the effect of modeling error, the class of models used in model updating assumes all ten elements have the same bending stiffness, and therefore, the corresponding minimization becomes a single variable problem.

Considering only the first three modes and the vertical DOFs, the objective function given in Equation (6.13) is employed in the model updating problem in this example. Use `gold()` to identify the model that fits the measured modal parameters.

SOLUTION

The objective function that calculates the discrepancy between the measured and model-predicted natural frequencies and mode shapes (only the first three modes and vertical DOFs) is implemented as a MATLAB function as `sbeam_jcal()` (see Program 6.5). The function is defined on Line 1. The measured data is loaded into the function on Lines 2 to 3, where `nfexp` and `msexp` are the measured natural frequencies and mode shapes, respectively. The function `sbeam_exe()` is called to calculate the model-predicted natural frequencies (`nf`) and mode shapes (`ms`) based on the given model (`theta`). Note that θ is a scalar (not a vector) in this example (as the stiffnesses of all 10 elements are assumed to be the same). The modal assurance criterion (`MAC`) matrix of the measured and model-predicted mode shapes is calculated on Line 5. The measured and model-predicted mode shapes are matched by the MAC value on Lines 6 to 8. The objective function value `J` is calculated on Lines 10 to 14 in a for-loop that goes through all the considered modes (i.e., three in this example, and it is equal to the length of the natural frequency vector `nfexp`).

Program 6.5: sbeam_jcal() for calculating the objective function value of the simple beam in this example.

```
Line 1.     function J=sbeam_jcal(theta)
Line 2.     S=load('sbeam_data');
Line 3.     nfexp=S.nfexp; msexp=S.msexp;
Line 4.     [nf,ms,node]=sbeam_exe(theta);
Line 5.     MAC=(msexp'*ms).^2;
Line 6.     [maxMAC,indx]=max(MAC');
Line 7.     nf=nf(indx);
Line 8.     ms=ms(:,indx);
Line 9.     J=0;
Line 10.    for ii=1:length(nfexp)
Line 11.       V1=msexp(:,ii);
Line 12.       V2=ms(:,ii);
Line 13.      J=J+((nfexp(ii)-nf(ii))/nfexp(ii))^2+(1-(V1'*V2)^2);
Line 14.    end
Line 15.    end
```

Once the measured data is available and the objective function is ready, the model updating can be easily carried out by calling the gold() function as:

```
>> [optx,infor] = gold('sbeam_jcal',0.5,0.05,0.0001,1)

locating upper and lower bound ...
iteration              x          y
      2.0000      0.6309      0.1212
      3.0000      0.7618      0.0449
      4.0000      0.9736      0.0007
      5.0000      1.3163      0.0722
```

```
locating optimal point ...
iteration       xLow       xUp
        1.0000     0.7618     1.1045
        2.0000     0.8927     1.1045
        3.0000     0.8927     1.0236
        4.0000     0.9427     1.0236
        5.0000     0.9736     1.0236
        6.0000     0.9736     1.0045
        7.0000     0.9736     0.9927
        8.0000     0.9809     0.9927
        9.0000     0.9854     0.9927
       10.0000     0.9854     0.9899
       11.0000     0.9854     0.9882
       12.0000     0.9864     0.9882
       13.0000     0.9864     0.9875
       14.0000     0.9869     0.9875
       15.0000     0.9871     0.9875
       16.0000     0.9871     0.9874
       17.0000     0.9872     0.9874
       18.0000     0.9873     0.9874
optx =
        0.9873
infor =
        0.0030     5.0000     18.0000
```

The command line shows that the initial trial, the initial step size, and the tolerance of iteration are set to 0.5, 0.05, and 0.0001, respectively. The iteration history is shown, where phase 1 takes five steps, and phase 2 takes 18 steps. The identified minimum is at $\theta = 0.9873$ with the objective function value of 0.0030. The objective function is plotted together with the identified minimum in Figure 6.24. It is clear from the figure that the identified minimum is accurate. The updated model is employed to calculate the modal parameters and matched to those from the measurement in Figure 6.25. The matching between the measured (exp) and updated (cal) modal parameters is very good.

Note that the number of iterations in phase 1 can be reduced if a large step size Δ is used (e.g., 1). But if Δ is too large, the initial interval, I, will be large, and the number of required iterations in phase 2 will be increased.

6.2.3.2 Polynomial approximation method

The polynomial approximation method is one of the most effective techniques for identifying the minimum of a single-variable objective function. The procedure is to evaluate the function at several points and then fit a polynomial to those known points. The minimum of this polynomial is then theoretically determined. The method is very simple and fast. If the objective function is approximated by a quadratic function, only three function evaluations are needed. The tradeoff is, of course, the accuracy of the identified minimum. If the objective function is highly non-linear or its behavior is very different from a polynomial function, the approximation will be very poor. As there is no guarantee of the accuracy of the identified minimum, additional function evaluations are usually needed to check if the objective function values at a close neighborhood of the identified minimum are matched to the corresponding polynomial function values. If their discrepancies are large, a higher order of polynomial may be required.

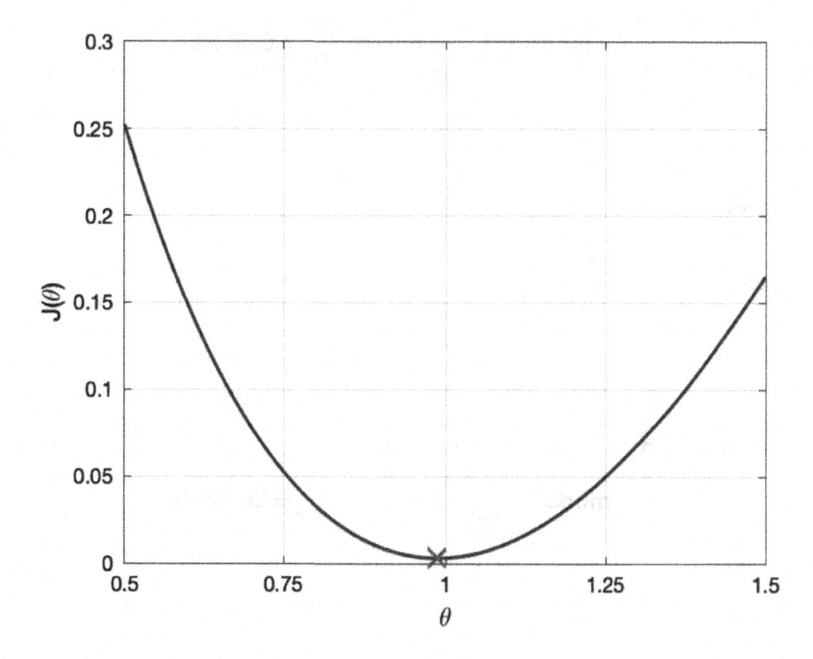

Figure 6.24 The objective function together with the identified minimum.

The use of the quadratic approximation method in identifying the minimum of the objective function $J(\theta)$ is given below as a reference.

$$\tilde{J}(\theta) = a_0 + a_1\theta + a_2\theta^2 \tag{6.75}$$

where $\tilde{J}(\theta)$ is the quadratic equation to approximate the objective function. As there are three unknown coefficients a_0, a_1, and a_2, at least three conditions are required to solve them. The objective function values at three different points (i.e., θ_1, θ_2, and θ_3) are evaluated, and it is assumed that the objective function has the same values as the approximated function at these three points.

$$\tilde{J}(\theta_1) = a_0 + a_1\theta_1 + a_2\theta_1^2 = J(\theta_1) \tag{6.76}$$

$$\tilde{J}(\theta_2) = a_0 + a_1\theta_2 + a_2\theta_2^2 = J(\theta_2) \tag{6.77}$$

$$\tilde{J}(\theta_3) = a_0 + a_1\theta_3 + a_2\theta_3^2 = J(\theta_3) \tag{6.78}$$

This is a set of three equations with three unknowns. Several solution methods are available for solving this kind of problem. If MATLAB is available, the matrix method is strongly recommended. The set of linear equations is expressed in the matrix form:

$$\begin{bmatrix} 1 & \theta_1 & \theta_1^2 \\ 1 & \theta_2 & \theta_2^2 \\ 1 & \theta_3 & \theta_3^2 \end{bmatrix} \begin{Bmatrix} a_0 \\ a_1 \\ a_2 \end{Bmatrix} = \begin{Bmatrix} J(\theta_1) \\ J(\theta_2) \\ J(\theta_3) \end{Bmatrix} \Rightarrow \mathbf{\Theta a} = \mathbf{J} \tag{6.79}$$

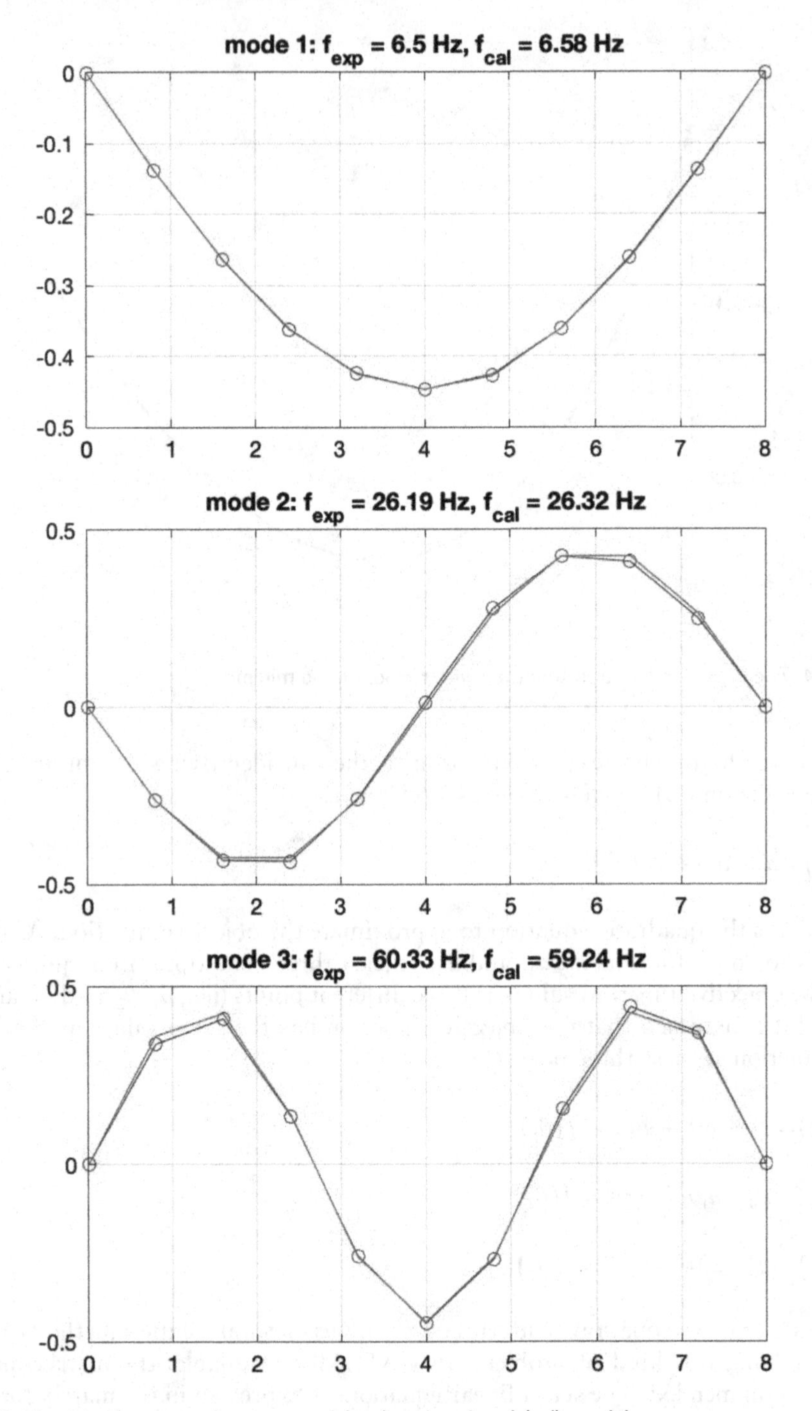

Figure 6.25 Matching between the measured (exp) and updated (cal) modal parameters.

The unknown coefficient vector, a, can be solved as:

$$\mathbf{a} = \Theta^{-1}\mathbf{J} \tag{6.80}$$

With the calculated coefficients, the objective function can now be approximated by the quadratic equation in Equation (6.75). The optimum can then be analytically calculated by obtaining the first derivative of the quadratic function with respect to the design variable:

$$\frac{d\tilde{J}(\theta)}{d\theta} = a_1 + 2a_2\theta \tag{6.81}$$

The optimum can be identified by assigning the first derivative to zero:

$$\theta^* = -\frac{a_1}{2a_2} \tag{6.82}$$

The second derivative must be positive for this solution to be a minimum:

$$\frac{d^2\tilde{J}(\theta)}{d\theta^2} = 2a_2 > 0 \tag{6.83}$$

6.2.3.3 Iterative quadratic approximation method

There are two main difficulties with the quadratic approximation method. First, the selection of the three points for calculating the quadratic function for the approximation of the objective function is arbitrary. On the one hand, the selected points have significant influences on the accuracy of the method, but on the other hand, there is no guideline for these selections. Second, the accuracy of the approximation is always an issue. The iterative quadratic approximation method was developed to address these two difficulties. First, the first phase of the golden-section method is used to identify the three required points for calculating the quadratic equation. Second, the accuracy of the quadratic approximation method is improved in an iterative manner.

Similar to the golden-section method, the algorithm for the iterative quadratic approximation method is divided into two phases, and phase 1 is exactly the same as that of the golden-section method (and it is not repeated here).

Three points (i.e., θ_{i-2}, θ_{i-1}, and θ_i) can be identified at the end of phase 1 of the algorithm, as shown in Figure 6.22. The lower and upper bounds are $\theta_L = \theta_{i-2}$ and $\theta_U = \theta_i$, respectively, and the intermediate point is defined as $\theta_M = \theta_{i-1}$.

Phase 2:

1. Evaluate the objective function at these three points to get $J(\theta_L)$, $J(\theta_M)$, and $J(\theta_U)$.
2. Calculate the unknown coefficients (i.e., a_0, a_1, and a_2) of the quadratic equation by solving the following set of three linear equations in matrix form as shown in Equations (6.79) and (6.80):

$$\begin{bmatrix} 1 & \theta_L & \theta_L^2 \\ 1 & \theta_M & \theta_M^2 \\ 1 & \theta_U & \theta_U^2 \end{bmatrix} \begin{Bmatrix} a_0 \\ a_1 \\ a_2 \end{Bmatrix} = \begin{Bmatrix} J(\theta_L) \\ J(\theta_M) \\ J(\theta_U) \end{Bmatrix} \Rightarrow \Theta\mathbf{a} = \mathbf{J} \Rightarrow \mathbf{a} = \Theta^{-1}\mathbf{J}$$

3. Determine the minimum of the quadratic equation as shown in Equation (6.82):

$$\bar{\theta} = -\frac{a_1}{2a_2}$$

4. Depending on the relative position of $\bar{\theta}$ and the relative value of $J(\bar{\theta})$, the new interval of the next iteration is defined.
   a. If $\theta_M < \bar{\theta}$ and $J(\bar{\theta}) > J(\theta_M)$, the new interval is from θ_L to $\bar{\theta}$ as shown in Figure 6.26(a).
      i. Update $\theta_U = \bar{\theta}$ and $J(\theta_U) = J(\bar{\theta})$.
   b. If $\theta_M < \bar{\theta}$ and $J(\theta_M) > J(\bar{\theta})$, the new interval is from θ_M to θ_U as shown in Figure 6.26(b).
      i. Update $\theta_L = \theta_M$ and $J(\theta_L) = J(\theta_M)$.
      ii. Update $\theta_M = \bar{\theta}$ and $J(\theta_M) = J(\bar{\theta})$.
   c. If $\bar{\theta} < \theta_M$ and $J(\bar{\theta}) > J(\theta_M)$, the new interval is from $\bar{\theta}$ to θ_U as shown in Figure 6.26(c).
      i. Update $\theta_L = \bar{\theta}$ and $J(\theta_L) = J(\bar{\theta})$.
   d. If $\bar{\theta} < \theta_M$ and $J(\theta_M) > J(\bar{\theta})$, the new interval is from θ_L to θ_M as shown in Figure 6.26(d).
      i. Update $\theta_U = \theta_M$ and $J(\theta_U) = J(\theta_M)$.
      ii. Update $\theta_M = \bar{\theta}$ and $J(\theta_M) = J(\bar{\theta})$.
5. Repeat steps 2 to 4 until the change in the estimated minimum of the quadratic equation in two successive iterations is sufficiently smaller. That is $|d\bar{\theta}| < \delta$, where δ is the pre-defined tolerance of iteration.
6. Phase 2 is completed. Return the identified minimum $\theta^* = \theta$ to the user.

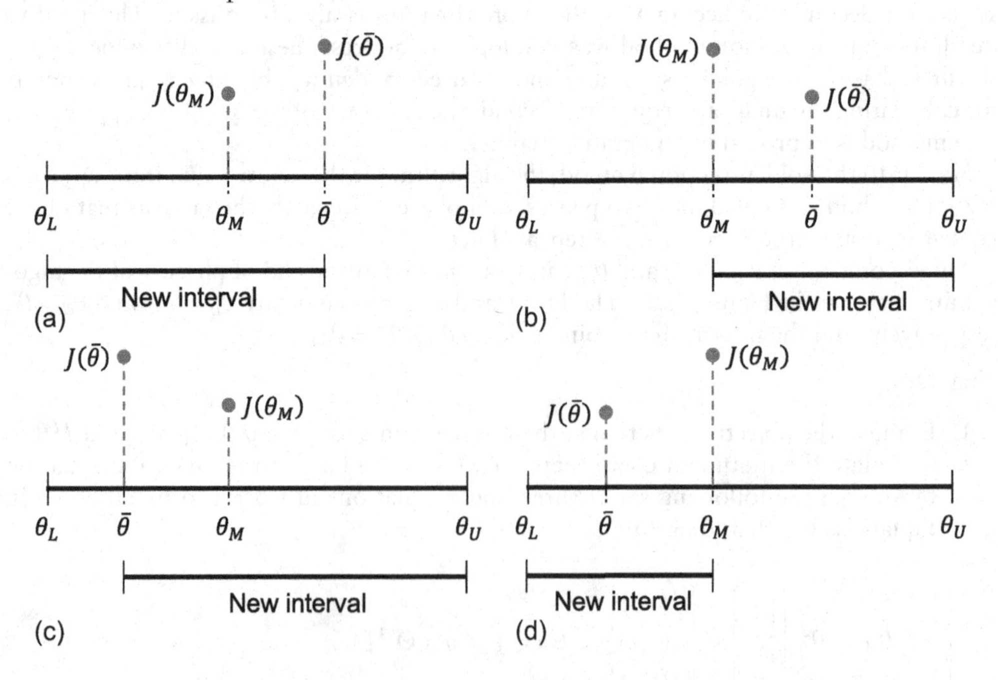

Figure 6.26 Defining new interval in phase 2 of the iterative quadratic approximate method.

It must be pointed out that only one function evaluation is required in each iteration (at step 4). The iterative quadratic approximation method is also easy to code, especially with MATLAB. Program 6.6 shows the self-developed MATLAB function, `polyapp()`, for implementing the iterative quadratic approximation method (phase 2 only).

```
Program 6.6: polyapp() for single-variable unconstrained minimization
by the iterative quadratic approximation method (phase 2 only).
```

```
function [x,f,lcount]=polyapp(funname,Ivect,tol,maxitn,dispindx)
if nargin < 1, funname= []; end
if nargin < 2, Ivect = []; end
if nargin < 3, tol = []; end
if nargin < 4, maxitn = []; end
if nargin < 5, dispindx = []; end
if isempty(funname), error('funname is missing ...'); end
if isempty(Ivect), error('Ivect is missing ...'); end
if isempty(tol), tol = 1e-5; end
if isempty(maxitn), maxitn = 1000; end
if isempty(dispindx), dispindx = 1; end
if ~isstr(funname), error('funname is not string ...'); end
if length(Ivect) ~= 3, error('length(Ivect) ~= 3 ...'); end
% the 3 given points must be different
if Ivect(1)==Ivect(2), error('Ivect(1)==Ivect(2) error ...'); end
if Ivect(1)==Ivect(3), error('Ivect(1)==Ivect(3) error ...'); end
if Ivect(2)==Ivect(3), error('Ivect(2)==Ivect(3) error ...'); end
xl = Ivect(1); xi = Ivect(2); xu = Ivect(3);
fl = feval(funname,xl);
fi = feval(funname,xi);
fu = feval(funname,xu);
chkout = 0;
lcount = 0;
while chkout == 0
  lcount = lcount + 1;
  if dispindx == 1
    disp([num2str(lcount) ': f(' num2str(xl) ')=' num2str(fl) , ...
        ', f(' num2str(xi) ')=' num2str(fi) ', f(' num2str(xu) , ...
        ')=' num2str(fu)]);
  end
  if fi > fu & fi > fl
    error('bad interval defined by xl and xu ...');
  end;
  % calculate the polynomial (quadratic) coefficients
  a2 = 1/(xu-xi)*((fu-fl)/(xu-xl)-(fi-fl)/(xi-xl));
  a1 = (fi-fl)/(xi-xl)-a2*(xl+xi);
  a0 = fl-a1*xl-a2*xl^2;
  % xbar is the minimum location for the fitted quadratic function
  xbar = -a1/2/a2;
  if xbar > xu | xbar < xl
    error('xbar lays outside xl and xu ...')
  end
  fbar = feval(funname,xbar);
  if a2 <= 0
    error('a2 <= 0 error ... it is not a minimum point ...');
```

```
  end;
  xindx = (xi < xbar);
  findx = (fi < fbar);
  if xindx & findx
    xu = xbar; fu = fbar;
  elseif xindx & ~findx
    xl = xi;  fl = fi;
    xi = xbar; fi = fbar;
  elseif ~xindx & findx
    xl = xbar; fl = fbar;
  elseif ~xindx & ~findx
    xu = xi;  fu = fi;
    xi = xbar; fi = fbar;
  end
  if lcount > 1
    % start from the 2nd iteration, check for convergence
    dx = abs(xbar-xbarold);
    if dx < tol
      chkout=1; % iteration stop when tolerance criteria is satisfied
    end
    if lcount >= maxitn
      % iteration stop when maximum number of iterations is achieved
      disp('program stop as number of iteration > maxitn ...')
      if dx > tol
        disp('tolerance is not achieved!!')
        disp('you may need to run the minimization again ...')
      else
        disp('tolerance is achieved!!')
        disp('you don''t need to re-run the minimization ...')
      end
      chkout=1;
    end
  end
  xbarold = xbar;
end
x=xbar;
f=fbar;
if dispindx == 1
  disp(['optimal x = ' num2str(x)])
  disp(['optimal f = ' num2str(f)])
end
end
```

Example 6.3

Repeat the model updating problem in Example 6.2 (i.e., the simple beam case study) by using the iterative quadratic approximation method with a tolerance of 0.0001. It is assumed that the first three points can be obtained from phase 1 of the golden-section method in Example 6.2 (see iteration steps 3, 4, and 5 in phase 1). That is the starting points are (0.7918, 0.9736, 1.3163).

SOLUTION

The objective function coded in `sbeam _ jcal()` as given in Program 6.5 is employed again. In this example, phase 2 of the iterative quadratic approximation method is used to replace the second phase of the golden-section method. The function polyapp() in Program 6.6 is used for model updating as follows:

```
>> [optx,f,lcount]=polyapp('sbeam_jcal',[0.7918 0.9736
1.3163],0.0001,[],1);

1: f(0.7918)=0.03571, f(0.9736)=0.003125, f(1.3163)=0.074729
2: f(0.7918)=0.03571, f(0.9736)=0.003125, f(1.0038)=0.0031875
3: f(0.9736)=0.003125, f(0.98749)=0.0029799, f(1.0038)=0.0031875
4: f(0.9736)=0.003125, f(0.98735)=0.0029798, f(0.98749)=0.0029799
optimal x = 0.98729
optimal f = 0.0029798
```

The identified minimum is at 0.98729 with an objective function value of 0.0029798. This is basically the same as that obtained by the golden-section method in Example 6.2. Only 4 iteration steps are required for the iterative quadratic approximation method to calculate the minimum of this objective function with an accuracy of 0.0001, while the second phase of the golden-section method takes 18 iteration steps (see Example 6.2). In general, the iterative quadratic approximation method is more efficient than the golden-section method.

6.2.3.4 MATLAB function: fminbnd()

With the MATLAB Optimization Toolbox, one can use the function fminbnd() to search the unconstrained minimum of a single-variable function. The syntax of fminbnd() is:

```
[X,FVAL]=fminbnd(MYFUN,X1,X2)
```

where X and FVAL are the minimum and the corresponding objective function value, respectively; MYFUN is the name of M-file containing the objective function; X1 and X2 are the lower and upper bounds of the design variable.

Example 6.4

Solve the single-variable minimization problem in Example 6.2 (the same in Example 6.3) using fminbnd().

SOLUTION

The objective function sbeam_jcal() is minimized as follows:

```
>> [optx,fval]=fminbnd('sbeam_jcal',0,10)

optx =
      0.9873
fval =
      0.0030
```

Here, 0 and 10 are given as the lower and upper bounds of the design variable. The answer is, of course, the same as that obtained by the two self-developed MATLAB functions gold() and polyapp(). In order to compare the efficiency of fminbnd() with the self-developed MATLAB functions, a few lines of code are added to the sbeam _ jcal() function to record the total number of function evaluations required in model updating. Program 6.7 shows the modified function sbeam _ jcalv2(). The global variable mycounter is defined on Line 2. The value of this global variable (the value can be accessed in both the function and also the workspace) is increased by unity at the end of the objective function on Line 14. Finally, the current number of function evaluations is displayed on Line 15.

Program 6.7: sbeam_jcalv2() modified from sbeam_jcal() for counting the number of function evaluations during model updating.

```
Line 1.     function J=sbeam_jcalv2(theta)
Line 2.     global mycounter
Line 3.     S=load('sbeam_data');
Line 4.     nfexp=S.nfexp; msexp=S.msexp;
Line 5.     [nf,ms,node]=sbeam_exe(theta);
Line 6.     MAC=(msexp'*ms).^2;
Line 7.     [maxMAC,indx]=max(MAC');
Line 8.     nf=nf(indx); ms=ms(:,indx);
Line 9.     J=0;
Line 10.    for ii=1:length(nfexp)
Line 11.       V1=msexp(:,ii); V2=ms(:,ii);
Line 12.       J=J+((nfexp(ii)-nf(ii))/nfexp(ii))^2+(1-(V1'*V2)^2);
Line 13.    end
Line 14.    mycounter=mycounter+1;
Line 15.    disp(mycounter);
Line 16.    end
```

Since all local variables within a function will be discarded after the completion of the function, the total number of function evaluations cannot be stored as a local variable. Therefore, a global variable mycounter is employed. The mycounter global variable must be initialized before the function sbeam_jcalv2() is called. In order to have a fair comparison, the interval [0.4 1.4] is used for fminbnd() to carry out the model updating as follows:

```
>> global mycounter; mycounter = 0;

>> [optx,fval]=fminbnd('sbeam_jcalv2',0.4,1.4)
     1
     2
     3
     4
     5
     6
     7
     8
optx =
    0.9873
fval =
    0.0030
```

The number of function evaluations is 8 for fminbnd().With the modified objective function sbeam_jcalv2(), gold() is re-run with an initial trial at 0.5, and ployapp() is re-run with the three points at [0.4 1 1.4]. It turns out that polyapp() needs 12 function evaluations and gold() uses 43 function evaluations. It is very clear that the total number of function evaluations for fminbnd() i.e., 8 is the smallest.

6.2.4 Multivariable unconstrained optimization

The standard form of a multivariable unconstrained optimization problem can be obtained by extracting Equation (6.44) and Equation (6.48) from a general multivariable constrained optimization problem as defined in Section 6.2.1:

Minimize:

$$J(\boldsymbol{\theta})$$

with design variables:

$$\boldsymbol{\theta} = \{\theta_1, \theta_2, \ldots, \theta_{N_\theta}\}^T$$

Most existing deterministic numerical optimization algorithms follow a similar solution procedure. That is: (1) define an initial trial; (2) find out the "best" direction; (3) find out the optimum along this "best" direction; and (4) repeat steps 3 and 4 until specified convergency criteria are achieved (this may involve many iterations). The optimum identified in the final iteration will be treated as the solution of the algorithm stops. Different algorithms have a different definition for the "best" direction in step 3, and so have different ways to find out this "best" direction. In step 4, the optimum along the "best" direction can be identified by any one of the single-variable optimization algorithms as introduced in Section 6.2.3.

To clearly present the different values of design variables in different iterations, $\boldsymbol{\theta}^{[0]}$ is defined as the initial trial of the design variables; and $\boldsymbol{\theta}^{[k]}$ is defined as the values of design variables at the k-th iteration, for $k = 1,2,3,\ldots$. The number of required iterations is unknown for most existing numerical optimization algorithms. Other quantities, whose values will change during the iterations can be defined in the same manner. The general procedure of a deterministic numerical optimization algorithm is as follows:

1. Set the iteration counter $k = 0$, and define the initial trial of design variables $\boldsymbol{\theta}^{[k]}$ by the user.
2. Calculate the "best" direction (or it is usually called the "search direction"), $\mathbf{d}^{[k+1]}$, which is a vector with the same dimension as the design variables, $\boldsymbol{\theta}$. It must be pointed out that the main difference among different optimization algorithms is their ways of determining the search direction.
3. For a given starting point and a search direction, all points along this direction can be expressed in terms of a single variable, say α, as follows:

$$\boldsymbol{\theta}(\alpha) = \boldsymbol{\theta}^{[k]} + \alpha \mathbf{d}^{[k+1]} \tag{6.84}$$

If the search direction vector $\mathbf{d}^{[k+1]}$ is normalized to have unit length, the variable α represents the step size. The optimum along the search direction can be identified by solving the single-variable optimization problem with the design variable, α, and objective function:

$$J(\boldsymbol{\theta}(\alpha)) = J\left(\boldsymbol{\theta}^{[k]} + \alpha \mathbf{d}^{[k+1]}\right) = J(\alpha) \tag{6.85}$$

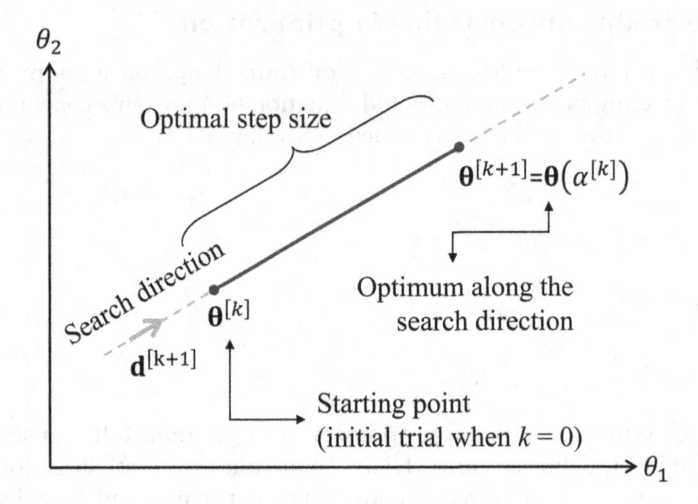

Figure 6.27 A two-variable example to illustrate the *k*-th iteration step.

After solving this single-variable optimization problem, the identified optimum along the search direction, $\boldsymbol{\theta}\left(\alpha^{[k]}\right)$, can be obtained, where $\alpha^{[k]}$ is the optimal step size in the k-th iteration and is equal to the distance from the starting point and the optimum along the search direction. This concept is illustrated in Figure 6.27 with a two-variable example.

4. The optimum along the search direction in the k-th iteration, $\boldsymbol{\theta}\left(\alpha^{[k]}\right)$, is considered as the starting point in the $(k + 1)$-th iteration:

$$\boldsymbol{\theta}^{[k+1]} = \boldsymbol{\theta}^{[k]} + \alpha^{[k]}\mathbf{d}^{[k+1]} \tag{6.86}$$

5. Increase the iteration counter by 1 (i.e., $k = k+1$), and repeat steps 2 to 4 until the specified convergency criteria are achieved. Some commonly used convergency criteria are listed below for references:
   a) The iteration is converged if the optimal step size, $\alpha^{[k]}$, is smaller than the pre-defined tolerance;
   b) The difference between objective function values from two successive iterations is smaller than the pre-defined tolerance;
   c) The number of maximum iterations is achieved.

Based on the above procedure, an N-dimensional unconstrained optimization problem can be resolved into many one-dimensional unconstrained optimization problems (in step 3 as illustrated in Figure 6.27) together with the calculation of the corresponding search directions (in step 2). The one-dimensional optimization problem was discussed in detail in Section 6.2.3. The main objective of this section is to calculate the search direction $\mathbf{d}^{[k+1]}$ in each iteration. Depending on the information required in calculating the search direction, the multivariable unconstrained optimization problem can be classified into three categories:

1. **Zero-order methods** require only the value of the objective function at different points. This type of method is usually reliable and easy to code. As only objective function values are needed, the methods usually can function even for discontinuous

functions. The tradeoff is that they usually require many iterations to converge, even for simple problems, resulting in very long computational time (and very high computational cost).

2. **First-order methods** require both the value and gradient of the objective function at different points. This type of method is usually more efficient compared to zero-order methods if the correct gradient information can be provided in a cost-effective way (e.g., by analytical methods). If the gradient information is approximated by a numerical method, the overall computational efficiency may not be necessarily better than that of the zero-order methods. As gradient information is important, their performance will be seriously affected if the first derivatives of the objective function are discontinuous.
3. **Second-order methods** require the value, gradient, and Hessian of the objective function at different points. The information of the second-order derivatives at a point allows a better description of the behavior of the objective function in the neighborhood of this point, and therefore, a better convergency rate is expected when compared to that of the zero- and first-order methods. However, the analytical solution of the Hessian matrix is usually very difficult, if not impossible, to obtain. If a numerical approximation is employed, the overall efficiency of this kind of method is questionable. This is because additional function evaluations are required in approximating the Hessian matrix. Furthermore, the error in the approximated Hessian matrix will certainly affect the accuracy of the calculated search direction.

These three types of methods will be introduced in the following sections.

6.2.4.1 *Zero-order: univariate search method*

The story starts with the easiest zero-order method, the univariable search method, in which the search direction is always parallel to one of the design variables. Each design variable will be considered in turn, and this will be repeated until the convergency criteria are achieved. For a general multivariable optimization problem with N_θ design variables in $\boldsymbol{\theta}$, the search direction $\mathbf{d}^{[k+1]}$ in the k-th iteration can be calculated as:

$$\mathbf{d}^{[k+1]}(i) = \begin{cases} 1 & \text{if } i = k + 1 - n_k N_\theta \\ 0 & \text{otherwise} \end{cases} \quad \text{for } i = 1,\ldots,N_\theta \tag{6.87}$$

where $\mathbf{d}^{[k+1]}(i)$ is the i-th element in the $\mathbf{d}^{[k+1]}$ vector; the integer n_k is given by:

$$n_k = \text{floor}\left(\frac{k}{N_\theta}\right) \tag{6.88}$$

where the operator $\text{floor}(\cdot)$ returns only the integer part of a number. Consider a simple case with only two design variables (i.e., $N_\theta = 2$). When $k = 0$, the search direction is:

$$\mathbf{d}^{[1]} = \begin{Bmatrix} 1 \\ 0 \end{Bmatrix} \tag{6.89}$$

That is, the algorithm will search from the initial trial, $\boldsymbol{\theta}^{[0]}$, in the direction along θ_1. The optimum along $\mathbf{d}^{[1]}$ and so as the optimal step size, $\alpha^{[0]}$, can be identified by solving the single-variable optimization problem with objective function $J\left(\boldsymbol{\theta}^{[0]}+\alpha\mathbf{d}^{[1]}\right)$. Then, the starting point for the next iteration can be calculated as:

$$\boldsymbol{\theta}^{[1]}=\boldsymbol{\theta}^{[0]}+\alpha^{[0]}\mathbf{d}^{[1]} \tag{6.90}$$

In the next iteration with $k = 1$, the search direction is:

$$\mathbf{d}^{[2]}=\begin{Bmatrix}0\\1\end{Bmatrix} \tag{6.91}$$

The algorithm will search the direction along θ_2. Similarly, the objective function $J\left(\boldsymbol{\theta}^{[1]}+\alpha\mathbf{d}^{[2]}\right)$ will be solved to find $\alpha^{[1]}$. Then, the starting point of the next iteration becomes:

$$\boldsymbol{\theta}^{[2]}=\boldsymbol{\theta}^{[1]}+\alpha^{[1]}\mathbf{d}^{[2]} \tag{6.92}$$

The iteration will continue until the convergence criteria are achieved. Note that there are only two different search directions for a two-variable case, and they will be employed repeatedly. That is $\mathbf{d}^{[1]}=\mathbf{d}^{[3]}=\mathbf{d}^{[5]}=\ldots$ and $\mathbf{d}^{[2]}=\mathbf{d}^{[4]}=\mathbf{d}^{[6]}=\ldots$.

Example 6.5

Identify the minimum of the Matyas function (shown below) by the univariate search method with an initial trial of $\{-2,-1\}^T$ and a tolerance of 1.0×10^{-8}.

$$J(\theta_1,\theta_2)=0.26\left(\theta_1^2+\theta_2^2\right)-0.48\theta_1\theta_2 \tag{6.93}$$

SOLUTION

The univariate search method is used to solve the two-variable minimization problem. The contour of the Matyas function in Equation (6.93) together with the search path (for all iterations) of the univariate search method is shown in Figure 6.28.

The minimization starts at the point $\{-2,-1\}^T$ and the iteration starts with the search direction $\{1,0\}^T$ (horizontal). When the algorithm obtains the minimum point along the $\{1,0\}^T$ direction, the second iteration starts along the $\{0,1\}^T$ direction (vertical). This continues until the convergence criterion is satisfied. In this example, the convergence criterion is that the difference in the objective function values for two successive iterations be smaller than the preset tolerance (i.e., 1×10^{-8}). It takes 85 iterations for the algorithm to complete the minimization process and the identified minimum is $\{-0.0011,-0.0012\}^T$ with the objective function value being 5.56×10^{-8}. The true minimum of this function is at $\{0,0\}^T$, with the objective function value being 0. The result from the univariate search method is reasonably close to the true solution.

The univariate search method is the simplest search technique. However, the search path is generally a series of zig-zag lines, as shown in Figure 6.28. As a result, it takes many iterations and a long time to converge. It is clear from Figure 6.28 that the "best" direction is about 45° from the horizontal. However, the algorithm cannot use it as the search direction. The convergence of the minimization process can be improved if the algorithm can calculate the search direction in a smarter way, e.g., based on previous search directions.

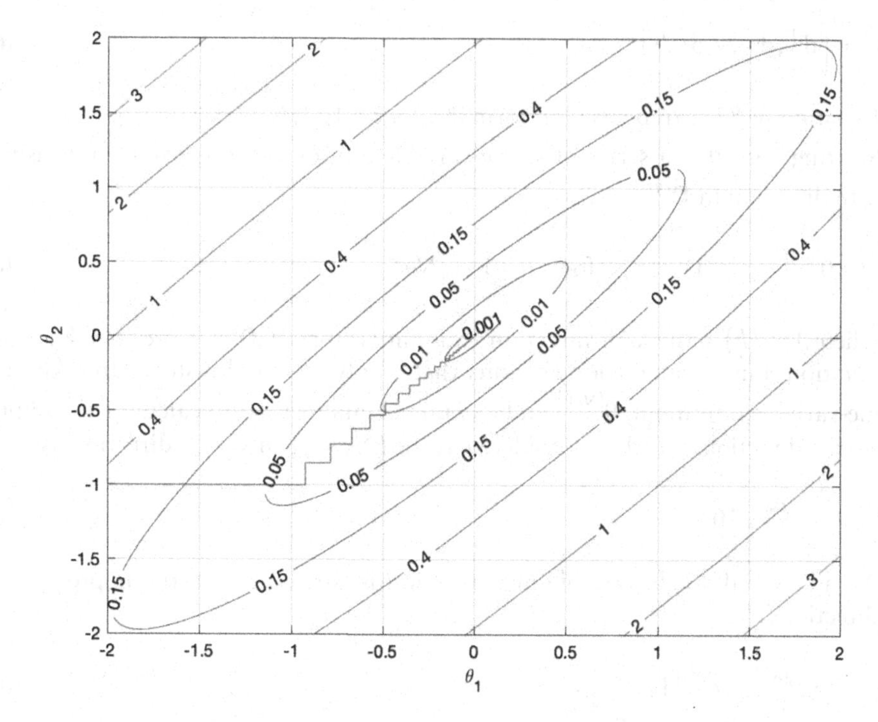

Figure 6.28 The contour of the Matyas function and the search path of the univariate search method.

6.2.4.2 Zero-order: conjugate direction method

The conjugate direction method can be considered as the significantly improved version of the univariate search method. The basic idea is to "memorize" a given number of search directions from the univariate search method and use this set of search directions to calculate an improved search direction. To implement this idea, a set of N_θ search directions is generated following the univariate search method. With a user-defined initial trial, a series of single-variable minimizations are carried out along all directions in the direction set one by one (this part is basically the same as the univariate search method). The improved search direction can then be calculated from the results of this series of single-variable minimizations (i.e., based on the set of search directions). A single-variable minimization is carried out along this improved search direction. If the convergence criteria are not achieved, this improved search direction will be added at the end of the set of N_θ search directions and an existing direction must be removed from the set (to keep the total number of directions in the set to be N_θ). There are different ways to decide which search direction is to be removed. In this section, the easiest way is employed, and that is to remove the first one in the set. For each direction in the new set, a single-variable minimization is carried out for calculating a new improved search direction. These processes are repeated until the pre-defined convergence criteria are achieved.

The detailed procedures of the conjugate direction method are as follows:

1. Initialize the iteration counter $g = 1$, and assign the user-defined initial trail to $\boldsymbol{\theta}^{[0]}$.
2. Generate the initial set of N_θ search directions (following the univariate search method) as:

$$\mathbf{D}^{\langle 1\rangle} = \left[\mathbf{d}^{[1]}, \mathbf{d}^{[2]}, \cdots \mathbf{d}^{[N\theta]}\right] \tag{6.94}$$

Equation (6.87) can be used to form $\mathbf{d}^{[i]}$, for $i = 1, \ldots, N_\theta$.

3. Starting with $\theta^{[0]}$, a series of single-variable minimization is carried out using the k-th direction in $\mathbf{D}^{\langle g\rangle}$:

$$\theta^{[k]} = \theta^{[k-1]} + \alpha^{[k-1]}\mathbf{D}^{\langle g\rangle}(k) \quad \text{for} \quad k = 1, \ldots, N_\theta \tag{6.95}$$

where $\mathbf{D}^{\langle g\rangle}(k)$ is the k-th direction (column vector) in $\mathbf{D}^{\langle g\rangle}$. Note that k is not the iteration counter but the loop counter that is only used in this step. After the last single-variable minimum, $\theta^{[N\theta]}$ can be obtained, and is used to calculate the improved search direction, which is considered as the $(N_\theta + g)$-th search direction as:

$$\mathbf{d}^{[N\theta+g]} = \theta^{[N\theta]} - \theta^{[0]} \tag{6.96}$$

4. An additional single-variable minimization is carried out in the improved search direction:

$$\theta^{[N\theta+1]} = \theta^{[N\theta]} + \alpha^{[N\theta]}\mathbf{d}^{[N\theta+g]} \tag{6.97}$$

The identified minimum along this improved direction is assigned to $\theta^{[0]} = \theta^{[N\theta+1]}$ for the next iteration.

5. If the convergence criteria are achieved, stop. Otherwise, increase the iteration counter $g = g + 1$, and the improved direction is added to the end of the direction set by removing the first direction:

$$\mathbf{D}^{\langle g\rangle} = \left[\mathbf{d}^{[g]}, \mathbf{d}^{[g+1]}, \cdots \mathbf{d}^{[N\theta+g-1]}\right] \tag{6.98}$$

6. Go to step 3.

It must be pointed out that the definition of an iteration step in the conjugate direction method is different from that of an iteration step in the univariate search method. In the univariate search method, an iteration step involves a single-variable minimization. However, in the conjugate direction method, an iteration step involves $N_\theta + 1$ single-variable minimizations. Therefore, comparing the performance of these two methods by looking at the number of iterations is meaningless. The number of function evaluations should be used.

The conjugate direction method works well in many applications, especially for lower-order objective functions with a small number of variables. However, many single-variable minimizations are required, which is computationally expensive for problems involving higher-order functions.

Example 6.6

Solve the minimization in Example 6.5 again by the conjugate direction method with an initial trial of $\{-2, -1\}^T$ and a tolerance of 1.0×10^{-8}. The objective function (Matyas function) is duplicated here for convenience.

$$J(\theta_1,\theta_2) = 0.26(\theta_1^2 + \theta_2^2) - 0.48\theta_1\theta_2$$

SOLUTION

The same initial trial and tolerance as in Example 6.5 are adopted. The contour of the Matyas function, together with the search path of the conjugate direction method (in red color), is shown in Figure 6.29. Only three iterations are required for the algorithm to converge, and three single-variable minimizations are needed in each iteration resulting in a total of 9 single-variable minimizations. This is a small number when compared to 85 single-variable minimizations for the univariate search method in Example 6.5. The identified minimum is $\{-0.2650, -0.2979\}^T \times 10^{-11}$, which is much closer to the true minimum at $\{0,0\}^T$ when compared to the univariate search method. At the minimum, the objective function value is 3.4383×10^{-25}, which is basically equal to zero.

Figure 6.29 clearly shows the efficiency of the conjugate direction method (when compared to the univariate search method). The first improved search direction, $\mathbf{d}^{[3]}$, only has little improvement. In the second iteration, the improved search direction, $\mathbf{d}^{[4]}$, leads the search path to a region extremely close to the global minimum of the objective function. The search path of the third iteration (the final one) is too short and cannot be clearly shown in the figure.

The conjugate directions method is one of the most efficient zero-order methods. In order to further improve the efficiency, one needs to consider the gradient information during the optimization process. That means one needs to use first-order methods.

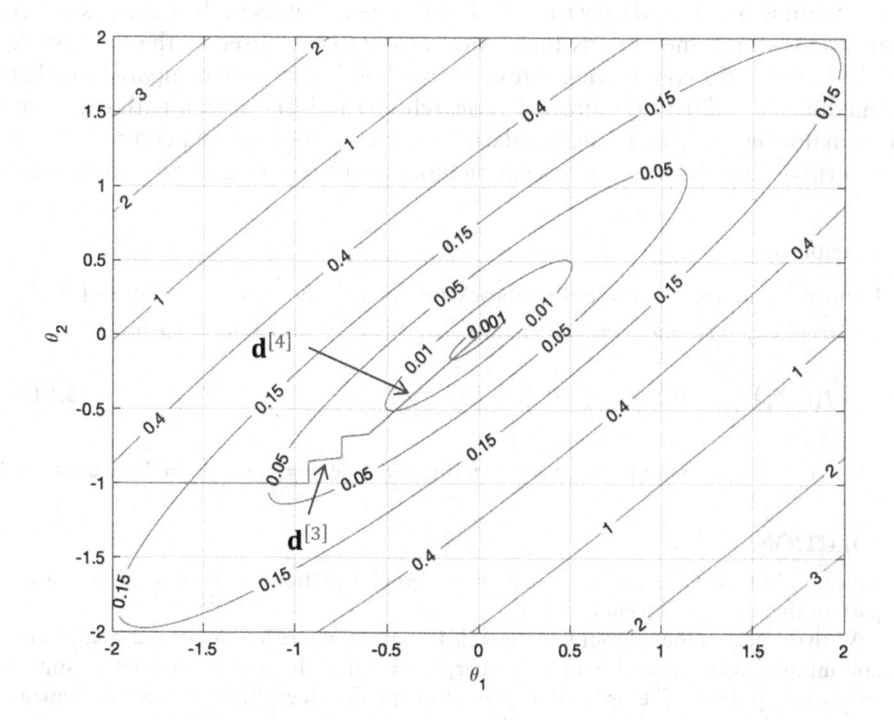

Figure 6.29 The contour of the Matyas function and the search path of the conjugate direction method.

6.2.4.3 First-order: steepest descent method

It must be pointed out that the formulation of the gradient vector either must be explicitly or implicitly given for using first-order methods. Otherwise, the numerical approximated gradient vector is required. This will certainly increase the required computational cost significantly, and, more importantly, the approximation error will certainly affect the performance of the algorithm.

The steepest descent method is probably the most well-known and yet the worst efficient first-order method (in terms of performance). The idea of this method is to select the search direction as the steepest descent direction, that is, the direction at which the rate of decrease for the function is the highest. As the gradient vector can be interpreted as the direction with the fastest rate of increase of the function, the opposite direction (i.e., negative of the gradient vector) is the steepest descent direction. The procedure of the steepest descent method is as follows:

1. Initialize the iteration counter $k = 1$, and assign the user-defined initial trail to $\theta^{[0]}$.
2. Carry out the following single-variable minimization:

$$\theta^{[k]} = \theta^{[k-1]} + \alpha^{[k-1]}\left[-\nabla J\left(\theta^{[k-1]}\right)\right] \tag{6.99}$$

   where $\nabla J\left(\theta^{[k-1]}\right)$ is the gradient vector of the objective function evaluated at $\theta^{[k-1]}$, $\alpha^{[k-1]}$ is the optimal step size in the k-th single-variable minimization.
3. If the convergence criteria are achieved, stop. Otherwise, increase the iteration counter $k = k + 1$ and go to step 2.

The algorithm is simple and easy to code. It is believed that its performance is better than the univariate search method, as first-order information is used in the steepest descent method. However, the convergence rate of the method is very poor compared to other first-order methods. Similar to the univariate search method, the search path of the method involves many zig-zag paths that significantly increase the required computational cost. The performance of the steepest descent method can be tested through a simple example.

Example 6.7

Identify the minimum of the Rosenbrock function as shown in Equation (6.100) by the steepest descent method with an initial trial $\{-4,3\}^T$ and a tolerance of 1.0×10^{-4}.

$$J(\theta_1,\theta_2) = (a-\theta_1)^2 + b\left(\theta_2 - \theta_1^2\right)^2 \tag{6.100}$$

where the true minimum is at $\{a,a^2\}^T$. In this example, $a = 1.5$ and $b = 2$ are used.

SOLUTION

Figure 6.30 shows the contour of the Rosenbrock function together with the search path of the steepest descent method.

With the help of the contour lines, it is clear that the algorithm carries out single-variable minimizations along lines that are perpendicular to the contour lines (at the initial or starting points). The optimal step sizes of the first few single-variable minimizations are relatively large. However, when the search path arrived at the region near the

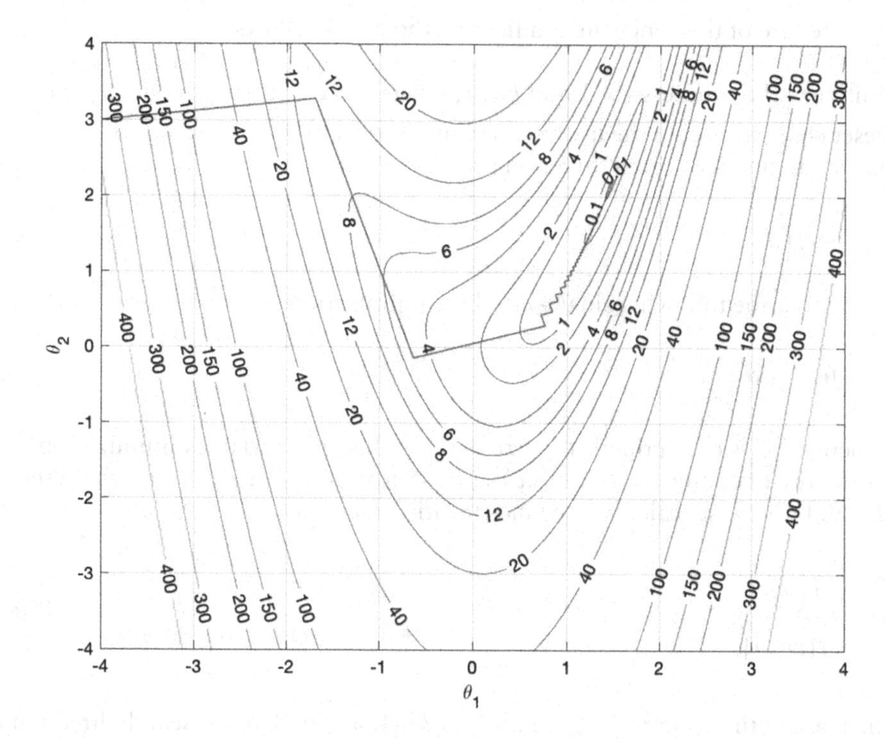

Figure 6.30 The contour of the Rosenbrock function and the search path of the steepest descent method.

minimum, the optimal step size becomes smaller and smaller in a zig-zag pattern similar to the univariate search method (zero-order). A total of 437 iterations are required for the algorithm to converge. The identified minimum is $\{1.4974, 2.2418\}^T$ with an objective function value of 7.1680×10^{-6}. The true minimum of the Rosenbrock function in this example is $\{1.5, 2.25\}^T$ and the objective function value is 0. The result from the steepest descent method can be improved by using a smaller tolerance, and the tradeoff is even more iterations and so an increase in the required computational cost.

6.2.4.4 First-order: conjugate gradient method

The conjugate gradient method modifies the search direction of the steepest descent algorithm, and this significantly increases the efficiency of the algorithm. Unlike the conjugate direction method, which requires the storage of N_θ single-variable minimizations to generate the improved search direction, the availability of the gradient information makes it possible to calculate the improved search direction without "memorizing" any past search direction. The reduction in the required memory and computational time makes this method very attractive.

In the first iteration, the steepest descent direction is used as the search direction. In subsequent iterations, the search direction is a conjugate direction, which is calculated based on the search direction in the previous iteration. The conjugate gradient method can minimize a quadratic objective function within N_θ iterations. However, the efficiency can be seriously reduced if the objective function is not quadratic. This problem can be solved by resetting the improved search direction to the steepest descent direction for every N_θ iterations. This reset process is extremely important in maintaining the performance of the conjugate gradient method.

The procedure of the conjugate gradient method is as follows:

1. Initialize the counters $g = 1$ and $k = 1$, where g is the iteration counter, and k is the reset counter. Assign the user-defined initial trial to $\boldsymbol{\theta}^{[0]}$.
2. Generate the first search direction as:

$$\mathbf{d}^{[1]} = -\nabla J\left(\boldsymbol{\theta}^{[0]}\right) \tag{6.101}$$

3. Carry out the following single-variable minimization:

$$\boldsymbol{\theta}^{[g]} = \boldsymbol{\theta}^{[g-1]} + \alpha^{[g-1]}\mathbf{d}^{[g]} \tag{6.102}$$

   where $\alpha^{[g-1]}$ is the optimal step size in the g-th single-variable minimization.
4. If the reset counter $k > N_\theta$, reset the modification factor $\beta = 0$ and the reset count $k = 0$. Otherwise, calculate the modification factor as:

$$\beta = \frac{\left\|\nabla J\left(\boldsymbol{\theta}^{[g]}\right)\right\|^2}{\left\|\nabla J\left(\boldsymbol{\theta}^{[g-1]}\right)\right\|^2} \tag{6.103}$$

5. Increase both counters $g = g + 1$ and $k = k + 1$, and update the search direction from the one in the previous iteration using the modification factor as:

$$\mathbf{d}^{[g]} = -\nabla J\left(\boldsymbol{\theta}^{[g-1]}\right) + \beta\mathbf{d}^{[g-1]} \tag{6.104}$$

6. If the convergence criteria are achieved, stop. Otherwise, go to step 3.

The algorithm is simple and easy to code.

Example 6.8

Solve the minimization problem in Example 6.7 again using the conjugate gradient method with an initial trial of $\{-4,3\}^T$ and a tolerance of 1.0×10^{-4}. The Rosenbrock function is shown below again for convenience.

$$J(\theta_1,\theta_2) = (a-\theta_1)^2 + b\left(\theta_2-\theta_1^2\right)^2$$

where the true minimum is at $\left\{a,a^2\right\}^T$. The same as the situation in Example 6.7, a = 1.5 and b = 2 are used.

SOLUTION

First, the problem is solved without considering the reset process. The search path is shown in Figure 6.31. In the first iteration (at the initial trial $\{-4,3\}^T$), the algorithm still starts in the steepest descent direction, that is, the search path is perpendicular to the contour line. It is clear from the figure that there are no zig-zag patterns. However,

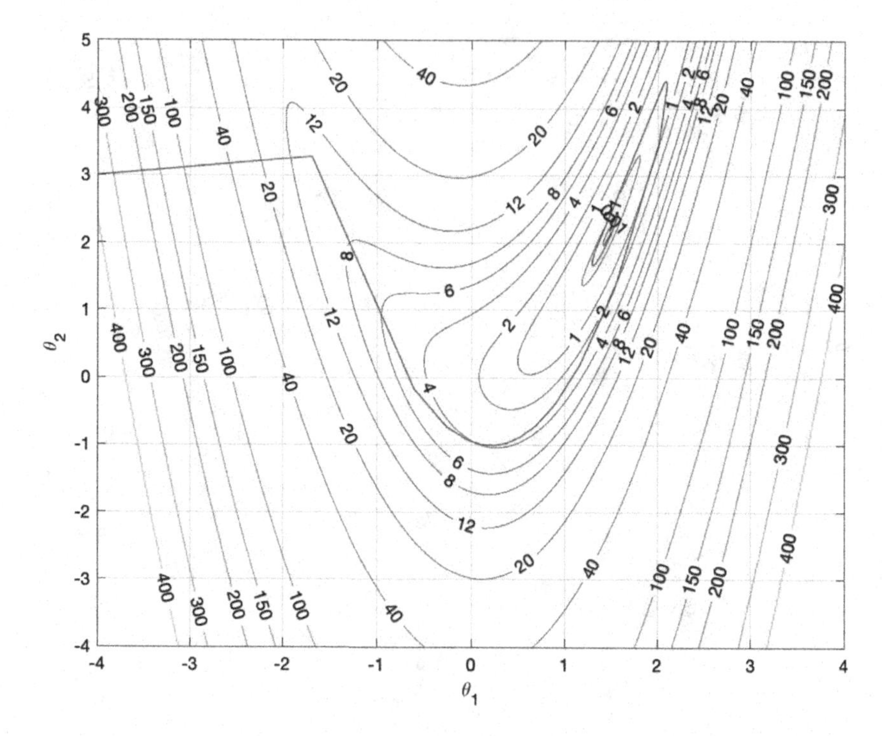

Figure 6.31 Minimization of the Rosenbrock function by the conjugate gradient method (without reset).

the efficiency of the algorithm is certainly reduced by the search path which follows the contour and spins about the minimum, as in Figure 6.31. This spinning effect is mainly due to the fact that the objective function is not quadratic. The total number of iterations is 54, and the identified minimum is at $\{1.4975, 2.2421\}^T$ with an objective value of 6.5693×10^{-6}. The result is not bad when compared to the steepest descent method in Example 6.7.

To overcome the spinning problem, the minimization problem is solved again with the reset process. The search path is shown in Figure 6.32. It is very clear from the figure that the spinning problem is solved. It takes 14 iterations for the algorithm to converge, and the identified minimum is at $\{1.4999, 2.2495\}^T$ with an objective value of 2.3293×10^{-8}. The improvement from the reset process can be clearly observed in this example.

6.2.4.5 Second-order: Newton–Raphson method

The basic idea of the Newton–Raphson method is to approximate the objective function as a quadratic function at an initial trial $\theta^{[0]}$ based on the gradient vector and Hessian matrix information.

$$J(\theta) \approx J\left(\theta^{[0]}\right) + \left(\theta - \theta^{[0]}\right)^T \nabla J\left(\theta^{[0]}\right) + \frac{1}{2}\left(\theta - \theta^{[0]}\right)^T \mathbf{H}\left(\theta^{[0]}\right)\left(\theta - \theta^{[0]}\right) \tag{6.105}$$

where the gradient vector, $\nabla J(\theta)$, and the Hessian matrix, $\mathbf{H}(\theta) = \nabla^2 J(\theta)$, are defined in Equations (6.50) and (6.53), respectively. The first derivative of it becomes:

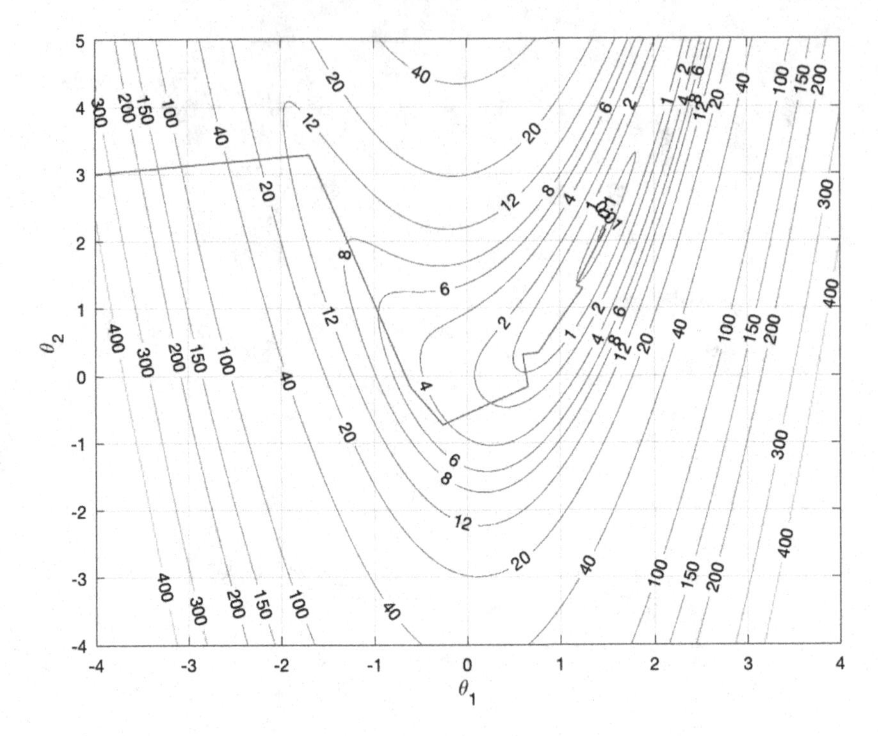

Figure 6.32 Minimization of the Rosenbrock function by the conjugate gradient method (with reset).

$$\nabla J(\theta) \approx \nabla J\left(\theta^{[0]}\right) + \mathbf{H}\left(\theta^{[0]}\right)\left(\theta - \theta^{[0]}\right) \tag{6.106}$$

At the minimum, θ^*, the first derivative is equal to zero:

$$\nabla J\left(\theta^{[0]}\right) + \mathbf{H}\left(\theta^{[0]}\right)\left(\theta^* - \theta^{[0]}\right) \approx 0 \tag{6.107}$$

Therefore:

$$\theta^* \approx \theta^{[0]} + \mathbf{H}\left(\theta^{[0]}\right)^{-1}\left[-\nabla J\left(\theta^{[0]}\right)\right] \tag{6.108}$$

If the objective function is quadratic, Equation (6.108) is not approximate but exact, and the minimum can be identified without iteration. As almost all objective functions in real applications are not quadratic, Equation (6.108) has to be implemented in an iterative manner through a series of single-variable minimizations.

$$\theta^{[k+1]} = \theta^{[k]} + \alpha^{[k]}\mathbf{H}\left(\theta^{[k]}\right)^{-1}\left[-\nabla J\left(\theta^{[k]}\right)\right] \tag{6.109}$$

where $\alpha^{[k]}$ is the optimal step size in the k-th iteration. It is clear that the search direction is calculated from the gradient and Hessian information.

$$\mathbf{d}^{[k+1]} = \mathbf{H}\left(\theta^{[k]}\right)^{-1}\left[-\nabla J\left(\theta^{[k]}\right)\right] \tag{6.110}$$

The inverse of the Hessian matrix is required in calculating the search direction. During the iteration, the Hessian matrix may become ill-conditioned (i.e., singular or almost singular). Under such a situation, the algorithm will be stopped with an error. Furthermore, the performance of the algorithm will be seriously affected if an accurate Hessian matrix cannot be provided.

The algorithm of the Newton–Raphson method is as follows:

1. Initialize the iteration counter $k = 0$, and assign the user-defined initial trail to $\theta^{[0]}$.
2. Evaluate the gradient vector, $\nabla J\left(\theta^{[k]}\right)$, and the inverse of the Hessian matrix, $\mathbf{H}\left(\theta^{[k]}\right)^{-1}$, for calculating the search direction as shown in Equation (6.110).
3. Carry out the following single-variable minimization:

$$\theta^{[k+1]} = \theta^{[k]} + \alpha^{[k]}\mathbf{H}\left(\theta^{[k]}\right)^{-1}\left[-\nabla J\left(\theta^{[k]}\right)\right] \tag{6.111}$$

4. If the convergence criteria are achieved, stop. Otherwise, increase the iteration counter $k = k + 1$ and go to step 2.

The algorithm is simple and easy to code. However, it is usually not easy to provide the Hessian information of the objective function.

Example 6.9

Solve the minimization problem in Example 6.7 (the Rosenbrock function) again using the Newton–Raphson method with an initial trial of $\{-4,3\}^T$ and a tolerance of 1.0×10^{-4}.

SOLUTION

Figure 6.33 shows the search path of the Newton–Raphson method in identifying the minimum of the Rosenbrock function. It is clear from the figure that there are no zig-zag patterns. The total number of iterations is 9, and the identified minimum is at $\{1.5,2.25\}^T$, which is exactly the same as the true minimum, with an objective value of 5.3306×10^{-22}, which is basically equal to zero. The result is very good when compared to all other algorithms introduced so far.

Example 6.10

Consider the set of measured (simulated) modal parameters of the simple beam in Example 6.2. The model for model updating is changed from 1D (assuming all 10 elements have the same stiffness in Example 6.2) to 2D by assuming the left and right parts of the beam have different bending stiffness as defined in Equation (6.30). That is, to update the model $\theta = \{\theta_L,\theta_R\}^T$, where θ_L and θ_R are the non-dimensional scaling factors for the bending stiffness of the left and right parts of the beam, respectively. Identify the updated model by using the following minimization algorithms to minimize the discrepancy between the measured and model-predicted natural frequencies and mode shapes (the first three modes and vertical DOFs only).

1. Univariate search.
2. Conjugate direction.

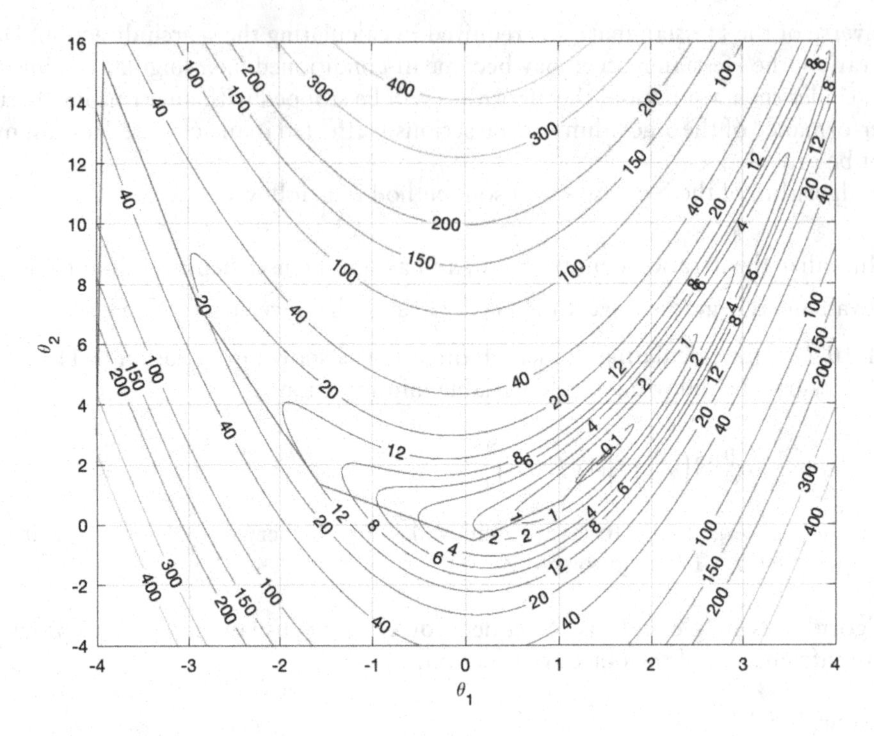

Figure 6.33 Minimization of the Rosenbrock function by the Newton–Raphson method.

3. Steepest descent.
4. Conjugate gradient.
5. Newton–Raphson.

For a fair comparison, all algorithms have the same initial trial as $\theta = \{0.6, 0.3\}^T$.

SOLUTION

The contour plot of the objective function as given in Equation (6.13) is presented in Figure 6.34 together with the search paths of the five numerical optimization algorithms, where the circle and cross markers show the initial trial and the identified optima. All algorithms return the same optimal model $\hat{\theta} = \{0.9815, 0.9931\}^T$, as expected. The number of iteration steps required for various algorithms is summarized below:

1. Univariate search: 6 iteration steps.
2. Conjugate direction: 12 iteration steps.
3. Steepest descent: 6 iteration steps.
4. Conjugate Gradient: 6 iteration steps.
5. Newton–Raphson: 6 iteration steps.

In this example, the gradient vector (used in the steepest descent and conjugate gradient method) and the Hessian matrix (used in the Newton–Raphson method) of the objective function are approximated by the finite difference method (see Section 6.2.5).

In this example, the univariate search method is more efficient (requires a smaller number of function evaluations) when compared to the conjugate direction method. This result is not the same as that in Example 6.5 and Example 6.6. Referring to the

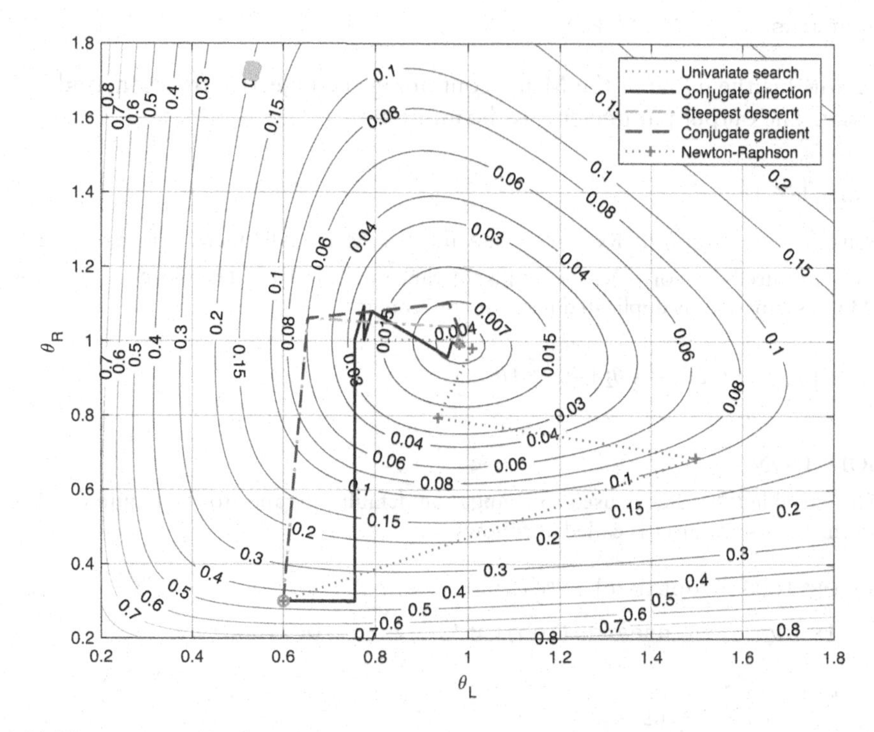

Figure 6.34 The contour plot of the objective function together with the search paths of different optimization algorithms.

contour lines in Figure 6.28, the objective function forms a long canyon at about 45° from the θ_1-axis. As the search directions for the univariate search method must be along either the θ_1 or θ_2-axis, the search path must follow a series of zig-zag lines. The resultant minimization process becomes very inefficient. Under such a situation, the conjugate direction method is the method of choice, as it can search a direction other than the θ_1 and θ_2 directions.

Referring to the contour lines in Figure 6.34, the objective function looks like a circular puddle. Thus, the minimization is efficient even if the search directions are along θ_L and θ_R. Due to the same reason, the efficiencies for both steepest descent and conjugate gradient methods are both good. Even though the number of iteration steps is small for both first-order methods, they are not necessarily more efficient when compared to the two zero-order methods, as additional function evaluations are needed in approximating the gradient vectors by the finite difference method. Note that more function evaluations are needed in approximating the Hessian matrix in the Newton–Raphson method. Thus, the Newton–Raphson method is less efficient than the steepest descent and conjugate gradient methods, even though the number of iteration steps is the same (i.e., 6) in this example.

6.2.4.6 *MATLAB function: fminsearch()*

The MATLAB function `fminsearch()` implements the Nelder–Mead Method (a kind of simplex method) to minimize a function following a direct search method; that is, only zero-order information is required. The Nelder–Mead method is believed to be one of the most efficient and popular optimization methods. The algorithm of the Nelder–Mead method is not given here. However, the syntax for using `fminsearch()` is given:

```
X = fminsearch(FNAME,X0)
```

where FNAME is the name of the M-file containing the objective function, and X0 is the initial trial. The output parameter X is the minimum.

Example 6.11

Solve the minimization in Example 6.5 again by the MATLAB function fminsearch() with an initial trial of $\{-2,-1\}^T$ and a tolerance of 1.0×10^{-8}. The objective function (Matyas function) is duplicated here for convenience.

$$J(\theta_1,\theta_2) = 0.26\left(\theta_1^2+\theta_2^2\right) - 0.48\theta_1\theta_2$$

SOLUTION

The variable OPTIONS is used to change the default tolerance to the required value. Then, fminsearch() is called as follows:

```
>> OPTIONS = optimset('TolX',1.e-8);

>> [x,fval]=fminsearch('objfun',[-2 -1],OPTIONS)
x =
   1.0e-08 *
   -0.1940   -0.0926
fval =
   3.3954e-19
```

where objfun is the file name of the objective function. It is clear from the results that the identified minimum is at $\{-0.1940,-0.0926\}^T \times10^{-8}$ with an objective value of 3.3954×10^{-19}. The total number of function evaluations is 145. The conjugate direction method solves the same problem with over 600 iterations.

6.2.5 Gradient and Hessian approximation using finite-difference

In most model updating problems, the objective function is the measure-of-fit function that usually involves solving an eigenvalue problem or a time-domain structural responses analysis. This kind of objective function cannot be expressed explicitly. Thus, the expression for gradient vector and Hessian matrix cannot be provided, and only zero-order methods can be used. One possible way for these complicated objective functions to use higher-order methods for minimization is to approximate the gradient and Hessian information by the finite-difference method.

Both the gradient vector and Hessian matrix can be approximated by the Taylor series expansion. A two-variable function $J(\theta_1, \theta_2)$ in Figure 6.35 is used as an example.

To simplify the formulation, the following definitions of various function values as shown in Figure 6.35 are given:

$$J_{-1,1} = J\left(\begin{Bmatrix}\theta_1-\Delta\theta_1\\ \theta_2+\Delta\theta_2\end{Bmatrix}\right) \quad J_{0,1} = J\left(\begin{Bmatrix}\theta_1\\ \theta_2+\Delta\theta_2\end{Bmatrix}\right) \quad J_{1,1} = J\left(\begin{Bmatrix}\theta_1+\Delta\theta_1\\ \theta_2+\Delta\theta_2\end{Bmatrix}\right)$$

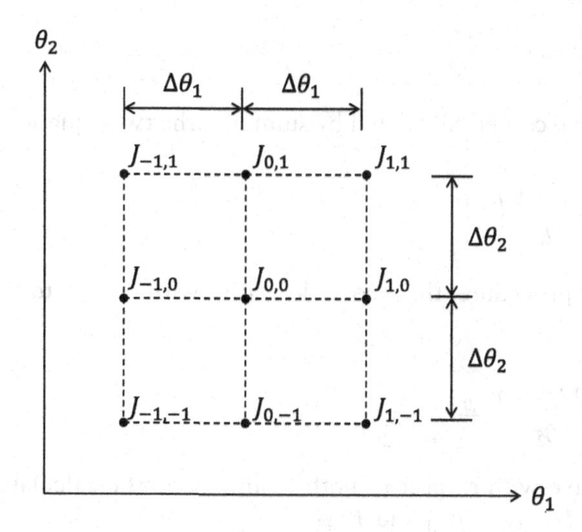

Figure 6.35 A example of a two-variable function.

$$J_{-1,0} = J\left(\begin{Bmatrix} \theta_1 - \Delta\theta_1 \\ \theta_2 \end{Bmatrix}\right) \quad J_{0,0} = J\left(\begin{Bmatrix} \theta_1 \\ \theta_2 \end{Bmatrix}\right) \quad J_{1,0} = J\left(\begin{Bmatrix} \theta_1 + \Delta\theta_1 \\ \theta_2 \end{Bmatrix}\right)$$

$$J_{-1,-1} = J\left(\begin{Bmatrix} \theta_1 - \Delta\theta_1 \\ \theta_2 - \Delta\theta_2 \end{Bmatrix}\right) \quad J_{0,-1} = J\left(\begin{Bmatrix} \theta_1 \\ \theta_2 - \Delta\theta_2 \end{Bmatrix}\right) \quad J_{1,-1} = J\left(\begin{Bmatrix} \theta_1 + \Delta\theta_1 \\ \theta_2 - \Delta\theta_2 \end{Bmatrix}\right)$$

where the perturbation vector is:

$$\Delta\theta = \begin{Bmatrix} \Delta\theta_1 \\ \Delta\theta_2 \end{Bmatrix} \tag{6.112}$$

Referring to Figure 6.35, the value of $J_{1,0}$ and $J_{-1,0}$ can be approximated by the Taylor series expansion at $J_{0,0}$ as:

$$J_{1,0} = J_{0,0} + \Delta\theta_1 \left.\frac{\partial J}{\partial \theta_1}\right|_{0,0} + \frac{\Delta\theta_1^2}{2!} \left.\frac{\partial^2 J}{\partial \theta_1^2}\right|_{0,0} + O\left(\Delta\theta_1^3\right) \tag{a1}$$

$$J_{-1,0} = J_{0,0} - \Delta\theta_1 \left.\frac{\partial J}{\partial \theta_1}\right|_{0,0} + \frac{\Delta\theta_1^2}{2!} \left.\frac{\partial^2 J}{\partial \theta_1^2}\right|_{0,0} + O\left(\Delta\theta_1^3\right) \tag{a2}$$

Calculating the difference between the two equations $a1$ and $a2$ and dividing by 2 (i.e., $(a1 - a2)/2$), the following expression can be obtained:

$$\left.\frac{\partial J}{\partial \theta_1}\right|_{0,0} = \frac{J_{1,0} - J_{-1,0}}{2\Delta\theta_1} \tag{6.113}$$

Following a similar procedure, the first derivative of the function with respect to θ_2 at $\{\theta_1, \theta_2\}^T$ can be obtained as:

$$\left.\frac{\partial J}{\partial \theta_2}\right|_{0,0} = \frac{J_{0,1} - J_{0,-1}}{2\Delta\theta_2} \tag{6.114}$$

The second derivative can be calculated by summing the two equations (i.e., $a1 + a2$):

$$\left.\frac{\partial^2 J}{\partial \theta_1^2}\right|_{0,0} = \frac{J_{0,1} - 2J_{0,0} + J_{0,-1}}{\Delta\theta_1^2} \tag{6.115}$$

Following a similar procedure, the second derivative with respect to θ_2 at $\{\theta_1, \theta_2\}^T$ can be obtained as:

$$\left.\frac{\partial^2 J}{\partial \theta_2^2}\right|_{0,0} = \frac{J_{1,0} - 2J_{0,0} + J_{-1,0}}{\Delta\theta_2^2} \tag{6.116}$$

The second derivative with respect to both θ_1 and θ_2 can be calculated from the Taylor series expansion of $J_{1,1}$, $J_{1,-1}$, $J_{-1,1}$ and $J_{-1,-1}$:

$$J_{1,1} = J_{0,0} + \Delta\theta_1 \left.\frac{\partial J}{\partial \theta_1}\right|_{0,0} + \Delta\theta_2 \left.\frac{\partial J}{\partial \theta_2}\right|_{0,0} + 2\frac{\Delta\theta_1\Delta\theta_2}{2!}\left.\frac{\partial^2 J}{\partial\theta_1\partial\theta_2}\right|_{0,0} + \ldots \tag{b1}$$

$$J_{1,-1} = J_{0,0} + \Delta\theta_1 \left.\frac{\partial J}{\partial \theta_1}\right|_{0,0} - \Delta\theta_2 \left.\frac{\partial J}{\partial \theta_2}\right|_{0,0} - 2\frac{\Delta\theta_1\Delta\theta_2}{2!}\left.\frac{\partial^2 J}{\partial\theta_1\partial\theta_2}\right|_{0,0} + \ldots \tag{b2}$$

$$J_{-1,1} = J_{0,0} - \Delta\theta_1 \left.\frac{\partial J}{\partial \theta_1}\right|_{0,0} + \Delta\theta_2 \left.\frac{\partial J}{\partial \theta_2}\right|_{0,0} - 2\frac{\Delta\theta_1\Delta\theta_2}{2!}\left.\frac{\partial^2 J}{\partial\theta_1\partial\theta_2}\right|_{0,0} + \ldots \tag{b3}$$

$$J_{-1,-1} = J_{0,0} - \Delta\theta_1 \left.\frac{\partial J}{\partial \theta_1}\right|_{0,0} - \Delta\theta_2 \left.\frac{\partial J}{\partial \theta_2}\right|_{0,0} + 2\frac{\Delta\theta_1\Delta\theta_2}{2!}\left.\frac{\partial^2 J}{\partial\theta_1\partial\theta_2}\right|_{0,0} + \ldots \tag{b4}$$

The four equations $(b1)$ to $(b4)$ can be operated as:

$$A(b1) + B(b2) + C(b3) + D(b4) = 2\frac{\Delta\theta_1\Delta\theta_2}{2!}\left.\frac{\partial^2 J}{\partial\theta_1\partial\theta_2}\right|_{0,0}$$

where A, B, C, and D are constant factors to be determined. The condition for the coefficient of $J_{0,0}$ to vanish:

$$A + B + C + D = 0$$

The condition for the coefficient of $\Delta\theta_1 \left.\frac{\partial J}{\partial \theta_1}\right|_{0,0}$ to vanish:

$$A + B - C - D = 0$$

The condition for the coefficient of $\Delta\theta_2 \left.\frac{\partial J}{\partial \theta_2}\right|_{0,0}$ to vanish:

$$A - B + C - D = 0$$

The condition for the coefficient of $2\frac{\Delta\theta_1\Delta\theta_2}{2!}\frac{\partial^2 J}{\partial\theta_1\partial\theta_2}\bigg|_{0,0}$ to be equal to unity:

$$A - B - C + D = 1$$

The four conditions result in a system of linear equations with the unknown coefficients A, B, C, and D. The system can be expressed in a matrix form as:

$$\begin{bmatrix} 1 & 1 & 1 & 1 \\ 1 & 1 & -1 & -1 \\ 1 & -1 & 1 & -1 \\ 1 & -1 & -1 & 1 \end{bmatrix}\begin{Bmatrix} A \\ B \\ C \\ D \end{Bmatrix} = \begin{Bmatrix} 0 \\ 0 \\ 0 \\ 1 \end{Bmatrix}$$

Therefore, the coefficients are $A = 1/4$, $B = -1/4$, $C = -1/4$, and $D = 1/4$, and one obtains:

$$\frac{J_{1,1} - J_{1,-1} - J_{-1,1} + J_{-1,-1}}{4} = \Delta\theta_1\Delta\theta_2 \frac{\partial^2 J}{\partial\theta_1\partial\theta_2}\bigg|_{0,0} \tag{6.117}$$

Finally,

$$\frac{\partial^2 J}{\partial\theta_1\partial\theta_2}\bigg|_{0,0} = \frac{J_{1,1} - J_{1,-1} - J_{-1,1} + J_{-1,-1}}{4\Delta\theta_1\Delta\theta_2} \tag{6.118}$$

Now the gradient vectors can be approximated by Equations (6.113) and (6.114). The Hessian matrix can be approximated by Equations (6.115), (6.116), and (6.118). Although the formulation is derived for a two-variable function, it can be easily extended to a general multivariable function with N_θ model parameters. The gradient vector as defined in Equation (6.50) (duplicated below for convenience):

$$\nabla J(\boldsymbol{\theta}) = \left\{\frac{\partial J(\boldsymbol{\theta})}{\partial\theta_1}, \frac{\partial J(\boldsymbol{\theta})}{\partial\theta_2}, \ldots, \frac{\partial J(\boldsymbol{\theta})}{\partial\theta_{N_\theta}}\right\}^T$$

where a general i-th term in it can be approximated as:

$$\frac{\partial J(\boldsymbol{\theta})}{\partial\theta_i} = \frac{J\left(\boldsymbol{\theta} + \mathbf{u}^{\langle i\rangle}\Delta\theta_i\right) - J\left(\boldsymbol{\theta} - \mathbf{u}^{\langle i\rangle}\Delta\theta_i\right)}{2\Delta\theta_i} \tag{6.119}$$

where $\mathbf{u}^{\langle i\rangle}$ is a vector with the same length as $\boldsymbol{\theta}$, and the k-th element of the $\mathbf{u}^{\langle i\rangle}$ vector is defined as:

$$\mathbf{u}^{\langle i\rangle}(k) = \begin{cases} 1 & \text{if } i = k \\ 0 & \text{otherwise} \end{cases} \tag{6.120}$$

Similarly, the Hessian matrix as defined in Equation (6.53) (duplicated below for convenience):

$$\nabla^2 J(\theta) = \mathbf{H}(\theta) \begin{bmatrix} \frac{\partial^2 J(\theta)}{\partial \theta_1^2} & \frac{\partial^2 J(\theta)}{\partial \theta_1 \partial \theta_2} & \cdots & \frac{\partial^2 J(\theta)}{\partial \theta_1 \partial \theta_{N_\theta}} \\ \frac{\partial^2 J(\theta)}{\partial \theta_2 \partial \theta_1} & \frac{\partial^2 J(\theta)}{\partial \theta_2^2} & \cdots & \frac{\partial^2 J(\theta)}{\partial \theta_2 \partial \theta_{N_\theta}} \\ \vdots & \vdots & \ddots & \vdots \\ \frac{\partial^2 J(\theta)}{\partial \theta_{N_\theta} \partial \theta_1} & \frac{\partial^2 J(\theta)}{\partial \theta_{N_\theta} \partial \theta_2} & \cdots & \frac{\partial^2 J(\theta)}{\partial \theta_{N_\theta}^2} \end{bmatrix}$$

The i-th diagonal element in the Hessian matrix can be approximated as:

$$\frac{\partial^2 J(\theta)}{\partial \theta_i^2} = \frac{J\left(\theta + \mathbf{u}^{(i)} \Delta\theta_i\right) - 2J(\theta) + J\left(\theta - \mathbf{u}^{(i)} \Delta\theta_i\right)}{\Delta\theta_i^2} \tag{6.121}$$

The off-diagonal element at the i-th row and j-th column can be approximated as:

$$\frac{\partial^2 J(\theta)}{\partial \theta_i \partial \theta_j} = \frac{J_{i,j} - J_{i,-j} - J_{-i,j} + J_{-i,-j}}{4\Delta\theta_i \Delta\theta_j} \tag{6.122}$$

where

$$J_{i,j} = J\left(\theta + \mathbf{u}^{(i)} \Delta\theta_i + \mathbf{u}^{(j)} \Delta\theta_j\right) \tag{6.123}$$

$$J_{i,-j} = J\left(\theta + \mathbf{u}^{(i)} \Delta\theta_i - \mathbf{u}^{(j)} \Delta\theta_j\right) \tag{6.124}$$

$$J_{-i,j} = J\left(\theta - \mathbf{u}^{(i)} \Delta\theta_i + \mathbf{u}^{(j)} \Delta\theta_j\right) \tag{6.125}$$

$$J_{-i,-j} = J\left(\theta - \mathbf{u}^{(i)} \Delta\theta_i - \mathbf{u}^{(j)} \Delta\theta_j\right) \tag{6.126}$$

By approximating the gradient vector and Hessian matrices using the finite-difference method, one may take advantage of higher-order methods with only zero-order information. Of course, the tradeoff is the additional computational cost for approximating the high-order information. Furthermore, the error in the approximated gradient vector and Hessian matrix may affect the efficiency of the higher-order algorithms.

6.2.6 Probabilistic optimization algorithms

A series of deterministic optimization algorithms are introduced in the previous sections. They are "deterministic" as the same minimum will be returned if the same initial trial (and the same set of algorithmic parameters) is provided for processing the optimization. Furthermore, this kind of algorithm aims at providing a single minimum even if the objective function has multiple minimums (or there are multiple points that give very similar objective function values). To address the non-uniqueness problem in optimization, probabilistic (or stochastic) optimization algorithms can be utilized. Unlike the deterministic methods, the probabilistic algorithms will usually return different results even when the same initial trial (and the same set of algorithmic parameters) is employed.

Probabilistic optimization algorithms are mainly zero-order methods, and they solve unconstrained optimization problems in the following standard form, which are extracted from Equations (6.44), (6.47), and (6.48).

Minimize:

$$J(\boldsymbol{\theta})$$

Subject to:

$$\theta_k^L \le \theta_k \le \theta_k^U \quad \text{for} \quad k = 1, \ldots, N_\theta$$

with design variables:

$$\boldsymbol{\theta} = \{\theta_1, \theta_2, \ldots, \theta_{N_\theta}\}^T$$

6.2.6.1 Random search

"Random search" refers to a wide range of methods, which search the parameter space of interest in a random manner. This kind of method is the most easily implemented, but it is also the most inefficient. One of the simplest methods of this kind is to randomly generate a given number of points in the parameter space and identify the one with the smallest objective function value. The procedure is as follows:

1. Define the number of points required, $g_{\max}$, and initialize the counter $g = 1$.
2. Generate the g-th random point as:

$$\boldsymbol{\theta}^{[g]} = \boldsymbol{\theta}^L + \boldsymbol{r}^T\left(\boldsymbol{\theta}^U - \boldsymbol{\theta}^L\right) \tag{6.127}$$

   where $\boldsymbol{r}$ is a column vector with length N_θ, the elements in $\boldsymbol{r}$ are independent random numbers generated from a uniform distribution with a value from 0 to 1; $\boldsymbol{\theta}^U = \{\theta_1^U, \theta_2^U, \ldots, \theta_{N_\theta}^U\}^T$ and $\boldsymbol{\theta}^L = \{\theta_1^L, \theta_2^L, \ldots, \theta_{N_\theta}^L\}^T$.
3. Calculate $J^{[g]} = J\left(\boldsymbol{\theta}^{[g]}\right)$.
4. If $g = 1$, assign $J^* = J^{[g]}$ and $\boldsymbol{\theta}^* = \boldsymbol{\theta}^{[g]}$. Otherwise:
   If $J^{[g]} < J^*$, assign $J^* = J^{[g]}$ and $\boldsymbol{\theta}^* = \boldsymbol{\theta}^{[g]}$.
5. Increase the counter $g = g + 1$.
6. If $g > g_{\max}$, stop. Otherwise, go to step 2.

When the algorithm stops, the identified minimum, $\boldsymbol{\theta}^*$, and the corresponding objective function value, J^*, can be obtained. Of course, this "minimum" is only the smallest one among all the generated random points. The result will be more reliable if a very large $g_{\max}$ is employed or if the parameters space is reduced (i.e., reducing the ranges between the upper and lower bounds). It must be pointed out that the algorithm returns not only a single minimum but all generated points, $\theta^{[g]}$, together with the corresponding objective function values, $J^{[g]}$. One can sort all points with an ascending order of the objective function value and investigate other points with similar objective function values as the minimum. If some points may be possible local or global minimums, a deterministic minimization can be carried out with each of these points as an initial trial.

The random search methods are very simple and easy to code. However, an extremely large value of g_{max} is needed to ensure the results are meaningful if the number of design variables, N_θ, is large. With simple modification in step 2 of the algorithm, the method can be used to handle discrete design variables.

6.2.6.2 Simulated annealing

Annealing in metallurgy is a heat treatment process that changes the mechanical properties of a material. It involves a very careful control of temperature that allows the formation of a better crystal structure resulting in a change in ductility and hardness of the material. During the process, an atom is allowed to go from the current energy state to a higher energy state (the chance of this depends on the temperature). The simulated annealing (SA) minimization algorithm makes use of this characteristic in "jumping" out from the local minimums.

Consider any one of the introduced deterministic minimization algorithms. The design point, $\theta^{[k]}$, in the k-th iteration certainly has an objective function value smaller than that in the $(k-1)$-th iteration, $\theta^{[k-1]}$. Points with higher objective function values will certainly be discarded during the minimization process. This rule makes the algorithms to be easily trapped by local minimums. The outstanding feature of simulated annealing is that a design point with a higher objective function value may still be accepted with a probability depending on the increase in objective function value. This reduces the chance for the algorithm to be trapped by local minimums.

The simulated annealing algorithm repeats a series of N_θ sample generating processes (each along the direction of a design variable) in multiple levels. The algorithm starts with a given temperature, T, and it will be reduced in successive levels by the temperature reduction factor, r_T (<1). A sample will be accepted (and replace the current design point) if its objective function value is lower than that of the current design point. The probability of accepting a sample with an objective function value higher than that of the current design point is:

$$p(\Delta J) = \exp\left(\frac{-\Delta J}{T}\right) \tag{6.128}$$

where ΔJ is the increase in objective function value, and T is the temperature at the corresponding level. At each temperature level, the series of N_θ sample generating process will be repeated in N_T temperature loops, and in each temperature loop, it will be repeated in N_c cycles. As a result, $N_T N_c N_\theta$ samples will be generated at each temperature level.

At the beginning of the algorithm (when the temperature is high), samples should be generated in a wide range that covers the entire parameter space of interest. Near the end of the algorithm (when the temperature is low), the range for generating samples must be small enough to ensure convergence. This is implemented by controlling the sampling region in each direction by:

$$\mathbf{R} = R_T\left(\theta^U - \theta^L\right) \tag{6.129}$$

where $\mathbf{R} = \{R_1, R_2, \ldots, R_{N_\theta}\}^T$, R_j for $j = 1,\ldots,N_\theta$ is the sampling range along the j-th design variable for sampling in each temperature level, and R_T is the scaling factor that controls the sampling ranges. In the first temperature level, $R_T = 1$ is employed. In each following temperature level, its value will be reduced by the range reduction factor, r_R (<1).

As samples are generated along each design variable (i.e., direction) in turn, if it is found that samples generated in a given direction are usually rejected, the corresponding sampling range should be reduced. Similarly, the sampling range should be increased if the acceptance rate for samples along the corresponding direction is high. This is implemented by defining the vector of acceptance rates $\mathbf{a} = \{a_1, a_2, \ldots, a_{N_\theta}\}^T$, where a_j for $j = 1,\ldots,N_\theta$ is the acceptance rate of the j-th design variable (direction). An initial value of unity is used. Each time a sample along a direction, say the j-th direction, is rejected in one of the N_c cycles, the corresponding acceptance rate will be reduced by:

$$a_j = a_j - \frac{1}{N_c} \tag{6.130}$$

The acceptance rates will be used to modify the sampling ranges after every N_c cycles. Then, the sampling rates are reset back to unity. The sampling range of the j-th direction is updated by the modification function (Belegundu & Chandrupatla 2011):

$$g(a) = \begin{cases} 1 + c\dfrac{a - 0.6}{0.4} & \text{if } a > 0.6 \\ \left(1 + c\dfrac{0.4 - a}{0.4}\right)^{-1} & \text{if } a < 0.4 \\ 1 & \text{otherwise} \end{cases} \tag{6.131}$$

where $c = 2$ is the recommended value. The modification function is shown in Figure 6.36. The range of the j-th design variable is modified as follows:

$$R_j = g(a_j)R_j \quad \text{for} \quad j = 1,\ldots,N_\theta \tag{6.132}$$

When all generated samples along a given direction are accepted, the corresponding sampling range will be increased by a factor of 3. Oppositely, if all generated samples along a given direction are all rejected, the sampling range will be reduced by a factor of 1/3.

The algorithm is relatively complicated when compared to the random search, and its procedure is given as follows:

1. Initialization:
   a) Initialize algorithmic constants:
      N_{max}: The maximum number of iterations (it is used to stop the algorithm if convergency cannot be achieved);

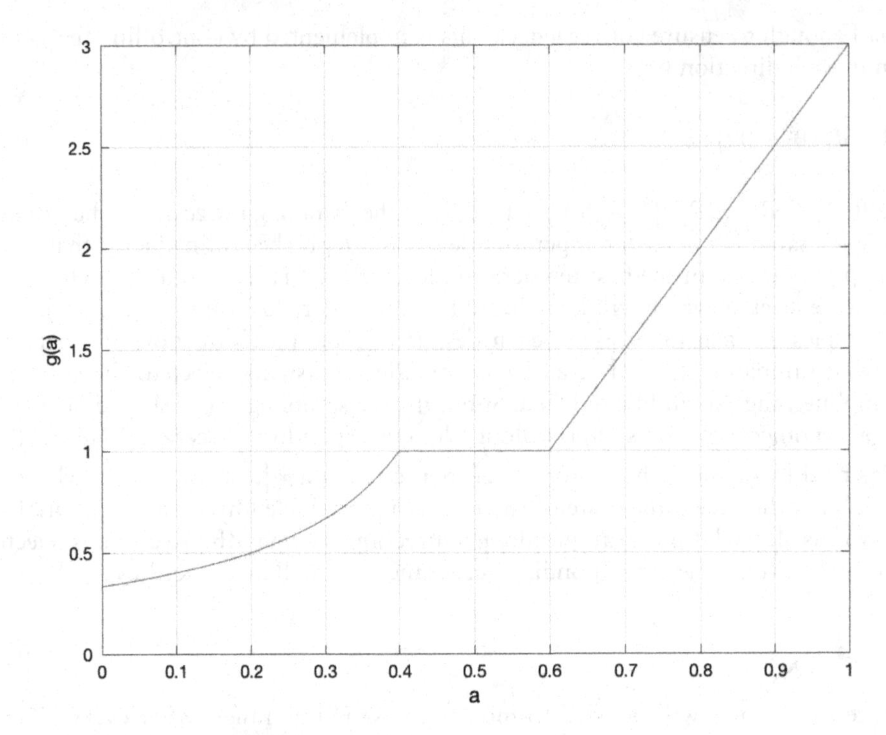

Figure 6.36 The modification function for adjusting the sample generation range based on the acceptance rate.

N_T: The number of temperature loops (the temperature is constant within a given temperature loop);
N_c: The number of cycle loops;
r_T: The temperature reduction factor;
r_R: The range reduction factor.

b) Initialize algorithmic variables (the initial value is assigned, and it will be updated in the iteration):
$R_T = 1$: The scaling factor that controls the sampling range of design variable;
T: The temperature.
$\theta = \theta^{[0]}$: The current design point;
$J = J(\theta)$: The objective function value of the current design point;
$J_{\min} = J$: The minimum objective function value (so far);
$\mathbf{R} = R_T\left(\theta^U - \theta^L\right)$: The sampling ranges of all design variables;
$\mathbf{a}$: The acceptance rates. A unit vector with length N_θ is assigned;
$i = 0$: The iteration counter.

2. Increase the iteration counter by unity, $i = i + 1$.
3. For $i_T = 1$ to N_T
   a) For $i_c = 1$ to N_c
      i. For $j = 1$ to N_θ
         1. Generate a sample along the j-th design parameter, θ^s:

$$\theta_k^s = \begin{cases} \theta_k + (2r-1)R_k & \text{for } k = j \\ \theta_k & \text{otherwise} \end{cases} \quad \text{for } k = 1,\ldots,N_\theta \tag{6.133}$$

where $\boldsymbol{\theta}^s = \{\theta_1^s, \theta_2^s, \ldots, \theta_{N\theta}^s\}^T$ is the generated sample, $\boldsymbol{\theta} = \{\theta_1, \theta_2, \ldots, \theta_{N\theta}\}^T$ is the current design point, $\mathbf{R} = \{R_1, R_2, \ldots, R_{N\theta}\}^T$ is the set of sampling ranges; and r is a random number generated from a uniform distribution with a value from 0 to 1.

2. If the generated sample is outside the feasible domain, it will be replaced by:

$$\theta_k^s = \begin{cases} \theta_k^L + r\left(\theta_k^U - \theta_k^L\right) & \text{for } k = j \\ \theta_k & \text{otherwise} \end{cases} \quad \text{for } k = 1, \ldots, N_\theta \tag{6.134}$$

3. Determine $J^s = J\left(\boldsymbol{\theta}^s\right)$.
4. If $J^s \le J$, then $\boldsymbol{\theta} = \boldsymbol{\theta}^s$ and $J = J^s$. The sample is accepted.
5. If $J^s > J$, then calculate:

$$p = \exp\left(\frac{J - J^s}{T}\right) \tag{6.135}$$

Generate a random number r from 0 to 1 following a uniform distribution.

a) If $r < p$, then $\boldsymbol{\theta} = \boldsymbol{\theta}^s$ and $J = J^s$. The sample is accepted.
b) If $r \ge p$, then the sample is rejected, and the corresponding acceptance rate is reduced by Equation (6.130).

6. If $J^s \le J_{\min}$, then $\boldsymbol{\theta}_{\min} = \boldsymbol{\theta}^s$ and $J_{\min} = J^s$.

ii. End For j

b) End For i_c
c) Update the sampling ranges $\mathbf{R}$ based on the acceptance rates $\mathbf{a}$ by Equation (6.132) with the modification function given in Equation (6.131).
d) Reset the acceptance rates $\mathbf{a}$ to a unit vector.

4. End For i_T
5. Reduce the temperature $T = r_T T$ and the scaling factor $R_T = r_R R_T$.
6. Reset the sampling ranges $\mathbf{R} = R_T\left(\boldsymbol{\theta}^U - \boldsymbol{\theta}^L\right)$.
7. Update the design point $\boldsymbol{\theta} = \boldsymbol{\theta}_{\min}$ and the corresponding objective function value $J = J_{\min}$.
8. If the convergency criteria are achieved or $i \ge N_{\max}$, stop. Otherwise, go to step 2.

The algorithm returns a single minimum together with the corresponding objective function value as output. With small modification by adding a variable to store all samples generated in the final iteration, the algorithm can provide samples that cover the important region of parameter space as another output on top of the identified minimum. Based on the experience of the author, SA is very robust and can provide valuable information about the important regions of the objective function in the parameter space of interest, especially when the optimization problem is non-unique.

6.3 CASE STUDIES

6.3.1 Model updating of a shear building model

This section starts with something relatively simple—a four-story shear building model as shown in Figure 6.37(a). As there are only four DOFs (considering vibration along the x-direction only as shown in the figure), it is not difficult to measure all the DOFs.

Furthermore, the level of modeling error is believed to be low as the structural system is made of steel (i.e., the modulus of elasticity is very accurate), and the bending stiffness of the slab is much higher than that of the columns (i.e., the shear building model assumptions as discussed in Section 2.2.1 are valid). Under small amplitude vibration, the system is certainly linear and elastic.

The model updating of the target four-story shear building model based on vibration data measured under laboratory conditions is considered here. First, the vibration data of the normal shear building model (without additional mass) is obtained for identifying the inter-story stiffness of the structure (i.e., stiffness identification). This model is considered as the baseline for the mass identification of the shear building models in other cases with additional mass(es) at the selected floor(s). Note that stiffness (or mass) identification is model updating by considering the inter-story stiffnesses (or floor masses) as uncertain model parameters.

This shear building model was built for the purpose of education in some structural dynamics courses (both undergraduate and postgraduate). The authors do not want to artificially damage the model, and therefore, this case study focuses on identifying the added masses instead of the reduction in stiffness due to damage.

6.3.1.1 Description of the structure and cases considered

Figure 6.37(a) shows the physical shear building model. The additional masses in one of the considered cases and the installed accelerometers are also shown in the figure. The entire model is made of steel (with modulus of elasticity $E = 200$ GPa and mass density $\rho = 8000\,\text{kg/m}^3$). Each floor is supported by four rectangular columns, which are

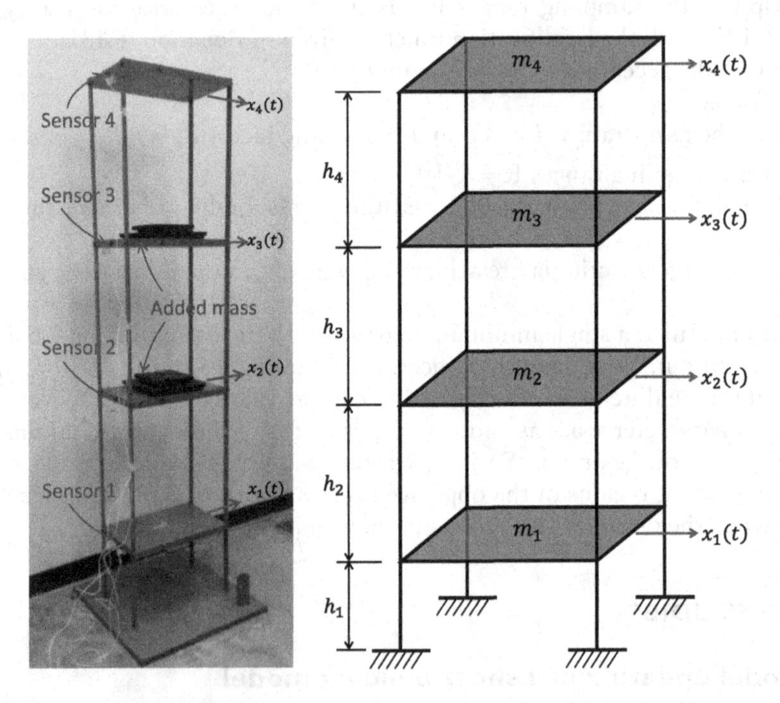

Figure 6.37 The four-story shear building model. (a) The physical model (with added masses and sensors). (b) The computer model.

continuous across all floors. The columns are fixed onto the floor slab through bolts, as shown in Figure 6.39. Under the assumption of a shear building model (as given in Section 2.2.1), the column-slab joints are rigid. However, it is believed that the real joint is semi-rigid in nature. The semi-rigid behavior of the column-slab joint will certainly reduce the inter-story stiffness that is calculated under the rigid joint assumption. This effect will be considered in the stiffness identification of the normal structure (without additional mass). The columns are steel strips each with width b and thickness (or depth) t. The dimensions of the columns are important in calculating the nominal value of the inter-story stiffness of the shear building model, and they are given in Table 6.5 together with the dimensions of the floor slab (important in calculating the floor masses). Figure 6.37(b) shows the corresponding shear building model (a 4-DOF system). The modeling will be discussed in detail in the next section.

A total of eight cases (including the normal case) are considered in this case study. The Case ID and the corresponding descriptions are given in Table 6.6. The NoMass case is the baseline without any additional mass. It is used as a reference for mass identification in other cases. A series of four single additional mass cases are considered in this study. They are FM2 (5 kg mass added to floor slab 2), TM2 (10 kg mass added to floor slab 2), FM3 (5 kg mass added to floor 3), and TM3 (10 kg mass added to floor 3). By comparing FM2 and TM2 (or FM3 and TM3), the effects of changing mass extent/magnitude can be

Table 6.5 Dimensions of the Shear Building Model

Column heights	
First story, h_1	278 mm
Second story, h_2	546 mm
Third story, h_3	546 mm
Fourth story, h_4	530 mm
Column cross-sectional dimensions	
Width, b	25.0 mm
Thickness (depth), t	5 mm
Floor slab dimensions	
Length, L	450 mm
Width, B	280 mm
Thickness, T	25 mm

Table 6.6 Cases Considered in the Vibration Tests of the Shear Building Model

Case ID	*Type*	*Description*
NoMass	No additional mass (normal)	This is the baseline (reference)
FM2	Single additional mass	5 kg additional mass at level 2
TM2		10 kg additional mass at level 2
FM3		5 kg additional mass at level 3
TM3		10 kg additional mass at level 3
FM2FM3	Double additional masses with the same extent	5 kg additional masses at levels 2 and 3
TM2TM3		10 kg additional masses at levels 2 and 3
FM2TM3	Double additional masses with different extents	5 kg additional mass at level 2 and 10 kg additional mass at level 3

studied. By comparing FM2 and FM3 (or TM2 and TM3), the effects of changing mass location can be observed. Three double additional masses cases are considered. Two of them have the same additional mass extent. They are FM2FM3 (5 kg mass at both floors 2 and 3) and TM2TM3 (10 kg masses at both floors 2 and 3). The final case has different extents for the additional masses i.e., FM2TM3 (5 kg mass added to floor 2 and 10 kg mass added to floor 3). Figure 6.37(a) shows the condition of TM2TM3. These comprehensive cases provide strong evidence in verifying the applicability of the model updating approach introduced in this chapter.

6.3.1.2 Impact hammer test and identified modal parameters

One of the most convenient (and cost-effective) methods for identifying the modal parameters of a structural system under laboratory conditions is the impact hammer test. With the impact excitation, the signal-to-noise ratio is usually high. As the spectrum of the impact force is flat for a given frequency range (depending on the hardness of the hammer head/tip), all modes within that frequency range should be excited. Note that the output-only method introduced in Chapter 3 can be used to identify the modal parameters from the measured time-domain acceleration responses from impact hammer tests. However, several important rules must be followed (to be discussed later in this section). Otherwise, a set of reliable modal parameters cannot be identified.

Four accelerometers are employed in this case study, and they are used to measure the impact hammer-induced acceleration responses of the four floor slabs along the horizontal x-direction as shown in Figure 6.37(a). Each sensor (i.e., accelerometer) is installed (by wax) on the edge of the corresponding floor slab along the center line of the slab (see Figure 6.38). An impact hammer is used to set the system in vibration through an impulse. The contact time between the impact hammer and the floor slab depends on the hardness of the hammer head. The higher the hardness, the shorter the contact time will be and the larger the resultant frequency range. Thus, a very hard hammer head is needed to excite a mode with very high natural frequency. In this case study, the natural frequencies of all four modes (this is a four-DOF system with only four modes) are very low. Therefore, a relatively soft hammer head is adopted (see the hammer head as shown in Figure 6.39) to increase the contact time and to allow more energy to contribute in

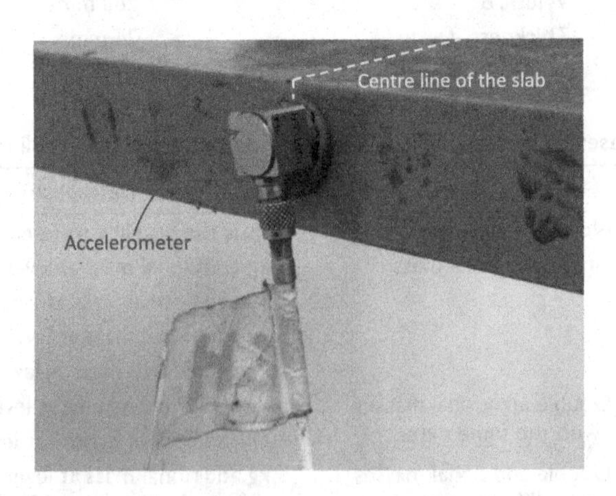

Figure 6.38 Installation of accelerometer for measuring the horizontal vibration of a floor slab.

exciting modes at a lower frequency range. The impact direction must be parallel to the DOF of the corresponding floor, especially for tests with multiple setups (when the number of available sensors is smaller than the number of targeted DOFs to be measured, the test has to be divided into more than one setup).

In this study, a sampling frequency of 2048 Hz is used and the measured duration is set to 1 min. The impact force is applied after the recording is started. The segment of measured responses before the impact is removed for the purpose of modal identification. Figure 6.40 shows an example of measured time-domain responses at DOF 4 (the top floor) under an impact force at DOF 2. The decay of vibration amplitude due to damping is typical for impact hammer test. Note that it is not necessary to measure the force input if the modal identification method in Chapter 3 is employed.

Each test (for a particular case) is divided into four parts, and each part corresponds to a given impact DOF (i.e., DOFs 1, 2, 3, and 4). Each part is repeated five times, and therefore, a total of 20 sets of time-domain data (each set consists of four channels for the four DOFs) are recorded in each test. Each set of data can be used to identify a set of modal parameters, and only the natural frequencies and mode shapes are of interest in this case study (for the purpose of model updating and mass identification).

To show the effects of impact DOF on the results of modal identification, the 20 sets of identified natural frequencies and mode shapes (5 for each impact DOF) in TM3 are shown in Figures 6.41 to 6.44, where the circle marker indicates the impact DOF, and the dashed vertical line shows the undeformed shape. Consider Figure 6.41 for impact at DOF 1 (floor 1), where the identified mode shapes for mode 1 from different trials are very different, especially at the impact DOF. Modes 2 and 3 are also not very good in a similar manner (not as bad as that in mode 1). However, the mode shapes for mode 4 from different trials are very good and overlapping each other (looks like a single mode shape). This can be explained by the fact that the impact DOF 1 has the largest amplitude of vibration for mode 4. However, this DOF has the smallest amplitude of vibration in all modes 1, 2, and 3. Thus, this impact DOF (i.e., DOF 1) result in mode 4 being

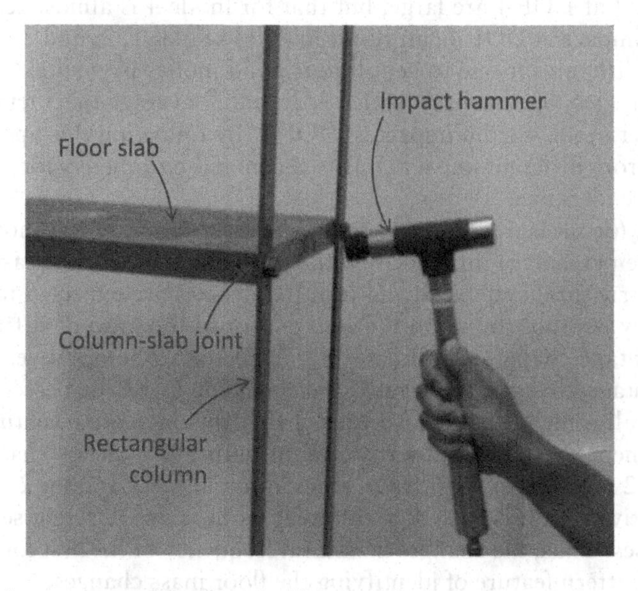

Figure 6.39 Impact hammer test of the shear building model.

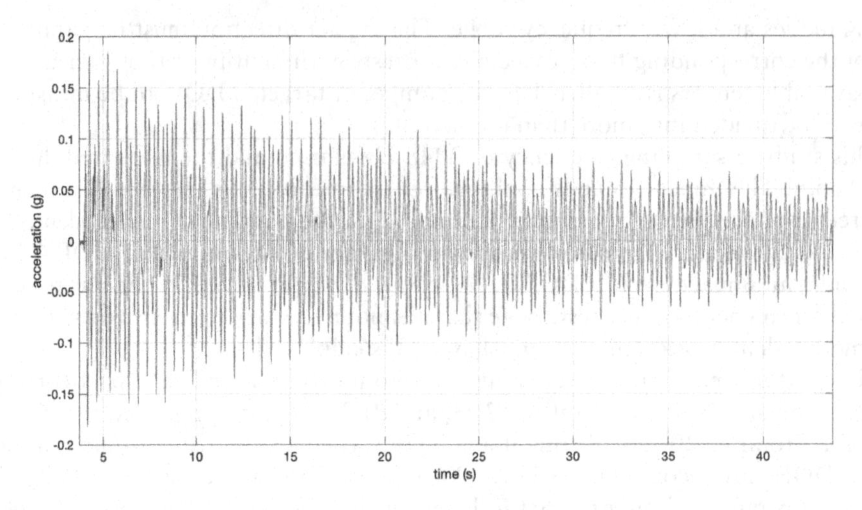

Figure 6.40 A typical measured acceleration from an impact hammer test.

the best. Note that the identified modal parameters of modes 1 to 3 with impact DOF at 1 should not be used. But, the identified modal parameters of mode 4 can be (and should be) used. Consider Figure 6.42 for impact at DOF 2. It is clear that the results are improved a lot even though mode 1 is still the worst. This is expected as the amplitude of vibration at DOF 2 (the impact DOF) is not the smallest in all modes. It is believed that this impact DOF (i.e., DOF 2) can result in very good modes 2 and 3. Next consider Figure 6.43 for impact at DOF 3. It can be observed that modes 1, 2, and 3 are good. However, mode 4 becomes very bad as the amplitude of vibration at DOF 3 is almost zero in this mode. It is believed that this impact DOF (i.e., DOF 3) can result in good modes 1, 2, and 3. Finally consider Figure 6.44 for impact at DOF 4. The situation is very similar to that in Figure 6.43 (impact at DOF 3). The amplitudes of vibration for modes 1, 2, and 3 at DOF 4 are large, but that for mode 4 is almost zero. Therefore, it is believed that impact at DOF 4 can result in good modes 1, 2, and 3. This illustration shows that it is difficult for one to get all four good modes by a single impact DOF. To obtain all four modes, one may get modes 1, 2, and 3 (average the 5 trials) from impact at DOF 4 and get mode 4 from impact at DOF 1. By following this approach (selecting the best modes from different impact DOFs), the modal parameters for all four modes in all eight cases are identified.

The identified (or measured) natural frequencies for all cases are shown in Table 6.7 together with the percentage differences with respect to the NoMass case. When masses are added to a structural system, the natural frequencies are expected to be reduced (as natural frequency is proportional to the square root of the ratio of stiffness over mass). Thus, the percentage changes in all cases are expected to be negative. Consider FM2, where all percentage changes are negative except mode 4. As the percentage change for mode 4 is so small (only 0.02%), it is believed that this increase in natural frequency is due to measurement noise. Consider the level of natural frequency changes, where it is clear that mode 2 is the most sensitive to the additional mass at floor 2, while mode 4 is the most insensitive one. It is clear that different modes have different sensitivity for different floor masses. Thus, the changes in natural frequency of the first four modes may be considered as a pattern feature of identifying the floor mass changes.

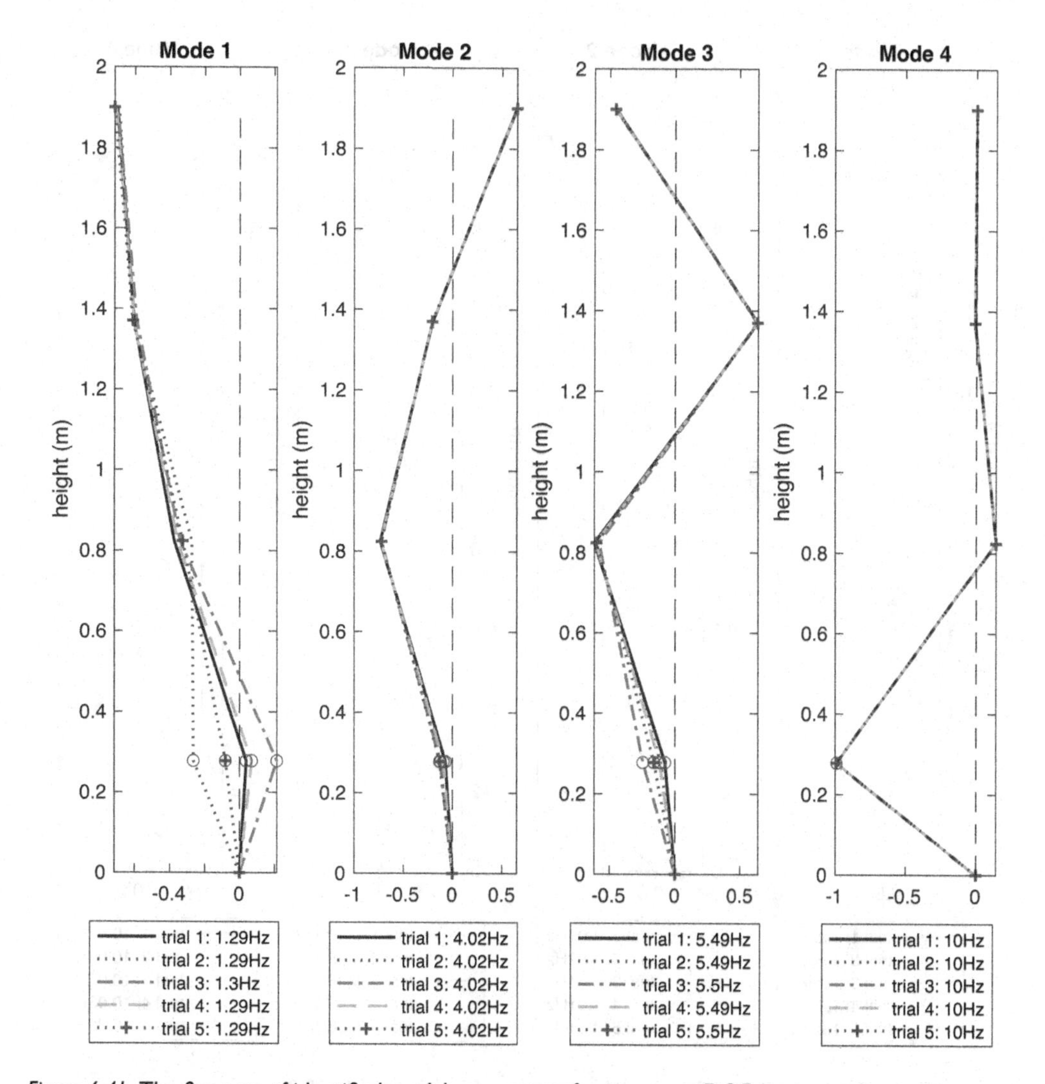

Figure 6.41 The five sets of identified modal parameters for impact at DOF 1.

Consider TM2 (double the additional mass at the same floor as FM2). The reductions in natural frequencies for mode 1 (from 1.31% to 3.06%), mode 2 (from 5.18% to 9.81%), and mode 3 (1.76% to 3.09%) are increased by ~2.3, ~1.9, and ~1.8 times, respectively; or one may say, the changes in natural frequencies are roughly double. Note that the percentage change in natural frequency for mode 4 is too small to be included in the discussion. It seems that when the additional mass is double, the pattern feature is double.

Consider FM3, where modes 1 and 3 (but not modes 2 and 4) are sensitive to the additional mass at floor 3. This pattern feature for mass at floor 3 is different from that for mass at floor 2. Comparing the percentage reductions in natural frequency from FM3 to TM3, mode 1 increases from 3.36% to 6.8%, mode 2 increases from 0.52% to 1.23%, and mode 3 increases from 4.95% to 8.8%. The increases for modes 1, 2, and 3 are ~2.0, ~2.4, and ~1.8 times, respectively, or roughly double.

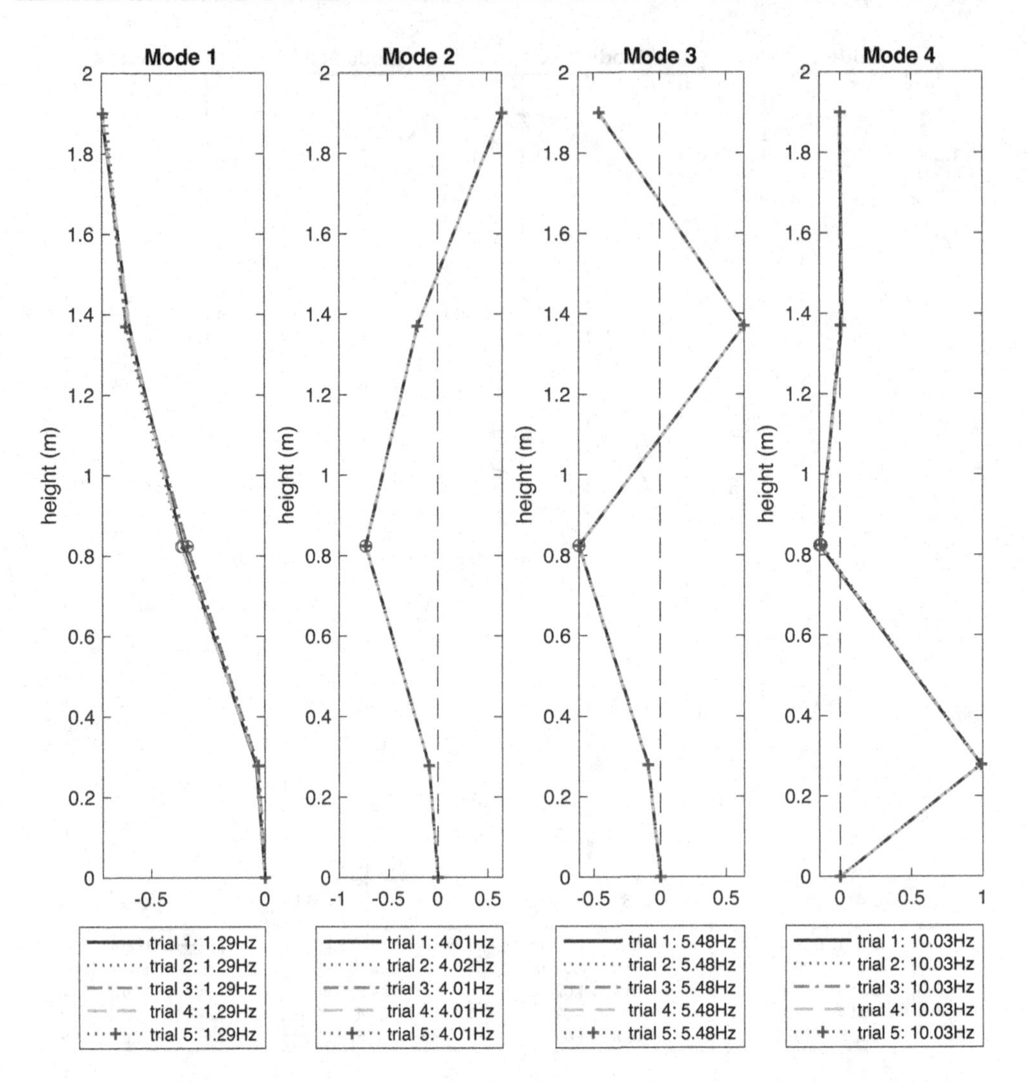

Figure 6.42 The five sets of identified modal parameters for impact at DOF 2.

Consider FM2FM3, where modes 1, 2, and 3 (but not mode 4) are sensitive. The percentage reductions in this case are similar to the sum of percentage reductions of FM2 and FM3. The reductions in FM2 for modes 1, 2, and 3 are 1.31%, 5.18% and 1.76%, respectively, while the reductions in FM3 for modes 1, 2, and 3 are 3.36%, 0.52%, and 4.95%, respectively. The sum for modes 1, 2, and 3 are 4.67%, 5.7%, and 6.71%, which are similar to the reduction in FM2FM3 of 4.73%, 5.60%, and 6.97%. When FM2FM3 is compared to TM2TM3, the reductions in natural frequencies are nearly double. If the modes are sorted according to their sensitivity in FM2FM3 and TM2TM3, the order of modes is 3, 2, 1, and 4.

Considering FM2TM3, the order of mode sensitivity is 3, 1, 2, and 4, which is different from that for FM2FM3 and TM2TM3. This reinforces the argument that the changes in natural frequency of the first four modes can be considered as a pattern feature for locating the floor of additional mass(es). The use of pattern features for the purpose of

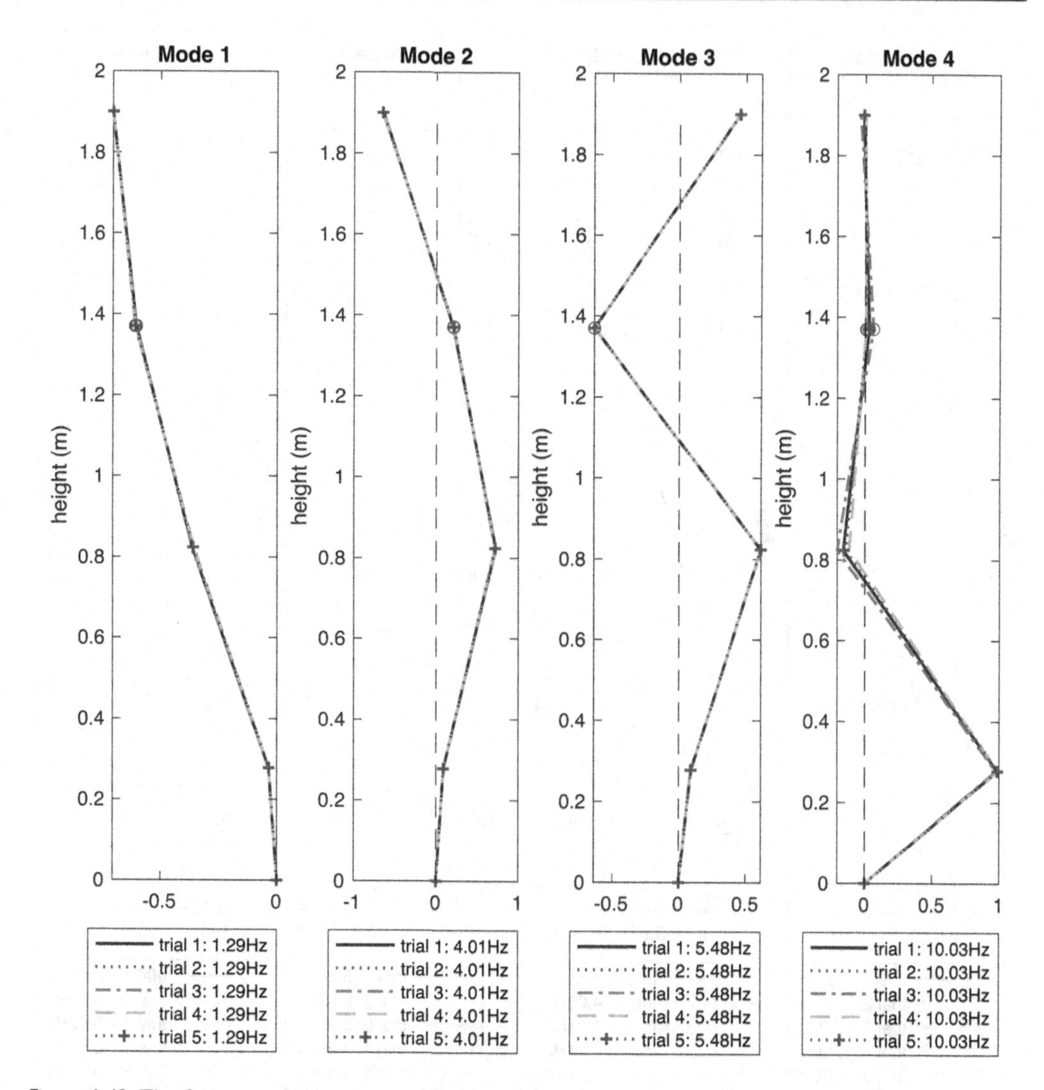

Figure 6.43 The five sets of identified modal parameters for impact at DOF 3.

mass identification (or damage detection) is not a new idea, and artificial neural networks (ANNs) are usually employed in the implementation of this idea. Interested readers are redirected to references (Lam & Lee 2005; Lam et al. 2006; Yuen & Lam 2006) for the use of ANN in structural damage detection. Although damage detection by pattern feature matching is not the main purpose of this case study, it clearly shows that the changes in natural frequency do contain valuable information about the location and extent of added mass(es).

The identified (or measured) mode shapes in all eight cases are grouped and presented in Figures 6.45 to 6.48. In Figure 6.45, the mode shapes for the NoMass (baseline), FM2, and TM3 cases are plotted together. From the figure, mode 1 is not very sensitive as only a very minor change in mode shape at DOF 2 is observed. Mode 2 and mode 3 are sensitive at DOF 3 and DOF 2, respectively. Mode 4 is not as sensitive as modes 2 and 3, but more

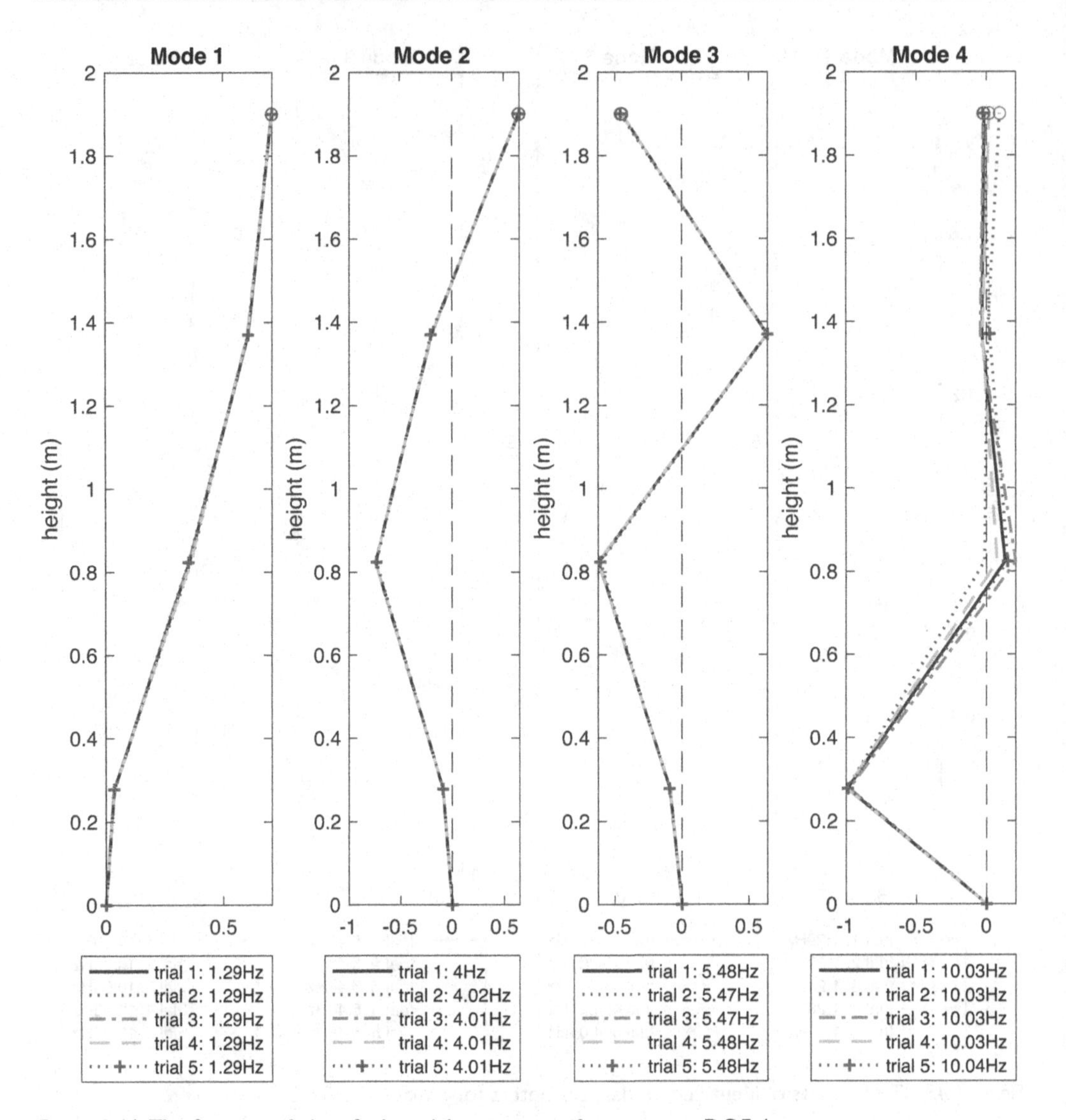

Figure 6.44 The five sets of identified modal parameters for impact at DOF 4.

sensitive to mode 1 especially at DOF 2. The results also show that the larger the mass extent, the larger the changes in mode shape will be. This can be clearly inferred from the changes in mode shape of mode 2 at DOF 3 (also of mode 3 at DOF 2).

From Figure 6.46 (FM3 and TM3), it is clear that modes 1, 2, and 4 are not sensitive to additional mass at floor 3. Only mode 3 is sensitive at DOFs 2, 3, and 4. The level of mode shape changes obviously depends on the level of mass change.

From Figure 6.47 (FM2FM3 and TM2TM3), modes 2 and 3 are very sensitive and mode 4 is sensitive, while mode 1 is less sensitive. Similarly, Figure 6.48 (FM2TM3) shows the changes in mode shapes due to the additional mass of 5 kg at floor 2 and 10 kg at floor 3. Modes 2 and 3 are very sensitive, while modes 1 and 4 are less sensitive.

It is very clear that the changes in natural frequencies and mode shapes contain valuable information about the location and extent of added mass(es).

Table 6.7 Measured Natural Frequency in Different Cases and the Corresponding Mass Induced Percentage Change

Case ID	*Mode 1 (Hz)*	*Mode 2 (Hz)*	*Mode 3 (Hz)*	*Mode 4 (Hz)*
NoMass	1.39	4.06	6.01	10.03
FM2	1.37 (–1.31%)	3.85 (–5.18%)	5.91 (–1.76%)	10.03 (0.02%)
TM2	1.34 (–3.06%)	3.66 (–9.81%)	5.83 (–3.09%)	10.01 (–0.18%)
FM3	1.34 (–3.36%)	4.04 (–0.52%)	5.71 (–4.95%)	10.08 (0.49%)
TM3	1.29 (–6.80%)	4.01 (–1.23%)	5.48 (–8.80%)	10 (–0.30%)
FM2FM3	1.32 (–4.73%)	3.84 (–5.60%)	5.59 (–6.97%)	10.05 (0.22%)
TM2TM3	1.26 (–9.35%)	3.64 (–10.43%)	5.27 (–12.34%)	9.94 (–0.87%)
FM2TM3	1.28 (–7.96%)	3.82 (–6.04%)	5.35 (–10.99%)	9.97 (–0.60%)

In summary, the rule of thumb for modal identification through impact hammer tests is as follows:

- Select an appropriate hammer head/tip with suitable hardness. If necessary, the test is repeated with different hammer heads.
- The impact direction should be parallel to the DOF. Any inclination angle may result in vibration in other unexpected directions. This may affect the accuracy of measured responses.
- Repeat the test with different impact DOFs to avoid missing modes or poorly identified modes. The identification result of a mode is expected to be poor if the mode shape value at the impact DOF is zero or very small.
- Repeat the test (at a given impact DOF) several times to ensure the reliability of the measured results. Discard outliers and calculate the average natural frequencies and mode shapes for the purpose of model updating.

The use of model updating technique in stiffness and mass identifications is demonstrated in the following sections.

6.3.1.3 Modeling of the four-story shear building model

The modeling method basically follows the one introduced in Section 2.2.1. The system stiffness matrix is considered first as:

$$\mathbf{K}=\begin{bmatrix} k_1+k_2 & -k_2 & 0 & 0 \\ -k_2 & k_2+k_3 & -k_3 & 0 \\ 0 & -k_3 & k_3+k_4 & -k_4 \\ 0 & 0 & -k_4 & k_4 \end{bmatrix} \tag{6.136}$$

where k_i, for $i=1,\ldots,4$, are the inter-story stiffnesses of the shear building at stories 1 to 4, respectively. As each floor is supported by four columns, the nominal value of the inter-story stiffness is defined as (see Section 2.2.1):

$$k_{0,i}=4\times\frac{12EI}{h_i^3},\quad \text{for } i=1,\ldots,4 \tag{6.137}$$

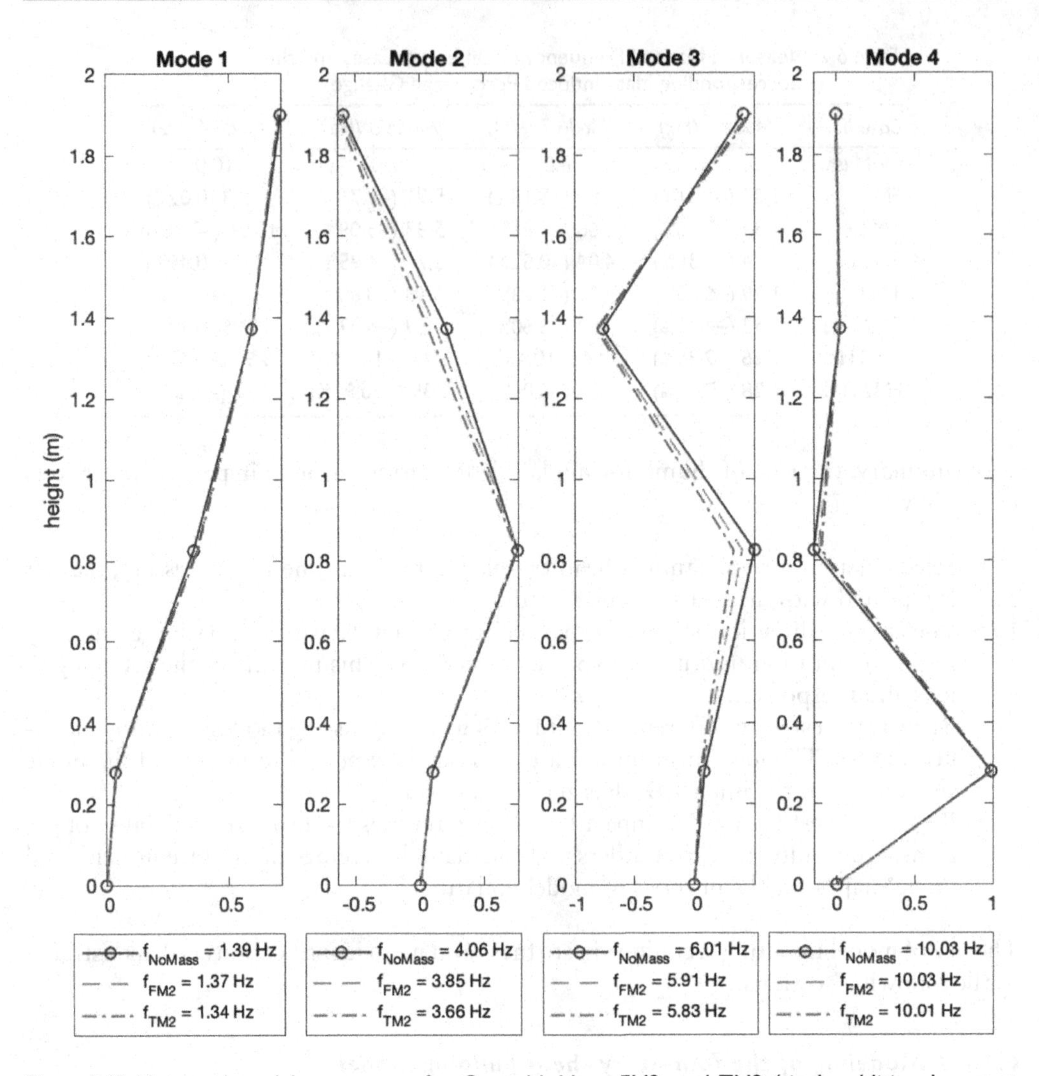

Figure 6.45 Measured modal parameters for Cases NoMass, FM2, and TM2 (single additional mass at floor 2).

where E is the modulus of elasticity and is equal to 200 GPa, I is the second moment of area of each column and is equal to $bt^3/12$, where the numerical values for b, t, h_i, for $i = 1,\ldots,4$, are given in Table 6.5. The nominal stiffness values are summarized in Table 6.8.

Note that the inter-story stiffness is considered as uncertain (i.e., treated as design variables in the minimization process) in the model updating of the NoMass case. This process is very important in identifying the baseline model for mass identification in other cases. In the model updating for the NoMass case, the numerical values of inter-story stiffness are given by:

$$k_i = \theta_{k,i} k_{0,i}, \quad \text{for } i = 1,\ldots,4 \tag{6.138}$$

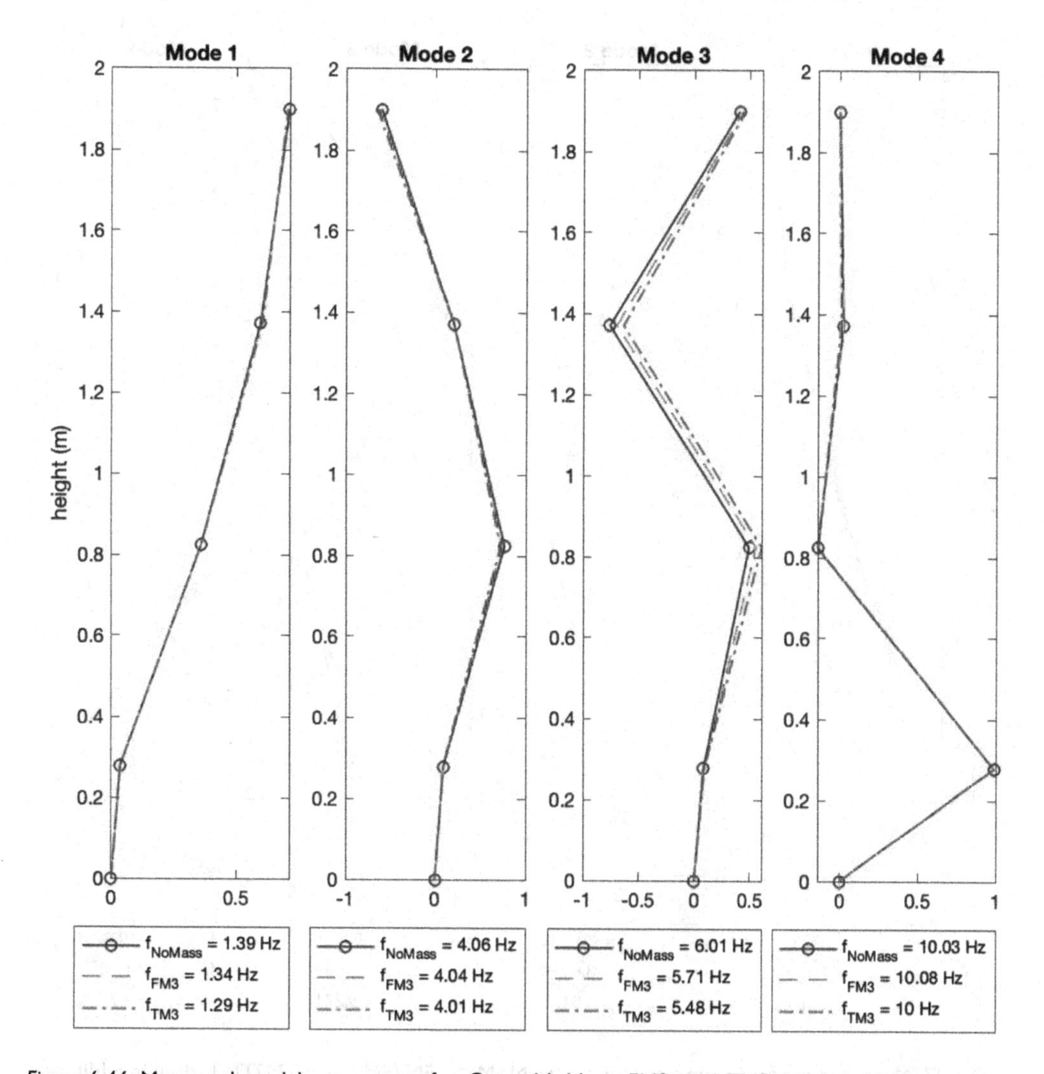

Figure 6.46 Measured modal parameters for Cases NoMass, FM3, and TM3 (single additional mass at floor 3).

where $\theta_{k,i}$, for $i = 1,\ldots,4$, are the non-dimensional scaling factors for the inter-story stiffness to be identified in the model updating process (NoMass). For a given $\boldsymbol{\theta}_k = \{\theta_{k,1}, \theta_{k,2}, \theta_{k,3}, \theta_{k,4}\}^T$, the inter-story stiffness can be calculated and the system stiffness matrix in Equation (6.136) can be obtained.

The system mass matrix is very simple, and it is expressed as (see Section 2.2.1):

$$\mathbf{M} = \begin{bmatrix} m_1 & 0 & 0 & 0 \\ 0 & m_2 & 0 & 0 \\ 0 & 0 & m_3 & 0 \\ 0 & 0 & 0 & m_4 \end{bmatrix} \tag{6.139}$$

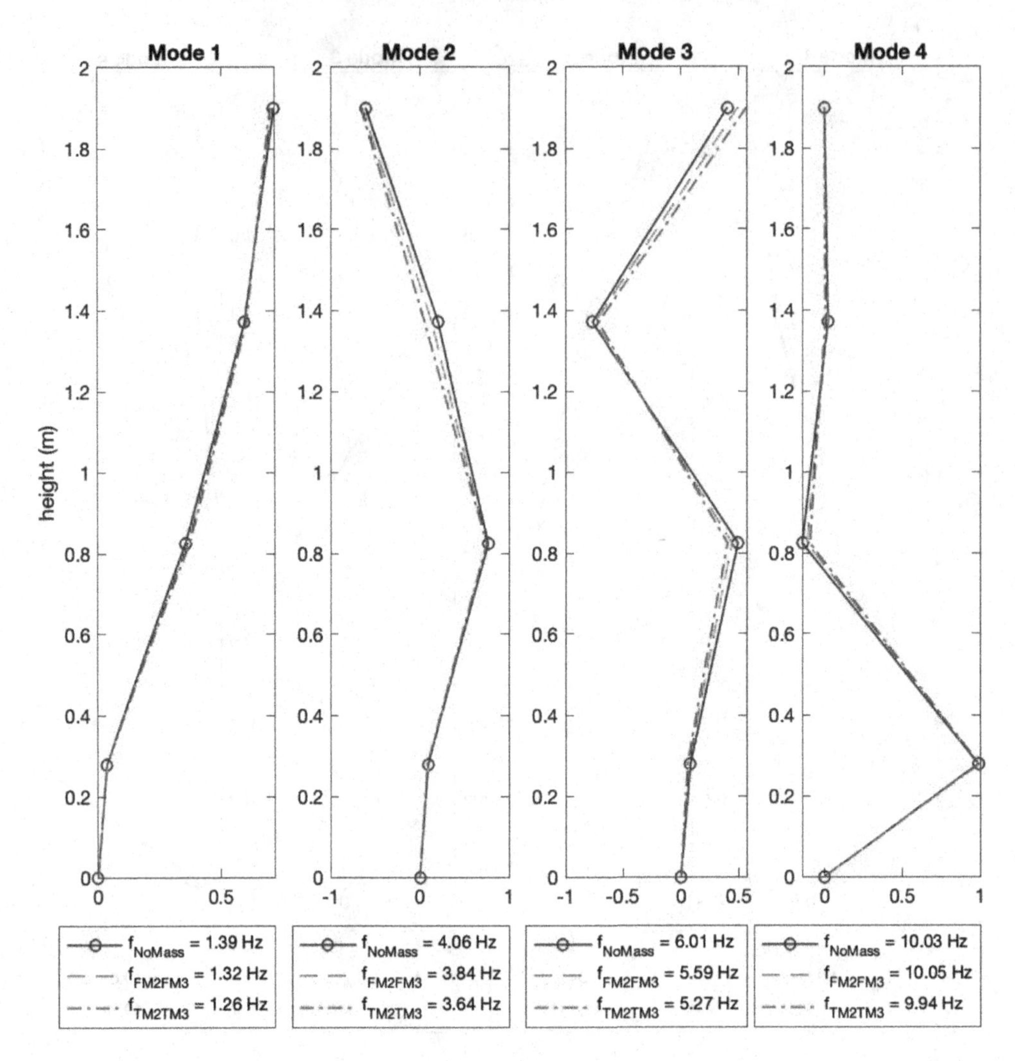

Figure 6.47 Measured modal parameters for Cases NoMass, FM2FM3, and TM2TM3 (double additional masses with the same extent).

where m_i, for $i = 1,\ldots,4$, are the total floor mass for floors 1 to 4. In all mass identification cases (i.e., all cases except NoMass), the total floor masses are considered as uncertain. Therefore, their values are calculated by:

$$m_i = \theta_{m,i} m_{0,i}, \quad \text{for } i = 1,\ldots,4 \tag{6.140}$$

where $\theta_{m,i}$, for $i = 1,\ldots,4$, are the non-dimensional scaling factors for the total floor masses to be identified in the mass identification cases. The column masses are lumped to the corresponding floors in the calculation of the total floor masses. For example, the nominal value of the total floor mass for floor 1, $m_{0,1}$ is calculated by summing the slab mass to half of the mass of the four columns at story 1 and half of the mass of the four columns at story 2. Note that the nominal value of the total floor mass for floor 4, $m_{0,4}$ is

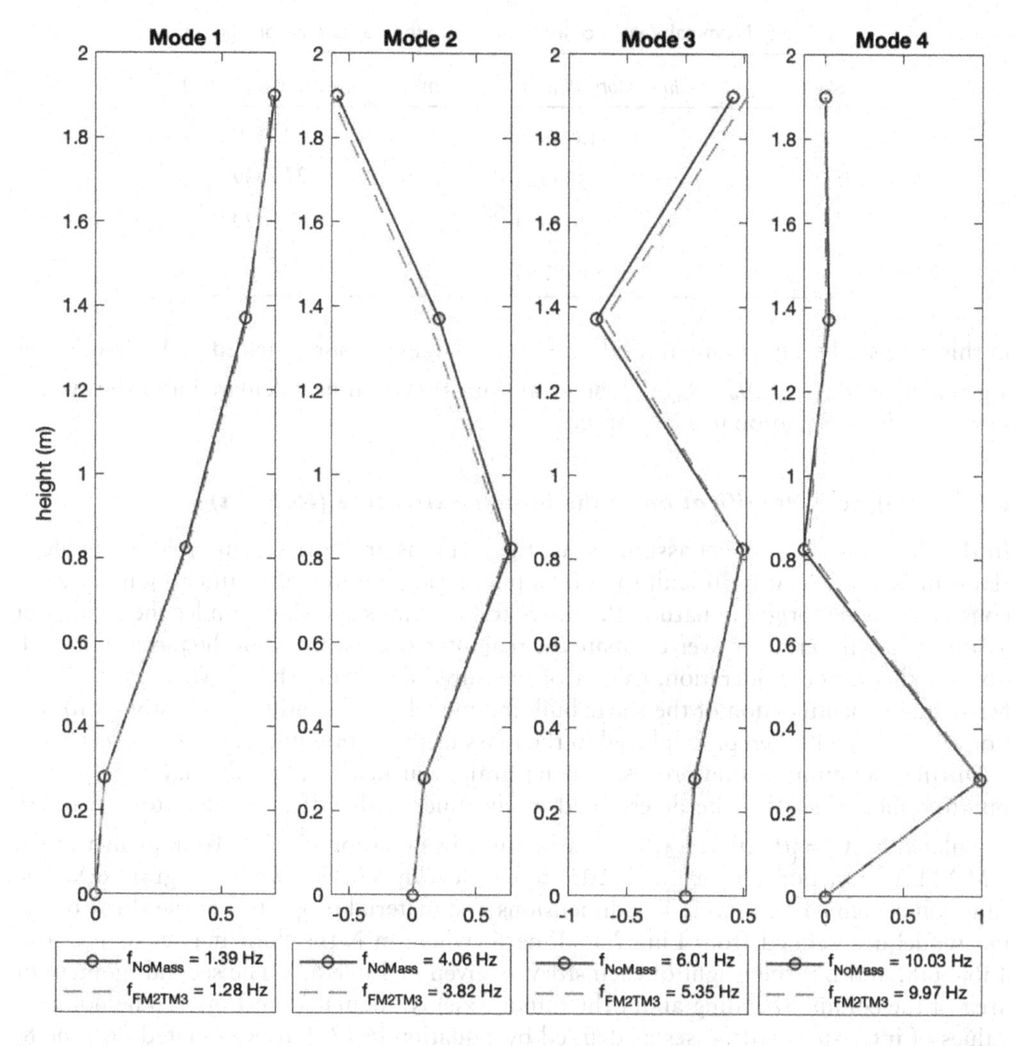

Figure 6.48 Measured modal parameters for cases NoMass and FM2TM3 (double additional masses with different extents).

equal to the sum of the slab mass and half of the mass of the four columns at story 4 (as there is no story 5). That is:

$$m_{0,i} = LBT\rho + 4\frac{h_i + h_{i+1}}{2}bt\rho, \quad \text{for } i = 1,\ldots,3 \tag{6.141}$$

and

$$m_{0,4} = LBT\rho + 4\frac{h_4}{2}bt\rho \tag{6.142}$$

where $LBT\rho$ is the mass of the floor slab. The numerical values of L, B, T, b, t, and h_i, for $i = 1,\ldots,4$, are given in Table 6.5; ρ is the mass density of steel and is equal to $8000\,\text{kg/m}^3$

Table 6.8 Nominal Values of Inter-Story Stiffnesses and Floor Masses

Story, i	*Inter-story stiffness, $k_{0,i}$ (N/m)*	*Floor mass, $m_{0,i}$ (kg)*
1	11.6361×10^4	26.8480
2	1.5359×10^4	27.3840
3	1.5359×10^4	27.3520
4	1.6792×10^4	26.2600

in this case study. The nominal total floor mass values are summarized in Table 6.8. For a given $\boldsymbol{\theta}_m = \{\theta_{m,1}, \theta_{m,2}, \theta_{m,3}, \theta_{m,4}\}^T$, the total floor mass can be calculated and the system mass matrix in Equation (6.139) can be obtained.

6.3.1.4 Stiffness identification of the baseline structure (NoMass)

In the shear building model assumptions, the columns are rigidly connected to the floor slabs (or beams). As it is difficult to build a pure rigid joint in reality and all joints can be considered as semi-rigid in nature, the inter-story stiffness calculated under the rigid joint assumption will certainly over-estimate the real inter-story stiffness of the physical model. To take this into consideration, the set of measured data from the NoMass case is used for stiffness identification of the shear building model. The identified inter-story stiffness (for all stories) will then be employed in the mass identification in the following sections.

During the minimization process, the natural frequencies and mode shapes of a shear building model needs to be determined many times with different inter-story stiffness calculated by $\boldsymbol{\theta}_k = \{\theta_{k,1}, \theta_{k,2}, \theta_{k,3}, \theta_{k,4}\}^T$ as defined in Equation (6.138). To implement this, a MATLAB function `shear_building()` is developed as shown in Program 6.8. The function is defined on Line 1. The dimensions and material properties of the shear building model are defined from Line 2 to Line 6, where m is the floor masses as given in Table 6.8, and `L` is the height of each story as given in Table 6.5. The second moment of area of the column (bending about the minor axis) is calculated on Line 7. The nominal values of inter-story stiffnesses as defined by Equation (6.137) are calculated on Line 8. The inter-story stiffnesses as defined by Equation (6.138) are calculated on Line 9. The system stiffness matrix is defined from Line 10 to Line 12. The system mass matrix is generated on Line 13. The eigenvalue and eigenvector of the shear building model as defined by `K` and `M` are calculated by the MATLAB built-in function `eig()` on Line 14, where `D` is a matrix with eigenvalues at the diagonal and `V` is the eigenvector matrix (each column is the mode shape of a given mode). The calculated eigenvalues are sorted in ascending order and assigned to the variable `evalue` on Line 15. The order of eigenvectors is rearranged to match that of the eigenvalues (ascending) and assigned to the variable `evector` on Line 16. The natural frequencies (`nf`) of the shear building model are calculated on Line 17, and the mode shapes (`ms`) are assigned on Line 18. Both `nf` and `ms` are the output parameters of the function (see Line 1).

Program 6.8: shear_building() to calculate the natural frequencies and mode shapes for a given model defined by $\boldsymbol{\theta}$.

```
Line 1.    function [nf,ms]=shear_building(theta)
Line 2.    m=[26.8480 27.3840 27.3520 26.2600];
Line 3.    L=[0.278 0.546 0.546 0.530];
```

```
Line 4.    b=0.025; % witdh of column
Line 5.    t=0.005; % thickness of column
Line 6.    E=200e9; % Young's modulus
Line 7.    I=b*t^3/12;
Line 8.    k=4*12*E*I./L.^3; % inter-story stiffness
Line 9.    k=k.*theta;
Line 10.   K=[
Line 11.   k(1)+k(2)     -k(2)         0             0
           -k(2)         k(2)+k(3)     -k(3)         0
           0             -k(3)         k(3)+k(4)     -k(4)
           0             0             -k(4)         k(4)
Line 12.   ];
Line 13.   M=diag(m);
Line 14.   [V,D]=eig(inv(M)*K);
Line 15.   [evalue,indx]=sort(diag(D));
Line 16.   evector=V(:,indx);
Line 17.   nf=evalue.^0.5/2/pi
Line 18.   ms=evector;
Line 19.   end
```

The objective function is implemented in MATLAB as `jcal()` in Program 6.9. The function is defined on Line 1. The function requires three input parameters. They are the model defined by θ_k (`theta`), the measured natural frequencies (`nfexp`) and mode shapes (`msexp`). The MATLAB function `shear_building()` as given in Program 6.8 is called on Line 2 to calculate the model-predicted natural frequencies and mode shapes. The measured mode shapes are already normalized to have unit length, and the model-predicted mode shapes are normalized in the same way from Line 3 to Line 6. As all mode shapes are normalized (unit length), the modal assurance criterion (MAC) value as defined in Equation (6.10) can be calculated as the square of the dot product between the two sets of vectors on Line 7. The model-predicted and measured mode shapes are matched by using the MAC value from Line 8 to Line 10. The objective function value as defined by Equation (6.13) is calculated from Line 11 to Line 14. The calculated `J` value is treated as the output parameters of this function as shown on Line 1.

Program 6.9: jcal() the objective function for stiffness identification of the shear building model.

```
Line 1.    function J=jcal(theta,nfexp,msexp)
Line 2.    [nf,ms]=shear_building(theta);
Line 3.    mm=ms'*ms;
Line 4.    for ii=1:length(nf)
Line 5.       ms(:,ii)=ms(:,ii)./mm(ii,ii)^0.5;
Line 6.    end
Line 7.    MAC=(msexp'*ms).^2;
Line 8.    [maxMAC,indx]=max(MAC');
Line 9.    nf=nf(indx);
Line 10.   ms=ms(:,indx);
Line 11.   J=0;
Line 12.   for ii=1:length(nfexp)
Line 13.      J=J+((nfexp(ii)-nf(ii))/nfexp(ii))^2+(1-mac(msexp(:,ii)
           ,ms(:,ii)));
Line 14.   end
Line 15.   end
```

After loading the experimental measured natural frequencies and mode shapes to the MATLAB workspace, one can use the MATLAB optimization toolbox to minimize the discrepancy between the measured and model-predicted modal parameters (i.e., the objective function in Equation (6.13)) to identify the inter-story stiffnesses as follows.

```
>> theta0=[1 1 1 1]; thetaL=theta0./5; thetaU=theta0*5;

>> [theta,j] = fmincon('jcal',theta0,[],[],[],[],thetaL,thetaU,[],[],n
fexp,msexp);
        0.0006    0.8008   0.7052    0.7523    0.7563
```

The initial trial of the minimization together with the upper and lower bounds are defined on the first line. The upper bound is defined as 5 times the initial trial, while the lower bound is defined as 0.2 times the initial trial. The factors 5 and 0.2 are fixed by trial and error. In general, a smaller search region is preferred to ensure a fast convergency. If the optimal solution touches the bound, the minimization should be repeated with a released search region. That is, to use a larger factor for the upper bound and/or a smaller factor for the lower bound. If necessary, different factors can be applied to different parameters.

The MATLAB function `fmincon()` is called to implement the minimization of the objective function defined in the function `jcal()`. It must be pointed out that the first input parameter of `jcal()` (i.e., `theta` on Line 1 of Program 6.9) is passed through the second input of `fmincon()`, while other input parameters of `jcal()` (i.e., `nfexp` and `msexp` on Line 1 of Program 6.9) are passed after the tenth input parameters of `fmincon()`. The last (third) line of the above code shows the results of the optimization process, where `j`=0.0006 is the objective function value at optimal; and `theta`=[0.8008 0.7052 0.7523 0.7563] are the optimal scaling factors of inter-story stiffness for stories 1, 2, 3 and 4, respectively. As the objective function value is very small, it is believed that the matching between the measured and model-predicted modal parameters is very good. The results show that the scaling factors take a value of around 0.7 to 0.8. It is believed that the effect of semi-rigid joint in this case can be reflected by a 20% to 30% reduction in inter-story stiffness. It is found that the reduction in inter-story stiffness of the first story (~20%) is relatively low when compared to the reduction of other stories (~30%). This is very likely caused by the fact that the height of the first story is relatively short when compared to other stories.

With this optimal model, the matching between the measured and updated modal parameters is shown in Figure 6.49. It is clear from the figure that the measured (with subscript exp) and model-predicted (with subscript cal) natural frequencies are very close to each other. Furthermore, the mode shapes are almost overlapping. This confirms the small objective function value of 0.0006 obtained. This model is treated as the baseline for mass identification in the subsequent sections.

6.3.1.5 Mass identification

With the baseline system identified in the previous section, mass identification can be carried out with measured data from different cases. The same objective function `jcal()` (i.e., Program 6.9) is used in this section for mass identification by modifying the `shear _ building()` function (i.e., Program 6.8) by replacing Line 9 with the following two lines:

```
k=k.*[0.8008 0.7052 0.7523 0.7563];

m=m.*theta;
```

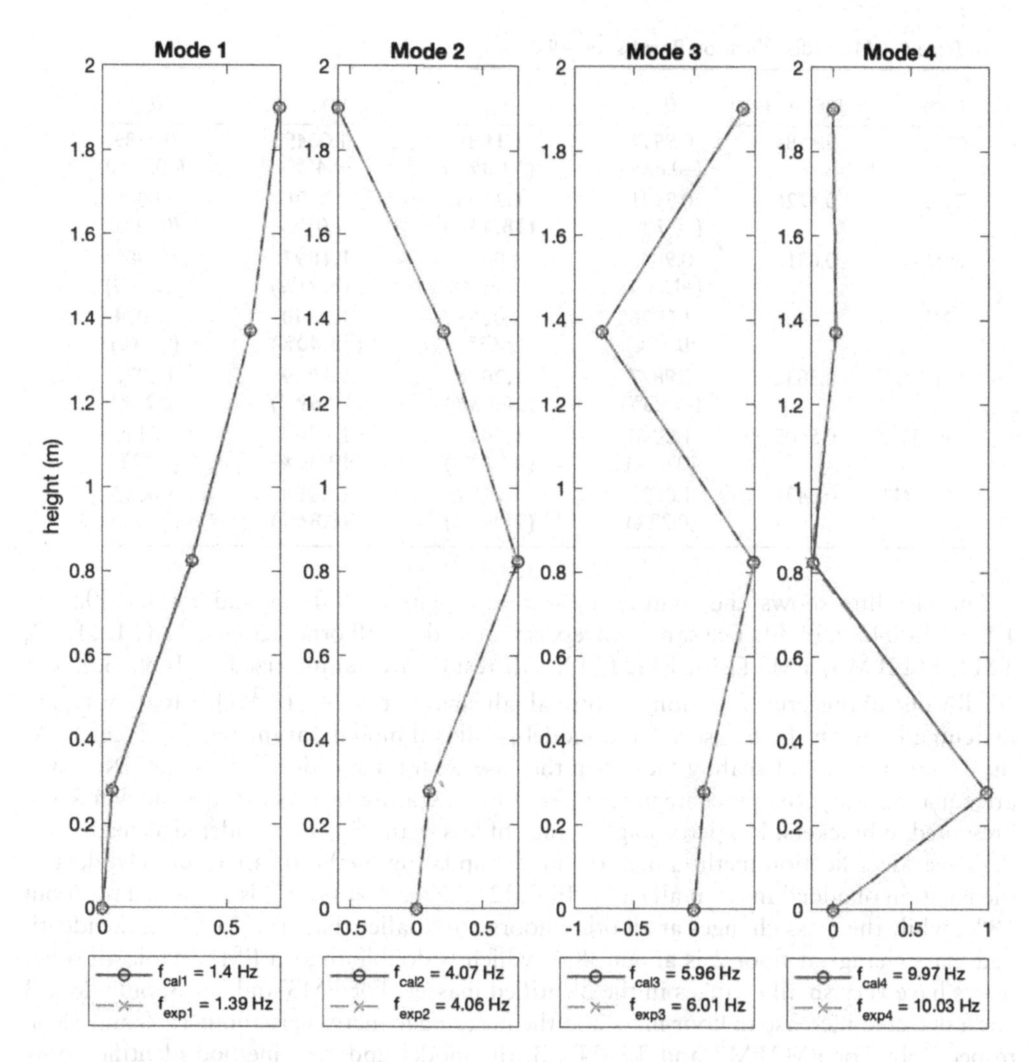

Figure 6.49 Matching between measured and updated modal parameters in NoMass.

The first line reduces the inter-story stiffnesses of the shear building model to consider the effect of semi-rigid column-slab joints. The scaling factors employed here are determined by stiffness identification in the previous section using the set of data from the NoMass case. The second line applies the uncertain parameters vector $\boldsymbol{\theta}_m = \{\theta_{m,1}, \theta_{m,2}, \theta_{m,3}, \theta_{m,4}\}^T$ as defined in Equation (6.140).

By using the same initial trial, upper, and lower bounds as before, the minimization can be carried out using the `fmincon()` MATLAB function. The FM2 case is considered first by loading the set of measured modal parameters in the working space, and executing the following lines of codes.

```
theta0=[1111];thetaL=theta0./5;thetaU=theta0*5;
[theta,j]=fmincon('jcal',theta0,[],[],[],[],thetaL,thetaU,[],[],nfexp,
msexp);
        0.0005      0.9917     1.1889    1.0045   0.9989
```

Table 6.9 Mass Identification Results for All Cases

Case	J ($\times 10^{-3}$)	$\theta_{m,1}$	$\theta_{m,2}$	$\theta_{m,3}$	$\theta_{m,4}$
FM2	0.4584	0.9917 (−0.83%)	**1.1889** **(18.89%)**	1.0045 (0.45%)	0.9989 (−0.11%)
TM2	0.4726	0.9921 (−0.79%)	**1.3843** **(38.43%)**	1.0101 (1.01%)	1.0081 (0.81%)
FM3	0.6312	0.9893 (−1.07%)	1.0087 (0.87%)	**1.1897** **(18.97%)**	0.9989 (−0.11%)
TM3	0.6401	1.0026 (0.26%)	1.0263 (2.63%)	**1.3840** **(38.40%)**	1.0024 (0.24%)
FM2FM3	0.5832	0.9877 (−1.23%)	**1.2003** **(20.03%)**	**1.1969** **(19.69%)**	1.0022 (0.22%)
TM2TM3	0.5605	1.0040 (0.40%)	**1.4077** **(40.77%)**	**1.4008** **(40.08%)**	1.0149 (1.49%)
FM2TM3	0.5481	1.0022 (0.22%)	**1.2192** **(21.92%)**	**1.3884** **(38.84%)**	1.0068 (0.68%)

The last line shows the minimization results with `j`=0.0005 and `theta`=[0.9917 1.1889 1.0045 0.9989]. The same process is repeated for all other cases (i.e., TM2, FM3, TM3, FM2FM3, TM2TM3, FM2TM3). All results are summarized in Table 6.9. The small value of objective function at optimal (all smaller than 1×10^{-3}) clearly shows good matching between the measured and model-predicted modal parameters in all cases. As the nominal values of scaling factors in the case with no additional mass (i.e., NoMass) are equal to unity, the percentage changes of mass scaling factors can be calculated and presented in brackets. If a percentage change of less than $\pm 3\%$ is considered as negligible, the mass identification method (i.e., the model updating method) can accurately identify the location of added mass in all cases. In FM2, the mass at floor 2 is increased by about 19%, while the mass changes at all other floors are smaller than 1%. In TM2, the identified mass change at floor 2 is about 38%, which is double that in FM2, while all other floors have very small changes in the identified masses. For FM3 and TM3, only floor 3 has a detected increase in floor mass and the percentage increase is about 19% and 38%, respectively. For FM2FM3 and TM2TM3, the model updating method identified only floors 2 and 3 with detectable increase in floor masses. Finally in FM2TM3, the method detects the mass increases at floors 2 and 3 with about 22% and 39%, respectively. The location(s) of added masses in all cases can be successfully detected by model updating.

The nominal values of floor masses are given in Table 6.8. With this information, the identified additional masses are summarized in Table 6.10. When the mass identification results are compared to the exact masses added in each case (in Table 6.6), the model updating results are very accurate with the identified added mass being a little overestimated. For locations without additional mass, the identified mass changes are all less than 1 kg.

Based on the identified model in each case, the updated modal parameters can be calculated to confirm the performance of the identified models. The updated natural frequencies in all cases are calculated and presented in Table 6.11. With the measured natural frequencies in Table 6.7, the percentage differences can be calculated and summarized in the brackets. The matchings in natural frequencies in all cases are, in general, good. However, the updated natural frequencies in mode 1 are overestimated with higher than

Table 6.10 Identified Additional Mass for All Cases (kg)

Case	*Floor 1*	*Floor 2*	*Floor 3*	*Floor 4*
FM2	–0.22	**5.17**	0.12	–0.03
TM2	–0.21	**10.52**	0.28	0.21
FM3	–0.29	0.24	**5.19**	–0.03
TM3	0.07	0.72	**10.50**	0.06
FM2FM3	–0.33	**5.49**	**5.39**	0.06
TM2TM3	0.11	**11.16**	**10.96**	0.39
FM2TM3	0.06	**6.00**	**10.62**	0.18

Table 6.11 Updated Natural Frequencies in All Cases (Hz)

Case	*Mode 1*	*Mode 2*	*Mode 3*	*Mode 4*
FM2	1.39 (1.21%)	3.85 (0.06%)	5.85 (–0.97%)	10.00 (–0.32%)
TM2	1.36 (1.37%)	3.66 (–0.06%)	5.76 (–1.11%)	9.99 (–0.23%)
FM3	1.36 (1.23%)	4.04 (0.00%)	5.67 (–0.70%)	10.03 (–0.55%)
TM3	1.31 (1.40%)	4.00 (–0.24%)	5.44 (–0.75%)	9.96 (–0.44%)
FM2FM3	1.34 (1.25%)	3.84 (–0.01%)	5.54 (–0.94%)	10.02 (–0.33%)
TM2TM3	1.28 (1.41%)	3.64 (–0.11%)	5.21 (–1.20%)	9.93 (–0.14%)
FM2TM3	1.29 (1.36%)	3.81 (–0.25%)	5.30 (–0.89%)	9.95 (–0.24%)

1% but lower than 2% in all cases. In terms of fitting the measured natural frequencies, mode 1 is the worst. The fitting in mode 2 is very good with the maximum percentage difference being less than 0.3%. The fitting in mode 3 is not as good as that in mode 2, but better than mode 1, with the maximum percentage difference at 1.2%. The fitting in mode 4 is also very good with the highest percentage difference being 0.55%.

The matchings in mode shapes are very good, as the measured mode shapes almost overlap with the corresponding updated mode shapes. For example, Figure 6.50 and Figure 6.51 show the matchings between the measured and updated mode shapes in FM2 (single additional mass) and FM2TM3 (double additional masses with different extents). It is clear that the matchings are almost perfect. The matchings in other cases are very similar and are not shown here.

6.3.2 Joint damage detection of a two-story steel frame by model updating

In this section, a two-story steel frame is employed as an example to illustrate the use of optimization algorithms in model updating by minimizing the discrepancy between the measured modal parameters, such as natural frequencies and mode shapes, and the model-predicted ones. In this case study, the vibration test was carried out on the undamaged steel frame to obtain the acceleration responses at different measured DOFs of the structure. The modal parameters were then identified from the power spectral density data, which was calculated from the measured acceleration responses based on the method introduced in Chapter 3. Artificial joint damages were then simulated to the joints of the steel frame case by case. For each damaged case, the modal parameters were obtained and recorded for the purpose of structural joint damage detection following the

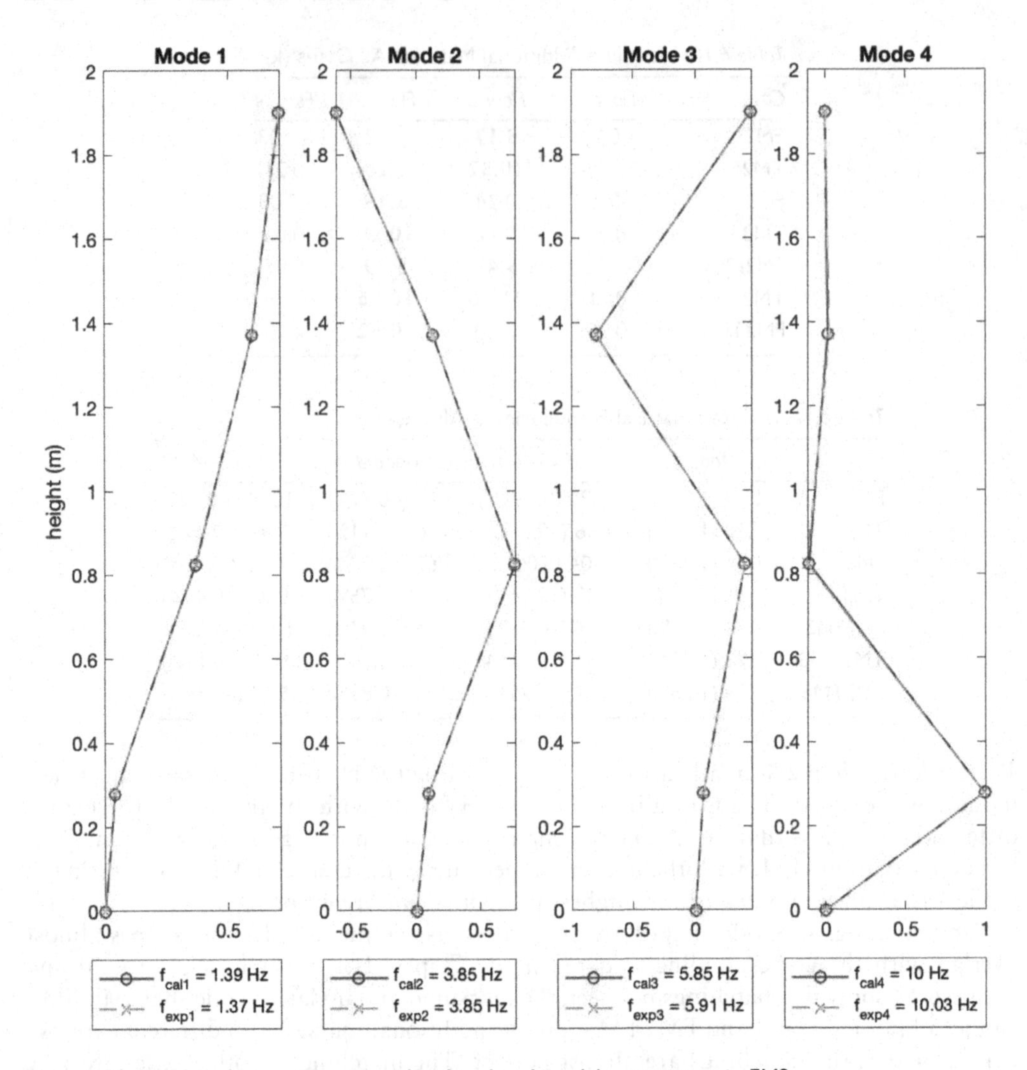

Figure 6.50 Matching between measured and updated modal parameters in FM2.

model updating approach. By the model updating method introduced in this chapter, the finite element model of the undamaged structure and those of the structure under various damaged cases were identified. By comparing the identified models from different cases, the locations and extents of the simulated damages in different damaged cases can be detected. This is one of the most popular applications of structural model updating in the field of structural health monitoring (SHM). In this case study, only the identified natural frequencies and mode shapes are considered in the model updating and damage detection processes without any prior information on the damage location.

The step-by-step procedure includes (1) measuring the acceleration responses of the target steel frame under ambient vibration (wind-induced vibration); (2) identifying the modal parameters by the method introduced; (3) formulating the model (i.e., system stiffness and mass matrices) of the target structure; and (4) updating the model based on the identified natural frequencies and mode shapes.

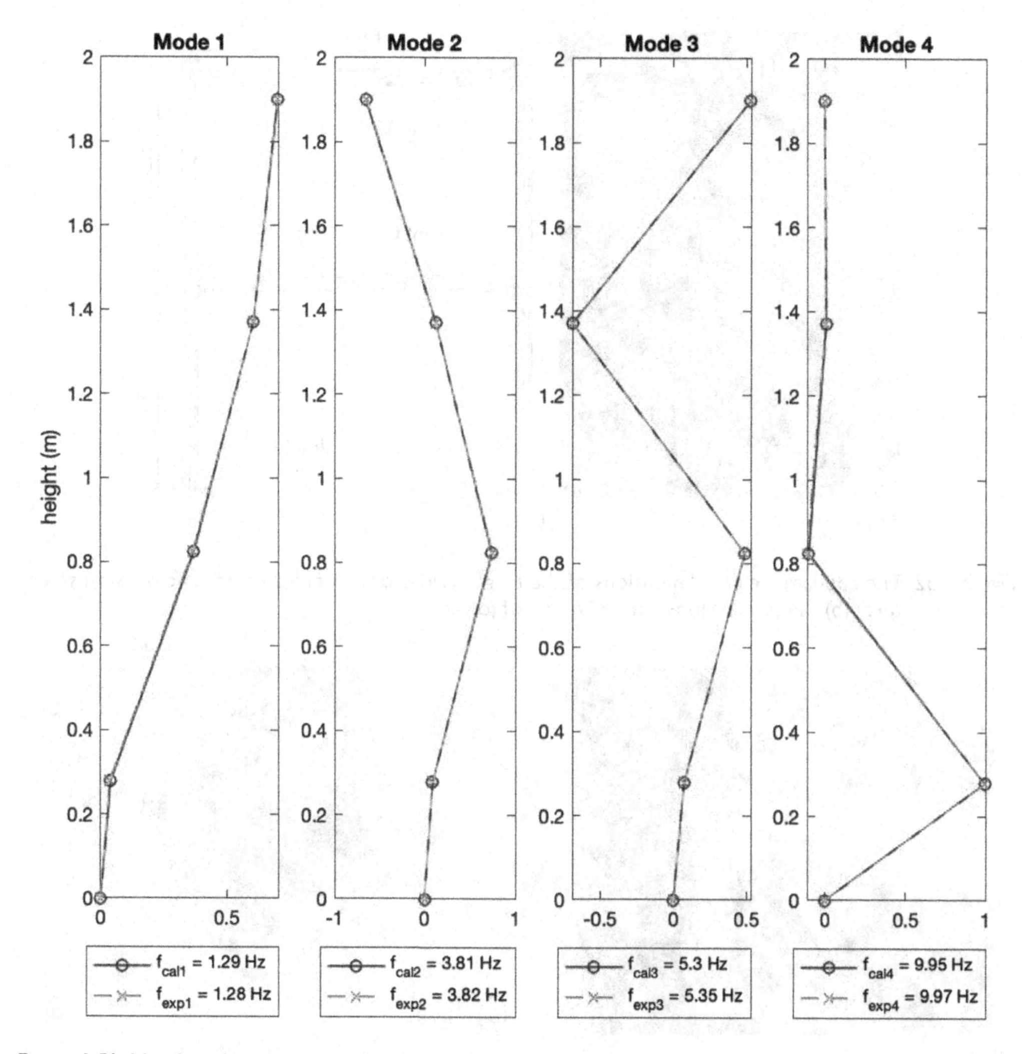

Figure 6.51 Matching between measured and updated modal parameters in FM2TM3.

6.3.2.1 Description of the structure

The two-story one-bay steel frame was built in the Heavy Structures Testing Laboratory of City University of Hong Kong as shown in Figure 6.52(a). The bay width is about 2 m and the height of the steel frame is about 2.5 m. The detailed dimensions of the frame are shown in Figure 6.52(b). As the main purpose of this case study is the detection of joint damages through model updating, it is important to label all joints. There are six joints on this frame, they are the four beam-column connections A (upper left), B (upper right), C (lower left), and D (lower right), and the two column-base connections E (left support) and F (right support) as indicated in Figure 6.52(b).

All beam-column connections were formed by fixing the top and bottom flanges of the beam to the flange of the column by angles. The web of the beam is also fixed to the flange of the column by angles as shown in Figure 6.53(a). This is a typical moment resisting joint as the moment can be transferred from the beam to the column through the tension

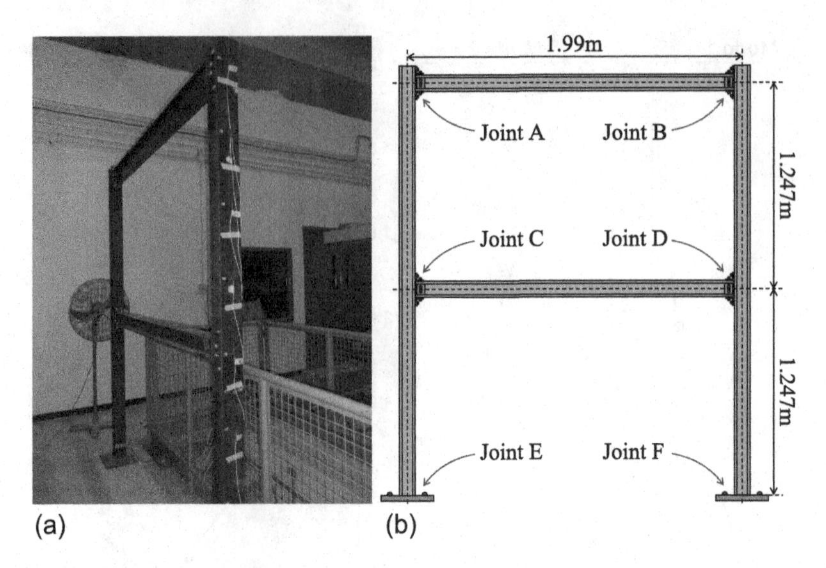

Figure 6.52 The appearance and dimensions of the target two-story steel frame. (a) The physical structure; (b) The dimensions and definition of joints.

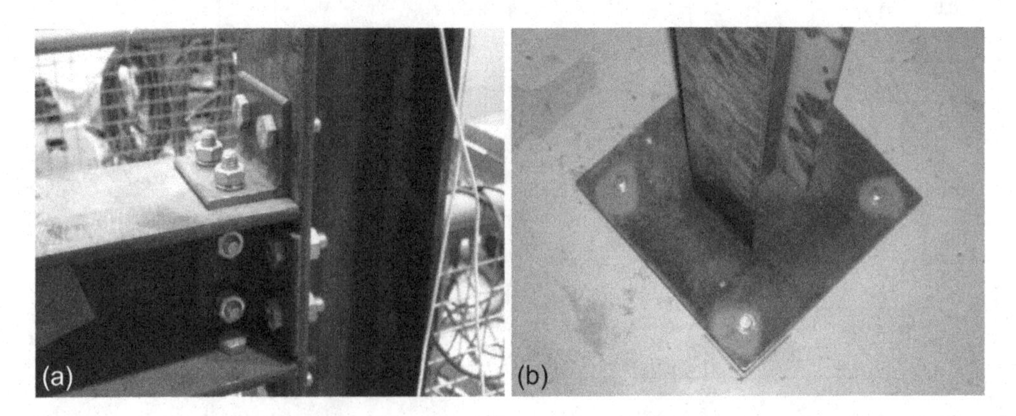

Figure 6.53 The joint and support conditions of the target two-story steel frame. (a) Beam-column connection. (b) Column-base connection.

and compression forces on the beam flanges through the angles. That is, the tension and compression forces on the top and bottom flanges of the beam (the beam moment is mainly generated by these components of internal member forces) is transferred to the top and bottom angles through the shear forces on the bolts connecting the angles and the beam. These shear forces are then transferred to the tension and compression forces on the bolts connecting the angles and the column. Finally, these tension and compression forces will generate moments on the column. Although this load path is not as effective as that of the welded connection, the behavior of this kind of joint is closer to a rigid connection from the design viewpoint. The beam shear can be effectively transferred through the bolts (connecting the web of beam) to the angles on both sides of the web, and from these two angles, through the bolts (connecting the flange of the column) to the column as axial force. The two column-base connections are formed by welding the column end to a thick steel plate, which is then fixed onto the rigid floor in the laboratory through four strong bolts at the four corners of the steel plate as shown in Figure 6.53(b). As the spacing of bolts is larger than the depth of the column, the column moment can be transferred to the ground through the tension and compression forces on the bolts. From the design viewpoint, this kind of connection is also considered closer to a rigid support.

From the analysis viewpoint, both beam-column and column-base connections are modeled as semi-rigid. This can be done by using a rotational spring with a given rotational stiffness (unit Nm/rad) to capture the moment-rotation behavior of the connection. When the rotational stiffness is high, a large moment is required to produce a small rotation, and the joint behavior is closer to a rigid connection. When rotational stiffness is very small, a small moment can induce a large rotation, and the joint behaves like a pinned connection. As a result, the identification of the rotational stiffness of various joints of the steel frame through model updating technique can help in detecting joint damages on the structure. Here, the reduction in the rotational stiffness of a joint is considered as joint damage.

Both the beams and columns are made of standard steel sections W$4\times4\times13$ (imperial) with a cross-sectional area of 2471 mm^2 and second moment of area about the major axis (y) 475.9×10^4 mm^4, and the moduli of elasticity and mass density are equal to 190 GPa and 19.3 kg/m, respectively. The important member properties (i.e., the cross-sectional and material properties) of all steel members are summarized in Table 6.12.

The values of rotational stiffness for both the beam-column connections and the column-base connections are difficult to obtain. In general, people believe that the rotational stiffness of the beam-column connection is smaller than that of the column-base connection (the support should be stiffer). In this study, a nominal value of rotational stiffness of 1×10^6 Nm/rad is employed for both types of connections. The purpose of this case study is to identify the rotational stiffnesses of all joints (i.e., joints A to F) in different cases based on the sets of measured vibration data.

6.3.2.2 *Modeling of the two-story steel frame*

For the purpose of structural damage detection, the class of models to be employed is important. Here the class of models refers to the discretization of the physical model to form the finite element model and its parameterization. In general, the higher the complexity of the model class (i.e., the larger the number of uncertain model parameters included in the model updating process), the more powerful of the model class is in fitting

Table 6.12 Cross-Sectional and Material Properties of the Target Structure

b
t_w
h
t_f

Height, h (mm):	106
Width, b (mm):	103
Web thickness, t_w (mm):	7.1
Flange thickness, t_f (mm):	8.8
Cross-sectional area, A (mm^2):	2471
Second moment of area about the major axis y, I_y (mm^4):	475.9×10^4
Modulus of elasticity E, (GPa):	190
Mass density m, (kg/m)	19.3

the measured responses. However, more information will be required in identifying all uncertain parameters to an acceptable level of accuracy with such a model class. The concept is similar to the fitting of a set of data by least-squares regression. The higher the order of the polynomial (that is the class of models), the higher the capability of the regression in fitting complicated data trends. However, the number of unknown coefficients increases as the order of polynomial increases. As a result, more data points are needed in calculating all coefficients. Some people have the impression that a model class with higher complexity is always good, as it can fit more complicated data trends. This is not necessarily true in model updating, especially for the purpose of damage detection, in which the reliability of the identified model parameters is essential. For a given set of measured data, the higher the model complexity, the higher the (posterior) uncertainties associated with the identified model parameters. It is true that the identified model can fit the measured data very well, but it is very likely that the sensitivity of the model-predicted responses with respect to the value of some model parameters is very low. As a result, the uncertainties on the identified model parameters are very high, and the results of damage detection are believed to be unreliable. In this case study, the class of models is selected based on trial-and-error following a deterministic approach (as the situation considered in this chapter). The problem is best addressed by following the probabilistic approach through Bayesian model class selection (Lam et al. 2017; Adeagbo et al. 2021), as the uncertainties induced by modeling error and measurement noise are involved. The use of the probabilistic approach in model updating will be discussed in detail in Chapter 7.

The two-story steel frame in this case study is modeled as a two-dimensional (2D) frame with 32 nodes (30 movable nodes and 2 nodes at the two supports) and 32 2D frame elements (16 elements for columns and another 16 elements for beams). The finite element model with the numbering of nodes and elements is shown in Figure 6.54(a) and Figure 6.54(b), respectively. Each of the 30 movable nodes has three DOFs (i.e., the lateral DOF in the global x-direction; the vertical DOF in the global y-direction; and the rotational DOF in the global z-direction (out of plane follows the right-hand rule)). The

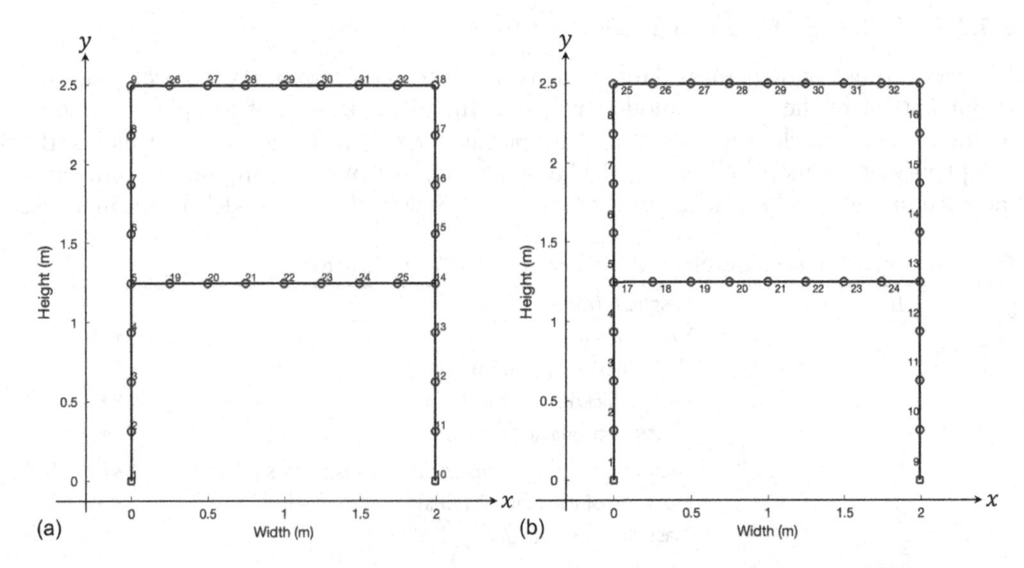

Figure 6.54 The computer model of the two-story frame with node and element numbers. (a) Node numbers. (b) Element numbers.

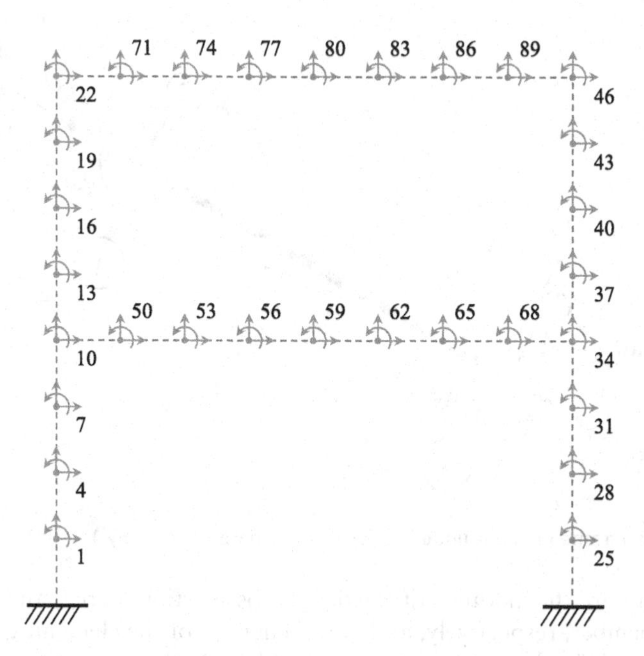

Figure 6.55 The DOFs of the computer model (the numbers shown are the measured DOFs).

numbering of DOFs is shown in Figure 6.55, in which the numbers shown are the measured DOFs in the series of vibration tests. The arrows show the positive directions of the corresponding DOFs.

Figure 6.56 is a typical 2D frame element employed in this case study. DOFs 1 and 4 correspond to the axial displacement (i.e., $d_{ix'}$ and $d_{jx'}$) and force (i.e., $q_{ix'}$ and $q_{jx'}$) along x'-direction at the starting node (i.e., node i) and ending node (i.e., node j), respectively. DOFs 2 and 5 correspond to the shear displacement (i.e., $d_{iy'}$ and $d_{jy'}$) and force (i.e., $q_{iy'}$ and $q_{jy'}$) along the y'-direction at node i and node j, respectively. DOFs 3 and 6 correspond to the rotation (i.e., $d_{iz'}$ and $d_{jx'}$) and bending moment (i.e., $q_{iz'}$ and $q_{jz'}$) along the z'-direction at nodes i and j, respectively. For the completeness of this book, the force-displacement relationship and the formulation of the local element stiffness matrix are given as:

$$\begin{Bmatrix} q_{ix'} \\ q_{iy'} \\ q_{iz'} \\ q_{jx'} \\ q_{jy'} \\ q_{jz'} \end{Bmatrix} = \begin{bmatrix} \frac{EA}{L} & 0 & 0 & -\frac{EA}{L} & 0 & 0 \\ 0 & \frac{12EI}{L^3} & \frac{6EI}{L^2} & 0 & -\frac{12EI}{L^3} & \frac{6EI}{L^2} \\ 0 & \frac{6EI}{L^2} & \frac{4EI}{L} & 0 & -\frac{6EI}{L^2} & \frac{2EI}{L} \\ -\frac{EA}{L} & 0 & 0 & \frac{EA}{L} & 0 & 0 \\ 0 & -\frac{12EI}{L^3} & -\frac{6EI}{L^2} & 0 & \frac{12EI}{L^3} & -\frac{6EI}{L^2} \\ 0 & \frac{6EI}{L^2} & \frac{2EI}{L} & 0 & -\frac{6EI}{L^2} & \frac{4EI}{L} \end{bmatrix} \begin{Bmatrix} d_{ix'} \\ d_{iy'} \\ d_{iz'} \\ d_{jx'} \\ d_{jy'} \\ d_{jz'} \end{Bmatrix} \Rightarrow \mathbf{q} = \mathbf{k'd} \quad (6.143)$$

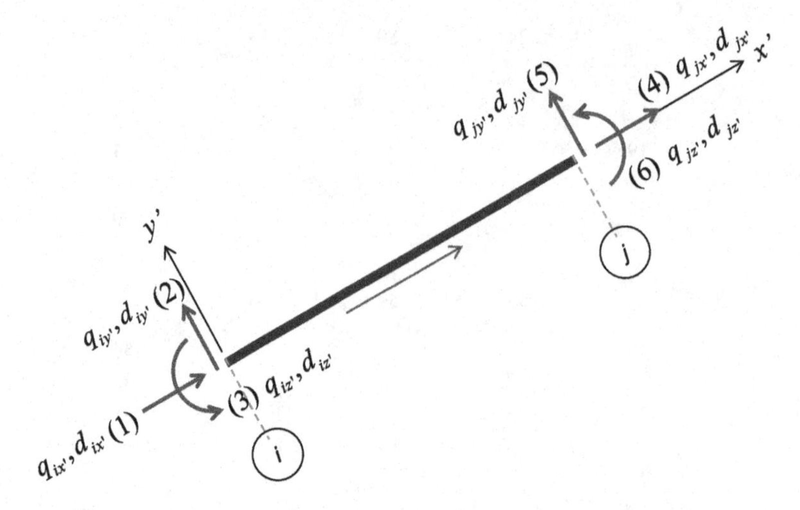

Figure 6.56 A typical frame element under its local coordinate system ($x'y'$).

where E, A, and I are the modulus of elasticity, cross-sectional area, and second moment of area of the member, respectively, and L is the length of the element; **q** and **d** represent the local force and displacement vectors, and **k**′ is the local element stiffness matrix, which is, of course, symmetrical. The local element stiffness matrix (and the local element mass matrix) must be transformed from the local coordinate system to the global coordinate system to form the global element stiffness matrix before the assembling process to form the system stiffness matrix (and also the system mass matrix). This requires the transformation of both the displacement and force vectors. The displacement transformation matrix, which is used to transform the global displacement vector to the local displacement vector **d**, is given as:

$$\begin{Bmatrix} d_{ix'} \\ d_{iy'} \\ d_{iz'} \\ d_{jx'} \\ d_{jy'} \\ d_{jz'} \end{Bmatrix} = \begin{bmatrix} \cos\phi & \sin\phi & 0 & 0 & 0 & 0 \\ -\sin\phi & \cos\phi & 0 & 0 & 0 & 0 \\ 0 & 0 & 1 & 0 & 0 & 0 \\ 0 & 0 & 0 & \cos\phi & \sin\phi & 0 \\ 0 & 0 & 0 & -\sin\phi & \cos\phi & 0 \\ 0 & 0 & 0 & 0 & 0 & 1 \end{bmatrix} \begin{Bmatrix} D_{ix} \\ D_{iy} \\ D_{iz} \\ D_{jx} \\ D_{jy} \\ D_{jz} \end{Bmatrix} \Rightarrow \mathbf{d} = \mathbf{TD} \tag{6.144}$$

where ϕ in Equation (6.144) is the angle from the global x-axis to the local x'-axis as shown in Figure 6.57(a). $\mathbf{D} = \{D_{ix} \quad D_{iy} \quad D_{iz} \quad D_{jx} \quad D_{jy} \quad D_{jz}\}^T$ is the global displacement vector with D_{ix} to D_{jz} being the displacements along the DOFs 1 to 6, respectively, under the global coordinate system as shown in Figure 6.57(b); **T** is the displacement transformation matrix. The transpose of the displacement transformation matrix is the force transformation matrix $\mathbf{T}^T$ that transforms the local force vector to the global force vector as:

$$\mathbf{Q} = \mathbf{T}^T\mathbf{q} \tag{6.145}$$

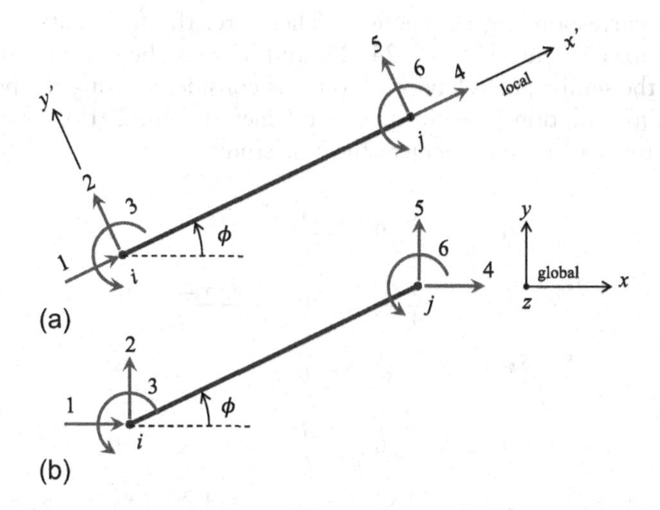

Figure 6.57 Transformation of the local coordinate to the global coordinate system. (a) Local coordinate system. (b) Global coordinate system.

where $\mathrm{Q} = \{Q_{ix} \quad Q_{iy} \quad Q_{iz} \quad Q_{jx} \quad Q_{jy} \quad Q_{jz}\}^T$ is the global force vector with Q_{ix} to Q_{jz} being forces along the DOFs 1 to 6, respectively, under the global coordinate system as shown in Figure 6.57(b).

Substituting Equation (6.144) into Equation (6.143), one obtains:

$$\mathrm{q} = \mathrm{k'TD} \tag{6.146}$$

Then substituting Equation (6.146) into Equation (6.145), one obtains:

$$\mathrm{Q} = \mathrm{T}^T\mathrm{k'TD} \Rightarrow \mathrm{Q} = \mathrm{kD} \tag{6.147}$$

where $\mathrm{k} = \mathrm{T}^T\mathrm{k'T}$ defines the relationship between the global forces and global displacements, and it is the global element stiffness matrix.

In this case study, the consistent mass matrix is employed. The typical element mass matrix, **m**, is given as (Paz & Leigh 2004):

$$\mathrm{m} = \frac{mL}{420}\begin{bmatrix} 140 & 0 & 0 & 70 & 0 & 0 \\ 0 & 156 & 22L & 0 & 54 & -13L \\ 0 & 22L & 4L^2 & 0 & 13L & -3L^2 \\ 70 & 0 & 0 & 140 & 0 & 0 \\ 0 & 54 & 13L & 0 & 156 & -22L \\ 0 & -13L & -3L^2 & 0 & -22L & 4L^2 \end{bmatrix} \tag{6.148}$$

where m is the mass density in kg/m and L is the length of the element. The transformation for the element stiffness matrix can be used for the element mass matrix. The element stiffness matrix in Equation (6.143) can be used only for normal frame elements but not for elements with rotational spring at one or both ends (for modeling the semi-rigid behavior of the corresponding connection). Therefore, the formulation of the element stiffness matrix for elements 1, 9, 17, 24, 25, and 32 must be considered differently. In this case study, the semi-rigid behavior of joints is considered using the popular element stiffness matrix formulation given in reference (Chen & Lui, 2018). The formulation is provided below for readers to implement the case study.

$$\mathbf{k}=\frac{EI}{L}\begin{bmatrix} \frac{A}{I} & 0 & 0 & -\frac{A}{I} & 0 & 0 \\ 0 & \frac{s_{ii}+2s_{ij}+s_{jj}}{L^2} & \frac{s_{ii}+s_{jj}}{L} & 0 & -\frac{s_{ii}+2s_{ij}+s_{jj}}{L^2} & \frac{s_{ij}+s_{jj}}{L} \\ 0 & \frac{s_{ii}+s_{ij}}{L} & s_{ii} & 0 & -\frac{s_{ii}+s_{ij}}{L} & s_{ij} \\ -\frac{A}{I} & 0 & 0 & \frac{A}{I} & 0 & 0 \\ 0 & -\frac{s_{ii}+2s_{ij}+s_{jj}}{L^2} & -\frac{s_{ii}+s_{ij}}{L} & 0 & \frac{s_{ii}+2s_{ij}+s_{jj}}{L^2} & -\frac{s_{ij}+s_{jj}}{L} \\ 0 & \frac{s_{ij}+s_{jj}}{L}L & s_{ij} & 0 & -\frac{s_{ij}+s_{jj}}{L} & s_{jj} \end{bmatrix} \tag{6.149}$$

where:

$$\begin{aligned} s_{ii} &= \frac{1}{R}\left(4+\frac{12EI}{L}\frac{1}{k_L}\right) \\ s_{jj} &= \frac{1}{R}\left(4+\frac{12EI}{L}\frac{1}{k_R}\right) \\ s_{ij} &= \frac{2}{R} \end{aligned} \tag{6.150}$$

where R is defined as:

$$R=\left(1+\frac{4EI}{L}\frac{1}{k_L}\right)\left(1+\frac{4EI}{L}\frac{1}{k_R}\right)-\left(\frac{EI}{L}\right)^2\frac{4}{k_L k_R} \tag{6.151}$$

where k_L and k_R are the rotational stiffness of the spring at the left (starting node i) and right (ending node j) of the element. As mentioned in the previous section, all joints are modeled as semi-rigid connections with a nominal rotational stiffness $k_0 = 1\times10^6$ Nm/rad. The rotational stiffness for each joint can be calculated as:

$$k_i = \theta_i k_0 \quad \text{for} \quad i = 1,\ldots,6 \tag{6.152}$$

where k_i, for $i = 1,\ldots,6$, are the rotational stiffness for joints A to F as defined in Figure 6.52(b), respectively. After transformation, both the element stiffness matrices and element mass matrices can be assembled to form the system stiffness (**K**) and mass matrices (**M**).

6.3.2.3 Vibration test

One of the most important parts of structural model updating is to measure a set of informative vibration data under ambient vibration. Even though the signal-to-noise ratio for forced vibration tests is usually higher than that for ambient vibration tests, most structural model updating cases for the purpose of damage detection are carried out through ambient vibration tests. Property owners will most likely not allow the application of external forces onto their possibly damaged structures. Therefore, ambient vibration test is almost the only choice. The measured vibration data must be informative, such that information can be extracted from it for the purpose of damage detection. In other words, the type of measured quantity together with the location and direction (i.e., the DOF) for measuring this quantity must be sensitive to the target type of damage. As the purpose of this case study is to identify the rotational stiffnesses at different joints (i.e., detecting joint damages), it would be good if the rotation at the joints can be measured. However, it is difficult to measure the rotation to an acceptable accuracy under ambient vibration. An acceptable (and also commonly used) way is to measure the acceleration at different DOFs of the target structural system. Unlike the measurement of displacement and velocity, a reference is not required in measuring acceleration responses. All that is needed are accelerometers, which are popular and cost-effective. During the vibration test, an electric fan (as shown at the back of Figure 6.52(a)) was employed to induce wind load on the frame, so as to increase the signal-to-noise ratio.

Eleven accelerometers are available to the authors for this case study under laboratory conditions. As the axial stiffness of beams and columns are not the target of this study, only lateral vibrations are measured for the two columns, and only vertical vibrations are measured for the two beams. The numbered DOFs in Figure 6.55 show the measured DOFs to be considered in the vibration test. The total number of measured DOFs is 30 (i.e., 16 DOFs on the two columns and 14 DOFs on the two beams). As only 11 sensors are available, the test is divided into four setups with 11 sensors/channels per setup. Setups 1, 2, 3, and 4 are selected to measure the vibration of the left column, upper beam, lower beam, and right column, respectively. To link the measured data from different setups, it is important to have at least one reference channel between two different setups. Otherwise, the four sets of data become independent. That is, when the left column is vibrating to the left, one does not know if the right column is vibrating to the left or right, and it is unclear if the beams are vibrating upward or downward. The sensor locations (and directions) together with the channel numbers for all four setups are shown in Figure 6.58 (the arrows show the positive direction of sensors). Consider the first setup in Figure 6.58(a), sensors/channels 1 to 8 measure the horizontal acceleration responses of the left column with three reference sensors. Reference sensors 9, 10, and 11 link the vibration of the left column to the vibrations of the right column, the upper beam and the lower beam, respectively. After setup 1, sensors 2, 3, 4, 5, 6, and 7 are moved to the upper beam for measuring the beam vibration in setup 2 (in Figure 6.58(b)), while the positions for sensors 1, 8, 9, 10, and 11 remain unchanged. To move from setup 2 to 3, sensors 2, 3, 4, 5, 6, and 7 are moved to the lower beam (see Figure 6.58(c)). Finally in setup 4 (see Figure 6.58(d)), sensors 2 to 8 are moved to the right column.

The installations of sensors for various setups are shown in Figure 6.59(a) to (d) for the reader's reference. For setups 1 and 4, it is clear from Figures 6.59(a) and (d) that the sensors were installed on the left surface of the left column and right surface of the right column. These match with the arrow directions as shown in Figures 6.58(a) and (d).

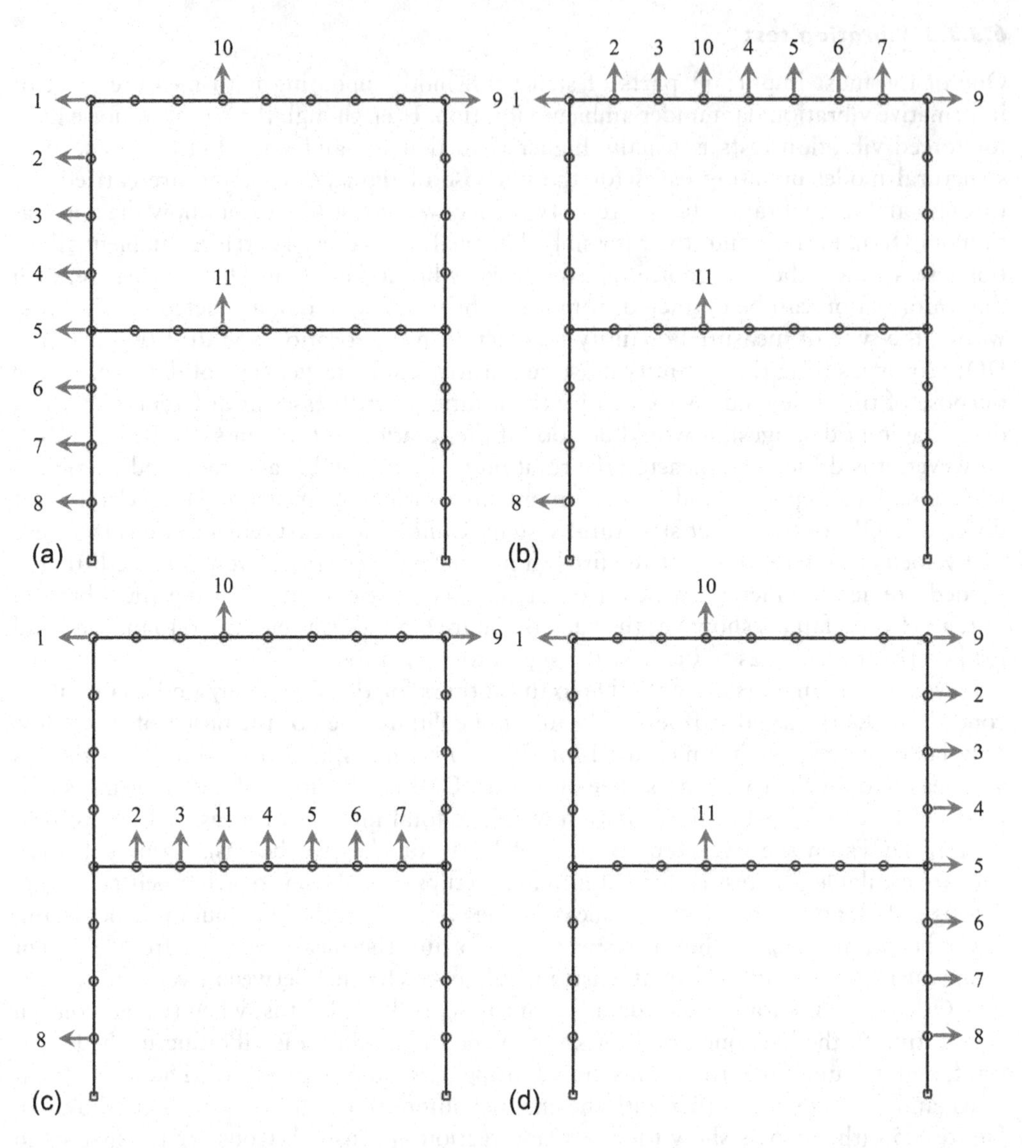

Figure 6.58 The sensor locations and directions (with channel numbers). (a) Setup 1. (b) Setup 2. (c) Setup 3. (d) Setup 4.

Similarly, Figures 6.59(b) and (c) show that the sensors were installed on the top surface of both beams, and these match with the arrow directions as shown in Figures 6.58(b) and (c).

For each setup, the measurement duration is 20 mins with a sampling frequency of 2048 Hz. Therefore, 2,457,600 data points were recorded in each channel. Figure 6.60 shows an example of measured acceleration response from sensor 1 (at channel 1) in setup 1 for time from 30 s to 50 s (20 s duration). The maximum acceleration during the period shown in Figure 6.60 is about 4.2×10^{-3} g.

A total of three cases are considered in this case study: the undamaged case (UD), the single-damage case with the damage at joint B, and the double-damage case with joints A and D being damaged (see Table 6.13).

Figure 6.59 Sensor installation on different components of the two-story steel frame. (a) Left column. (b) Upper beam. (c) Lower beam. (d) Right column.

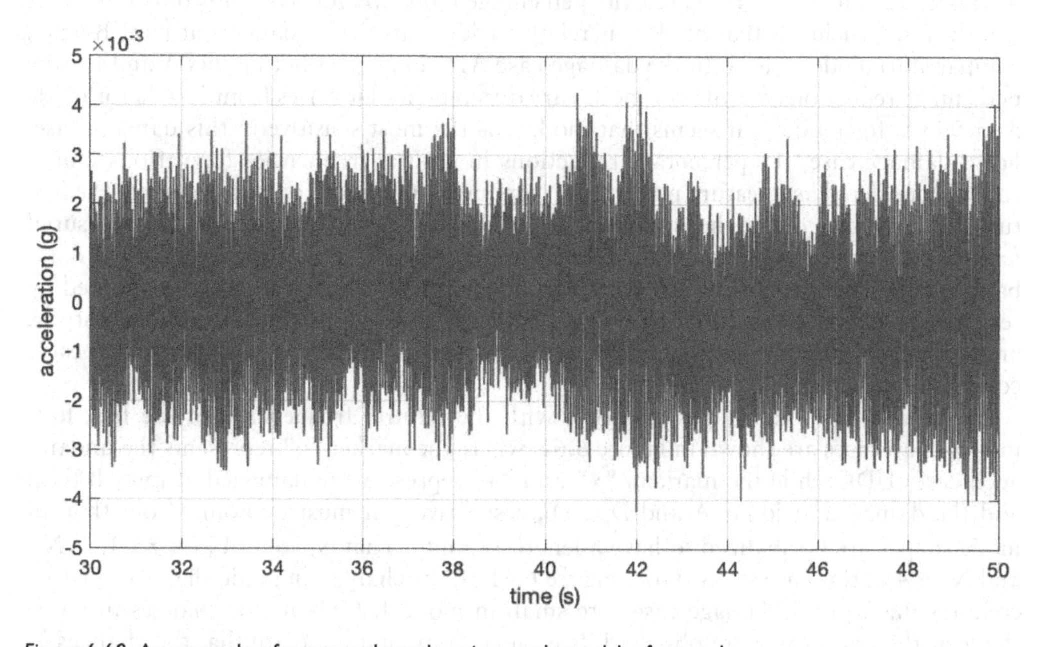

Figure 6.60 An example of measured acceleration at channel 1 of setup 1.

6.3.2.4 Modal identification

Following the method introduced in Chapter 3, the modal parameters of the steel frame in each case are identified. In this study, only the identified natural frequencies and mode shapes are employed for the purpose of model updating and damage detection. The identified natural frequencies are summarized in Table 6.14. The percentage reductions in natural frequencies for the different damage cases are given in parentheses. As joint

Table 6.13 Cases Considered in the Vibration Tests of the Two-Story Frame

Case	*Case ID*	*Remarks*
Undamaged	UD	This case is used as the baseline reference for damage detection of other cases.
Damaged at joint B	DB	This case is a single-damage case.
Damaged at joints A and D	AD	This case is a double-damage case.

Table 6.14 Identified Natural Frequencies of the First Four Modes in All Cases

Case	*Mode 1*	*Mode 2*	*Mode 3*	*Mode 4*
UD (Hz)	16.78	69.85	109.14	114.87
DB (Hz)	16.41 (–2.21%)	68.28 (–2.25%)	106.70 (–2.24%)	114.77 (–0.09%)
AD (Hz)	15.91 (–5.18%)	68.72 (–1.62%)	107.55 (–1.46%)	111.12 (–3.26%)

damage reduces the rotational stiffness of the system, reductions in natural frequencies are expected. The extent of natural frequency reduction for a given mode shows the sensitivity of that mode to the damage. Consider damage case DB (damage at joint B) as an example, where the percentage reductions for modes 1 to 3 are very similar and are equal to about 2.2% to 2.3%. However, the percentage reduction for mode 4 is only 0.09%. It can thus be concluded that mode 4 is relatively less sensitive to damage at joint B when compared to modes 1 to 3. In the damage case AD (damage at both joints A and D), the percentage reductions for all four modes are different, and it varies from 1.46% for mode 3 to 5.18% for mode 1. It seems that mode 1 is the most sensitive in this damage case. For a damage case, the percentage reductions in natural frequencies for various modes can be considered as a feature pattern. By computer simulation, one can generate the feature patterns for a comprehensive database of damage cases. By matching the measured feature pattern to the model-predicted feature patterns of all damage cases in the database, the damage case of the structure can be identified. This idea has been adopted by researchers (Lam & Lee 2005; Lam et al. 2006; Yuen & Lam 2006). In this case study, another approach (i.e., minimizing the error in natural frequencies and mode shapes) is considered.

The measured mode shapes (together with the natural frequencies) of the first four modes in all cases are shown in Figure 6.61, where the marker "o" represents the undamaged case (UD), while the markers "x" and "+" represent the damaged at joint B (DB) and the damaged at joints A and D (AD), respectively. It must be pointed out that all mode shapes are normalized to have a length of unity. That is, $\|\hat{\psi}_r\| = 1$, for $r = 1,\ldots,N_m$ and $N_m = 4$ in this case study. From Figure 6.61(a), the changes in mode shape at the two columns due to both damage cases are small in mode 1. Only minor changes in mode shape at the two beams are observed. It is clear from Figure 6.61(b) that the changes in mode shape at the upper beam due to both damage cases are very large. Considering mode 3 in Figure 6.61(c), the changes in mode shape at the lower beam are large. Unlike the situations in modes 1 and 2, the changes in mode shape at the two columns are also observable in this mode. In Figure 6.61(d), the changes in mode shape at the top beam are not small. However, it is difficult to conclude the damaged joint by observing the mode shape changes.

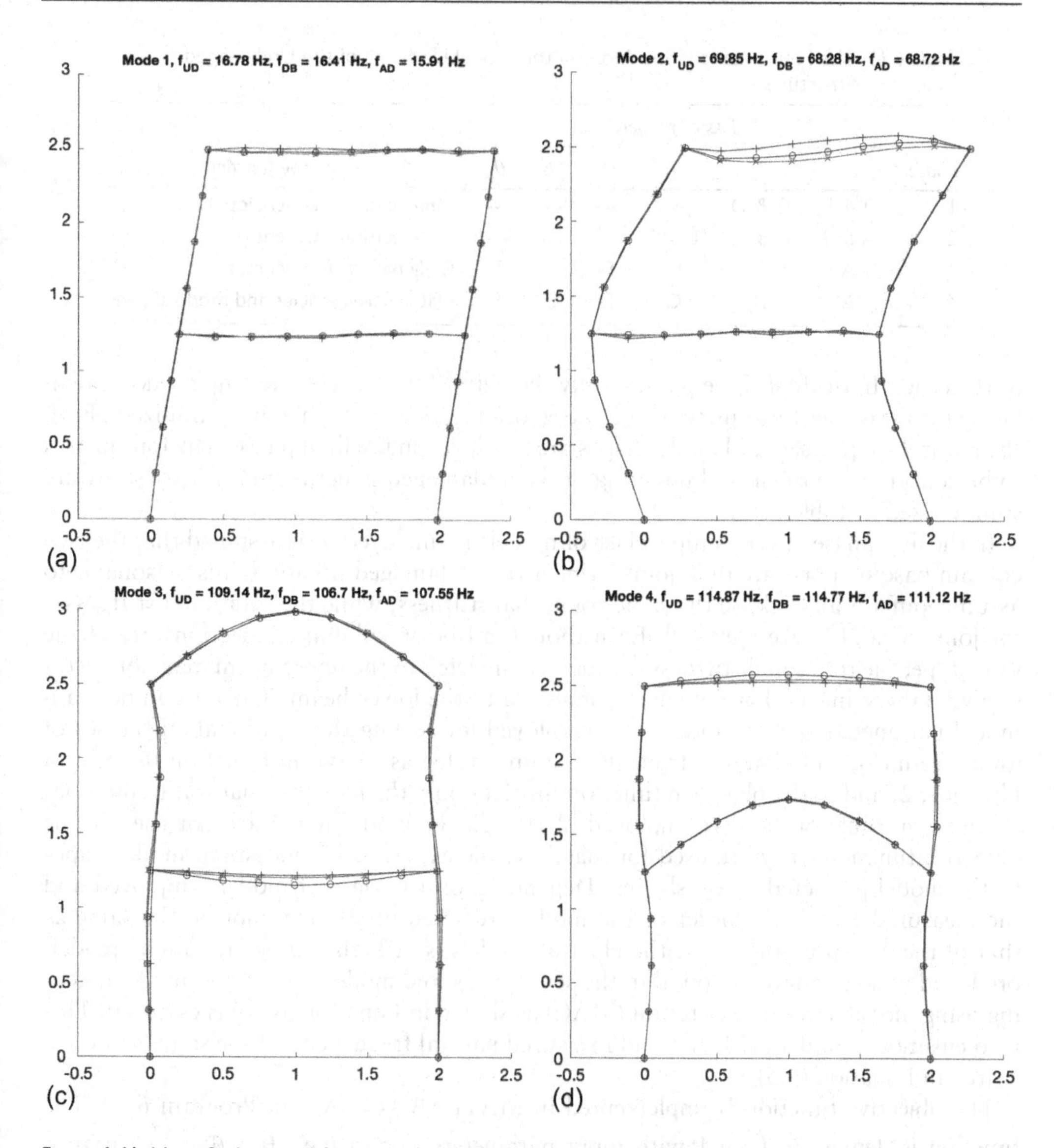

Figure 6.61 Identified natural frequency and mode shapes of the first four modes (o: undamaged (UD); x: damaged at joint B (DB); +: damaged at joints A and D (AD)). (a) Mode 1. (b) Mode 2. (c) Mode 3. (d) Mode 4.

6.3.2.5 Model updating of the undamaged structure

Most existing structural damage detection methods based on model updating require a baseline (reference), which is obtained from a set of measured data from the undamaged (intact) structure. In this section, a "representative" model of the undamaged structure is determined by model updating utilizing the set of natural frequencies and mode shapes identified in the previous section. To illustrate the effects on the model parameterization (i.e., the class of models) on the results of model updating, the model updating is carried out in four phases with the complexity of the model class being increased from one phase

Table 6.15 The Four Phases Considered in the Model Updating of the Undamaged Structure

	Class of models						
Phase	θ_1	θ_2	θ_3	θ_4	θ_5	θ_6	*Objective function*
1	A & B	C & D	---	---	---	---	Only natural frequencies
2	A & B	C & D	E & F	---	---	---	Only natural frequencies
3	A	B	C	D	E	F	Only natural frequencies
4	A	B	C	D	E	F	Natural frequencies and mode shapes

to the other. In the first three phases, only the natural frequencies are employed in calculating the measure-of-fit function (i.e., the error/objective function to be minimized). Both the natural frequencies and mode shapes are employed in the final phase. The four phases to be considered in the model updating of the undamaged structure in this case study are summarized in Table 6.15.

In the first phase, a very simple class of models is employed. It is assumed that the two column-base supports are rigid joints. Under the undamaged situation, it is reasonable to assume joints A and B have the same rotational stiffness, while the rotational stiffnesses for joints C and D are identical. Even though all beam-column connections are of the same type, the rotational stiffness for joints connected to the upper beam may not necessarily be the same as that for joints connected to the lower beam. Thus, two uncertain model parameters (i.e., θ_1 and θ_2) are employed for scaling the rotational stiffnesses of joints A and B, and C and D from the nominal value as shown in Equation (6.152). In Phases 1, 2, and 3, the objective function involves only the identified natural frequencies as given in Equation (6.5) is employed. Although the mode shapes are not used in the objective function, they are used for matching the experimental measured mode shapes to the model-predicted mode shapes. Depending on the class of models employed and the measured DOFs, the order of the model-predicted modes may not be the same as that of the measured modes. If the class of models is of high complexity, many model-predicted modes cannot be found in the set of measured modes. Therefore, mode matching using modal assurance criterion (MAC) as shown in Equation (6.10) is essential. This is to ensure the model-predicted and measured natural frequencies of the same mode are paired in Equation (6.5).

The objective function is implemented in MATLAB as shown in Program 6.10. The function is defined in Line 1 with input parameters `theta` (i.e., $\boldsymbol{\theta} = \{\theta_1, \theta_2\}^T$), `nfexp` (i.e., the vector of experimental measured natural frequencies), `msexp` (i.e., the matrix of experimental measured model shapes), and `mdof` (i.e., the vector of measured DOFs). In Line 2, the self-developed MATLAB function `twostoryframe_chenwf()` is called to calculate the natural frequencies and mode shapes of the two-story steel frame structure based on the given values of $\boldsymbol{\theta}$, where `nf` and `ms` are the model-predicted natural frequencies and mode shapes, respectively. The unmeasured DOFs in the model-predicted mode shapes are removed from the matrix `ms` on Line 3. The model-predicted mode shapes are normalized to have unit length from Lines 4 to 7. Similarly, the measured mode shapes are normalized from Lines 8 to 11. The MAC matrix (i.e., the MAC values of all model-predicted mode shapes vs. all measured mode shapes) is calculated on Line 12, where the i-th row and j-th column of the MAC matrix is the MAC value corresponding to the i-th experimental mode shape and the j-th model-predicted mode shape. The largest value on each row of the MAC matrix is found on Line 13. The vector indx shows the matching between the measured and model-predicted mode shapes. On Lines 14 and 15, the model-predicted natural frequency vector (nf) and mode shape matrix (ms) are rearranged to follow those of the measured ones. Finally, the J_f value in Equation (6.5) is calculated from Lines 16 to 19.

Program 6.10: jcalnf() the objective function employed only natural frequencies.

```
Line 1.   function J=jcalnf(theta,nfexp,msexp,mdof)
Line 2.   [nf,ms]=twostoryframe_chenwf(theta);
Line 3.   ms=ms(mdof,:);
Line 4.   mm=ms'*ms;
Line 5.   for ii=1:length(nf)
Line 6.      ms(:,ii)=ms(:,ii)./mm(ii,ii)^0.5;
Line 7.   end
Line 8.   mmexp=msexp'*msexp;
Line 9.   for ii=1:length(nfexp)
Line 10.     msexp(:,ii)=msexp(:,ii)./mmexp(ii,ii)^0.5;
Line 11.  end
Line 12.  MAC=(msexp'*ms).^2;
Line 13.  [maxMAC,indx]=max(MAC');
Line 14.  nf=nf(indx);
Line 15.  ms=ms(:,indx);
Line 16.  J=0;
Line 17.  for ii=1:length(nfexp)
Line 18.    J=J+((nfexp(ii)-nf(ii))/nfexp(ii))^2;
Line 19.  end
Line 20.  end
```

In this case study, the MATLAB optimization toolbox is employed to minimize the objective function as given in Program 6.10. The function fmincon() is adopted. To ensure the reliability of the numerical minimization result, the minimization process is repeated many times (say 100 times, in this case study) with different randomly generated initial trials. The optimization results (i.e., the identified model together with the objective function value) are recorded in Table 6.16 (Phase 1). The identified rotational stiffness for joints A and B is 0.3587×10^6 Nm/rad, while that for joints C and D is 0.9923×10^6 Nm/rad. The objective function minimum value is 0.0263, which is the sum of squares of natural frequency fractional error for all modes. To see the error contribution from different

Table 6.16 **The Identified Optimal Model Together with the Objective Function Value in Different Phases**

	Class of models						
Phase	θ_1	θ_2	θ_3	θ_4	θ_5	θ_6	*Objective function value*
1	0.3587	0.9923	---	---	---	---	0.0263
2	1.2646	1.4510	1.8733	---	---	---	3.0556×10^{-4}
3	0.3103	4.2854	2.6739	0.7284	2.4146	2.6809	1.6474×10^{-9}
4	1.0134	0.8366	1.6397	1.7428	2.8383	1.6570	0.0127

modes, the model-predicted natural frequencies of the optimal model in phase 1 are given in Table 6.17 (Phase 1). It is clear that the percentage errors in natural frequencies are not small. The largest percentage error is 11.68% in mode 3, and the smallest one is 4.74% in mode 4. As the model class in phase 1 is very simple (with only two uncertain model parameters), it is believed that the fitting can be improved by raising the complexity of the model class (in later phases). As mode shapes are not used in calculating the objective function, it is interesting to see the matchings between the model-predicted and measured mode shapes (see Figure 6.62). Consider mode 1 in Figure 6.62(a), where the mode shape matching at the two columns is relatively poor. Consider the column deflection near the support, where the experimental measured one ("+") is larger than the model-predicted one ("x"). Similar observations can be made in modes 2, 3, and 4 in Figure 6.62(b), (c), and (d), respectively. As the lateral deflection shape of a column depends very much on the support condition, it is believed that the assumption of rigid connection at the support may not represent the real situation as reflected in the measured mode shape. With the rigid support assumption, the class of models in this phase overestimate the overall stiffness of the system. To reduce the stiffness of the system in matching the measured natural frequencies (which shows the real stiffness of the system), the estimated rotational stiffnesses are believed to be underestimated in this phase. With this observation, the rotational stiffness of column-base connections at joints E and F are considered as uncertain parameters in the class of models in phase 2 of this study.

The model class in phase 2 with three uncertain model parameters is more complex compared to that in phase 1. The additional parameter θ_3 is used to scale the rotational stiffness of joints E and F from the nominal value. Similar to the procedure in phase 1, the minimization is carried out with different random initial trials. The identified model together with the objective function value are given in Table 6.16 (Phase 2). The result obtained is consistent with the previous discussion that the rotational stiffnesses in phase 1 are underestimated. The identified rotational stiffness values for joints A, B, C, and D in phase 2 are higher than those in phase 1. The identified rotational stiffness for joints F and G is 1.8733×10^6 Nm/rad. Although this rotational stiffness is higher than that of the beam-column connections, it is far from being a rigid connection. It is also clear from Table 6.16 that the objective function value of phase 2 drops significantly when compared to that in phase 1, implying that the fitting in natural frequencies has improved a lot. Referring to Table 6.17 (Phase 2), most percentage errors in natural frequencies are smaller than 1% except for mode 2 (1.26%). Figure 6.63 compared the measured and model-predicted mode shapes in phase 2. It is clear that the matching of mode shapes at the two columns improves a lot when compared to the situation in phase 1 (compare to Figure 6.62). The consideration of the semi-rigid behavior of the column-base connection significantly reduces the effect of modeling error in the class of models. Another

Table 6.17 Model-Predicted Natural Frequencies of the Updated Model in All Phases

Case	*Mode 1*	*Mode 2*	*Mode 3*	*Mode 4*
Measured	16.78	69.85	109.14	114.87
Phase 1	17.61 (4.95%)	76.08 (8.92%)	96.39 (−11.68%)	109.43 (−4.74%)
Phase 2	16.66 (−0.72%)	70.73 (1.26%)	108.28 (−0.79%)	115.61 (0.64%)
Phase 3	16.78 (0.00%)	69.85 (0.00%)	109.13 (−0.01%)	114.87 (0.00%)
Phase 4	16.94 (0.95%)	69.90 (0.07%)	105.29 (−3.53%)	117.32 (2.13%)

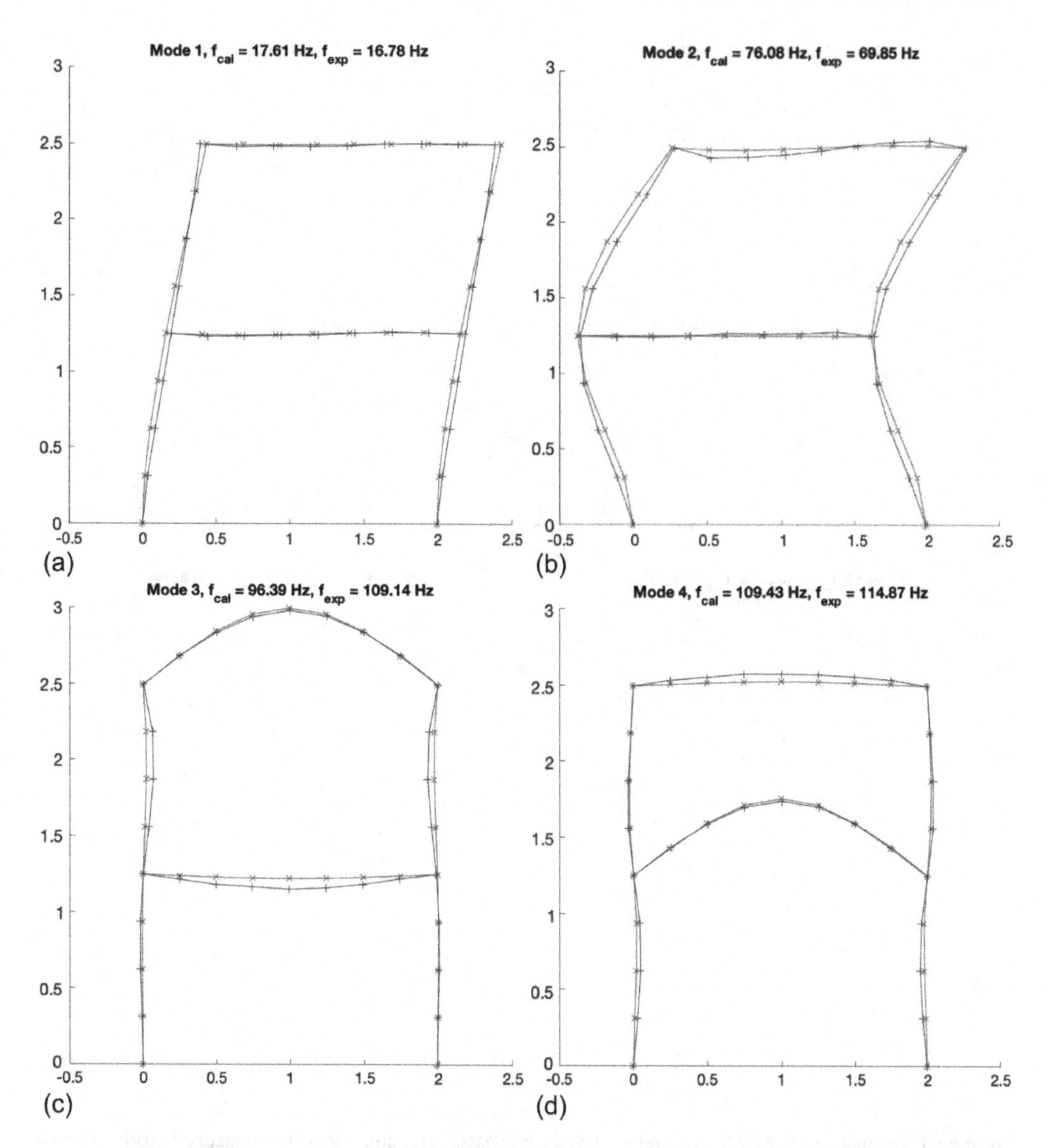

Figure 6.62 Comparison between model-updated (only natural frequencies) and measured modal parameters in phase 1 ("+" represents experimental measured, and "x" represents model-predicted). (a) Mode 1. (b) Mode 2. (c) Mode 3. (d) Mode 4.

important assumption in both phases 1 and 2 is that the steel frame is symmetrical about the center line. That is, the rotational stiffnesses for the left and right joints at the same level are assumed to be the same. This assumption is important to reduce the number of uncertain model parameters when it is known that the system is undamaged. It is a common trick to quickly get a reasonable initial model. This symmetrical assumption is not considered in phase 3.

In phase 3, the complexity of model class is increased as the number of uncertain model parameters is doubled from 3 (in phase 2) to 6. That is, θ_1 to θ_6 are used to scale the

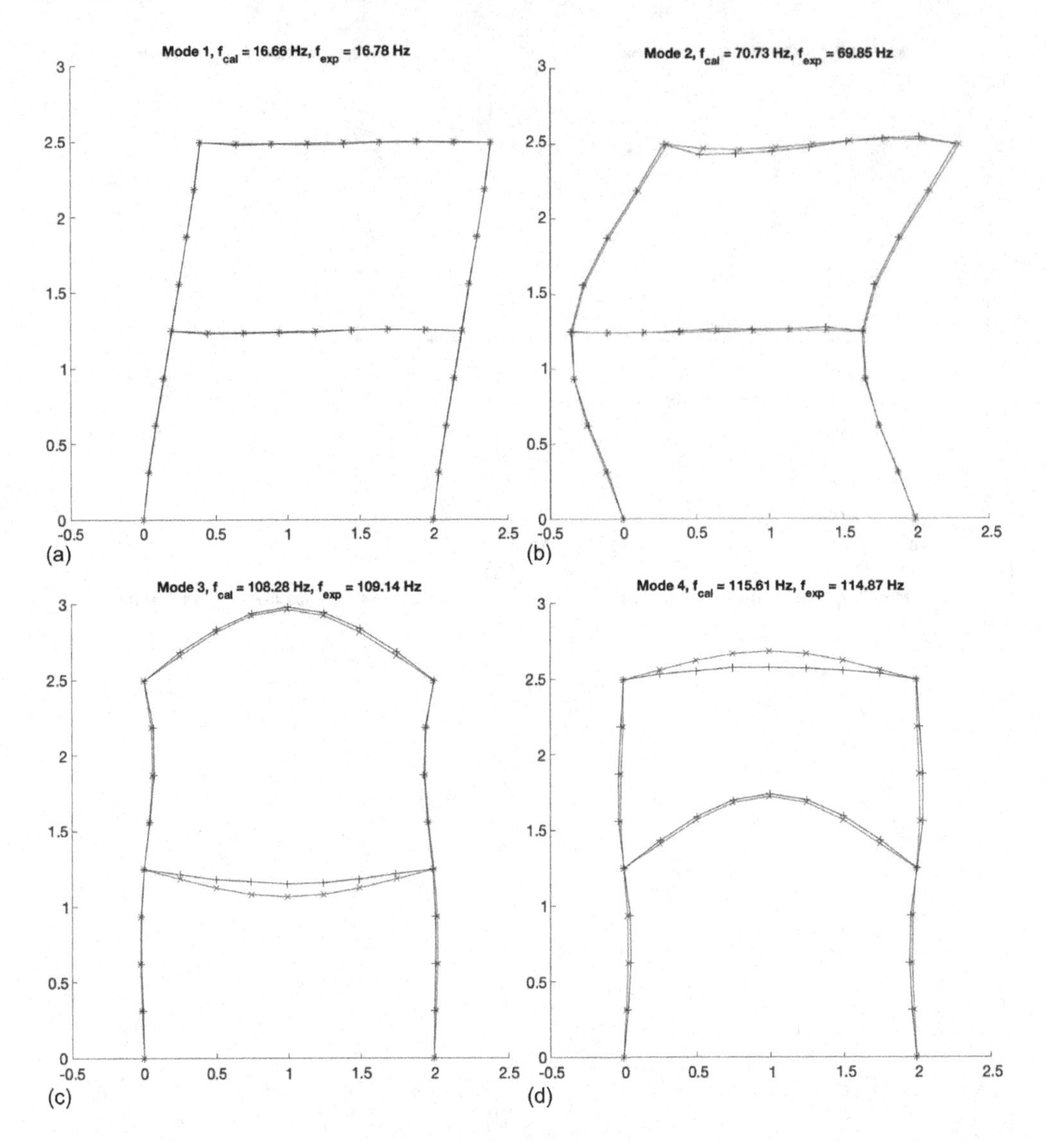

Figure 6.63 Comparison between model-updated (only natural frequencies) and measured modal parameters in the undamaged case (θ_1 for joints A and B, θ_2 for joints C and D, and θ_3 for joints E and F). (a) Mode 1. (b) Mode 2. (c) Mode 3. (d) Mode 4.

rotational stiffness at joints A to F, respectively. As the number of uncertain parameters is increased, the time required to solve the numerical optimization problem also increased. The model updating results are shown in Table 6.16 (Phase 3). It is a bit odd that the identified stiffness for joint A is very different from that for joint B; similarly for joints C vs. D and joints E vs. F. For an undamaged structure, such a difference in joint stiffness is questionable. It seems that the identified model is very good as the objective function value is extremely low (almost equal to zero) showing that the fitting of measured and model-predicted natural frequencies is almost perfect. This viewpoint is reinforced by looking at the percentage error in natural frequencies in Table 6.17 (Phase 3). The percentage errors

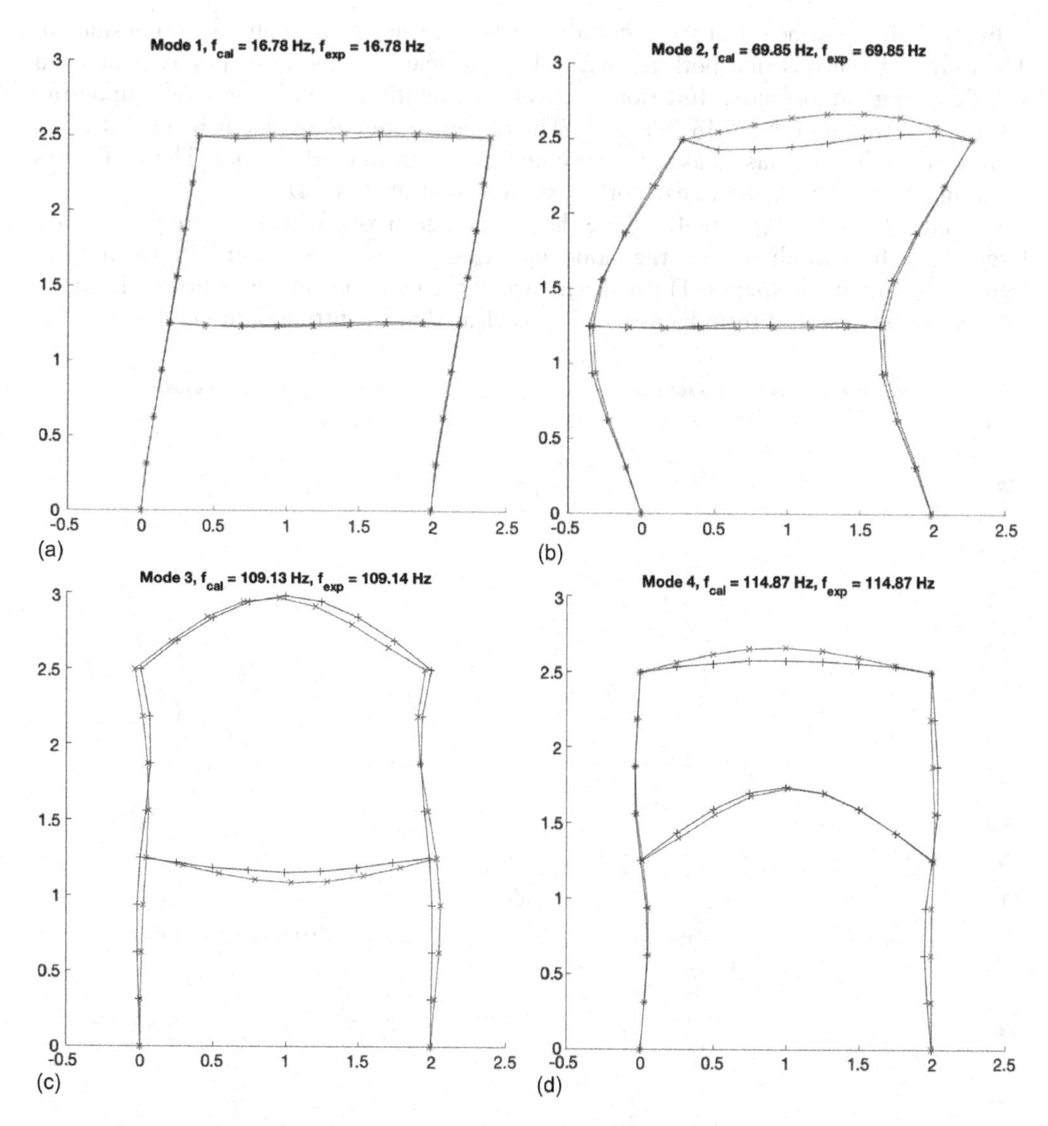

Figure 6.64 Comparison between model-updated (only natural frequencies) and measured modal parameters in the undamaged case (θ_1 to θ_6 for joints A to F, respectively). (a) Mode 1. (b) Mode 2. (c) Mode 3. (d) Mode 4.

are basically equal to zero, except the 0.01% for mode 3. This expectation is completely destroyed by looking at the matching between the measured and model-predicted mode shapes in Figure 6.64. The matching is very bad in general, especially for the upper beam of mode 2; and the mode shape for mode 3 seems shifted. This is, in fact, understandable, as mode shapes are not included in the calculation of the objective function. This phase clearly illustrates the situation when the class of models is complex but the amount of measured information is limited. The identified model is good to fit the measured data that was presented in the updating process, but the prediction ability of this model is questionable. Finally, the mode shape information is considered in formulating the objective function in phase 4.

In phase 4, the same set of six uncertain model parameters as in phase 3 is considered. The only difference is that both the natural frequencies and mode shapes are employed in calculating the objective function as given in Equation (6.13). The model updating results are given in Table 6.16 (Phase 4). The objective function value is increased when compared to that of phase 3 as the error contribution from mode shapes. The stiffnesses for joints A and B are close to each other, similar to joints C vs. D.

Consider Table 6.17 (phase 4), where the percentage errors for natural frequencies are larger than those in phase 3, as the model updating process needs to fit both the natural frequencies and mode shapes. The performance of the identified model in fitting the mode shapes can be observed from Figure 6.65. It is clear that the fitting of mode shapes is the

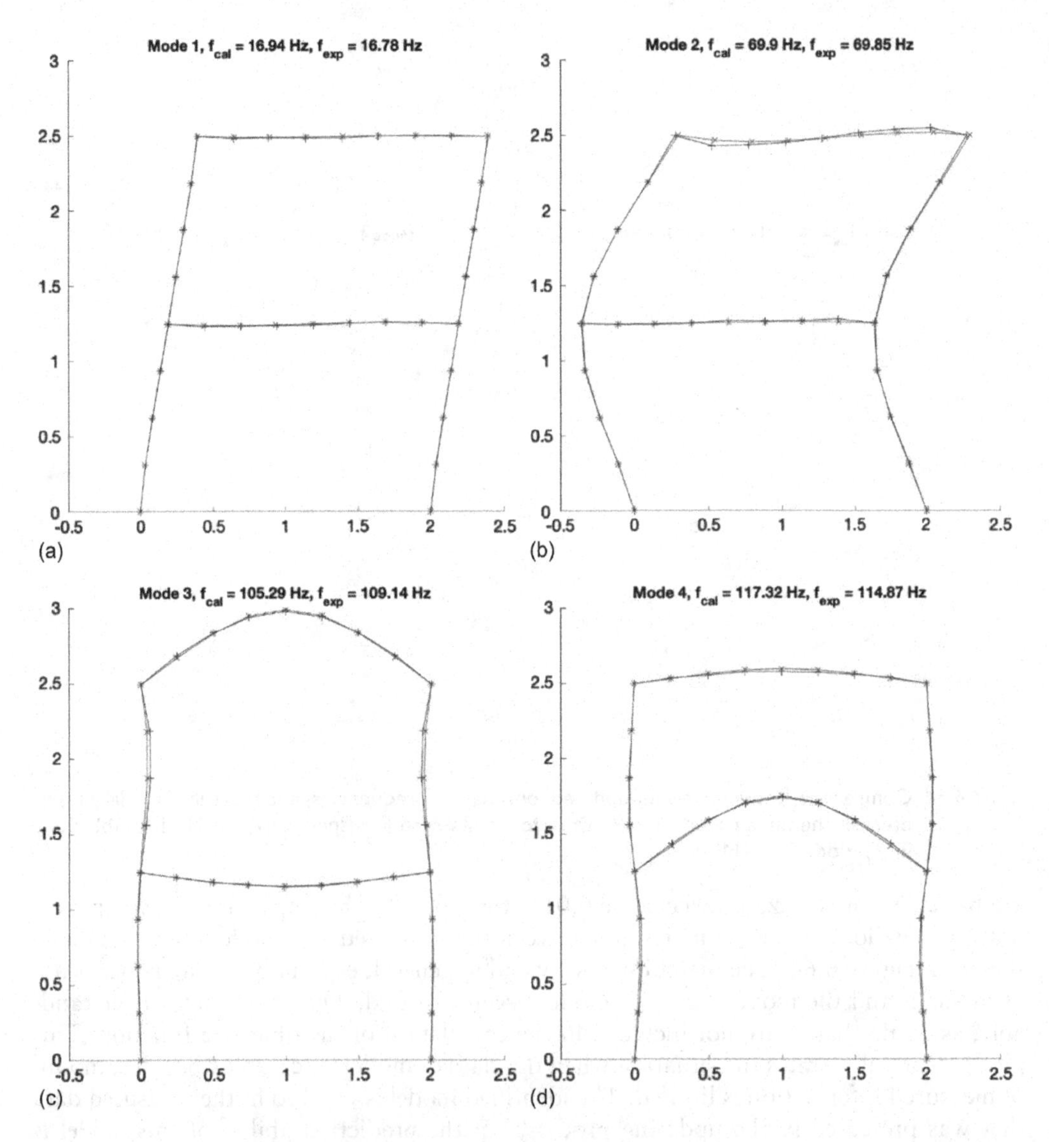

Figure 6.65 Comparison between model-updated (both natural frequencies and mode shapes) and measured modal parameters in the undamaged case (θ_1 to θ_6 for joints A to F, respectively). (a) Mode 1. (b) Mode 2. (c) Mode 3. (d) Mode 4.

best among all considered phases (as expected). This identified model will be employed as the baseline for joint damage detection in the next section.

6.3.2.6 Model updating in damage case DB

In this damage case, joint damage is simulated to joint B by loosening the screws of the top angle as shown in Figure 6.66. It is clear from the figure that the set of screws connecting the top flange of the beam to the angle and the set of screws connecting the angle to the flange of the beam are loosened. This will effectively cut off the load path from the top flange of the beam to the flange of the column. As a result, the bending moment on the beam cannot be effectively transferred to the column, thus reducing the rotational stiffness of joint B.

By using the same class of models in phase 4 of the undamaged case (i.e., six uncertain model parameters to scale the rotational stiffness of the six joints), model updating is carried out with the consideration of both the natural frequencies and mode shapes in the objective function. The optimal model in phase 4 (i.e., the last row in Table 6.16) is employed as the initial trial in the model updating of the damaged case DB. The optimal model identified is given in Table 6.18 (Case DB). The percentage differences of the six identified rotational stiffness between the case UD and DB are calculated and given in parentheses in Table 6.18. It is clear from the table that the percentage changes for rotational stiffness at joints A, C, D, E, and F are all smaller than 8%, while the rotational stiffness at joint B is reduced by about 47%. Based on this result, it is realistic to conclude that joint B is damaged in this case with almost half of the rotational stiffness lost. This simulated joint damage can be detected by the model updating method.

Note that the objective function value (i.e., the J value) in case DB (damage at joint B) is in the same order of magnitude as that in case UD (undamaged). It is believed that the

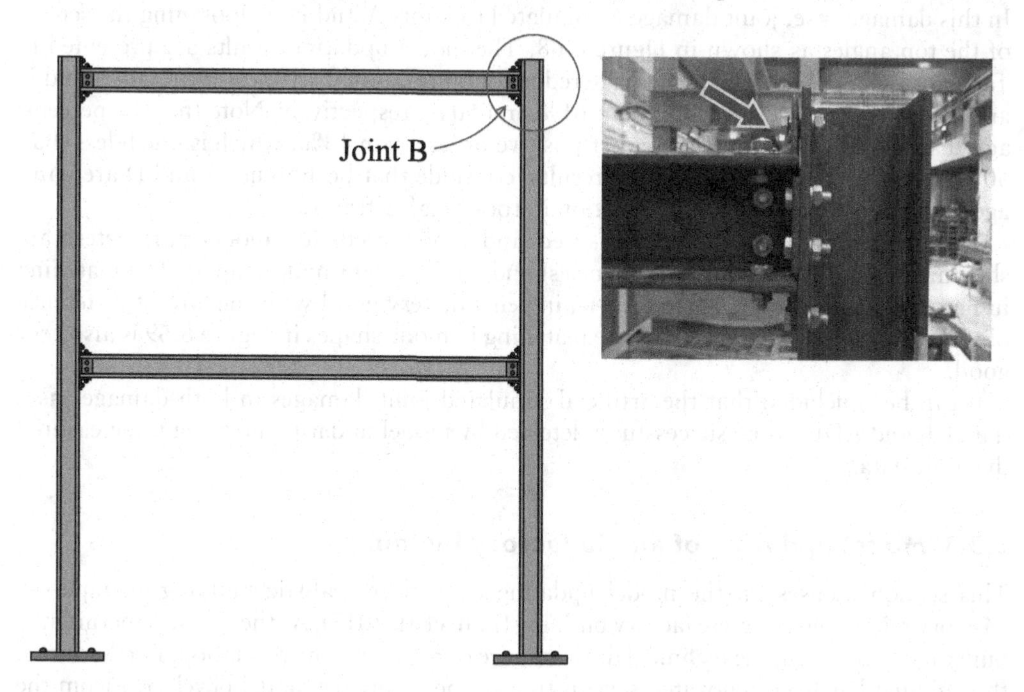

Figure 6.66 Simulated joint damage in case DB.

Table 6.18 Model Updating Results for All Undamaged and Damaged Cases

Case	*Joint A*	*Joint B*	*Joint C*	*Joint D*	*Joint E*	*Joint F*	*J value*
UD	1.0134	0.8366	1.6397	1.7428	2.8383	1.6570	0.0127
DB	0.9382 (−7.42%)	**0.4454 (−46.76%)**	1.5466 (−5.68%)	1.8623 (6.86%)	3.0034 (5.82%)	1.5628 (−5.68%)	0.0152
AD	**0.3905 (−61.47%)**	0.8074 (−3.49%)	2.1803 (32.97%)	**0.8658 (−50.32%)**	2.4860 (−12.41%)	2.0116 (21.40%)	0.0199

Table 6.19 Measured vs. Model-Predicted Natural Frequencies in Case DB

	Mode 1	*Mode 2*	*Mode 3*	*Mode 4*
Measured	16.41	68.28	106.70	114.77
Model-predicted	16.54 (0.79%)	68.14 (−0.21%)	102.06 (−4.35%)	117.08 (2.01%)

matching between the measured and model-predicted modal parameters is reasonable, as confirmed by the results in Table 6.19. The percentage differences between the measured and model-predicted natural frequencies are very small with the largest being about 4%. Figure 6.67 shows the matching between the measured and model-predicted mode shapes (of the updated model) in this case. The matching is very good in all modes. The damage detection result for the single damage situation in this case is very satisfactory. The double damage case is considered in the next section.

6.3.2.7 Model updating in damage case AD

In this damage case, joint damage is simulated to joints A and D by loosening the screws of the top angles as shown in Figure 6.68. The model updating results are presented in Table 6.18 (Case AD). The percentage reduction in rotational stiffness for joints A and D are very large and are equal to about 61% and 50%, respectively. Note that the percentage changes for other joints are either positive or less than 13%, which is much less than 50%. Based on the model updating results, conclude that both joints A and D are damaged with approximately 50% reduction in rotational stiffness.

The matchings between the measured and model-predicted model parameters are shown in Table 6.20 (natural frequencies) and Figure 6.69 (mode shapes). The matching in natural frequencies in Table 6.20 is, in general, very good with the largest difference observed at mode 3 at about 6%. The matching in mode shapes in Figure 6.69 is also very good.

It can be concluded that the artificial simulated joint damages in both damage cases (i.e., DB and AD) can be successfully detected by model updating utilizing the measured dynamic data.

6.3.3 Model updating of an old factory building

This section focuses on the model updating of a more realistic building example—a 14-story reinforced concrete factory building (Lam et al. 2017). As the target structure was built many years ago, the technical drawings are of relatively low resolution. Furthermore, the building has been renovated several times. The computer model developed from the

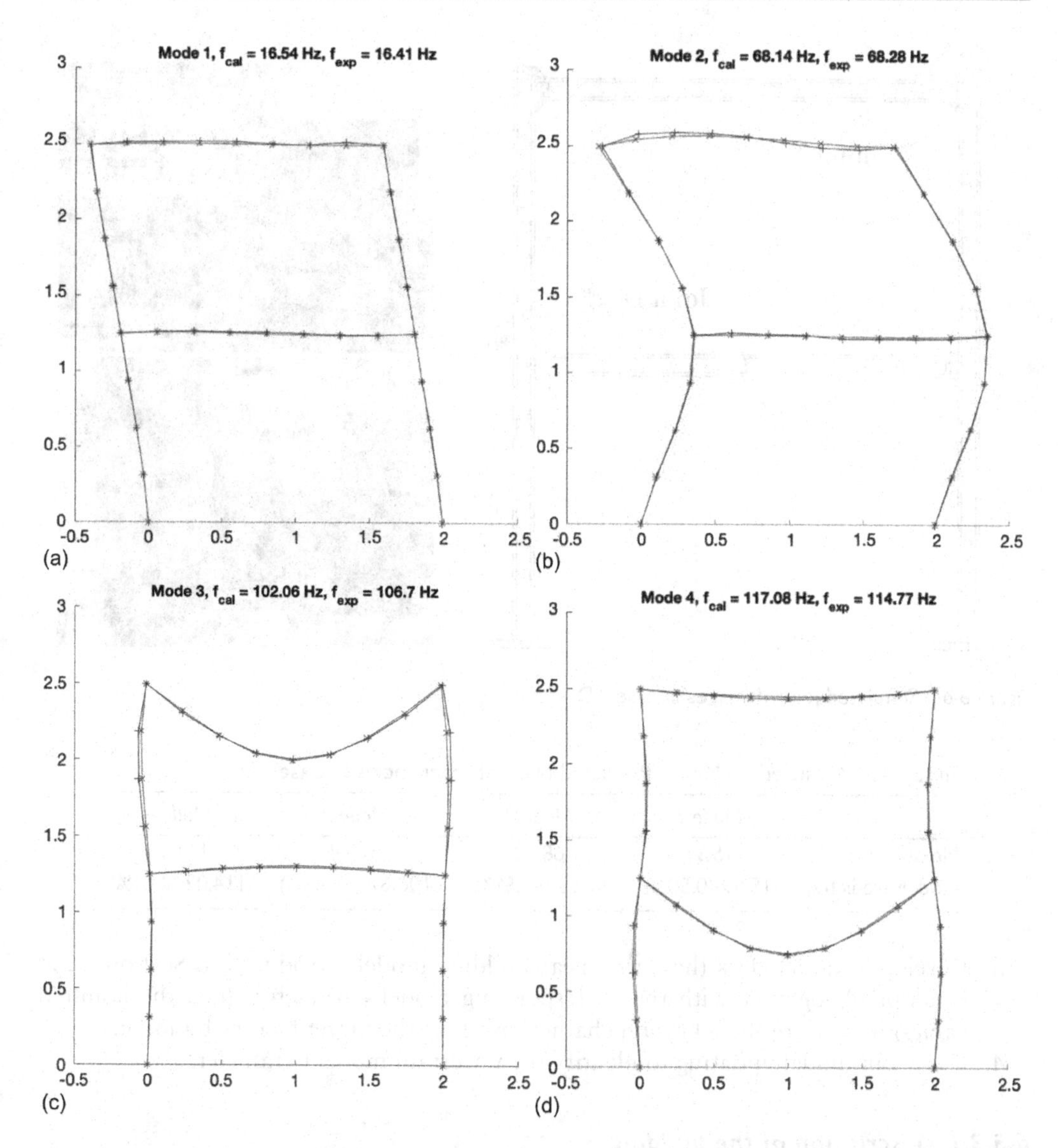

Figure 6.67 Comparison between model-updated and measured modal parameters in damaged case DB (θ_1 to θ_6 for joints A to F, respectively). (a) Mode 1. (b) Mode 2. (c) Mode 3. (d) Mode 4.

available structural drawings consists of uncertainties. The main purpose of this case study is to identify the inter-story stiffness along the global x (i.e., East-West, EW), y (i.e., North-South, NS), and z (i.e., Torsional, T) directions based on a set of measured vibration data obtained from field test. To achieve this purpose, the following objectives are considered:

1. Design and implement the field test to measure the ambient vibration of the building at pre-defined DOFs (i.e., measurement locations and directions).
2. Identify a set of modal parameters in particular the natural frequencies and mode shapes based on the measured ambient vibration data.

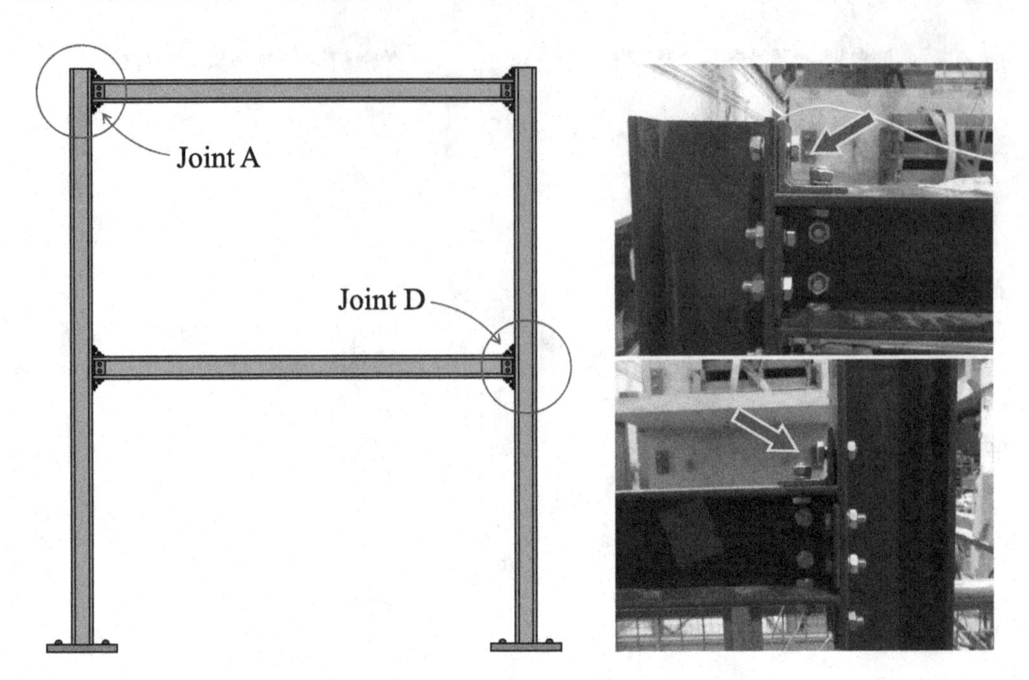

Figure 6.68 Simulated joint damages in case AD.

Table 6.20 Measured vs. Model-Predicted Natural Frequencies in Case AD

	Mode 1	*Mode 2*	*Mode 3*	*Mode 4*
Measured	15.91	68.72	107.55	111.12
Model-predicted	15.99 (0.50%)	67.63 (−1.59%)	100.67 (−6.40%)	114.02 (2.61%)

3. Develop a model class (here the shear building model introduced in Section 2.2.1 is adopted) together with the corresponding model parameters (i.e., the nominal values) to capture the vibration characteristics of the target factory building.
4. Carry out model updating to identify the uncertain model parameters.

6.3.3.1 Description of the building

The target structure in this case study is the 14-story old factory building with a reduction in floor area from 4/F to 5/F as shown in Figure 6.70. This reinforced concrete building was built sometime around 1970, and the drawings follow the imperial units (i.e., foot and inch). The units are converted to SI units in this case study, and Figure 6.71 shows the re-plotted typical floor plan for 5/F to 13/F with mm as the unit. It must be pointed out that the floor plan given in Figure 6.71 is re-plotted with reference to the old technical drawings and information collected from several site inspections. This is the best the authors can do; however, it may not be very accurate. Referring to the figure, there are three types of columns and three types of beams. Their dimensions are summarized in Table 6.21. This information is important for the estimation of the nominal values of the inter-story stiffness and floor mass in the modeling of the factory building.

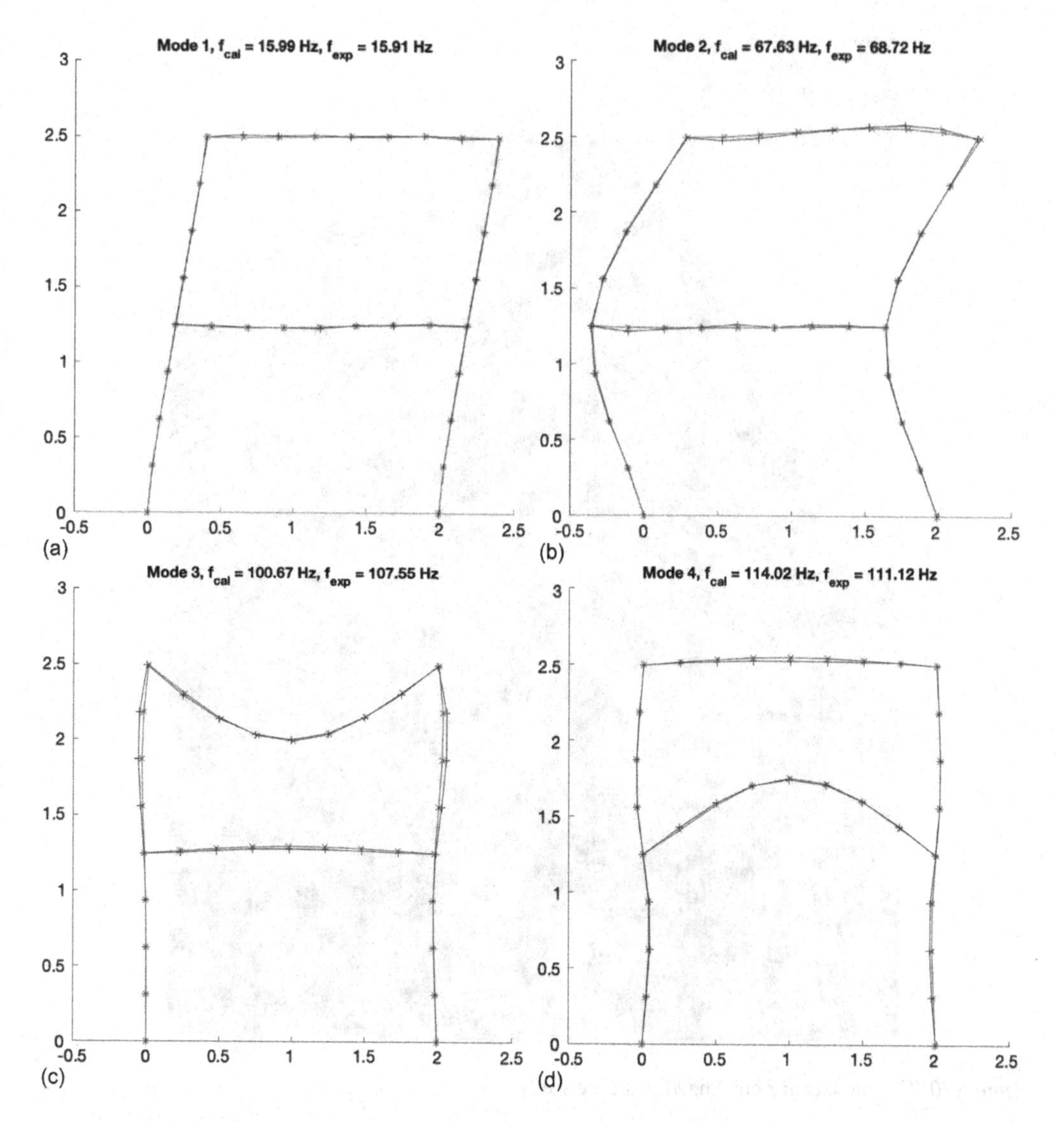

Figure 6.69 Comparison between model-updated and measured modal parameters in damaged case AD (θ_1 to θ_6 for joints A to F, respectively). (a) Mode 1. (b) Mode 2. (c) Mode 3. (d) Mode 4.

Based on the available information, reinforced concrete rigid frame is the lateral load resistant system. That is, the horizontal load, such as wind load, is mainly resisted by the double bending of the beams and columns. The floor plan is basically a rectangle with dimensions $32.77\ \text{m} \times 29.18\ \text{m}$ (almost a square). As the distribution of columns is very uniform for all floors, the stiffness center and the geometric center are very close to each other. It is expected that the factory building can be modeled as three independent shear building models, i.e., the three 14-story shear building models for vibration along the NS, the EW, and the torsional directions. Note that bending along the NS direction (i.e., the global y-axis) means bending about the global x-axis.

To identify the translational vibration modes (e.g., the vibration along the EW or NS directions), a minimum of one triaxial accelerometer per floor is good enough.

Figure 6.70 The old factory building in this case study.

Nonetheless, at least three triaxial accelerometers per floor are needed to capture the torsional vibration modes. As shown in the floor plan in Figure 6.71, there are three staircases (i.e., Staircases A, B, and C) in this factory building. It is convenient to carry out the vibration measurement on each floor in these three staircases without affecting the normal operation of the factory building. The measurements from the three staircases must be linked, and this can be done on the roof of the building. Figure 6.72 shows the roof of the factory building with the locations of the three staircases. It is clear from the figure that there are two reinforced concrete rooms (one on the left and one on the right). They are used as the entrances to the staircases, the lift machine room, and also as supports for water tanks. Owing to the lack of a structural drawing, the mass of the roof floor is highly uncertain. Thus, the mass of the roof floor (i.e., 14/F) will be considered as one of the uncertain model parameters in the model updating process.

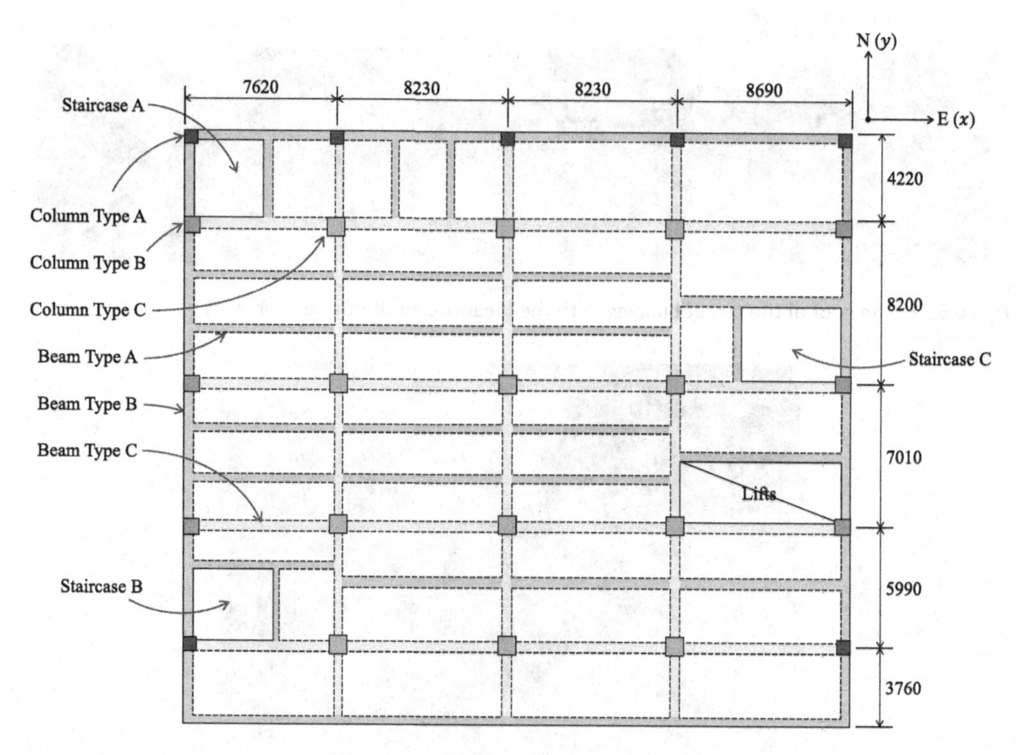

Figure 6.71 Typical floor plan for 5/F to 13/F together with the global coordinate system.

Table 6.21 Dimensions of Various Types of Columns and Beams in Figure 6.71

Type	*Dimensions*	*Remarks*
Column Type A	610 × 610	Small size columns usually at corners or edges of the building.
Column Type B	760 × 760	Middle size columns usually at edges of the building.
Column Type C	860 × 860	Large size columns used as internal columns.
Beam Type A	300 × 710	Small size beams usually used as secondary and external beams.
Beam Type B	360 × 760	Middle size beams usually used as secondary and external beams.
Beam Type C	480 × 760	Large size beams used as internal main beams.

6.3.3.2 Field test and modal identification

For ambient vibration tests, high sensitivity accelerometers must be used (not the same as the small size accelerometers employed in the previous case studies). Figure 6.73 shows the six triaxial accelerometers (each with a 30m cable) employed in this field test. Each sensor outputs three channels of vibration data corresponding to the local x-, y-, and z-directions (x and y are horizontal, and z is vertical). During the field test, all sensors are carefully orientated to point towards the north direction. The measured signals are transferred from the sensors to the self-developed data acquisition system (i.e., the console) through cables as shown in Figure 6.74. The self-developed data acquisition system consists of a series of analog to digital (A/D) cards and a car battery (as electricity

Figure 6.72 The roof of the target building with the locations of the three staircases.

Figure 6.73 The six triaxial accelerometers employed in the field test.

supply is usually unavailable in field tests), and it is controlled by a laptop with LabVIEW (National Instruments 2003) installed. The laptop is used to set up the A/D parameters, such as the sampling frequency and duration together with the sensitivity factors for each sensor. It is also used to record the measured vibration data, and perform real-time data analysis using MATLAB. In this field test, the sampling frequency and the recording duration are set to 2048 Hz and 30 mins, respectively. Note that the sampling frequency and duration adopted are much higher than necessary. In Figure 6.73, the six accelerometers were placed very close to each other, and a set of 10 mins vibration data was recorded for on-site calibration to ensure the accuracy of the measured signals and the performance of all sensors. With these six sensors, six measurement points can be tested simultaneously.

As discussed in the previous section, the vibration measurement is focused on the three staircases. For each staircase, 15 measurement points are required (one per floor, including G/F). Owing to the importance in linking the three sets of measured data from the three staircases, one additional measurement point is set on the roof (i.e., 14/F). As a result, a total of 46 (= 15 × 3 + 1) measurement points are needed. With six available triaxial accelerometers, the entire field test must be divided into multiple setups (at least one overlapped measurement point in each setup as a reference to link the measured data from different setups). Finally, a total of 12 setups are required (4 setups for each staircase). The locations of all six sensors in all setups are summarized in Table 6.22. In the table, the labels "A," "B," and "C" represent Staircase A, B, and C, respectively. The

Figure 6.74 The console for controlling and recording measured signals from the six sensors.

label "Roof" represents the sensor location on the roof (not belonging to any staircase). The superscript "R" represents the reference for linking different setups. For example, "R1,5,9" for Sensors 3 to 6 in Setups 1, 5, and 9 means that Sensors 3 to 6 are reference sensors to link the measured data from Setups 1, 5, and 9.

In Setup 1, Sensors 1, 2, and 3 are placed in Staircase A, while Sensors 4, 5, and 6 are placed in Staircase B, the roof, and Staircase C, respectively. Setup 1 is mainly used to link measurement of Staircase A to other staircases. After Setup 1, Sensors 3, 4, 5, and 6 are moved to floors 11, 10, 9, and 8, respectively, for Setup 2. Note that Sensors 1 and 2 are not moved, so they can be used as references to link data in Setups 1 and 2. After Setup 2, Sensors 1, 2, 3, and 4 are moved to floors 5, 4, 7, and 6, respectively, for Setup 3. Sensors 5 and 6 are kept in the same locations to serve as references to link data from Setups 2 and 3. As each setup is linked to other setups through more than two references, the measured partial mode shapes from different setups can be combined to form the global mode shape with high accuracy. Interested readers are redirected to Lam et al (2017) for the detailed measurement plan. As an example, the measured time-domain response in Setup 1, Sensor 1, Channel 1 (along EW direction) is plotted in Figure 6.75. The vibration level is under 4×10^{-4} g, which is low, as this is an ambient vibration test. For measurement of a given staircase, sensors at different floors should be placed along the same vertical line. Figure 6.76 shows the alignment of sensors in one of the setups.

Based on the modal identification method introduced in Chapter 3, the modal parameters are identified. The six identified (or denoted as measured) natural frequencies are summarized in Table 6.23. Note that the modes are arranged according to their vibration directions (i.e., translational along north-south (global y), translational along east-west (global x), and torsional (global z)). For each direction, two modes are identified. Therefore, the model updating of each shear building model can be done using information from two modes. It is clear that the order of modes are NS, EW, and T (for both first and second modes) implying that the lateral stiffness (or the inter-story stiffness of the shear building) along the NS direction is lower than that along the EW direction. Furthermore, both lateral stiffness values are smaller than the torsional stiffness.

Table 6.22 Sensor Locations in Various Setups and the Corresponding Reference Sensors

Setup	*Sensor 1*	*Sensor 2*	*Sensor 3*	*Sensor 4*	*Sensor 5*	*Sensor 6*	*Remarks*
1	A, 13/F$^{(R1,2)}$	A, 12/F$^{(R1,2)}$	A, 14/F$^{(R1,5,9)}$	B, 14/F$^{(R,1,5,9)}$	Roof, 14/F$^{(R1,5,9)}$	C, 14/F$^{(R1,5,9)}$	Linking
2	A, 13/F$^{(R1,2)}$	A, 12/F$^{(R1,2)}$	A, 11/F	A, 10/F	A, 9/F$^{(R2,3)}$	A, 8/F$^{(R2,3)}$	Staircase A
3	A, 5/F$^{(R3,4)}$	A, 4/F$^{(R3,4)}$	A, 7/F	A, 6/F	A, 9/F$^{(R2,3)}$	A, 8/F$^{(R2,3)}$	Staircase A
4	A, 5/F$^{(R3,4)}$	A, 4/F$^{(R3,4)}$	A, 3/F	A, 2/F	A, 1/F	A, G/F	Staircase A
5	B, 13/F$^{(R5,6)}$	B, 12/F$^{(R5,6)}$	A, 14/F$^{(R1,5,9)}$	B, 14/F$^{(R,1,5,9)}$	Roof, 14/F$^{(R1,5,9)}$	C, 14/F$^{(R1,5,9)}$	Linking
6	B, 13/F$^{(R5,6)}$	B, 12/F$^{(R5,6)}$	B, 11/F	B, 10/F	B, 9/F$^{(R6,7)}$	B, 8/F$^{(R6,7)}$	Staircase B
7	B, 5/F$^{(R7,8)}$	B, 4/F$^{(R7,8)}$	B, 7/F	B, 6/F	B, 9/F$^{(R6,7)}$	B, 8/F$^{(R6,7)}$	Staircase B
8	B, 5/F$^{(R7,8)}$	B, 4/F$^{(R7,8)}$	B, 3/F	B, 2/F	B, 1/F	B, G/F	Staircase B
9	C, 13/F$^{(R9,10)}$	C, 12/F$^{(R9,10)}$	A, 14/F$^{(R1,5,9)}$	B, 14/F$^{(R,1,5,9)}$	Roof, 14/F$^{(R1,5,9)}$	C, 14/F$^{(R1,5,9)}$	Linking
10	C, 13/F$^{(R9,10)}$	C, 12/F$^{(R9,10)}$	C, 11/F	C, 10/F	C, 9/F$^{(R10,11)}$	C, 8/F$^{(R10,11)}$	Staircase C
11	C, 5/F$^{(R11,12)}$	C, 4/F$^{(R11,12)}$	C, 7/F	C, 6/F	C, 9/F$^{(R10,11)}$	C, 8/F$^{(R10,11)}$	Staircase C
12	C, 5/F$^{(R11,12)}$	C, 4/F$^{(R11,12)}$	C, 3/F	C, 2/F	C, 1/F	C, G/F	Staircase C

Note: A = Staircase A; B = Staircase B; C = Staircase C; R = the reference sensor to link different setups.

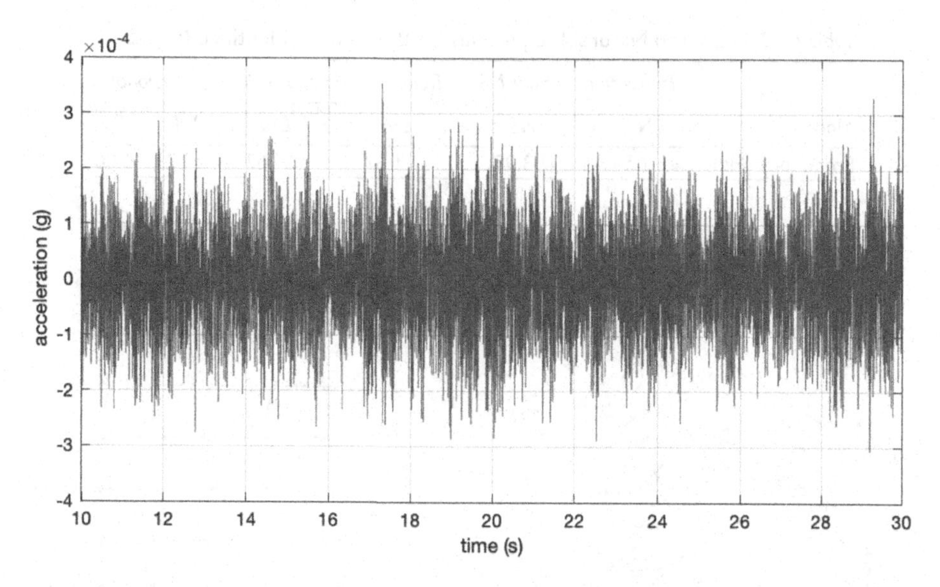

Figure 6.75 A sample of measured time-domain acceleration.

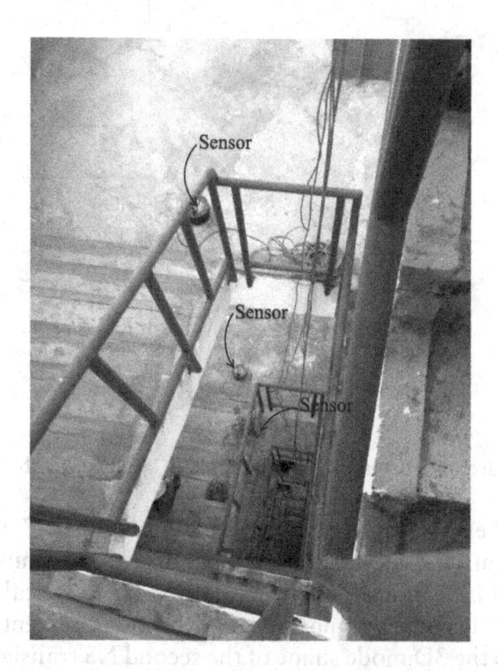

Figure 6.76 The alignment of sensors for vibration measurement in a staircase.

The measured mode shapes are shown in Figures 6.77 to 6.82. In Figure 6.77(a), the three-dimensional view of the mode shape of the first translational NS mode is shown in solid lines, while the undeformed building is shown in dashed lines. The three vertical lines represent the three staircases. The ground-level coordinates for Staircases A, B, and C are (0, 0), (0,–21.71), and (24.28,–8.56), respectively. To clearly show a floor, the three measurement points (in the three staircases) of a given floor are connected by straight

Table 6.23 Measured Natural Frequencies for the First Six Identified Modes

	Translational about NS		*Translational about EW*		*Torsional*	
Mode	*NS 1*	*NS 2*	*EW 1*	*EW 2*	*T 1*	*T 2*
Frequency (Hz)	1.51	3.99	1.67	4.67	2.39	6.00

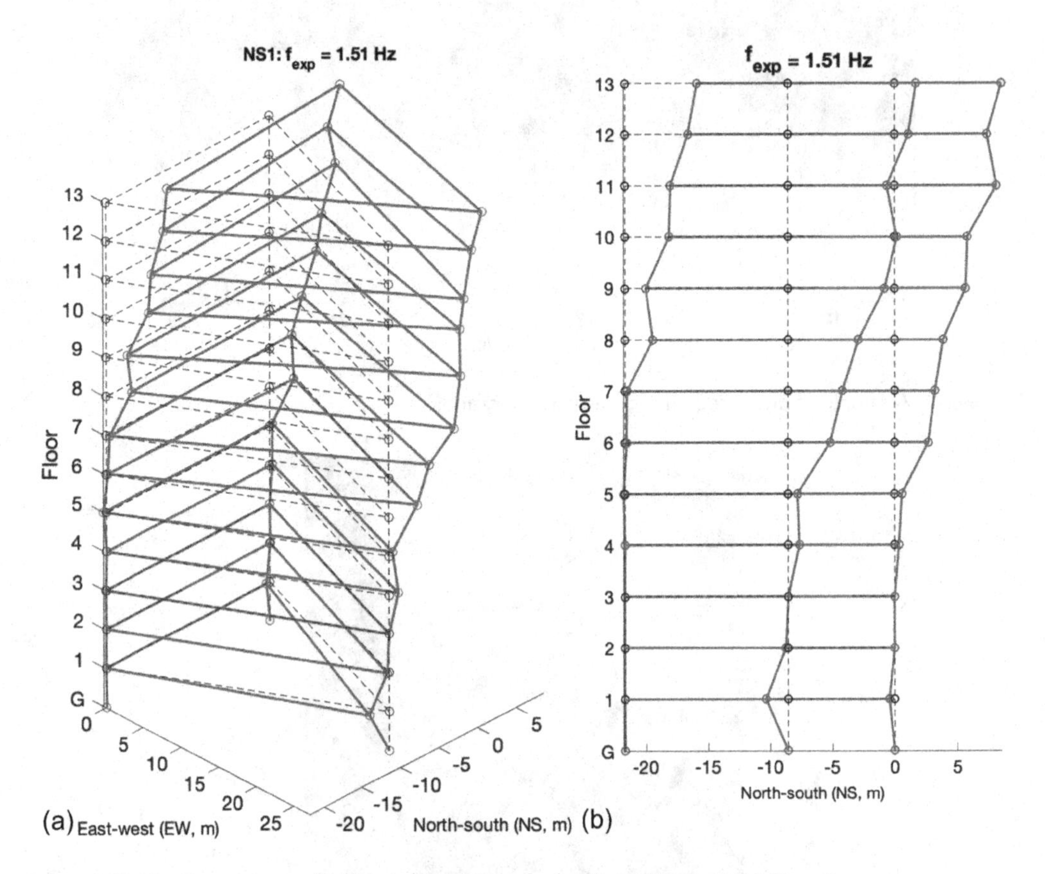

Figure 6.77 The first measured NS translational mode (NS1). (a) 3D view. (b) Elevation.

lines (to show a triangle). From the elevation view in Figure 6.77(b), it is clear that this mode is the first translational mode along the NS direction. It must be pointed out that the mode shape at the 14/F is not shown in the figure as the vertical alignment of the measurement points at 14/F is not the same as that of the measurement points at other floors.

Figure 6.78(a) shows the 3D mode shape of the second NS translational mode. It is clear from the elevation view in Figure 6.78(b) that this is a second bending mode with the nodal point at about 11/F. Figure 6.79 and Figure 6.80 show the first and second translational EW modes, respectively. The mode shapes look similar to those in Figure 6.77 and Figure 6.78, but the vibration direction is along EW instead of NS. Figure 6.81 shows the first torsional mode of the factory building. To clearly show the torsional effect, the plan view is plotted in Figure 6.81(b). To go from the bottom to the top, the mode shape rotates in a clockwise direction. Finally, the second torsional mode is shown in Figure 6.82. To go from the bottom to the top, the mode shape first rotates counterclockwise and then

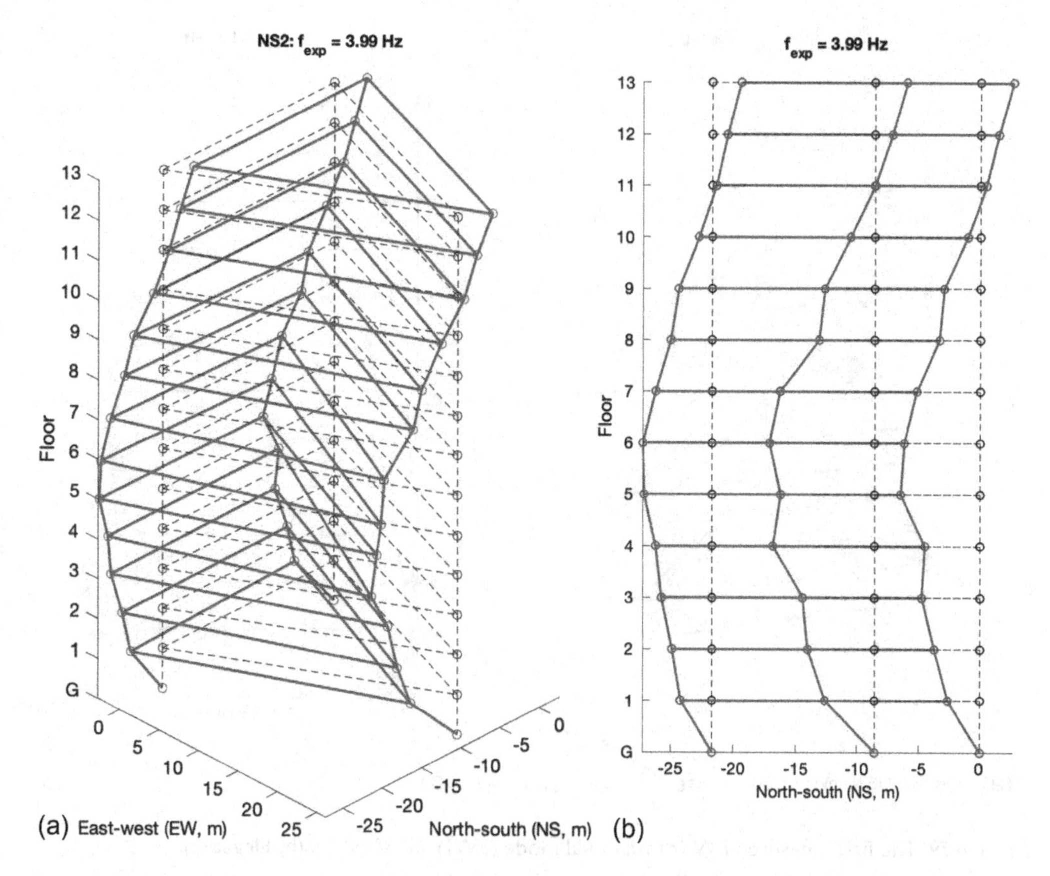

Figure 6.78 The second measured NS translational mode (NS2). (a) 3D view. (b) Elevation.

clockwise. Based on the plan view in Figure 6.81(b) and Figure 6.82(b), the center of rotation is approximately at the centroid of the triangle.

6.3.3.3 Modeling of the factory building

The factory building is modeled by three shear building models (denoted as the NS model, the EW model, and the T model), and their model updatings are carried out utilizing measured modal parameters along the NS, EW, and T directions, respectively.

It is assumed that all beam-column connections are rigid, and the nominal value of the translational inter-story stiffness about the global x-axis (i.e. along the NS direction) can be calculated as:

$$k_{0,i}^{(x)} = \sum_{j=1}^{N_{c,i}} k_{j,i}^{(x)} = \sum_{j=1}^{N_{c,i}} \frac{12EI_{j,i}^{(x)}}{h_i^3} \quad \text{for} \quad i = 1,\ldots,14 \tag{6.153}$$

where h_i is the story height at the i-th story, and is equal to 3 m for all stories in this case study; E is the modulus of elasticity of concrete, and it is assumed to be 30×10^9 GPa; $I_{j,i}^{(x)}$ is the second moment of area for bending about the x-axis of the j-th column on the i-th

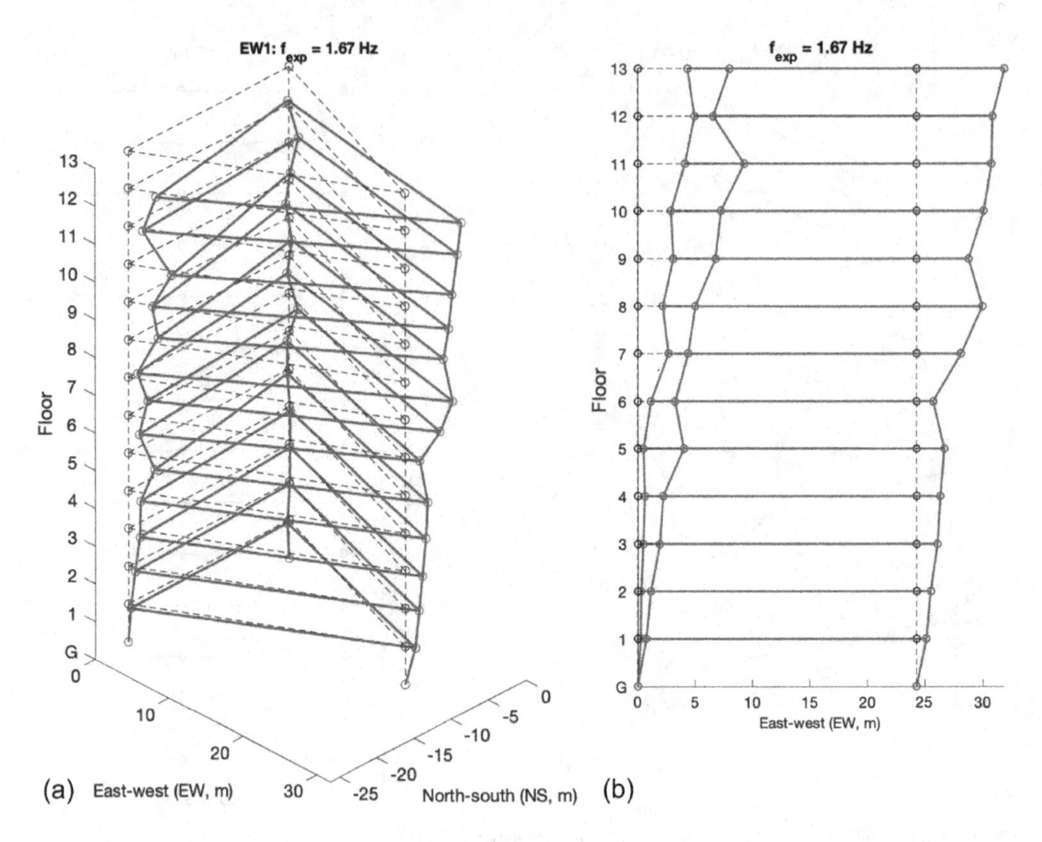

Figure 6.79 The first measured EW translational mode (EW1). (a) 3D view. (b) Elevation.

story; $N_{c,i}$ is the total number of columns on the i-th story; and $k_{j,i}^{(x)}$ represents the lateral stiffness about the x-axis of the j-th column on the i-th story. Similarly, the nominal value of the translational inter-story stiffness about the global y-axis (i.e, along the EW direction) can be expressed as:

$$k_{0,i}^{(y)} = \sum_{j=1}^{N_{c,i}} k_{j,i}^{(y)} = \sum_{j=1}^{N_{c,i}} \frac{12EI_{j,i}^{(y)}}{h_i^3} \quad \text{for} \quad i = 1,\ldots,14 \tag{6.154}$$

where $k_{j,i}^{(y)}$ and $I_{j,i}^{(y)}$ are the lateral stiffness and second moment of area for bending about the global y-axis for the j-th column on the i-th story. It must be pointed out that the values of $k_{0,i}^{(x)}$ and $k_{0,i}^{(y)}$ in this case study are the same as the cross-sections for all columns and are square with the same second moment of area for bending in both principal axes.

To calculate the inter-story torsional stiffness, the stiffness center for the i-th story $(x_{k,i}, y_{k,i})$ is required. It can be calculated as:

$$x_{k,i} = \frac{\sum_{j=1}^{N_{c,i}} x_{j,i} k_{j,i}^{(y)}}{\sum_{j=1}^{N_{c,i}} k_{j,i}^{(y)}} \quad \text{and} \quad y_{k,i} = \frac{\sum_{j=1}^{N_{c,i}} y_{j,i} k_{j,i}^{(x)}}{\sum_{j=1}^{N_{c,i}} k_{j,i}^{(x)}} \quad \text{for} \quad i = 1,\ldots,14 \tag{6.155}$$

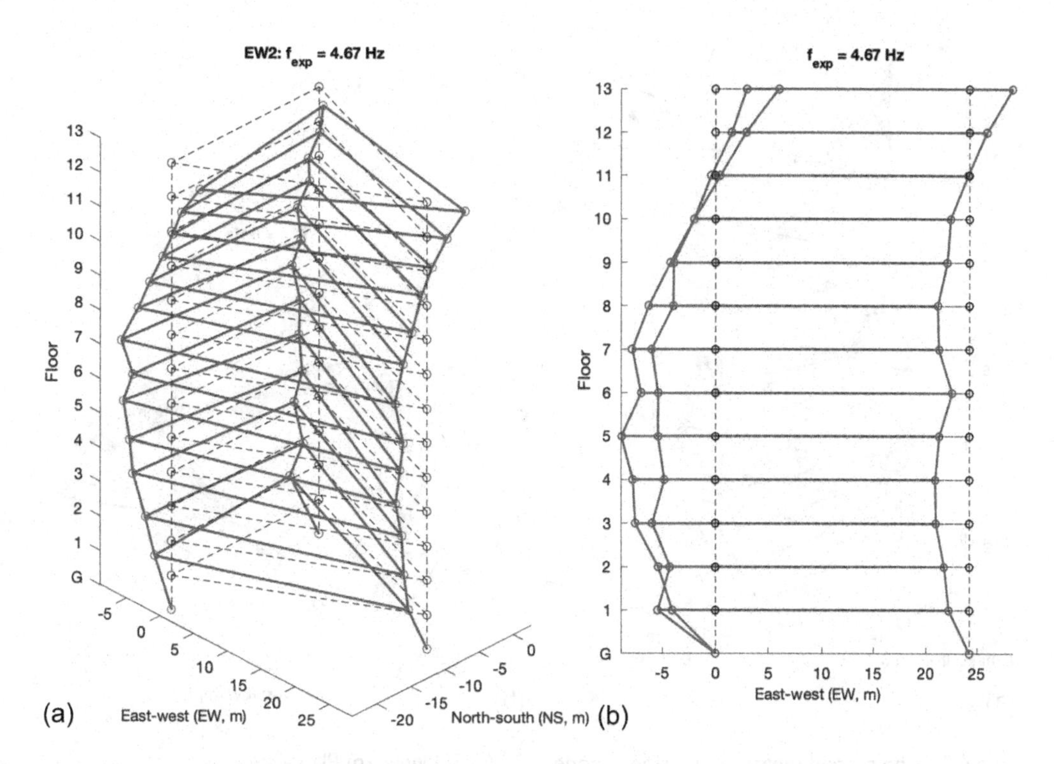

Figure 6.80 The second measured EW translational mode (EW2). (a) 3D view. (b) Elevation.

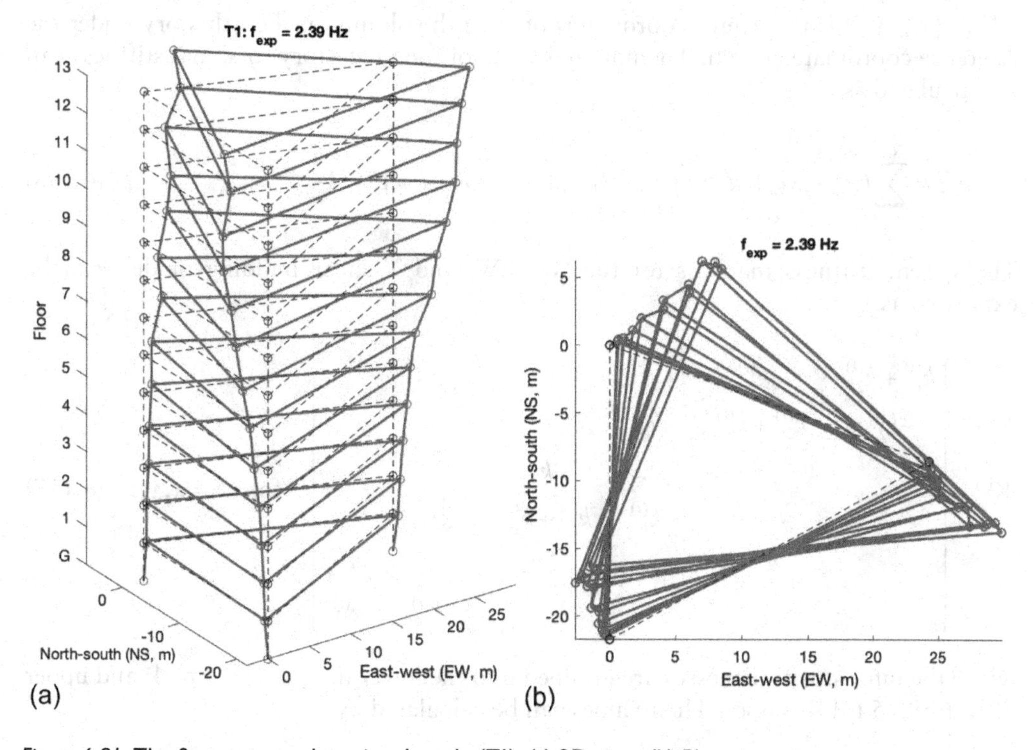

Figure 6.81 The first measured torsional mode (T1). (a) 3D view. (b) Plan view.

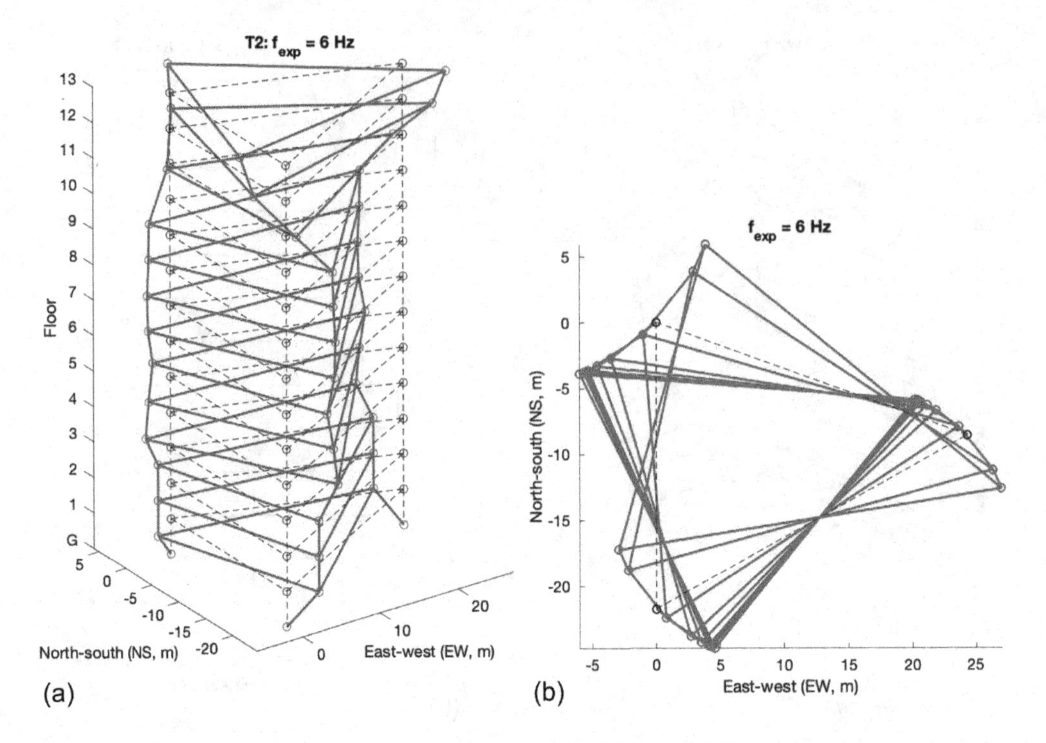

Figure 6.82 The second measured torsional mode (T2). (a) 3D view. (b) Plan view.

where $(x_{j,i}, y_{j,i})$ is the center coordinates of the j-th column on the i-th story under the reference coordinate system. The nominal value of the inter-story torsional stiffness can be calculated as:

$$k_{0,i}^{(z)} = \sum_{j=1}^{N_{c,i}} \left(x_{j,i} - x_{k,i}\right)^2 k_{j,i}^{(y)} + \left(y_{j,i} - y_{k,i}\right)^2 k_{j,i}^{(x)} \quad \text{for} \quad i = 1,\ldots,14 \tag{6.156}$$

The system stiffness matrices for the NS, EW, and T shear building models can be expressed as:

$$\mathbf{K}^{(i)} = \begin{bmatrix} k_1^{(i)} + k_2^{(i)} & -k_2^{(i)} & & & & \\ -k_2^{(i)} & k_2^{(i)} + k_3^{(i)} & \ddots & & & \\ & \ddots & \ddots & -k_j^{(i)} & & \\ & & -k_j^{(i)} & k_j^{(i)} + k_{j+1}^{(i)} & \ddots & \\ & & & \ddots & \ddots & -k_{14}^{(i)} \\ & & & & -k_{14}^{(i)} & k_{14}^{(i)} \end{bmatrix} \quad \text{for} \quad i = x, y, z \tag{6.157}$$

where the inter-story stiffnesses are grouped into the lower (i.e., stories 1 to 4) and upper (i.e., stories 5 to14) classes. Their values can be calculated by:

$$k_j^{(i)} = \theta_L^{(i)} k_{0,L}^{(i)} \quad \text{for} \quad i = x,y,z \quad \text{and} \quad j = 1,\ldots,4 \tag{6.158}$$

and

$$k_j^{(i)} = \theta_U^{(i)} k_{0,U}^{(i)} \quad \text{for} \quad i = x,y,z \quad \text{and} \quad j = 5,\ldots,14 \tag{6.159}$$

where $k_{0,L}^{(i)}$ and $k_{0,U}^{(i)}$, for $i = x,y,z$, are the nominal values of inter-story stiffness for lower and upper stories, respectively; $\theta_L^{(i)}$ and $\theta_U^{(i)}$, for $i = x,y,z$, are the non-dimensional scaling factor to be updated in the model updating process. Figure 6.83 considers the shear building model NS as an example to show the parameterization.

The system mass matrices for the NS and EW shear building models can be expressed as:

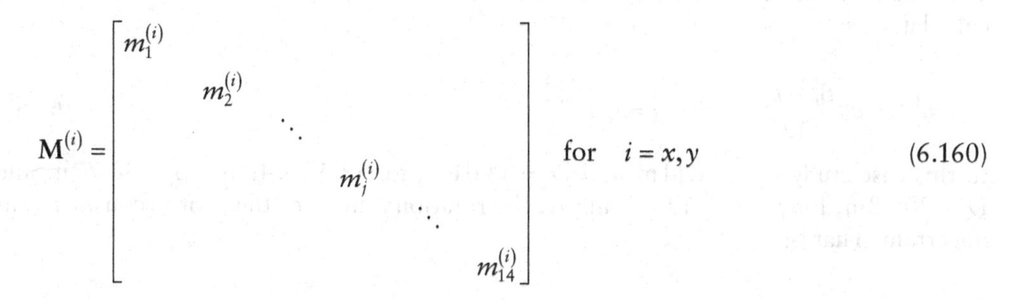

$$\mathbf{M}^{(i)} = \begin{bmatrix} m_1^{(i)} & & & & \\ & m_2^{(i)} & & & \\ & & \ddots & & \\ & & & m_j^{(i)} & \\ & & & & \ddots \\ & & & & & m_{14}^{(i)} \end{bmatrix} \quad \text{for} \quad i = x,y \tag{6.160}$$

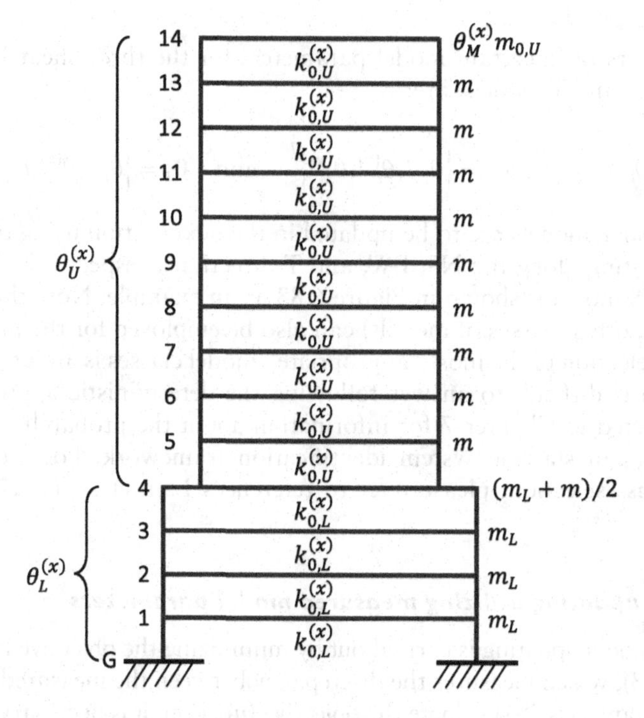

Figure 6.83 Parameterization of the shear building model NS.

where $m_j^{(i)} = m_L$, for $j = 1,\ldots,3$; $m_j^{(i)} = m$, for $j = 5,\ldots,13$; $m_j^{(i)} = (m_L + m)/2$, for $j = 4$. As the floor area is reduced at the fourth floor, the corresponding floor mass is assumed to be the average of the masses of the third and fifth floors. Owing to the relatively high uncertainty, the estimated floor mass at the roof (i.e., $m_{14}^{(i)}$) is also considered as an uncertain model parameter in model updating. That is:

$$m_{14}^{(i)} = \theta_M^{(i)} m_{0,U} \quad \text{for} \quad i = x, y \tag{6.161}$$

where $m_{0,U}$ is the nominal value of the roof mass. In this study, $m_L = 1294787\,\text{kg}$, $m = 955151\,\text{kg}$, and $m_{0,U} = 1206436\,\text{kg}$ are adopted. These values consider not only the mass of the floor slab but also the column and beam masses (assuming half of the column mass is lumped to the upper floor and another half is lumped to the lower floor). The rotational mass of the j-th rectangular floor slab with dimensions B_j and D_j can be calculated as:

$$m_j^{(z)} = m_j \frac{B_j^2 + D_j^2}{12} \quad \text{for} \quad j = 1,\ldots,13 \tag{6.162}$$

In this case study, $B_j = 38.11\,\text{m}$ and $D_j = 34.01\,\text{m}$, for $j = 1,\ldots,4$; and $B_j = 32.77\,\text{m}$ and $D_j = 29.18\,\text{m}$, for $j = 5,\ldots,14$. Similarly, the rotational mass of the roof is considered as uncertain. That is:

$$m_{14}^{(z)} = \theta_M^{(z)} m_{0,U} \frac{B_{14}^2 + D_{14}^2}{12} \tag{6.163}$$

Finally, the vectors of uncertain model parameters for the three shear building models (i.e., the NS, WE, and T models) are:

$$\boldsymbol{\theta}_{\text{NS}} = \left\{\theta_L^{(x)}, \theta_U^{(x)}, \theta_M^{(x)}\right\}^T, \boldsymbol{\theta}_{\text{EW}} = \left\{\theta_L^{(y)}, \theta_U^{(y)}, \theta_M^{(y)}\right\}^T \quad \text{and} \quad \boldsymbol{\theta}_{\text{T}} = \left\{\theta_L^{(z)}, \theta_U^{(z)}, \theta_M^{(z)}\right\}^T \tag{6.164}$$

The NS, EW, and T models are to be updated in the next section using measured modal parameters vibrating along the NS, EW, and T directions, respectively. The parameterization of the NS model is shown in Figure 6.83 as an example. Note that other parameterizations (i.e., other classes of models) can also be employed for the purpose of model updating. The selection of the most "appropriate" model classes is an interesting research question, which is difficult to answer following the deterministic approach. Interested readers are directed to Chapter 7 for information about the probabilistic approach following the Bayesian statistic system identification framework. For more information about model class selection, please refer to references Lam et al. (2017) and Adeagbo et al. (2021).

6.3.3.4 Model updating utilizing measured modal parameters

In this section, model updating is carried out by minimizing the objective function as given in Equation (6.13), which measures the discrepancy between the measured and model-predicted modal parameters. To evaluate the objective function, it is necessary to calculate the natural frequencies and mode shapes of the shear building model for a given set of model parameters as defined in Equation (6.164). Three MATLAB functions are developed to calculate the modal parameters of the NS, EW, and T models. As an example, the MATLAB function for the NS model `factory_buildingX()` is shown in Program 6.11.

The function is defined on Line 1. It gets the vector of uncertain parameter (`theta`) as input and gives the natural frequencies (`nfx`) and mode shapes (`msx`) as output. On Line 2, the self-developed function `properties_cal()` is called to provide the numerical values of the structural parameters including `mL`, `m` and `m0U` as defined in Equations (6.160) and (6.161). The variables `fR` and `fRL` are the factors for converting the floor mass to rotational mass. That is $\left(B_j^2 + D_j^2\right)/2$ in Equation (6.162). fR is for stories 5 to 14 (i.e., $j = 5,\ldots,14$), while `fRL` is for stories 1 to 4 (i.e., $j = 1,\ldots,4$). The variables `k0x` and `k0xL` are the nominal value of the inter-story stiffness (for bending about the x-axis) as given in Equation (6.153). `k0x` is for stories 5 to 14, and `k0xL` is for stories 1 to 4. `k0y` and `k0yL` are the nominal inter-story stiffness for bending about the y-axis, and they are not used in this function (the `properties_cal()` function is used in both NS and EW models). The inter-story stiffness is calculated on Line 3 by multiplying the nominal value to the uncertain scaling parameters (i.e., `theta` as defined in Equation (6.164)). The system stiffness matrix for the NS model (i.e., `Kx`) is formed by the sub-function `myK-cal()` on Line 2. Note that the function `myKcal()` is defined on Lines 23 to 32 (after the main function `factory_buildingX()`). Line 5 implements Equation (6.161) to calculate the floor mass at story 14 (i.e., `mU`). The system mass matrix (`Mx`) is formed on Lines 6 and 7. With the system stiffness and mass matrices, the natural frequencies and mode shapes are calculated by the sub-function `myeign()`, which is defined on Lines 33 to 38 (at the end of the program). As only the first two modes are needed, all modes higher than 2 are discarded from `nfx` and `ms` on Line 9. The mode shapes in `ms` consist of 14 DOFs (corresponding to the lateral vibration for floors 1 to 14), and it is converted to the arrangement of the measured mode shapes (`msx`) on Lines 10 to 17. The measured mode shape consists of 84 DOFs corresponding to the lateral vibration for floors 13 to G along the global x-axis and y-axis for the three staircases (i.e., $14 \times 2 \times 3$). For example, the measured mode shape of a given mode ($\boldsymbol{\Psi}$) can be expressed as:

$$\Psi = \left\{\varphi^{(A)}, \varphi^{(B)}, \varphi^{(C)}\right\}^T \tag{6.165}$$

where $\varphi^{(i)}$, for $i = A,B,C$, are the row vectors of measured mode shapes for Staircases A, B, and C, respectively, and they are defined as:

$$\varphi^{(i)} = \left\{\varphi_{x,13}^{(i)}, \varphi_{y,13}^{(i)}, \varphi_{x,12}^{(i)}, \varphi_{y,12}^{(i)}, \ldots, \varphi_{x,1}^{(i)}, \varphi_{y,1}^{(i)}, \varphi_{x,G}^{(i)}, \varphi_{y,G}^{(i)}\right\} \quad \text{for} \quad i = A,B,C \tag{6.166}$$

where $\varphi_{x,j}^{(i)}$ and $\varphi_{y,j}^{(i)}$ are the mode shape value along the global x and y-axis, respectively, on the j-th floor in the i-th staircase. Finally, the mode shapes in `msx` are normalized to have unit length on Lines 18 to 21. The MATLAB functions `factory_buildingY()` (for EW model) and `factory_buliidingZ()` (for T model) are in a similar format, and they are not repeated here.

Program 6.11: `factory_buildingX()` for calculating the modal parameters of the NS model.

```
Line 1.    function [nfx,msx]=factory_buildingX(theta)
Line 2.    [mL,m,m0U,fR,fRL,k0x,k0y,k0xL,k0yL]=properties_cal;
Line 3.    kx(1:4)=k0xL*theta(1); kx(5:14)=k0x*theta(2);
```

```
Line 4.    Kx=myKcal(kx);
Line 5.    mU=m0U*theta(3);
Line 6.    Vm=[mL mL mL (mL+m)/2 m m m m m m m m m mU];
Line 7.    Mx=diag(Vm);
Line 8.    [nfx,ms]=myeign(Kx,Mx);
Line 9.    nfx=nfx(1:2); ms=ms(:,1:2);
Line 10.   msx=zeros(84,2);
Line 11.   c=1;
Line 12.   for ii=1:13
Line 13.       msx(c,:)=ms(14-ii,:);
Line 14.       msx(c+28,:)=ms(14-ii,:);
Line 15.     msx(c+2*28,:)=ms(14-ii,:);
Line 16.     c=c+2;
Line 17.   end
Line 18.   mm=msx'*msx;
Line 19.   for ii=1:2
Line 20.     msx(:,ii)=msx(:,ii)./mm(ii,ii).^0.5;
Line 21.   end
Line 22.   end
Line 23.   function K=myKcal(k)
Line 24.   K=zeros(length(k));
Line 25.   K(1,1)=k(1);
Line 26.   for ii=2:length(k)
Line 27.       K(ii-1,ii-1)=K(ii-1,ii-1)+k(ii);
Line 28.       K(ii-1,ii)=K(ii-1,ii)-k(ii);
Line 29.       K(ii,ii-1)=K(ii-1,ii);
Line 30.     K(ii,ii)=K(ii,ii)+k(ii);
Line 31.   end
Line 32.   end
Line 33.   function [nf,ms]=myeign(K,M)
Line 34.   [V,D]=eig(inv(M)*K);
Line 35.   [d,indx]=sort(diag(D));
Line 36.   nf=d.^0.5/2/pi;
Line 37.   ms=V(:,indx);
Line 38.   end
```

For the T model, the mode shape obtained from solving the eigenvalue problem consists of 14 DOFs (corresponding to the rotation at the stiffness center on floors 1 to 14). It must be transformed to the displacement mode shape along the global x- and y-axes for Staircases A, B, and C. Figure 6.84 shows a typical floor with the locations of stiffness center together with the three staircases. Assuming that the rotational mode shape at a particular floor is β (a very small angle in radians), the corresponding displacement at the three staircases can be calculated based on the rigid floor diaphragm assumption as:

$$V_i = R_i\beta \quad \text{for} \quad i = A, B, C \tag{6.167}$$

where R_A, R_B, and R_C are the distances from the stiffness center to Staircases A, B, and C, respectively. Note that the displacements V_A, V_B, and V_C are perpendicular to the lines from the stiffness center as shown in Figure 6.84. With the coordinates of the stiffness center and the three staircases, the angles α_A, α_B, and α_C can be calculated. The displacement components along the global x- and y-axes can be determined as:

$$V_{i,x} = -R_i\beta\sin\alpha_i \quad \text{and} \quad V_{i,y} = R_i\beta\cos\alpha_i \quad \text{for} \quad i = A, B, C \tag{6.168}$$

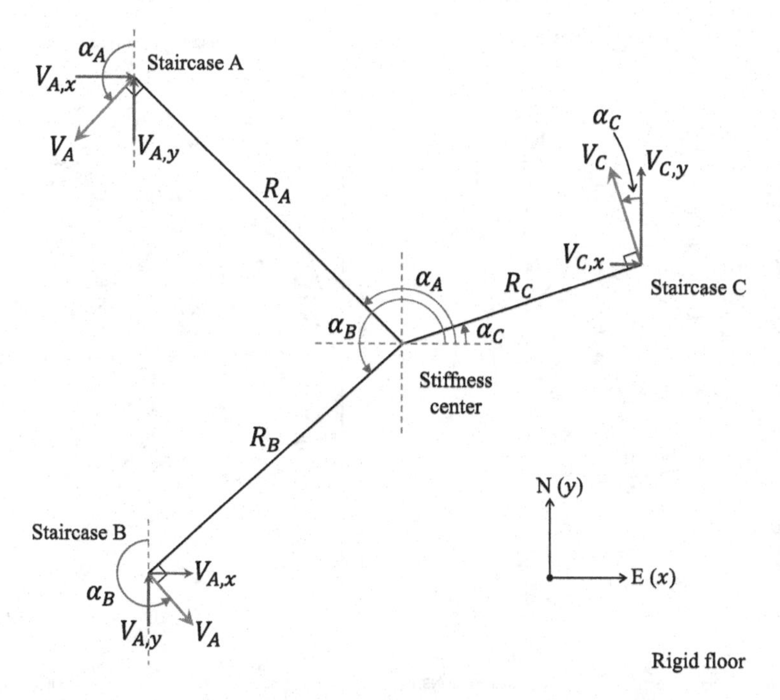

Figure 6.84 Converting the rotation at stiffness center to the displacement at the three staircases.

Table 6.24 Model Updating Results of the Three Shear Building Models

Model	*J function*	θ_L	θ_U	θ_M
NS	0.4582	1.2789	0.7762	2.1380
EW	0.5187	1.4464	1.1282	2.2627
T	0.4208	3.1321	1.2612	1.7775

With the transformation, the model-predicted rotational mode shapes (from the T model) can be directly compared to the measured ones as shown in Figures 6.81 and 6.82.

By minimizing the discrepancy between the measured and model-predicted natural frequencies and mode shapes, model updating of the NS, EW, and T models is carried out. The results are summarized in Table 6.24. Consider the model updating results of the NS model. The identified inter-story stiffness for the lower stories (~1.28) is about 28% higher than the nominal value, while that for the upper stories (~0.78) is about 22% lower than the nominal value. The identified roof mass is two times that of the nominal value. It can be concluded that the mass estimation (i.e., the nominal) value is seriously underestimated. For the EW model, the identified lower and upper floor stiffness values are both higher than the nominal values. The estimated roof mass is consistent with the one identified by the NS model (also about two times). The nature of the T model is different from that of the NS and EW models, as the identified inter-story rotational stiffness for lower stories is much higher than the nominal value, and that for upper stories is just a little higher than the nominal value.

The updated models are used to calculate the modal parameters of the system, and the updated modal parameters of the NS, EW, and T models are plotted in Figure 6.85,

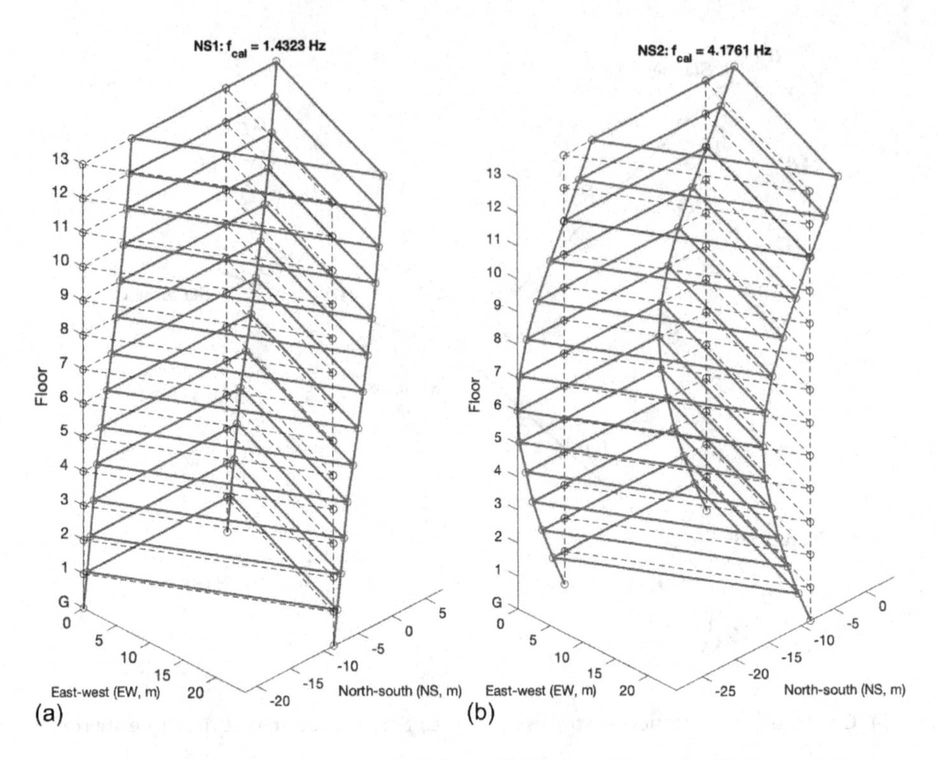

Figure 6.85 Updated modal parameters of the NS model. (a) First NS mode. (b) Second NS mode.

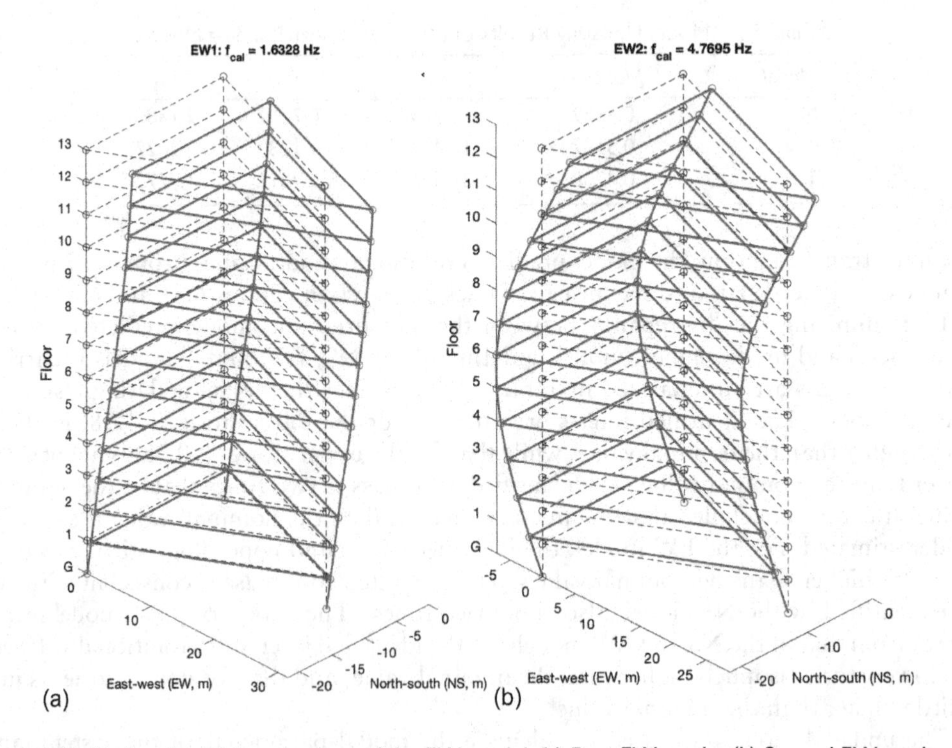

Figure 6.86 Updated modal parameters of the EW model. (a) First EW mode. (b) Second EW mode.

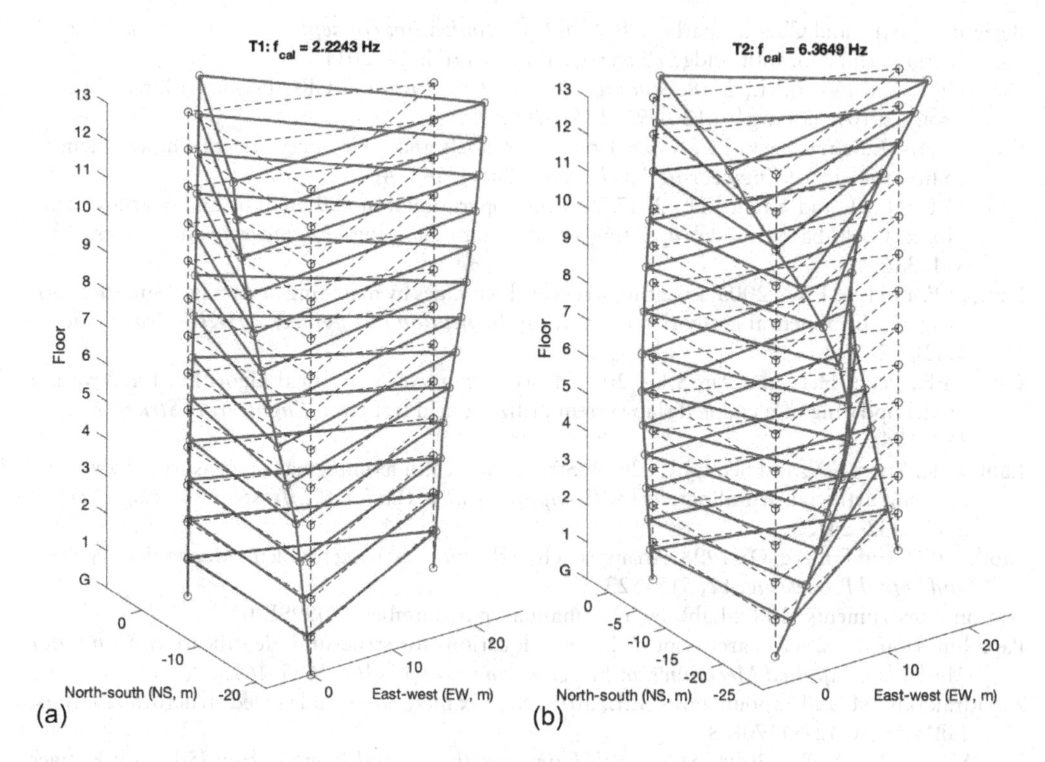

Figure 6.87 Updated modal parameters of the T model. (a) First T mode. (b) Second T mode.

Figure 6.86, and Figure 6.87, respectively. When the updated mode shapes are compared to the measured ones in Figure 6.77 to Figure 6.82, they are reasonably close to each other. Of course, the model-predicted mode shapes are much smoother when compared to the measured ones.

The series of case studies are arranged from indoor laboratory test to field test (i.e., from low to high level of measurement noise) and from simple to complex structural model (i.e., from low to high degree of modeling error). Referring to the matchings between the measured and the model-predicted responses from the updated model, it is believed that the uncertainties associated with the model updating results become higher and higher. All case studies in this chapter follow the deterministic approach in model updating through numerically minimizing the discrepancy between the measured and model-predicted modal parameters. This approach is relatively simple in formulation but the model updating results may be misleading when the associated uncertainty is high. The best approach to address the uncertainty problem in model updating is to follow the probabilistic approach, and it will be introduced in the next chapter.

REFERENCES

Adeagbo, M.O., Lam, H.F. and Hu, Q., 2021. On the selection of the most plausible non-linear axial stress-strain model for railway ballast under different impulse magnitudes. *Structural Health Monitoring*. https://doi.org/10.1177/14759217211033968.

Beck, J.L. and Katafygiotis, L.S., 1998. Updating models and their uncertainties – Bayesian statistical framework. *Journal of Engineering Mechanics*, 124(4), 455–461.

Belegundu, A.D. and Chandrupatla, T.R., 2011. *Optimization concepts and applications in engineering* (2nd ed.). Cambridge University Press, March 28, 2011.

Chen, W.F. and Lui, E.M., 2018. *Stability design of steel frames*. CRC Press, Taylor & Francis Group, https://doi.org/10.1201/9781351076852.

Katafygiotis, L.S. and Beck, J.L., 1998. Updating models and their uncertainties – model identifiability. *Journal of Engineering Mechanics*, 124(4), 463–467.

Lam, H.F., Hu, J. and Yang, J.H., 2017. Bayesian operational modal analysis and Markov chain Monte Carlo-based model updating of a factory building. *Engineering Structures*, 132, 314–336.

Lam, H.F. and Lee, E.W., 2005. Locating structural damages by matching of damage signatures utilizing artificial neural networks. *Hong Kong Institution of Engineers (HKIE) Transactions*, 12(2), 22–29.

Lam, H.F., Peng, H.Y. and Au, S.K., 2014. Development of a practical algorithm for Bayesian model updating of a coupled slab system utilizing field test data. *Engineering Structures*, 79, 182–194.

Lam, H.F., Yuen, K.V. and Beck, J.L., 2006. Structural health monitoring via measured ritz vectors utilizing artificial neural networks. *Computer-Aided Civil and Infrastructure Engineering*, 21(4), 232–241.

Moller, P.W. and Friberg, O., 1998. An approach to the mode pairing problem. *Mechanical Systems and Signal Processing*, 12, 515–523.

National Instruments, 2003. LabView user manual, part number 320999E-01.

Papadimitriou, C., 2005. Pareto optimal sensor locations for structural identification. *Computer Methods in Applied Mechanics and Engineering*, 194(12–16), 1655–1673.

Papadrakakis, M. and Sapountzakis, E.J., 2018. Matrix methods for advanced structural analysis, ISBN 978-0-12-811708-8.

Paz, M. and Leigh, W., 2004. *Structural dynamics, theory and computation* (5th ed.). Kluwer Academic Publisher.

Vanik, M.W., Beck, J.L. and Au, S.K., 2000. Bayesian probabilistic approach to structural health monitoring. *Journal of Engineering Mechanics*, 126(7), 738–745.

Yuen, K.V. and Lam, H.F., 2006. On the complexity of artificial neural networks for smart structures monitoring. *Engineering Structures*, 28(7), 977–984.

Chapter 7

Bayesian model updating based on Markov chain Monte Carlo

7.1 DETERMINISTIC MODEL BASED ON THE EIGENVALUE PROBLEM

As shown in Chapter 6, there are mainly two steps for formulating Bayesian model updating, namely, deterministic modeling and stochastic embedding. A deterministic model is first developed in this section, and a probabilistic model is then chosen for the error of the deterministic model, i.e., stochastic embedding. The mathematical model is the finite element model. Consider the error of the eigenvalue problem equation:

$$\mathbf{r}_m = \left(\mathbf{K}(\boldsymbol{\theta}) - \omega_m^2 \mathbf{M}\right)\boldsymbol{\phi}_m \tag{7.1}$$

where $\mathbf{r}_m \in R^{N_d}$ denotes the error of the m-th mode; N_d is the number of DOFs of the finite element model; $\mathbf{K}(\boldsymbol{\theta}) \in R^{N_d \times N_d}$ is the stiffness matrix whose dependence on uncertain model parameters $\boldsymbol{\theta} \in R^{N_p}$ is emphasized; $\mathbf{M} \in R^{N_d \times N_d}$ is the mass matrix, which is assumed to be known; and ω_m and $\boldsymbol{\phi}_m \in R^{N_d}$ are the actual natural frequency and mode shape of the m-th mode, respectively. They are unknown and different from the ones of the finite element model, which is the reason why the error $\mathbf{r}_m$ exists. The eigenvalue problem is considered because excitations of full-scale structures are almost impossible or very costly to measure, and measured responses are not convenient to be used as data in model updating. Experimental modal parameters, i.e., natural frequencies and mode shapes, are usually used as data. The objective function to be minimized is thus written by considering the errors of N_m modes:

$$J(\boldsymbol{\alpha}) = \sum_{m=1}^{N_m}\left(\mathbf{r}_m^T\mathbf{r}_m + \left(\omega_m - \hat{\omega}_m\right)^2 + \left\|\mathbf{L}\boldsymbol{\phi}_m - \hat{\boldsymbol{\varphi}}_m\right\|^2\right) \tag{7.2}$$

where apart from the model parameters $\boldsymbol{\theta}$ (related to stiffness), the actual natural frequencies and mode shapes $\{\omega_m, \boldsymbol{\phi}_m\}$ are also treated as uncertain parameters to be identified, so the uncertain parameter vector $\boldsymbol{\alpha}$ contains both $\boldsymbol{\theta}$ and $\{\omega_m, \boldsymbol{\phi}_m\}$; $\hat{\omega}_m$ and $\hat{\boldsymbol{\varphi}}_m \in R^{N_o}$ are the experimental natural frequencies and mode shapes, respectively, and N_o is the number of measured DOFs ($N_o < N_d$); the normalization of mode shapes will be discussed later; $\mathbf{L}$ is the selection matrix that selects the mode shapes at the N_o measured DOFs corresponding to $\hat{\boldsymbol{\varphi}}_m$, and it contains only 0s and 1s; and $\|\;\|$ denotes the Euclidean norm. Note that the last two terms in Equation (7.2) work as the constrain that the experimental natural frequencies and mode shapes are the actual ones of the target structure.

DOI: 10.1201/9780429445866-7

The advantage of the above deterministic model is that it is not required to solve the eigenvalue problem of the large-scale finite element model. However, the direct optimization of the high-dimensional global mode shape $\boldsymbol{\phi}_m$ (compared to the experimental one $\hat{\boldsymbol{\varphi}}_m$ that is only the partial observation of $\boldsymbol{\phi}_m$) is almost impossible. An efficient method is proposed to solve this problem. The basic idea is to express the global mode shape given other uncertain parameters and the finite element model. Assuming that the model parameters are known $\boldsymbol{\theta} = \breve{\boldsymbol{\theta}}$, taking the actual natural frequency to be the experimental one $\omega_m = \hat{\omega}_m$ and taking the eigenvalue problem equation error to be zero, the following system of linear equations is obtained:

$$\left(\mathbf{K}\left(\breve{\boldsymbol{\theta}}\right) - \hat{\omega}_m^2\mathbf{M}\right)\boldsymbol{\phi}_m = \mathbf{B}\boldsymbol{\phi}_m = 0 \tag{7.3}$$

where $\mathbf{B} = \left(\mathbf{K}\left(\breve{\boldsymbol{\theta}}\right) - \hat{\omega}_m^2\mathbf{M}\right)\boldsymbol{\phi}_m$. To normalize $\boldsymbol{\phi}_m$, a reference DOF r is selected and then the mode shape at DOF r is normalized to be 1. The DOF numbering corresponds to the global mode shapes. This reference DOF should also be contained in the measured DOFs, so that the experimental mode shapes can have the same normalization as the global ones. Rearranging $\boldsymbol{\phi}_m$ into two blocks with one block corresponding to the reference DOF and the other corresponding to the non-reference DOFs, then rearranging $\mathbf{B}$ into a block matrix accordingly gives

$$\begin{bmatrix} b_{rr} & \mathbf{B}_{rb} \\ \mathbf{B}_{br} & \mathbf{B}_{bb} \end{bmatrix}\begin{bmatrix} \phi_{r,m} \\ \boldsymbol{\phi}_{b,m} \end{bmatrix} = \begin{bmatrix} b_{rr} & \mathbf{B}_{rb} \\ \mathbf{B}_{br} & \mathbf{B}_{bb} \end{bmatrix}\begin{bmatrix} 1 \\ \boldsymbol{\phi}_{b,m} \end{bmatrix} = 0 \tag{7.4}$$

where $b_{rr} \in R$, $\mathbf{B}_{rb} \in R^{1\times(N_d-1)}$, $\mathbf{B}_{br} \in R^{(N_d-1)\times 1}$, $\mathbf{B}_{bb} \in R^{(N_d-1)\times(N_d-1)}$, and $\boldsymbol{\phi}_{b,m} \in R^{(N_d-1)}$; $\phi_{r,m}$ has been taken to be 1 following the normalization scheme. The second block line of the above equation will give the mode shapes at all the non-reference DOFs:

$$\mathbf{B}_{br} + \mathbf{B}_{bb}\boldsymbol{\phi}_{b,m} = 0 \tag{7.5}$$

$$\boldsymbol{\phi}_{b,m} = -\mathbf{B}_{bb}^{-1}\mathbf{B}_{br} \tag{7.6}$$

Based on a given value of the model parameters $\breve{\boldsymbol{\theta}}$, the finite element model, and the experimental natural frequencies, the global mode shapes can be obtained. However, it is not recommended to use Equation (7.6) to calculate the global mode shapes because the size of $\mathbf{B}_{bb}$ is large for full-scale structures and taking its inverse directly is not efficient. Instead, one should solve the system of linear equations Equation (7.5). Efficient methods such as the LU decomposition method, least-squares method, and conjugate gradients squared method are available to do so.

The calculated global mode shapes can then be used to update the natural frequencies. The way of calculating the natural frequencies is similar to the one for the mode shapes. We want to make use of the experimental mode shapes. To do this, note that both the calculated global mode shapes and the experimental ones are normalized such that they have the "1" element at the same DOF; and for the optimal case, the experimental mode shapes are just selected from a particular part of the global mode shapes. Therefore, $\hat{\boldsymbol{\varphi}}_m$ is substituted into $\boldsymbol{\phi}_m$ for the corresponding measured DOFs, while the mode shape values at the unmeasured DOFs of $\boldsymbol{\phi}_m$ are still kept as the ones calculated by Equation (7.5). Taking the eigenvalue problem equation error to be zero and denoting the global mode shape after substitution by $\breve{\phi}_m$ gives

$$\mathbf{K}\left(\breve{\theta}\right)\breve{\phi}_m = \omega_m^2\mathbf{M}\breve{\phi}_m \tag{7.7}$$

Pre-multiplying $\breve{\phi}_m^T$ at both sides of the above equation gives the updated natural frequency

$$\omega_m = \sqrt{\frac{\breve{\phi}_m^T\mathbf{K}\left(\breve{\theta}\right)\breve{\phi}_m}{\breve{\phi}_m^T\mathbf{M}\breve{\phi}_m}} \tag{7.8}$$

The calculated ω_m and $\breve{\phi}_m$ are substituted back into Equation (7.1) and then Equation (7.2) to calculate the objective function error.

The above deterministic model also includes the actual natural frequencies and mode shapes of the target structure related to the experimental ones as uncertain parameters, but they can be conveniently calculated based on the stiffness and mass matrices and the experimental modal parameters, so neither direct optimization of the actual modal parameters nor the eigenvalue problem of the whole finite element model is required. Efficiency is ensured. This deterministic model can thus be integrated with any optimization algorithm for updating.

7.2 DETERMINISTIC MODEL UPDATING

It is known from the previous section that if the finite element model of the target structure is built and given experimental modal data and a particular model parameter vector $\breve{\theta}$, the objective function can be evaluated. By using any optimization algorithm to repeatedly generate samples of $\breve{\theta}$, the objective function can then be continuously evaluated until convergence, therefore model updating is achieved. An optimization algorithm usually needs an initial point to start. To avoid the updated model being trapped at a local optimum, it is proposed that multiple initial points are randomly generated, and then the optimization algorithm starts at each of the generated points to minimize the objective function. Among all the optimized models, the one with the smallest objective function value is taken to be the optimal model. The proposed deterministic model updating method thus contains two layers. The outside layer generates all the initial uncertain parameter vectors which the optimization algorithm of the inside layer uses as the starting points. The inside layer optimizes at each starting point based on the proposed deterministic model. The proposed model updating method is summarized in the following algorithm.

Algorithm 7.1: Deterministic model updating based on the error of the eigenvalue problem equation

Outside layer

1. Set the number of initial parameter vectors N_p. Set the range for each component of $\boldsymbol{\theta}$, $[r, r]$, within which each component of $\boldsymbol{\theta}$ will be generated with a uniform distribution. N_a initial vectors of $\boldsymbol{\theta}$ are then generated within $[r, r]$ with a uniform distribution.

2. Construct the mass matrix $\mathbf{M}$.
3. For each initial vector $\boldsymbol{\theta}_i$, $i = 1,2,\cdots,N_a$, do the minimization with the Nelder-Mead optimization algorithm in the Inside layer (e.g., the MATLAB function *fminsearch* can be used). Among the N_a optimizations, the one with the smallest objective function value gives the optimal model.

Inside layer

A. Starting at $\boldsymbol{\theta}_i$, set $\breve{\theta} = \theta_i$.
B. Construct $\mathbf{K}\left(\breve{\theta}\right)$, and then form $\mathbf{B}$ with $\hat{\omega}_m$ and $\mathbf{M}$.
C. Choose a reference DOF d_m from the measured DOFs for the m-th mode (different reference DOFs can be chosen for different modes). Rearrange $\mathbf{B}$ and ϕ_m according to Equation (7.4). Solve Equation (7.5) to get $\phi_{b,m}$, and thus ϕ_m.
D. Normalize $\hat{\varphi}_m$ such that its component at DOF d_m is unity, and replace the part of ϕ_m that is related to the measured DOFs with the normalized $\hat{\varphi}_m$ to obtain $\breve{\phi}_m$.
E. Calculate the updated natural frequency ω_m using Equation (7.8).
F. Substitute $\mathbf{K}\left(\breve{\theta}\right)$, ω_m, and ϕ_m into Equation (7.2) to calculate the error for the m-th mode.
G. Repeat steps B to F for the N_m modes to obtain the objective function value $J(\boldsymbol{\alpha})$ at $\breve{\theta}$.
H. Use the optimization algorithm to update $\breve{\theta}$ and repeat steps B to G until convergence.

7.2.1 Model updating of a scaled transmission tower

A scaled transmission tower (see Figure 7.1) was employed to illustrate the proposed deterministic model updating method based on the eigenvalue problem. It was studied by the authors (Lam & Yang 2015; Yang & Lam 2018). This tower has eight levels. It consists of four main columns, beams, braces, and cross-arm members at Levels 6 and 8. The columns, beams, and cross-arm members are made of steel and are welded to connect at each joint. The braces are made of aluminum. Steel plates are welded at each beam-column joint, and the braces are bolted at these steel plates. Figure 7.2 shows the front view, the side view, and the floor plans of this tower, and the dimensions can be seen.

A dynamic test was conducted on one column of the tower to get natural frequencies and mode shapes of this column for model updating. This test setup simulates the practical situation where only partial mode shapes can be obtained (in this case, the mode shapes of one column), but the system mode shapes are not available. The detailed sensor setup is shown in Figure 7.3. Dynamic responses along x- and y-directions were to be measured, so that the translational and torsional modes could be identified. There were only nine uniaxial accelerometers at the time of the test, so the test was divided into two setups. In Setup 1 (see the picture on the left of Figure 7.3), Sensors 1 to 8 were installed on the target column along the x-direction. The arrows indicate the positive direction of the sensors. In Setup 2 (see the picture on the right of Figure 7.3), Sensors 1 to 8 were installed along the y-direction. A reference sensor is needed to combine the mode shapes of these two setups. Because the sensors in the two setups were installed along two perpendicular directions, the reference sensor was installed such that the angle between the sensor direction and the x-direction is about 45° (see the arrows in Figure 7.4 for the direction of the reference sensor), so that the reference sensor contained vibration along both x- and y-directions for combining mode shapes of Setups 1 and 2. Figure 7.4(a)

Figure 7.1 The scaled transmission tower.

shows how the sensors were installed along the x- and y-directions. The positive directions of the sensors are indicated by the arrows. Figure 7.4(b) shows the reference sensor installed at the top level together with a sensor installed along the y-direction.

In the dynamic test, acceleration responses of the target column were first measured for Setup 1 and then for Setup 2. The natural frequencies and mode shapes were identified from the measured data of Setups 1 and 2, respectively. The mode shapes identified from the two setups were assembled into the global mode shapes using the reference sensor. The average was taken for the natural frequencies of the same mode in the two setups to produce the natural frequency for model updating. The details of modal analysis of the tower column are not included, as the focus here is model updating. Five modes were identified for model updating. They are the first translational modes along the y- and x-directions, respectively, a torsional mode, and the second translational modes along the y- and x-directions, respectively (see the model updating results in Figure 7.6 later).

Model updating was conducted for the transmission tower using the proposed Algorithm 7.1. In this case, the uncertain parameters include three structural parameters, the natural frequencies, and global mode shapes of the five identified modes (from the measured accelerations). The assignment of the three structural parameters in $\boldsymbol{\theta}$ is that $\boldsymbol{\theta}(1)$ is used to scale the stiffness of levels 1 to 3 of the tower, $\boldsymbol{\theta}(2)$ is used to scale the stiffness of levels 4 to 5 of the tower, and $\boldsymbol{\theta}(3)$ is used for levels 6 to 8. Note that $\boldsymbol{\theta}$ is a scaling coefficient to scale stiffness values. Multiple initial $\boldsymbol{\theta}$s were generated within the range [0, 2] to use as starting points of the deterministic optimization. Among these optimizations,

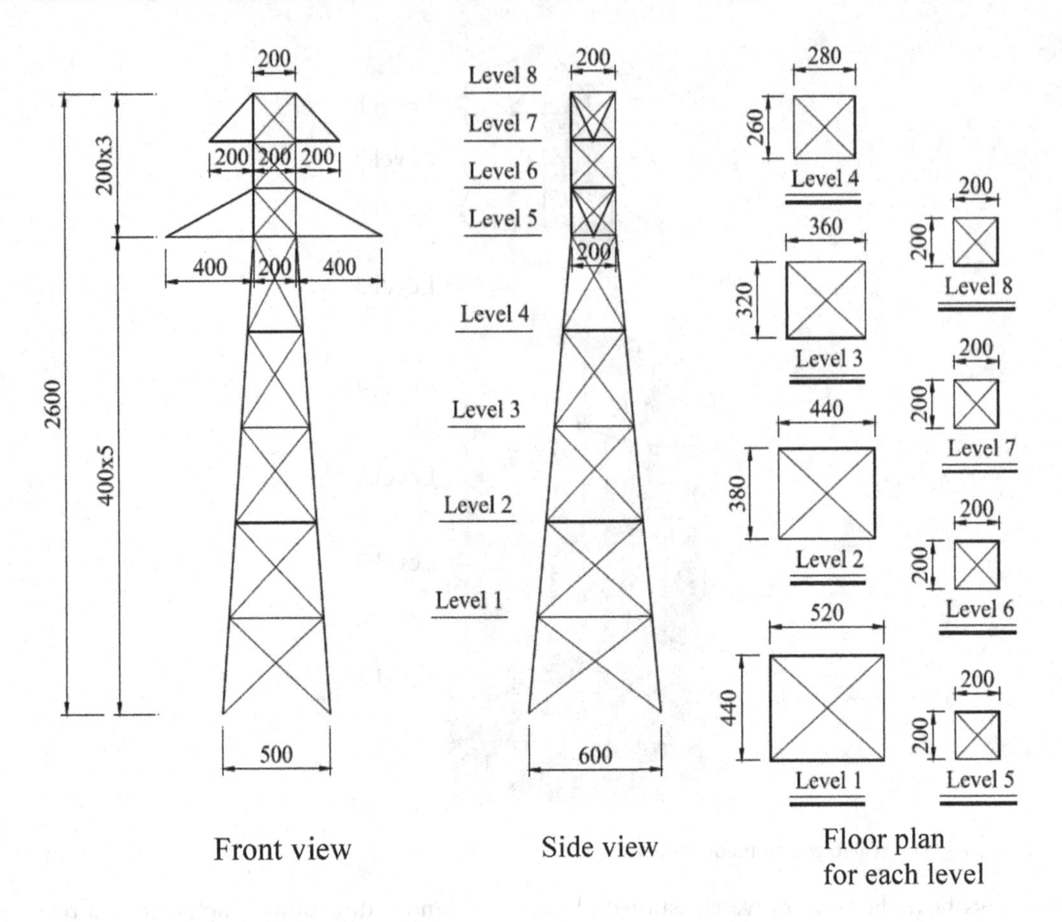

Figure 7.2 Dimensions of the transmission tower.

27 different optima were obtained and shown in Figure 7.5. The vertical axis corresponds to the normalized objective function value of each optimum, which is normalized such that all the objective function values of these optima sum to 1. The horizontal axes correspond to $\boldsymbol{\theta}$. It can be seen that starting from different points, $\boldsymbol{\theta}$ may arrive at different local optima. The optimal model is obtained with the $\hat{\boldsymbol{\theta}}$ that produces the smallest objective function value among the 27 optima.

To check the performance of the optimal model of the tower, modal parameters of the tower were calculated using the optimal $\hat{\boldsymbol{\theta}}$. The experimental natural frequencies of the 5 modes are 35.37 Hz, 40.75 Hz, 67.99 Hz, 92.03 Hz, and 133.18 Hz, while the model-predicted natural frequencies are 38.29 Hz, 44.31 Hz, 67.65 Hz, 92.16 Hz, and 128.42 Hz. The matching between the experimental and the model-predicted natural frequencies is good. Figure 7.6 compares the experimental and model-predicted mode shapes. Each column of this figure corresponds to one mode. The upper row corresponds to the mode shapes along the x-direction and the lower row corresponds to the mode shapes along the y-direction. The lines with circles correspond to the undeformed column. The lines with squares correspond to the measured mode shapes. The black lines with stars correspond to the calculated mode shapes. The matching for the mode shapes of the column is also

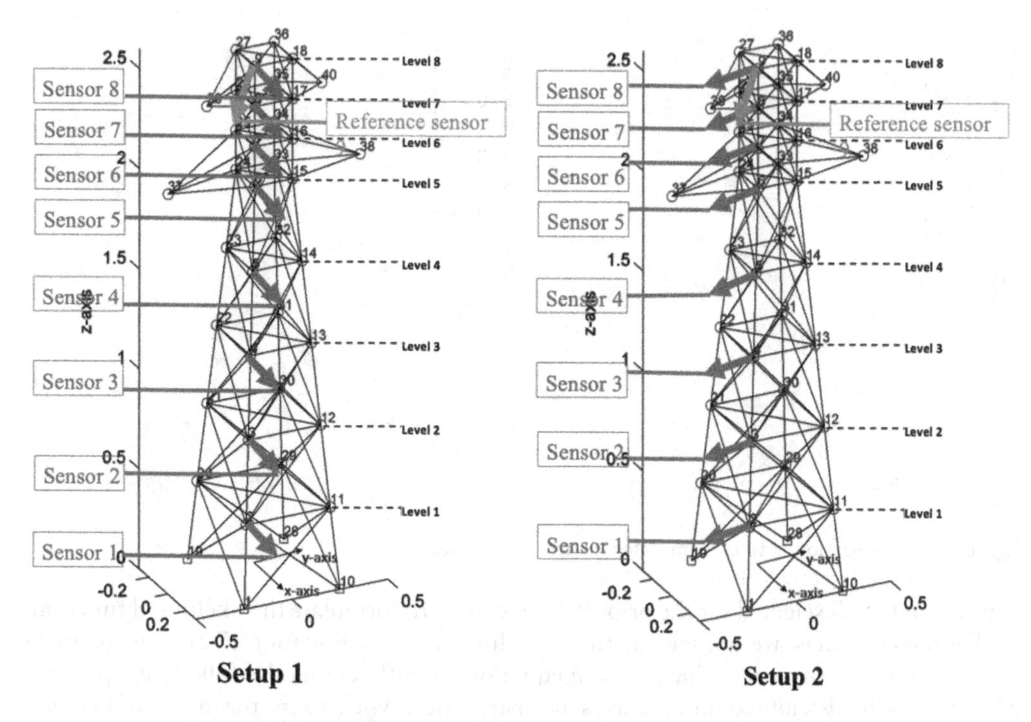

Figure 7.3 Sensor setup for the column of the transmission tower.

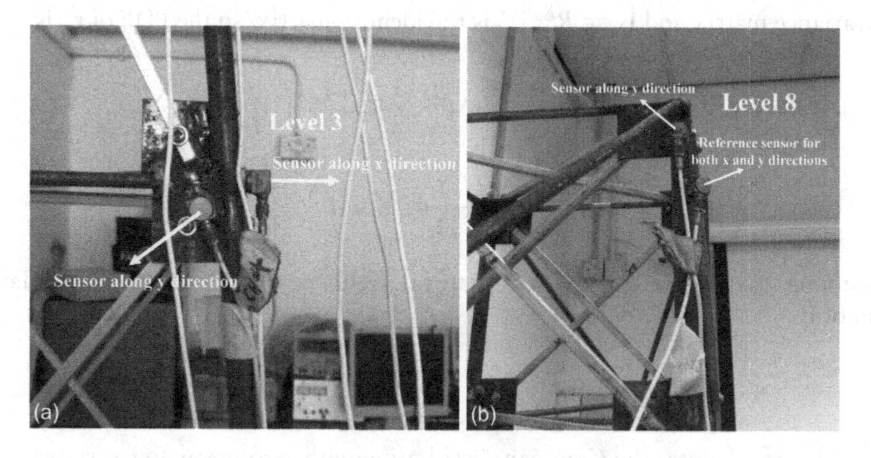

Figure 7.4 Sensor installation: (a) sensors along *x*- and *y*-directions; (b) the reference sensor.

good. Figure 7.7 shows the identified global mode shapes corresponding to the ones in Figure 7.6. It can be seen that these global mode shapes are reasonable.

7.3 BAYESIAN UPDATING: THE POSTERIOR PDF

To quantify the uncertainties given measured data D and the chosen model class $\mathcal{M}_P$, the posterior PDF is derived in this section following Bayes's theorem (see Chapter 6).

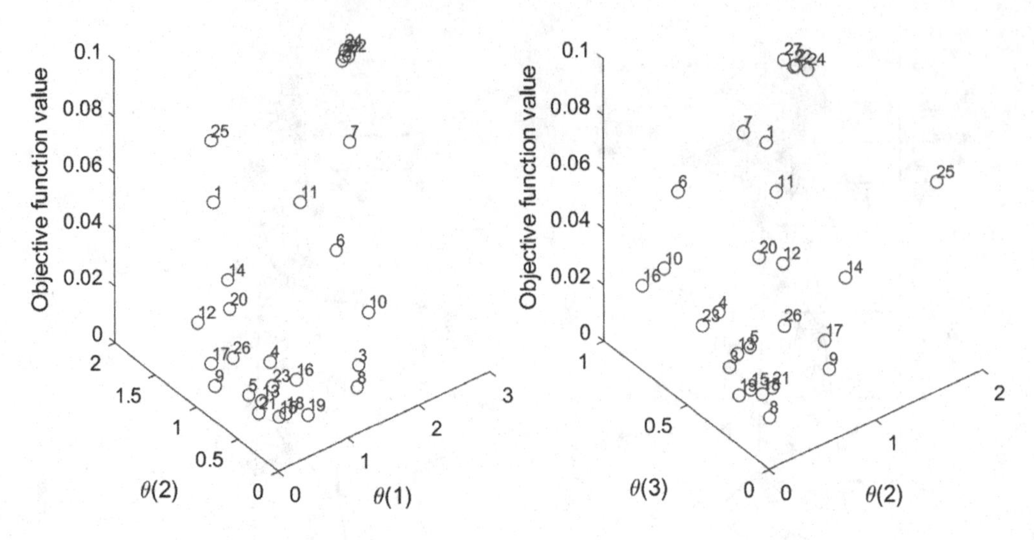

Figure 7.5 Optimization results from different starting points.

A uniform PDF is selected for the prior PDF $p(\boldsymbol{\alpha}|\mathcal{M}_P)$. To formulate the likelihood function, probabilistic models are needed for the modeling errors in Equation (7.2). It is assumed that the errors of the eigenvalue problem equations of different modes follow independent and identically distributed (i.i.d.) Gaussian distributions with a zero mean and a diagonal covariance matrix with equal diagonal terms $\mathbf{C}_m = \sigma_1^2 \mathbf{I}_{N_d}$, where σ_1^2 is the diagonal term of the covariance matrix and $\mathbf{I}_{N_d} \in R^{N_d \times N_d}$ is the identity matrix, so the PDF of $\mathbf{r}_m$ is

$$p\left(\mathbf{r}_m \middle| \boldsymbol{\theta}, \omega_m, \boldsymbol{\phi}_m, \mathcal{M}_P\right) = \frac{1}{(2\pi)^{N_d/2} \left|\mathbf{C}_m\right|^{1/2}} \exp\left(-\frac{1}{2}\mathbf{r}_m^T \mathbf{C}_m^{-1} \mathbf{r}_m\right)$$
$$= \left(2\pi\sigma_1^2\right)^{-N_d/2} \exp\left(-\frac{1}{2}\sigma_1^{-2} \mathbf{r}_m^T \mathbf{r}_m\right) \tag{7.9}$$

By assuming that the error of each natural frequency follows the same Gaussian distribution

$$e_{\omega_m} = \omega_m - \hat{\omega}_m \sim \mathcal{N}\left(0, \sigma_2^2\right) \tag{7.10}$$

where $\mathcal{N}\left(0, \sigma_2^2\right)$ denotes the Gaussian distribution with zero mean and variance σ_2^2, the PDF of the natural frequency of each mode is

$$p\left(\hat{\omega}_m \middle| \omega_m, \mathcal{M}_P\right) = \frac{1}{\sqrt{2\pi\sigma_2^2}} \exp\left(-\frac{\left(\omega_m - \hat{\omega}_m\right)^2}{2\sigma_2^2}\right) \tag{7.11}$$

Similarly, by assuming that the error vector of each mode shape follows the same multivariate Gaussian distribution

$$e_{\boldsymbol{\phi}_m} = \mathbf{L}\boldsymbol{\phi}_m - \hat{\boldsymbol{\varphi}}_m \sim \mathcal{N}\left(0, \sigma_3^2 \mathbf{I}_{N_o}\right) \tag{7.12}$$

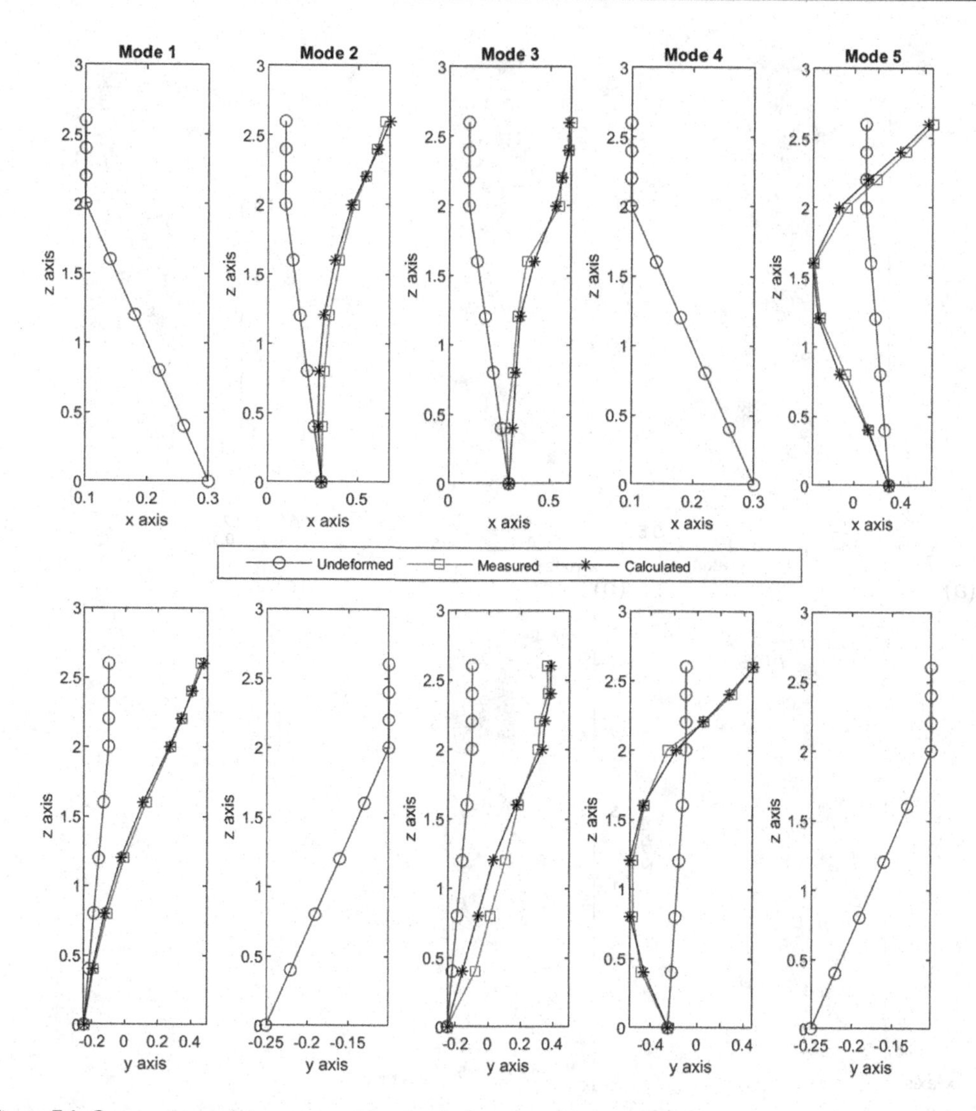

Figure 7.6 Comparison of the measured and calculated mode shapes of the tower column.

where a diagonal covariance matrix with equal diagonal elements is used. The PDF of the mode shape of each mode is thus

$$p\left(\hat{\varphi}_m \middle| \phi_m, \mathcal{M}_P\right) = \left(2\pi\sigma_3^2\right)^{-N_o/2} \exp\left(-\frac{1}{2}\sigma_3^{-2}\left(\mathbf{L}\phi_m - \hat{\varphi}_m\right)^T\left(\mathbf{L}\phi_m - \hat{\varphi}_m\right)\right) \tag{7.13}$$

The likelihood function is obtained by considering the PDFs of all the modes, where the independence is assumed among different parameters and different modes,

$$\begin{aligned} p\left(\hat{\omega}, \hat{\varphi} \middle| \alpha, \mathcal{M}_P\right) &= \prod_{m=1}^{N_m} p\left(\mathbf{r}_m \middle| \theta, \omega_m, \phi_m, \mathcal{M}_P\right) p\left(\hat{\omega}_m \middle| \omega_m, \mathcal{M}_P\right) p\left(\hat{\varphi}_m \middle| \phi_m, \mathcal{M}_P\right) \\ &= \left(2\pi\right)^{-\frac{1}{2}N_m(N_d+N_o+1)} \left(\sigma_1^2\right)^{-\frac{1}{2}N_m N_d} \left(\sigma_2^2\right)^{-\frac{1}{2}N_m} \left(\sigma_3^2\right)^{-\frac{1}{2}N_m N_o} \exp\left(-\frac{1}{2}J(\alpha)\right) \end{aligned} \tag{7.14}$$

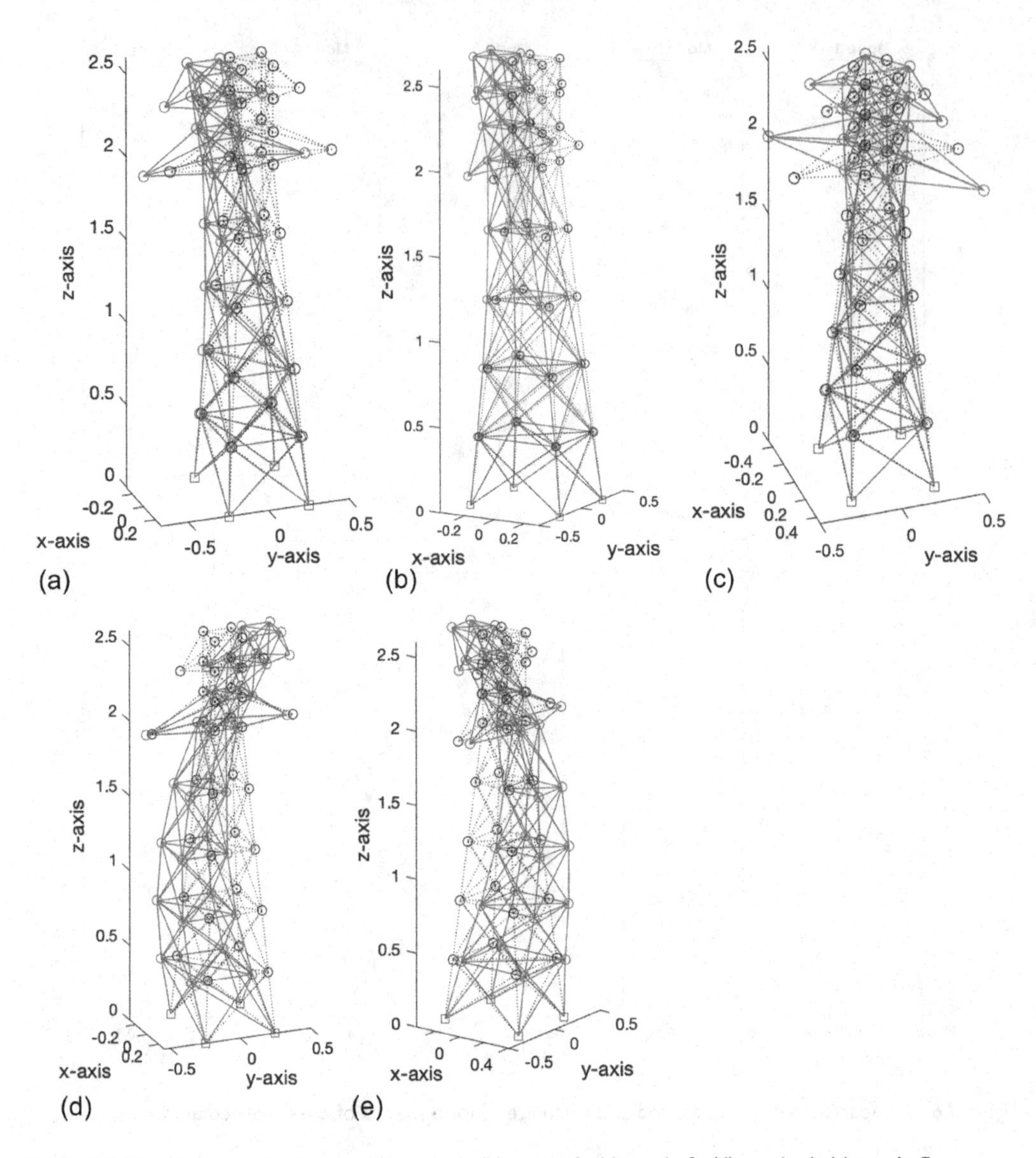

Figure 7.7 The global mode shapes: (a) mode 1; (b) mode 2; (c) mode 3; (d) mode 4; (e) mode 5.

where

$$J(\alpha)=\sum_{m=1}^{N_m}\left(\frac{\mathbf{r}_m^T\mathbf{r}_m}{\sigma_1^2}+\frac{\left(\omega_m-\hat{\omega}_m\right)^2}{\sigma_2^2}+\frac{\left(\mathbf{L}\phi_m-\hat{\varphi}_m\right)^T\left(\mathbf{L}\phi_m-\hat{\varphi}_m\right)}{\sigma_3^2}\right) \tag{7.15}$$

$\hat{\omega}$ contains all the experimental natural frequencies $\hat{\omega}=\left[\hat{\omega}_1,\hat{\omega}_2,\cdots,\hat{\omega}_{N_m}\right]^T$ and $\hat{\varphi}=\left[\hat{\varphi}_1^T,\hat{\varphi}_2^T,\cdots,\hat{\varphi}_{N_m}^T\right]^T$ contains all the experimental mode shapes. The posterior PDF of the uncertain parameters based on the error of the eigenvalue problem equation is

$$p(\alpha|\hat{\omega},\hat{\varphi},\mathcal{M}_P)=\frac{p(\alpha|\mathcal{M}_P)p(\hat{\omega},\hat{\varphi}|\alpha,\mathcal{M}_P)}{p(\alpha|\mathcal{M}_P)} \tag{7.16}$$

where $p(\alpha|\mathcal{M}_P)=\int_\Theta p(\alpha|\mathcal{M}_P)p(\hat{\omega},\hat{\varphi}|\alpha,\mathcal{M}_P)\mathrm{d}\alpha$ is a normalizing constant; in this case, the data are the experimental modal parameters $\mathbf{D}=\{\hat{\omega}_m,\hat{\varphi}_m : m=1,2,\cdots,N_m\}$ and the model class $\mathcal{M}_P$ includes both the finite element model and the probabilistic model. Absorbing all the constants, including the uniform prior PDF and the normalizing constant, into a factor c changes the posterior PDF to

$$p(\alpha|\hat{\omega},\hat{\varphi},\mathcal{M}_P)=cp(\hat{\omega},\hat{\varphi}|\alpha,\mathcal{M}_P) \tag{7.17}$$

For convenience, the negative log-likelihood function is given here by taking the natural logarithm of the likelihood function (Equation (7.14)):

$$\begin{aligned} L(\alpha)&=-\ln\left(p(\hat{\omega},\hat{\varphi}|\alpha,\mathcal{M}_P)\right)\\ &=\frac{1}{2}\left(N_m(N_d+N_o+1)\ln 2\pi+N_mN_d\ln\sigma_1^2+N_m\ln\sigma_2^2+N_mN_o\ln\sigma_3^2+J(\alpha)\right) \end{aligned} \tag{7.18}$$

7.4 IMPORTANCE SAMPLING

In many cases of science and engineering we need to calculate the integral with the following form

$$E_p(a(\mathbf{y}))=\int a(\mathbf{y})p(\mathbf{y})\mathrm{d}\mathbf{y} \tag{7.19}$$

where $\mathbf{y}$ is a random variable with PDF $p(\mathbf{y})$, e.g., the posterior PDF in a Bayesian inference problem; $a(\mathbf{y})$ is the quantity of interest, e.g., the displacement or stress of a structure; $E_p(a(\mathbf{y}))$ is thus the expectation with respect to the PDF $p(\mathbf{y})$. To evaluate this integral in Bayesian updating problems, a full description of the posterior PDF is needed. However, identifying the posterior PDF is difficult. The integrand of the above equation is thus complicated, and analytical evaluation of the integral is almost impossible. Moreover, $\mathbf{y}$ is high-dimensional for practical problems, so the normal numerical integration will take a long time. In Bayesian updating, we need an efficient method to identify the posterior PDF and evaluate the high-dimensional integral (Equation (7.19)) at the same time.

When $p(\mathbf{y})$ is convenient to be used for sampling, i.i.d. samples $\{\mathbf{y}^{(i)} : i=1,2,\cdots,N_i\}$ can be sampled from $p(\mathbf{y})$, and Equation (7.19) is approximated by the samples:

$$\bar{a}\approx\frac{1}{N_i}\sum_{i=1}^{N_i}a\left(\mathbf{y}^{(i)}\right) \tag{7.20}$$

This is the standard Monte Carlo method. When the number of samples N_i is large enough, this approximation is accurate, according to the Law of Large Numbers. If the variance of $a(\mathbf{y})$ is finite, i.e.,

$$\sigma_a^2 = \int \left(a(\mathrm{y}) - E_p\left(a(\mathrm{y})\right)\right)^2 p(\mathrm{y})\mathrm{dy} < \infty \tag{7.21}$$

Central Limit Theorem shows that the PDF of the approximation error of Equation (7.20) is a Gaussian PDF:

$$\frac{\sqrt{N_i}\left(\bar{a} - E_p\left(a(\mathrm{y})\right)\right)}{\sigma_a} \sim \mathcal{N}(0,1) \tag{7.22}$$

Standard Monte Carlo is only applicable when the target PDFs $p(\mathrm{y})$ are the ones that are convenient to sample from, for example, lognormal PDF, exponential PDF, uniform PDF, and Gaussian PDF. However, the difficulty is that $p(\mathrm{y})$ is complicated in practice, e.g., the posterior PDF in Bayesian inference, so most of the time directly sampling from $p(\mathrm{y})$ is not possible, and standard Monte Carlo is not suitable for Bayesian updating.

Following the idea of sampling, importance sampling can be an alternative to the standard Monte Carlo method. Importance sampling is constructed by introducing another PDF in the above integral:

$$E_q\left(a(\mathrm{y})\right) = \int a(\mathrm{y})\frac{p(\mathrm{y})}{q(\mathrm{y})}q(\mathrm{y})\mathrm{dy} \tag{7.23}$$

where $q(\mathrm{y})$ is the importance PDF. It can be seen that the basic idea of importance sampling is to sample from another PDF instead of the original complicated one. $q(\mathrm{y})$ is usually chosen to be a PDF that is convenient to sample from. The ratio in the above equation is called the importance weight:

$$w(\mathrm{y}) = \frac{p(\mathrm{y})}{q(\mathrm{y})} \tag{7.24}$$

The importance weight works as a correction for the bias when sampling from a PDF that is different from the target one. In Bayesian updating, importance sampling can be defined as follows (Robert & Casella 2013; Yang 2015).

Definition. Importance sampling is a process that generates i.i.d. samples $\left\{\mathrm{y}^{(i)} : i = 1,2,\cdots,N_i\right\}$ with an importance PDF $q(\mathrm{y})$ to approximate the expectation in Equation (7.19) using the generated samples:

$$\bar{a} = \frac{1}{N_i}\sum_{i=1}^{N_i} w\left(\mathrm{y}^{(i)}\right)a\left(\mathrm{y}^{(i)}\right) \tag{7.25}$$

where $w\left(\mathrm{y}^{(i)}\right)$ is the importance weight evaluated at each sample.

Theorem 7.1

When the number of samples is large enough, the empirical average Equation (7.25) based on the samples from importance sampling converges almost surely to Equation (7.19), i.e.,

$$P\left(\lim_{N_i \to \infty} \bar{a} = E_p\left(a(\mathrm{y})\right)\right) = 1 \tag{7.26}$$

Proof

Using the Law of Large Numbers for Equation (7.25) gives

$$\lim_{N_i \to \infty} \frac{1}{N_i} \sum_{i=1}^{N_i} w\left(\mathrm{y}^{(i)}\right) a\left(\mathrm{y}^{(i)}\right) = E_q\left(w(\mathrm{y})a(\mathrm{y})\right) \tag{7.27}$$

By definition, we also have

$$E_q\left(w(\mathrm{y})a(\mathrm{y})\right) = \int w(\mathrm{y})a(\mathrm{y})q(\mathrm{y})\mathrm{dy} \tag{7.28}$$

where the expectation is with respect to $q(\mathrm{y})$ because the samples in Equation (7.27) are generated by $q(\mathrm{y})$ in importance sampling. Using the definition of importance weight in Equation (7.28) reads

$$\begin{aligned} E_q\left(w(\mathrm{y})a(\mathrm{y})\right) &= \int \frac{p(\mathrm{y})}{q(\mathrm{y})} a(\mathrm{y})q(\mathrm{y})\mathrm{dy} \\ &= \int a(\mathrm{y})p(\mathrm{y})\mathrm{dy} \\ &= E_p\left(a(\mathrm{y})\right) \end{aligned} \tag{7.29}$$

Substituting Equation (7.29) into Equation (7.27) then completes the proof. □

It has been shown that importance sampling can well approximate the posterior PDF and the high-dimensional integral (Equation (7.19)) given enough samples. However, the difficulty in Bayesian inference is that the normalizing constant of $p(y)$ is not available because it requires knowledge of $p(y)$ and the evaluation of a high-dimensional integral, both of which are extremely difficult to obtain beforehand. As a result, the importance weight cannot be calculated and the original definition of importance sample cannot be used in Bayesian inference. To handle this problem, in Bayesian inference the un-normalized importance weights are calculated:

$$w_n(\mathrm{y}) = \frac{p_n(\mathrm{y})}{q_n(\mathrm{y})} \tag{7.30}$$

where $p_n(\mathrm{y})$ and $q_n(\mathrm{y})$ are the un-normalized posterior and importance PDFs, respectively; $p(y) = cp_n(y)$ and $q(y) = c_q q_n(y)$; and c and c_q are the normalizing constants of the two PDFs. Note that the normalizing constant of the importance PDF is usually known, but it is still considered to be unknown here for generality, because there may be cases where using the un-normalized $q_n(\mathrm{y})$ is more convenient than using $q(y)$. In Bayesian inference, the approximation of Equation (7.19) is then done by considering a normalizing constant to account for the effects of using un-normalized PDFs:

$$\bar{a} = \frac{\sum_{i=1}^{N_i} w_n\left(\mathrm{y}^{(i)}\right) a\left(\mathrm{y}^{(i)}\right)}{\sum_{i=1}^{N_i} w_n\left(\mathrm{y}^{(i)}\right)} \tag{7.31}$$

We want to show that Equation (7.31) also converges to Equation (7.19) almost surely.

Theorem 7.2
When the number of samples is large enough, the empirical average Equation (7.31) calculated by importance sampling with the un-normalized posterior and importance PDFs converges almost surely to Equation (7.19), i.e.,

$$P\left(\lim_{N_i \to \infty} \frac{\sum_{i=1}^{N_i} w_n\left(\mathbf{y}^{(i)}\right) a\left(\mathbf{y}^{(i)}\right)}{\sum_{i=1}^{N_i} w_n\left(\mathbf{y}^{(i)}\right)} = E_p\left(a(\mathbf{y})\right) \right) = 1 \tag{7.32}$$

Proof
Equation (7.31) is first transformed as follows:

$$\begin{aligned} \bar{a} &= \frac{\frac{1}{N_i}\sum_{i=1}^{N_i} w_n\left(\mathbf{y}^{(i)}\right) a\left(\mathbf{y}^{(i)}\right)}{\frac{1}{N_i}\sum_{i=1}^{N_i} w_n\left(\mathbf{y}^{(i)}\right)} \\ &= \frac{\frac{1}{N_i}\sum_{i=1}^{N_i} \left(\frac{p\left(\mathbf{y}^{(i)}\right)/c}{q\left(\mathbf{y}^{(i)}\right)/c_q} \right) a\left(\mathbf{y}^{(i)}\right)}{\frac{1}{N_i}\sum_{i=1}^{N_i} \frac{p\left(\mathbf{y}^{(i)}\right)/c}{q\left(\mathbf{y}^{(i)}\right)/c_q}} \\ &= \frac{\frac{1}{N_i}\sum_{i=1}^{N_i} w\left(\mathbf{y}^{(i)}\right) a\left(\mathbf{y}^{(i)}\right)}{\frac{1}{N_i}\sum_{i=1}^{N_i} w\left(\mathbf{y}^{(i)}\right)} \end{aligned} \tag{7.33}$$

It can be seen that the normalizing constants in the un-normalized importance weight are canceled, and Equation (7.31) can be rewritten using normalized importance weights with a normalization in the denominator. It has been proven by Theorem 7.1 that the nominator $\frac{1}{N_i}\sum_{i=1}^{N_i} w\left(\mathbf{y}^{(i)}\right) a\left(\mathbf{y}^{(i)}\right)$ converges to $E_p\left(a(\mathbf{y})\right)$, almost surely, with enough samples. We now study the denominator. The Law of Large Numbers gives

$$\lim_{N_i \to \infty} \frac{1}{N_i} \sum_{i=1}^{N_i} w\left(\mathbf{y}^{(i)}\right) = E_q\left(w(\mathbf{y})\right) \tag{7.34}$$

Expanding the right-hand side of the above equation by definition reads

$$\begin{aligned} E_q\left(w(\mathbf{y})\right) &= \int \frac{p(\mathbf{y})}{q(\mathbf{y})} q(\mathbf{y}) \mathrm{d}\mathbf{y} \\ &= \int p(\mathbf{y}) \mathrm{d}\mathbf{y} \\ &= 1 \end{aligned} \tag{7.35}$$

where the last equality is due to the theorem of total probability. Substituting Equation (7.35) into (7.34) gives

$$\lim_{N_i \to \infty} \frac{1}{N_i} \sum_{i=1}^{N_i} w\left(\mathbf{y}^{(i)}\right) = 1 \tag{7.36}$$

i.e., the denominator converges to unity almost surely. Using Theorem 7.1 and Equation (7.36) in Equation (7.33) completes the proof. □

With Theorem 7.2, importance sampling can also be used in Bayesian inference without knowing the normalizing constant of the posterior PDF. Difficulties still exist. In Bayesian inference, the posterior PDF is usually peaked and occupies only a small portion of the parameter space. This is what we hope for because the data are informative for the inference, making the posterior PDF be peaked in the parameter space. However, this poses a significant challenge for sampling because, most of the time, the samples will fall in the regions of small probability, unless an importance density that can well approximate the posterior PDF can be found; but it is hardly likely before importance sampling is done for identifying the posterior PDF. Under this situation the importance weights are close to zero and importance sampling becomes very inefficient. Markov chain Monte Carlo (MCMC) could be a good alternative for importance sampling. It will be introduced in the next section.

7.5 METROPOLIS–HASTINGS ALGORITHM

7.5.1 Overview

In this section, a widely applied algorithm, the Metropolis–Hastings (MH) algorithm, is introduced as a special case of MCMC methods for efficient applications in Bayesian model updating. Only implementing procedures and some important properties of the MH algorithm are discussed here, while thorough exposition about theoretical and applied aspects can be found in excellent references such as Robert and Casella 2013; Meyn and Tweedie 2012 and Owen 2013. The MH algorithm is partly based on the idea of importance sampling. Its development is mainly due to Metropolis et al. 1953 and Hastings 1970. Based on the Markov chain theory, the MH algorithm generates each sample conditional on the previous sample with some probability. All the samples constitute a Markov chain. Similar to the importance sampling, a proposal PDF is used to generate samples instead of using the complicated posterior PDF. The advantages of this algorithm are that information from other samples can be exploited to construct the proposal PDF and explore the parameter space, and the posterior PDF is only required to be known up to a multiplicative constant, so the normalizing constant that involves a high-dimensional integral need not be computed. The procedures of the MH algorithm are summarized in the following.

Algorithm 7.2: The MH algorithm

Conditional on the i-th sample $\mathbf{y}^{(i)}$,

1. Generate a candidate $\mathbf{Y}$ for the $(i + 1)$-th sample using the proposal PDF $\mathbf{Y} \sim q\left(\mathbf{y}|\mathbf{y}^{(i)}\right)$.
2. Take the $(i + 1)$-th sample $\mathbf{y}^{(i+1)}$ to be

$$\begin{cases} \mathbf{Y} \text{ with probability } r\left(\mathbf{Y}, \mathbf{y}^{(i)}\right); \\ \mathbf{y}^{(i)} \text{ with probability } 1 - r\left(\mathbf{Y}, \mathbf{y}^{(i)}\right) \end{cases}$$

where $r\left(\mathbf{Y}, \mathbf{y}^{(i)}\right)$ is called the acceptance probability:

$$r\left(\mathbf{Y}, \mathbf{y}^{(i)}\right) = \min\left(\frac{\dfrac{p(\mathbf{Y})}{q\left(\mathbf{Y}|\mathbf{y}^{(i)}\right)}}{\dfrac{p\left(\mathbf{y}^{(i)}\right)}{q\left(\mathbf{y}^{(i)}|\mathbf{Y}\right)}}, 1\right) \tag{7.37}$$

where $\min(\bullet, \bullet)$ gives the smaller argument of the two arguments.

In the MH algorithm, the proposal PDF $q(\mathbf{y})$ is chosen to be a PDF that is convenient to sample from. Conditional on the i-th sample, a candidate is first sampled using the proposal PDF. This candidate is then accepted as the $(i + 1)$-th sample with probability $r\left(\mathbf{Y}, \mathbf{y}^{(i)}\right)$, and therefore rejected with probability $1 - r\left(\mathbf{Y}, \mathbf{y}^{(i)}\right)$. If the candidate is rejected, the i-th sample is again used as the $(i + 1)$-th sample. It can be seen that when the ratio in Equation (7.37) is larger than 1, i.e., $p(\mathbf{Y})/q\left(\mathbf{Y}|\mathbf{y}^{(i)}\right) > p\left(\mathbf{y}^{(i)}\right)/q\left(\mathbf{y}^{(i)}|\mathbf{Y}\right)$, the candidate is always accepted.

7.5.2 Convergence of the algorithm

By repeatedly using the MH algorithm, multiple samples can be generated. These samples are generated from the proposal PDF that is different from the posterior PDF. We want to show that if the current sample distributes as the posterior PDF, then the next generated sample also distributes as the posterior PDF, so that when the number of generated samples is large enough, these samples are expected to well approximate the posterior PDF as well as the expectation in Equation (7.19). This property is useful in practice. For example, in Bayesian model updating, the first sample of the uncertain parameters in the MH algorithm can be chosen based on the initial finite element model. This sample is considered to distribute as the posterior PDF (it need not necessarily be in the region of high probability) because it is thought that the modeling errors of the initial finite element model built according to structural geometrical information will not be too large. The properties of the MH algorithm in this section are based on Yang (2015) and Robert and Casella (2013).

Theorem 7.3
If the current sample distributes as the posterior PDF, then the next sample generated by the MH algorithm also distributes as the posterior PDF.

Proof
The Markov transition kernel of the MH algorithm from $\mathbf{y}^{(i)}$ to $\mathbf{y}^{(i+1)}$ is formulated:

$$\begin{aligned} K\left(\mathbf{y}^{(i)}, \mathbf{y}^{(i+1)}\right) = {} & r\left(\mathbf{y}^{(i+1)}, \mathbf{y}^{(i)}\right) q\left(\mathbf{y}^{(i+1)}|\mathbf{y}^{(i)}\right) \\ & + \left(1 - \int r\left(\mathbf{y}^{(i+1)}, \mathbf{y}^{(i)}\right) q\left(\mathbf{y}^{(i+1)}|\mathbf{y}^{(i)}\right) d\mathbf{y}^{(i+1)}\right) \delta_{\mathbf{y}^{(i)}}\left(\mathbf{y}^{(i+1)}\right) \end{aligned} \tag{7.38}$$

where $\delta_{\mathbf{y}^{(i)}}$ is the Dirac mass in $\mathbf{y}^{(i)}$, so $\delta_{\mathbf{y}^{(i)}}\left(\mathbf{y}^{(i+1)}\right)=1$ only if $\mathbf{y}^{(i+1)}=\mathbf{y}^{(i)}$, otherwise $\delta_{\mathbf{y}^{(i)}}\left(\mathbf{y}^{(i+1)}\right)=0$. The second part of the above equation only exists when the candidate in the MH algorithm is rejected and the i-th sample is taken to be the $(i+1)$-th sample. The first part of the above equation corresponds to the situation where the candidate is accepted. As a result, this transition kernel describes the movement of a sample from the MH algorithm in two scenarios. In the first scenario, the candidate is generated by the proposal PDF $q\left(\mathbf{y}\middle|\mathbf{y}^{(i)}\right)$ and then accepted with probability $r\left(\mathbf{y}^{(i+1)},\mathbf{y}^{(i)}\right)$. In the second scenario, the candidate is rejected, where the plausibility of being rejected is quantified by the complement of the probability that the candidate is accepted (the integral in the above kernel). To prove this theorem, the detailed balance condition needs to be proved first:

$$p\left(\mathbf{y}^{(i)}\right)K\left(\mathbf{y}^{(i)},\mathbf{y}^{(i+1)}\right)=p\left(\mathbf{y}^{(i+1)}\right)K\left(\mathbf{y}^{(i+1)},\mathbf{y}^{(i)}\right) \tag{7.39}$$

where $p(\mathbf{y})$ is the stationary PDF of the Markov chain, the posterior PDF in this case. The detailed balance condition is formulated using the Markov transition kernel, and says that the probability of the state moving from $\mathbf{y}^{(i)}$ to $\mathbf{y}^{(i+1)}$ is equal to the probability of the state moving from $\mathbf{y}^{(i+1)}$ to $\mathbf{y}^{(i)}$. We will study the detailed balance condition for the two scenarios, respectively. If the candidate is accepted, using the first part of the Markov transition kernel gives

$$\begin{aligned}p\left(\mathbf{y}^{(i)}\right)K\left(\mathbf{y}^{(i)},\mathbf{y}^{(i+1)}\right)&=p\left(\mathbf{y}^{(i)}\right)r\left(\mathbf{y}^{(i+1)},\mathbf{y}^{(i)}\right)q\left(\mathbf{y}^{(i+1)}\middle|\mathbf{y}^{(i)}\right)\\&=p\left(\mathbf{y}^{(i)}\right)\min\left(\frac{\dfrac{p\left(\mathbf{y}^{(i+1)}\right)}{q\left(\mathbf{y}^{(i+1)}\middle|\mathbf{y}^{(i)}\right)}}{\dfrac{p\left(\mathbf{y}^{(i)}\right)}{q\left(\mathbf{y}^{(i)}\middle|\mathbf{y}^{(i+1)}\right)}},1\right)q\left(\mathbf{y}^{(i+1)}\middle|\mathbf{y}^{(i)}\right)\end{aligned} \tag{7.40}$$

where Equation (7.37) is used for the acceptance probability. Putting the first and third terms of the above equation into $\min(\bullet,\bullet)$ reads

$$\begin{aligned}p\left(\mathbf{y}^{(i)}\right)K\left(\mathbf{y}^{(i)},\mathbf{y}^{(i+1)}\right)&=\min\left(p\left(\mathbf{y}^{(i)}\right)\frac{\dfrac{p\left(\mathbf{y}^{(i+1)}\right)}{q\left(\mathbf{y}^{(i+1)}\middle|\mathbf{y}^{(i)}\right)}}{\dfrac{p\left(\mathbf{y}^{(i)}\right)}{q\left(\mathbf{y}^{(i)}\middle|\mathbf{y}^{(i+1)}\right)}}q\left(\mathbf{y}^{(i+1)}\middle|\mathbf{y}^{(i)}\right),p\left(\mathbf{y}^{(i)}\right)q\left(\mathbf{y}^{(i+1)}\middle|\mathbf{y}^{(i)}\right)\right)\\&=\min\left(p\left(\mathbf{y}^{(i+1)}\right)q\left(\mathbf{y}^{(i)}\middle|\mathbf{y}^{(i+1)}\right),p\left(\mathbf{y}^{(i)}\right)q\left(\mathbf{y}^{(i+1)}\middle|\mathbf{y}^{(i)}\right)\right)\end{aligned}$$

$$
\begin{aligned}
&= p\left(\mathbf{y}^{(i+1)}\right) q\left(\mathbf{y}^{(i)} \middle| \mathbf{y}^{(i+1)}\right) \min\left(1, \frac{p\left(\mathbf{y}^{(i)}\right) q\left(\mathbf{y}^{(i+1)} \middle| \mathbf{y}^{(i)}\right)}{p\left(\mathbf{y}^{(i+1)}\right) q\left(\mathbf{y}^{(i)} \middle| \mathbf{y}^{(i+1)}\right)}\right) \\
&= p\left(\mathbf{y}^{(i+1)}\right) \min\left(1, \frac{\dfrac{p\left(\mathbf{y}^{(i)}\right)}{q\left(\mathbf{y}^{(i)} \middle| \mathbf{y}^{(i+1)}\right)}}{\dfrac{p\left(\mathbf{y}^{(i+1)}\right)}{q\left(\mathbf{y}^{(i+1)} \middle| \mathbf{y}^{(i)}\right)}}\right) q\left(\mathbf{y}^{(i)} \middle| \mathbf{y}^{(i+1)}\right)
\end{aligned}
\tag{7.41}
$$

Using the definition of the acceptance probability and Markov transition kernel for the above equation reads

$$
\begin{aligned}
p\left(\mathbf{y}^{(i)}\right) K\left(\mathbf{y}^{(i)}, \mathbf{y}^{(i+1)}\right) &= p\left(\mathbf{y}^{(i+1)}\right) r\left(\mathbf{y}^{(i)}, \mathbf{y}^{(i+1)}\right) q\left(\mathbf{y}^{(i)} \middle| \mathbf{y}^{(i+1)}\right) \\
&= p\left(\mathbf{y}^{(i+1)}\right) K\left(\mathbf{y}^{(i+1)}, \mathbf{y}^{(i)}\right)
\end{aligned}
\tag{7.42}
$$

The detailed balance condition is thus proved for the first scenario.

It is trivial to prove the detailed balance condition for the second scenario. Using the second part of the Markov transition kernel gives

$$
\begin{aligned}
p\left(\mathbf{y}^{(i)}\right) K\left(\mathbf{y}^{(i)}, \mathbf{y}^{(i+1)}\right) &= p\left(\mathbf{y}^{(i)}\right)\left(1 - \int r\left(\mathbf{y}^{(i+1)}, \mathbf{y}^{(i)}\right) q\left(\mathbf{y}^{(i+1)} \middle| \mathbf{y}^{(i)}\right) d\mathbf{y}^{(i+1)}\right) \delta_{\mathbf{y}^{(i)}}\left(\mathbf{y}^{(i+1)}\right) \\
&= p\left(\mathbf{y}^{(i+1)}\right)\left(1 - \int r\left(\mathbf{y}^{(i)}, \mathbf{y}^{(i+1)}\right) q\left(\mathbf{y}^{(i)} \middle| \mathbf{y}^{(i+1)}\right) d\mathbf{y}^{(i)}\right) \delta_{\mathbf{y}^{(i+1)}}\left(\mathbf{y}^{(i)}\right) \\
&= p\left(\mathbf{y}^{(i+1)}\right) K\left(\mathbf{y}^{(i+1)}, \mathbf{y}^{(i)}\right)
\end{aligned}
\tag{7.43}
$$

The second equality is due to $\mathbf{y}^{(i+1)} = \mathbf{y}^{(i)}$ in this case, so one can replace the other and they can swap the order of each other. The detailed balance condition (Equation (7.39)) is thus proved.

Next, If the current sample $\mathbf{y}^{(i)}$ distributes as the posterior PDF, $\mathbf{y}^{(i)} \sim p(\mathbf{y})$, the PDF of the next sample is obtained by describing the transition from the current sample to the next:

$$
p_n\left(\mathbf{y}^{(i+1)}\right) = \int p\left(\mathbf{y}^{(i)}\right) K\left(\mathbf{y}^{(i)}, \mathbf{y}^{(i+1)}\right) d\mathbf{y}^{(i)}
\tag{7.44}
$$

Using the detailed balance condition in the above equation gives

$$
\begin{aligned}
p_n\left(\mathbf{y}^{(i+1)}\right) &= \int p\left(\mathbf{y}^{(i+1)}\right) K\left(\mathbf{y}^{(i+1)}, \mathbf{y}^{(i)}\right) d\mathbf{y}^{(i)} \\
&= p\left(\mathbf{y}^{(i+1)}\right) \int K\left(\mathbf{y}^{(i+1)}, \mathbf{y}^{(i)}\right) d\mathbf{y}^{(i)} \\
&= p\left(\mathbf{y}^{(i+1)}\right)
\end{aligned}
\tag{7.45}
$$

where $p\left(\mathbf{y}^{(i+1)}\right)$ can be taken out of the integrand because the integral is with respect to $\mathbf{y}^{(i)}$; $\int K\left(\mathbf{y}^{(i+1)},\mathbf{y}^{(i)}\right)d\mathbf{y}^{(i)}=1$ due to the theorem of total probability has been used. The PDF of the next sample is also the posterior PDF, and the theorem is thus proved. □

7.6 BAYESIAN UPDATING USING THE MH ALGORITHM AND THE EIGENVALUE PROBLEM ERROR

In this section, the MH algorithm is used to sample the posterior PDF derived based on the eigenvalue problem error (i.e., Equation (7.17)), so that the uncertainties can be quantified using the generated samples. An experimental case study on the shear building (see Chapter 5) is presented to illustrate the Bayesian method. Some implementation issues are first discussed.

7.6.1 The proposal PDF

The proposal PDF for the MH algorithm should be chosen as the one that is convenient to sample from. In this case, the proposal PDF is chosen as the lognormal PDF that always generates positive samples, because the quantity of interest are stiffness parameters and they must be positive. The definition of a lognormal PDF can be described: If a random variable y follows the lognormal distribution with parameters μ and κ, then the natural logarithm of y, $\ln(y)$, follows the normal distribution with mean μ and standard deviation κ, so μ is the mean of logarithmic values and κ is the standard deviation of logarithmic values. Each component of the uncertain parameter vector is sampled independently. For sampling the candidate of the j-th component of the uncertain parameter vector $\mathbf{y}_j$, μ is taken as $\ln\left(\mathbf{y}_j^{(i)}\right)$, where $\mathbf{y}_j^{(i)}$ denotes the j-th component of the current sample $\mathbf{y}^{(i)}$, so that the lognormal proposal PDF centers at the current sample; and κ is taken as the sample standard deviation calculated with the natural logarithm of the previous samples of $\mathbf{y}_j$.

When sampling for the first sample, the lognormal proposal PDF is set to center at the initial parameters of the finite element model. κ is determined by choosing the initial coefficient of variation (COV) of each uncertain parameter. It controls the spread of the proposal PDF. Note that the initial κ chosen this way will not affect the efficiency of the sampling method because it will be adaptively updated using generated samples later. When sampling for a general sample, the lognormal proposal PDF centers at the current sample, and its standard deviation parameter κ is calculated as the sample standard deviation of the natural logarithm of the previous samples. The implementation details are summarized as follows.

7.6.1.1 *Sampling for the first sample*

Let $l\left(\mu_j,\kappa_j\right)$ denote the lognormal PDF with parameter μ_j and κ_j for the j-th component of the uncertain parameter vector. For $j=1,2,\cdots,N_p$, take $\mu_j=\ln\left(\mathbf{y}_j^{(0)}\right)$, where $\mathbf{y}_j^{(0)}$ is the j-th component of the initial uncertain parameter vector $\mathbf{y}^{(0)}$; take $\kappa_j=\upsilon_0\mathbf{y}_j^{(0)}$ for convenience, where υ_0 is the COV set by the user, e.g., υ_0 can be 0.25.

7.6.1.2 *Sampling for a general sample*

For $j = 1,2,\cdots,N_p$, take $\mu_j = \ln\left(y_j^{(i)}\right)$; obtain κ_j as the sample standard deviation calculated by the previous N_1 samples:

$$\kappa_j = \frac{1}{N_1 - 1} \sum_{k=i-N_1+1}^{i} \left(\ln\left(y_j^{(k)}\right) - \bar{y}_j\right)^2 \tag{7.46}$$

where

$$\bar{y}_j = \frac{1}{N_1} \sum_{k=i-N_1+1}^{i} \ln\left(y_j^{(k)}\right) \tag{7.47}$$

Note that κ_j is updated only when every N_1 sample is generated – but not updated for each sample. This strategy gives time for the samples to explore the local area of the parameter space.

7.6.2 The prediction-error variances

In the MH algorithm, the prediction-error variances σ_1, σ_2, and σ_3 are needed to calculate the acceptance probability because the ratio of posterior PDF values is needed. In the following, their analytical formulas, given other uncertain parameters, are derived. The most probable value (MPV) of σ_1^2 is obtained by maximizing the posterior PDF, which is equivalent to minimizing the negative log-likelihood function L (Equation (7.18)). To do this, the first derivative of L with respect to σ_1^2 is calculated:

$$\frac{\partial L}{\partial \sigma_1^2} = \frac{1}{2}\left(N_m N_d - \left(\sigma_1^2\right)^{-2} \sum_{m=1}^{N_m} \mathbf{r}_m^T \mathbf{r}_m\right) \tag{7.48}$$

Setting the above equation to zero and solving for σ_1^2 gives the optimum of σ_1^2 based on other uncertain parameters:

$$\tilde{\sigma}_1^2 = \frac{\sum_{m=1}^{N_m} \mathbf{r}_m^T \mathbf{r}_m}{N_m N_d} \tag{7.49}$$

where $\mathbf{r}_m$ is given in Equation (7.1). The first derivative of L is calculated with respect to σ_2^2 and set to zero:

$$\frac{\partial L}{\partial \sigma_2^2} = \frac{1}{2}\left(N_m \left(\sigma_2^2\right)^{-1} - \left(\sigma_2^2\right)^{-2} \sum_{m=1}^{N_m} \left(\omega_m - \hat{\omega}_m\right)^2\right) = 0 \tag{7.50}$$

Solving for σ_2^2 gives

$$\tilde{\sigma}_2^2 = \frac{\sum_{m=1}^{N_m} \left(\omega_m - \hat{\omega}_m\right)^2}{N_m} \tag{7.51}$$

Similarly,

$$\frac{\partial L}{\partial \sigma_3^2} = \frac{1}{2}\left(N_m N_o \left(\sigma_3^2\right)^{-1} - \left(\sigma_3^2\right)^{-2} \sum_{m=1}^{N_m} \left(\mathbf{L}\boldsymbol{\phi}_m - \hat{\boldsymbol{\varphi}}_m\right)^T \left(\mathbf{L}\boldsymbol{\phi}_m - \hat{\boldsymbol{\varphi}}_m\right) \right) = 0 \tag{7.52}$$

$$\tilde{\sigma}_3^2 = \frac{1}{N_m N_o} \sum_{m=1}^{N_m} \left(\mathbf{L}\boldsymbol{\phi}_m - \hat{\boldsymbol{\varphi}}_m\right)^T \left(\mathbf{L}\boldsymbol{\phi}_m - \hat{\boldsymbol{\varphi}}_m\right) \tag{7.53}$$

With Equations (7.49), (7.51), and (7.53), the acceptance probability can be calculated given a generated sample. During the operation of the proposed MH algorithm, σ_1, σ_2, and σ_3 are calculated based on the current sample, and then they are used for generating N_1 samples. After N_1 samples are generated, the variance parameters are updated using the latest generated sample. This strategy is similar to that for the proposal PDF, i.e., the variance parameters are updated for every N_1 sample.

7.6.3 Summary of the proposed algorithm

Algorithm 7.3: Bayesian updating based on the MH algorithm

(1) Initialization

a. Set the sample number N_1 for updating the variance parameters, e.g., $N_1 = 300$; set the sample counter $N_2 = 0$; set the maximum number of samples N_3 to be generated.

b. Set the initial COV v_0, e.g., $v_0 = 0.25$; for $j = 1$ to N_p, take $\kappa_j = v_0 \theta_j^{(0)}$, where $\theta_j^{(0)}$ denotes the j-th initial structural uncertain parameter, and $\mu_j = \ln\left(\theta_j^{(0)}\right)$ for the lognormal proposal PDF.

c. Generate the first sample of the structural parameter $\boldsymbol{\theta}^{(1)}$. Each component of $\boldsymbol{\theta}^{(1)}$ is generated independently using the lognormal proposal PDF in the previous step:

$$\theta_j^{(1)} \sim l\left(\mu_j, \kappa_j\right)$$

d. Update the counter $N_2 = N_2 + 1$.

e. Repeat steps B to E of Algorithm 7.1 for $m = 1$ to N_m to get $\mathbf{K}\left(\boldsymbol{\theta}^{(1)}\right)$, $\omega_m\left(\boldsymbol{\theta}^{(1)}\right)$ and $\boldsymbol{\phi}_m\left(\boldsymbol{\theta}^{(1)}\right)$, where the dependence on the sample is emphasized.

f. Calculate σ_1, σ_2, and σ_3 using Equations (7.49), (7.51), and (7.53).

(2) Generate a general sample

For $i = 2$ to N_3, do the following.

The MH algorithm

Start

a. Sample a candidate $\mathbf{Y}$ of $\boldsymbol{\theta}^{(i)}$. For $j = 1$ to N_p, generate $\mathbf{Y}_j \sim q\left(y \middle| \theta_j^{(i-1)}\right) = l\left(\ln\left(\theta_j^{(i-1)}\right), \kappa_j\right)$, where $q\left(y \middle| \theta_j^{(i-1)}\right)$ denotes the lognormal proposal PDF conditional on $\theta_j^{(i-1)}$.

b. Take the i-th sample $\boldsymbol{\theta}^{(i)}$ to be

$$\begin{cases} \mathbf{Y} \text{ with probability } r\left(\mathbf{Y},\boldsymbol{\theta}^{(i-1)}\right); \\ \boldsymbol{\theta}^{(i-1)} \text{ with probability } 1-r\left(\mathbf{Y},\boldsymbol{\theta}^{(i-1)}\right) \end{cases}$$

where

$$r\left(\mathbf{Y},\boldsymbol{\theta}^{(i-1)}\right)=\min\left(\frac{p(\mathbf{Y})}{p\left(\boldsymbol{\theta}^{(i-1)}\right)}\frac{q\left(\boldsymbol{\theta}^{(i-1)}|\mathbf{Y}\right)}{q\left(\mathbf{Y}|\boldsymbol{\theta}^{(i-1)}\right)},1\right) \tag{7.54}$$

where the ratio of the posterior PDF values $\frac{p(\mathbf{Y})}{p\left(\boldsymbol{\theta}^{(i-1)}\right)}$ is calculated using Equations (7.14) and (7.15); $\frac{q\left(\boldsymbol{\theta}^{(i-1)}|\mathbf{Y}\right)}{q\left(\mathbf{Y}|\boldsymbol{\theta}^{(i-1)}\right)}$ is calculated using the lognormal proposal PDF.

End

c. Update the counter $N_2 = N_2 + 1$.
d. If $N_2 = N_1$, then
   update the variance parameters for the proposal PDF ($j = 1$ to N_p):

$$\kappa_j = \frac{1}{N_1 - 1}\sum_{k=i-N_1+1}^{i}\left(\ln\left(\theta_j^{(k)}\right)-\bar{\theta}_j\right)^2$$

$$\bar{\theta}_j = \frac{1}{N_1}\sum_{k=i-N_1+1}^{i}\ln\left(\theta_j^{(k)}\right)$$

   update σ_1, σ_2, and σ_3:
   Repeat steps B to E of Algorithm 7.1 for $m = 1$ to N_m to get $\mathbf{K}\left(\boldsymbol{\theta}^{(i)}\right)$, $\omega_m\left(\boldsymbol{\theta}^{(i)}\right)$, and $\boldsymbol{\phi}_m\left(\boldsymbol{\theta}^{(i)}\right)$, where the dependence on the current sample $\boldsymbol{\theta}^{(i)}$ is emphasized;
   Calculate σ_1, σ_2, and σ_3 using Equations (7.49), (7.51), and (7.53);
   Reset the counter $N_2 = 0$.
e. Repeat steps a to d until $i = N_3$.
f. **End For**

7.6.4 Bayesian model updating of the shear building

To demonstrate the proposed method, Bayesian model updating is conducted on the four-story shear building (see Chapter 5) in this section. The experimental modal parameters of this shear building, i.e., four modes of natural frequencies and mode shapes, have been identified based on measured accelerations under a stochastic wind excitation. The detailed discussions can be found in Chapter 5. These experimental modal parameters are used as data for Bayesian model updating. The shear building model (see Chapter 2 for the detailed derivation) is used for modeling. The uncertain parameter vector $\boldsymbol{\alpha}$ in this case thus contains the four inter-story stiffness parameters for stories 1 to 4 $\boldsymbol{\theta} \in R^4$ and the actual natural frequencies and mode shapes $\left\{\omega_m, \boldsymbol{\phi}_m \in R^4 : m = 1,2,\cdots,4\right\}$.

Algorithm 7.3 is used to sample from the posterior PDF. Note that only the structural parameters $\boldsymbol{\theta}$ are directly sampled, and the model-predicted modal parameters are

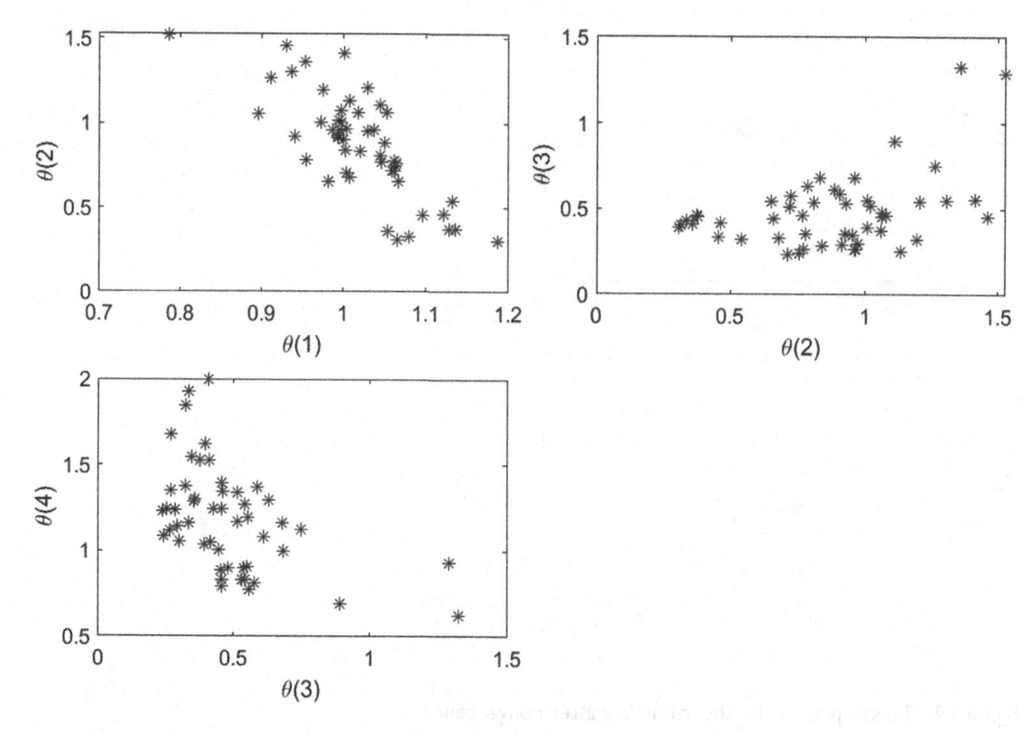

Figure 7.8 The first 500 samples of the uncertain structural parameters of the shear building.

calculated using the deterministic model based on the eigenvalue problem to facilitate computing the acceptance probability in the MH algorithm. In total, about 1500 samples of $\boldsymbol{\theta}$ were generated. Figure 7.8 shows the first 500 samples of the uncertain structural parameters generated by the MH algorithm, and Figure 7.9 shows the rest of the samples. By comparing these two figures, it can be seen that, at the early stage, the samples distribute in a relatively wide domain (see Figure 7.8). After the important region of the parameter space is found (the region of high probability), the samples converge in a narrower domain than that at the early stage. The samples after convergence (samples 500 to 1500 in this case) are used to approximate the posterior PDF, so not only the MPVs of the uncertain parameters can be obtained, but also their posterior uncertainties can be quantified. Because samples are generated from the posterior PDF, the MPV of $\boldsymbol{\theta}$ is approximated following the Monte Carlo method:

$$\begin{aligned}\hat{\boldsymbol{\theta}} &= \int \boldsymbol{\theta} p\left(\boldsymbol{\theta} \middle| \hat{\boldsymbol{\omega}}, \hat{\boldsymbol{\varphi}}, \mathcal{M}_P\right) d\boldsymbol{\theta} \\ &\approx \frac{1}{N_i} \sum_{i=1}^{N_i} \boldsymbol{\theta}^{(i)}\end{aligned} \tag{7.55}$$

where the MPV can be simply obtained by the sample average because each of these samples was generated with probability $p\left(\boldsymbol{\theta}^{(i)} \middle| \hat{\boldsymbol{\omega}}, \hat{\boldsymbol{\varphi}}, \mathcal{M}_P\right) d\boldsymbol{\theta}$, and their probability distribution has been considered. This approximation method is also applied for the modal

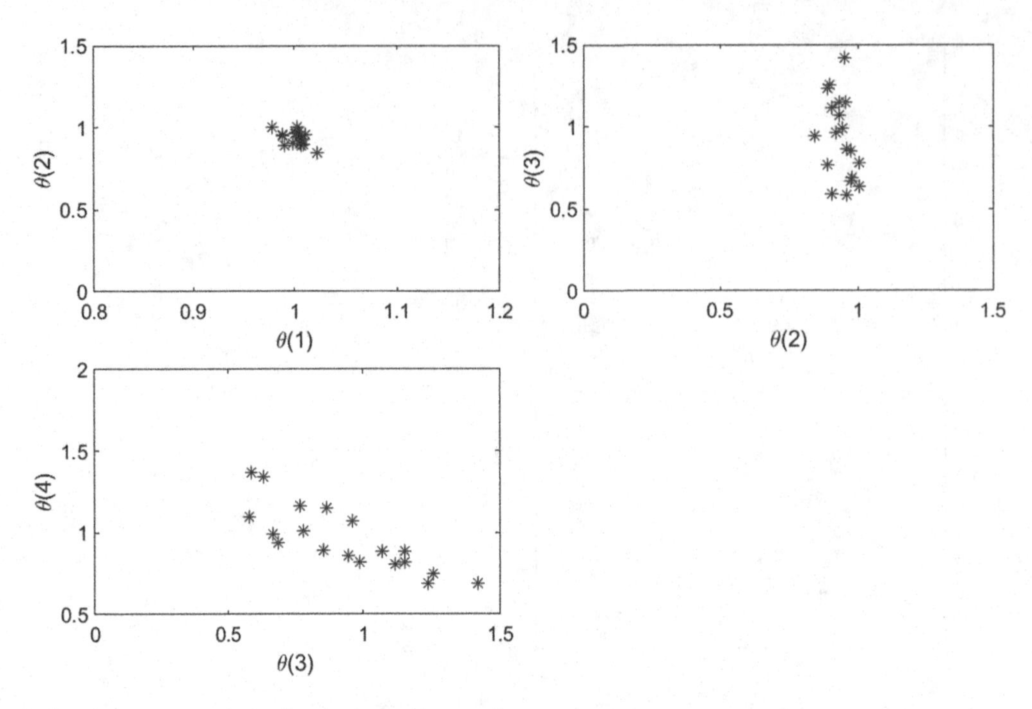

Figure 7.9 The samples of the shear building after convergence.

parameters. The MPV of the natural frequency of the m-th mode is also calculated using the samples:

$$\begin{aligned}\hat{\omega}_m(\theta) &= \int \omega_m(\theta) p\left(\theta \mid \hat{\omega}, \hat{\varphi}, \mathcal{M}_P\right) d\theta \\ &\approx \frac{1}{N_i} \sum_{i=1}^{N_i} \omega_m\left(\theta^{(i)}\right)\end{aligned} \tag{7.56}$$

where $p\left(\theta \mid \hat{\omega}, \hat{\varphi}, \mathcal{M}_P\right)$ denotes the posterior PDF of $\boldsymbol{\theta}$; $\omega_m\left(\theta^{(i)}\right)$ denotes the model-predicted natural frequency of the m-th mode calculated using the i-th sample $\theta^{(i)}$. Similarly, the MPV of the mode shape of the m-th mode is approximated by

$$\begin{aligned}\hat{\phi}_m(\theta) &= \int \mathbf{L}\phi_m(\theta) p\left(\theta \mid \hat{\omega}, \hat{\varphi}, \mathcal{M}_P\right) d\theta \\ &\approx \frac{1}{N_i} \sum_{i=1}^{N_i} \mathbf{L}\phi_m\left(\theta^{(i)}\right)\end{aligned} \tag{7.57}$$

where the mode shapes calculated by different samples $\phi_m\left(\theta^{(i)}\right)$ should have the same normalization and direction. The posterior uncertainties of $\boldsymbol{\theta}$ and the predicted modal parameters are also obtained using the samples. They are sample standard deviations:

$$s_\theta \approx \sqrt{\frac{1}{N_i - 1}\sum_{i=1}^{N_i}\left(\theta^{(i)} - \hat{\theta}\right)^2} \tag{7.58}$$

$$s_m \approx \sqrt{\frac{1}{N_i - 1}\sum_{i=1}^{N_i}\left(\omega_m\left(\theta^{(i)}\right) - \hat{\omega}_m(\theta)\right)^2} \tag{7.59}$$

$$\mathbf{s}_m \approx \sqrt{\frac{1}{N_i - 1}\sum_{i=1}^{N_i}\left(\mathrm{L}\phi_m\left(\theta^{(i)}\right) - \hat{\phi}_m(\theta)\right)^2} \tag{7.60}$$

where s_m is the sample standard deviation of the natural frequency of the m-th mode; note that all the operators in Equation (7.60) work component-wise for the mode shape vectors, so $\mathbf{s}_m \in R^{N_o}$ contains the standard deviations of the mode shapes at each measured DOF.

The MPVs of the mode shapes were calculated using the samples after convergence and compared with the experimental mode shapes (the solid lines) in Figure 7.10. The MPVs of the natural frequencies were compared in Table 7.1. It is seen that the MPVs of the modal parameters can match the experimental ones well.

The posterior uncertainties of the uncertain parameters and model-predicted modal parameters can also be obtained using the Bayesian method. Table 7.2 presents the posterior uncertainties of the identified structural parameters, natural frequencies, and mode shapes in terms of the coefficients of variation (COVs). The COV is defined as the ratio of the standard deviation to the MPV, and these quantifies can be calculated using Equations (7.56) to (7.60). The posterior uncertainties in terms of COV are in a normalized sense, so uncertainties of different parameters can be compared directly. It can be observed from Table 7.2 that the posterior uncertainties of the identified parameters are small. Moreover, given the measured data, the remaining uncertainties of the identified mode shapes are a bit larger than those of the identified natural frequencies. To see the posterior uncertainties of the mode shapes clearly, the posterior standard deviations of the identified mode shapes at each DOF were calculated, so the posterior uncertainties of the identified mode shapes can be graphically shown in Figure 7.11. In this figure, the lines with triangles show the lower and upper bounds of variation of the mode shapes, and the lines with circles show the MPVs.

7.7 BAYESIAN MODEL UPDATING USING A MULTI-LEVEL MCMC METHOD

The MCMC method discussed in the previous section is a single-level method. It explores the whole parameter space directly. For a complex posterior PDF in a high-dimensional parameter space, the single-level MCMC method may not be able to effectively generate samples in regions of high probability. In this section, a multi-level MCMC method is introduced to solve the above-mentioned problem. This method is suitable for solving unidentifiable problems with the posterior PDFs that lie in complex and extended manifolds.

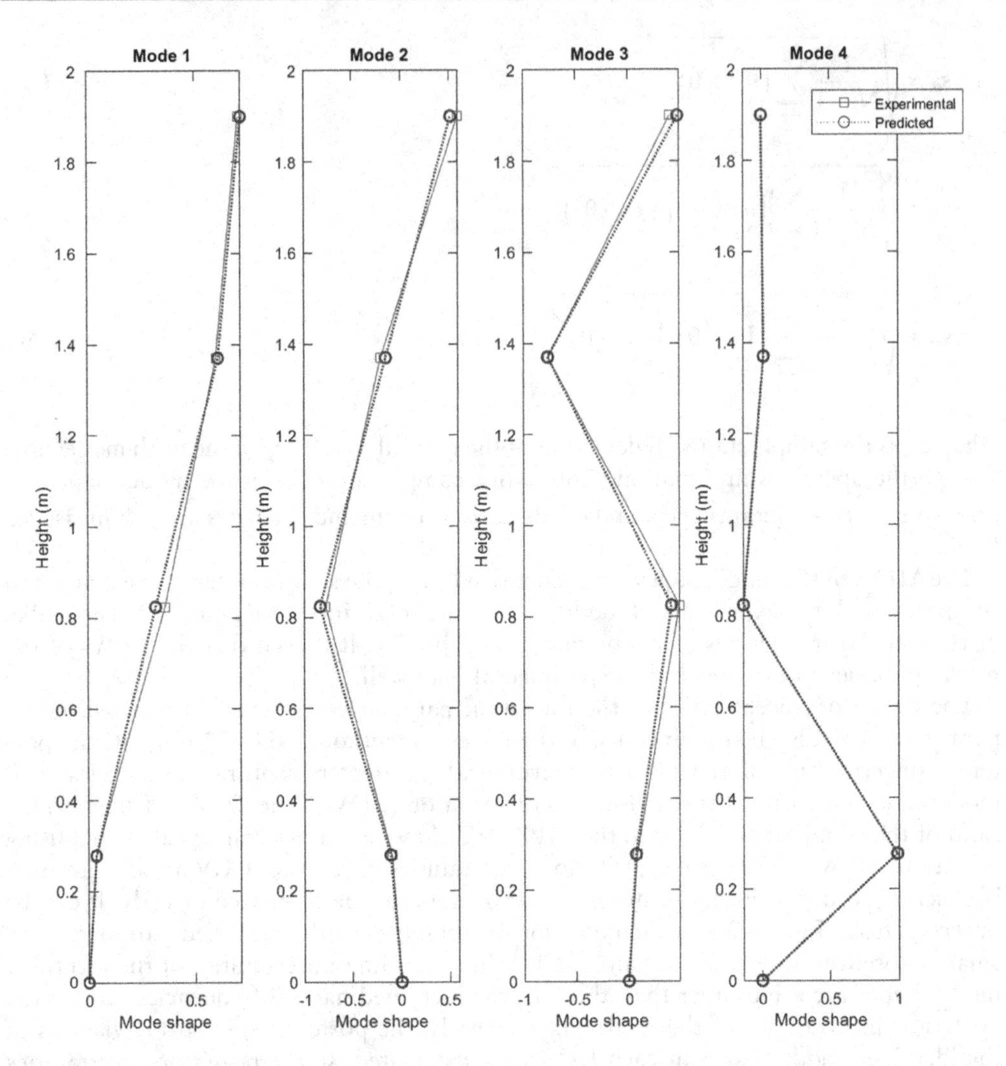

Figure 7.10 Comparison of the experimental and model-predicted mode shapes for the proposed MH algorithm.

Table 7.1 Comparison of the Experimental and Model-Predicted Natural Frequencies for the Proposed MH Algorithm

Mode	*1*	*2*	*3*	*4*
Experimental (Hz)	1.56	4.56	6.74	11.26
Predicted (Hz)	1.54	4.51	6.53	11.26

7.7.1 The posterior PDF for the multi-level MCMC method

The important regions (i.e., the regions of high probability) are determined by the variances of the prediction errors, but they are unknown before sampling is conducted; this is the main difficulty of sampling from the posterior PDF. The basic idea of the proposed multi-level MCMC method is to systematically construct a series of bridge PDFs

Table 7.2 The Posterior Coefficients of Variation of the Uncertain Parameters, Natural Frequencies, and Mode Shapes

Story	1	2	3	4
θ	0.0058	0.0303	0.2618	0.1257
Mode	1	2	3	4
ω	0.0304	0.0202	0.0437	0.0028
ϕ	0.0706	0.1006	0.2553	0.0008
	0.0870	0.0840	0.2266	0.0360
	0.0294	0.6067	0.0176	0.3229
	0.0088	0.1243	0.1967	0.2182

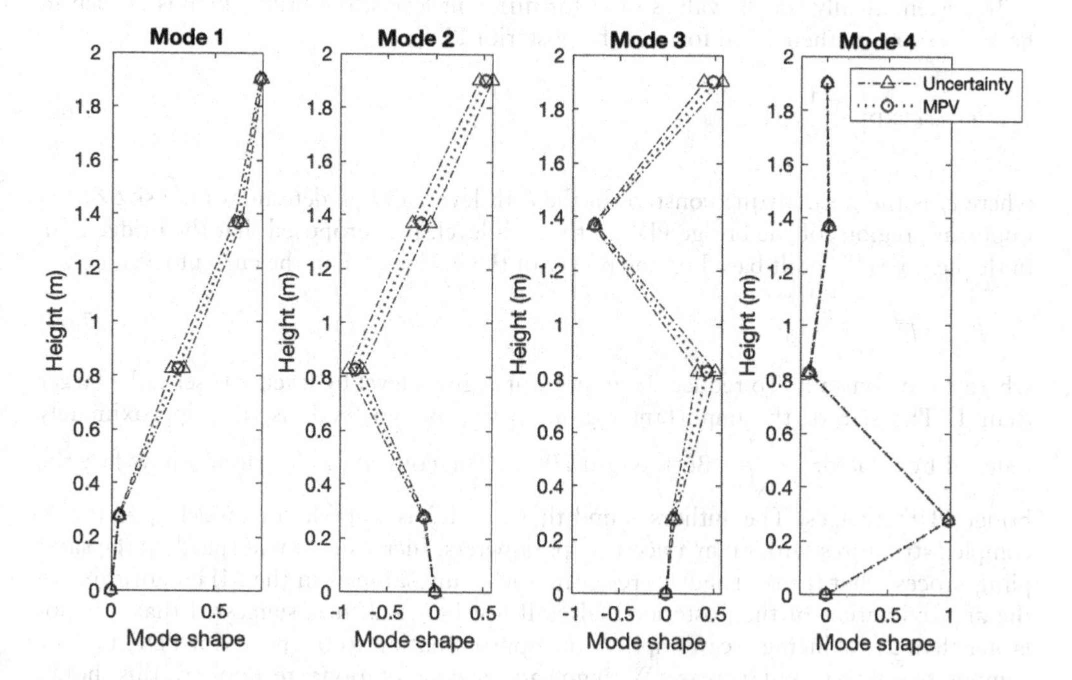

Figure 7.11 The posterior uncertainties of the identified mode shapes.

in multiple levels that bridge the gap between the initial PDF in the first sampling level and the posterior PDF in the final sampling level, such that the important regions of these PDF reduce gradually. By doing this, samples will be generated in the important regions of each intermediate PDF, and move gradually toward the important regions of the posterior PDF. The posterior PDF can generally be written in the following form:

$$p\left(\alpha \middle| \hat{\omega}, \hat{\varphi}, \mathcal{M}_P\right) = c\exp\left(-\frac{1}{2\gamma^2} J(\alpha)\right) \tag{7.61}$$

where γ^2 is unknown and is related to the important regions of the posterior PDF; $J(\boldsymbol{\alpha})$ is used again for convenience, but it does not include variance parameters here, so it is different from Equation (7.15). The multi-level MCMC method is based on the works of Beck and Au (2002), Yang (2015), and Lam et al. (2015). It is proposed in the method that the sampling is divided into multiple levels, in each of which one bridge PDF is constructed. In the first sampling level γ^2 is set to a large value, so that the bridge PDF of the first level has a large important region. γ^2 is reduced gradually in a controlled manner in the following levels to reduce the important regions of the bridge PDFs. Finally, the important regions of the bridge PDF reach those of the target posterior PDF, and the bridge PDF approximates the posterior PDF. In each level, sampling is conducted for the bridge PDF using the MH algorithm. A key factor for the proposed method to work well is how to change γ^2 in two successive levels. If γ^2 changes too much, the important regions will change much, so most of the generated samples will be rejected. If γ^2 changes too little, it will take a long time to reach the important regions of the posterior PDF. In both cases the sampling is inefficient.

To systematically set the values of γ^2 for different levels, the bridge PDF is chosen to have the same mathematical form as the posterior PDF:

$$p_k = c_k \exp\left(-\frac{1}{2\gamma_k^2} J(\alpha)\right) \tag{7.62}$$

where c_k is the normalizing constant in the k-th level, and γ_k^2 determines the size of the important regions of the bridge PDF in the k-th level. It is proposed that the bridge PDF in the next level is built based on the power of the bridge PDF in the current level, i.e.,

$$p_{k+1} \sim p_k^a \tag{7.63}$$

where a is a constant. To reduce the important regions level by level, a is set to be larger than 1. The size of the important regions between two levels is thus approximately reduced by a factor of $\frac{1}{\sqrt{a}}$ (Beck & Au 2002). The constant a determines how fast the bridge PDF changes. The authors found that $a = 1.2$ is suitable for model updating of complex structures with many uncertain parameters. Increasing a will speed up the sampling process, but it may result in rejecting too many samples in the MH algorithm, so the approximation of the posterior PDF will not be good. It is suggested that a is not larger than 2. Reducing a can improve the approximation to the posterior PDF, but the computational time will increase. With nowadays' computational technology, this should not be a big problem.

The next step is to re-formulate the posterior PDF based on the eigenvalue problem error to accommodate the general form in Equation (7.61). It can be seen in Equations (7.14), (7.15), and (7.17) that the original posterior PDF based on the eigenvalue problem error includes three variance parameters. To unify the three variance parameters into one, the fractional errors related to the eigenvalue problem, natural frequencies, and mode shapes are considered. The fractional error ranges from 0 to 1, so using fractional errors normalizes errors of different quantities, and using one variance parameter for different errors is possible. The fractional error of the eigenvalue problem is first considered. The eigenvalue problem error is a vector, and the modal assurance criterion (MAC) is used to derive the fractional error of the eigenvalue problem. MAC quantifies the degree of consistency between two vectors. Given vectors $\mathbf{a}, \mathbf{b} \in R^{N_1}$, MAC is defined as

$$\text{MAC}(\mathbf{a}, \mathbf{b}) = \frac{\left|\mathbf{a}^T \mathbf{b}\right|}{\left(\mathbf{a}^T \mathbf{a}\right)\left(\mathbf{b}^T \mathbf{b}\right)} \tag{7.64}$$

MAC(**a**, **b**) takes values between 0 and 1. MAC(**a**, **b**) = 0 means that **a** is orthogonal to **b**, and MAC(**a**, **b**) = 1 means that **a** and **b** are perfectly matched (parallel to each other). The complement of MAC, 1 – MAC, is thus treated as the quadratic form of the fractional error between two vectors, and its square root is interpreted as the fractional error between two vectors. For the case of the eigenvalue problem, according to Equation (7.1), the two vectors to be used for MAC calculation can be $\mathbf{K}(\theta)\breve{\phi}_m$ and $\hat{\omega}_m^2\mathbf{M}\breve{\phi}_m$, where the computation of $\breve{\phi}_m$ has been introduced in Algorithm 7.1. The fractional error of the eigenvalue problem is then written as

$$e_{\mathbf{r}_m} = \left(1 - \mathrm{MAC}\left(\mathbf{K}(\theta)\breve{\phi}_m, \omega_m^2\mathbf{M}\breve{\phi}_m\right)\right)^{\frac{1}{2}} \tag{7.65}$$

where the m-th mode is considered. The fractional error of the natural frequency of the m-th mode can be easily written as:

$$e_{\omega_m} = \frac{\hat{\omega}_m - \omega_m}{\hat{\omega}_m} \tag{7.66}$$

The fractional error of the mode shape of the m-th mode can also be similarly written using MAC:

$$e_{\phi_m} = \left(1 - \mathrm{MAC}\left(\mathbf{L}\phi_m, \varphi_m\right)\right)^{\frac{1}{2}} \tag{7.67}$$

It is then reasonable to assume that the fractional errors $e_{\mathbf{r}_m}$, e_{ω_m}, and e_{ϕ_m} follow the same Gaussian distribution $\mathcal{N}\left(0, \gamma^2\right)$ with zero mean and variance γ^2. The PDFs of the "experimental eigenvalue problem" (we invent this name to emphasize that the experimental modal parameters are used in the eigenvalue problem), the experimental natural frequency, and the experimental mode shape of the m-th mode can be formulated given the finite element model:

$$\begin{aligned} p\left(\mathbf{r}_m \middle| \theta, \omega_m, \phi_m, \mathcal{M}_P\right) &= p\left(e_{\mathbf{r}_m}\right) \\ &= \left(2\pi\gamma^2\right)^{-\frac{1}{2}} \exp\left(-\frac{1}{2\gamma^2}\left(1 - \mathrm{MAC}\left(\mathbf{K}(\theta)\breve{\phi}_m, \omega_m^2\mathbf{M}\breve{\phi}_m\right)\right)\right) \end{aligned} \tag{7.68}$$

$$p\left(\hat{\omega}_m \middle| \omega_m, \mathcal{M}_P\right) = p\left(e_{\omega_m}\right) = \left(2\pi\gamma^2\right)^{-\frac{1}{2}} \exp\left(-\frac{1}{2\gamma^2}\left(\frac{\hat{\omega}_m - \omega_m}{\hat{\omega}_m}\right)^2\right) \tag{7.69}$$

$$p\left(\hat{\varphi}_m \middle| \phi_m, \mathcal{M}_P\right) = p\left(e_{\phi_m}\right) = \left(2\pi\gamma^2\right)^{-\frac{1}{2}} \exp\left(-\frac{1}{2\gamma^2}\left(1 - \mathrm{MAC}\left(\mathbf{L}\phi_m, \hat{\varphi}_m\right)\right)\right) \tag{7.70}$$

The likelihood function based on these fractional errors can be obtained:

$$\begin{aligned} p\left(\hat{\omega}, \hat{\varphi} \middle| \alpha, \mathcal{M}_P\right) &= \prod_{m=1}^{N_m} p\left(\mathbf{r}_m \middle| \theta, \omega_m, \phi_m, \mathcal{M}_P\right) p\left(\hat{\omega}_m \middle| \omega_m, \mathcal{M}_P\right) p\left(\hat{\varphi}_m \middle| \phi_m, \mathcal{M}_P\right) \\ &= \left(2\pi\gamma^2\right)^{-\frac{3}{2}N_m} \exp\left(-\frac{1}{2\gamma^2} J(\alpha)\right) \end{aligned} \tag{7.71}$$

where

$$J(\alpha) = \sum_{m=1}^{N_m} \left(\left(1 - \mathrm{MAC}\left(\mathbf{K}(\theta)\breve{\phi}_m, \omega_m^2 \mathbf{M}\breve{\phi}_m\right)\right) + \left(\frac{\omega_m - \omega_m}{\omega_m}\right)^2 + \left(1 - \mathrm{MAC}\left(\mathbf{L}\phi_m, \varphi_m\right)\right) \right) \quad (7.72)$$

Assuming uniform prior PDFs for the uncertain parameters gives the posterior PDF to be used in the multi-level MCMC method:

$$p\left(\alpha \middle| \hat{\omega}, \hat{\varphi}, \mathcal{M}_P\right) = c\exp\left(-\frac{1}{2\gamma^2} J(\alpha)\right) \quad (7.73)$$

where the constants have been absorbed into c.

7.7.2 The sampling scheme for the multi-level MCMC method

Having derived the posterior PDF for the multi-level MCMC method, the bridge PDF in the k-th level can be set as Equation (7.62), with $J(\boldsymbol{\alpha})$ given by (7.72). According to Equation (7.63), the variances of two successive levels are related by the following equation

$$\gamma_{k+1}^2 = \frac{1}{a}\gamma_k^2 \quad (7.74)$$

That is, when the sampling moves from the current level to the next, the variance of the bridge PDF is reduced by a factor of $\frac{1}{a}$. This recursive relation requires setting up an initial variance. In the first sampling level, the variance of the bridge PDF is set to a relatively large value, in this case $\gamma_1^2 = 3$, so that p_1 covers a relatively large region to let samples well distribute in the parameter space. Note that the fractional error is in a percentage sense, so $\gamma_1^2 = 3$ means that the variance is 300%, which is a large value in terms of percentage. Having set up the initial variance, the variances in the following levels can be calculated using Equation (7.74). The important regions of the bridge PDFs thus reduce level by level, and the generated samples move from a wide region in the parameter space to the important regions of the posterior PDF.

Sampling is to be conducted in each level from the bridge PDF. The bridge PDF is only known up to a normalizing constant, so we cannot sample from it directly, but this mathematical form is suitable for the MH algorithm. The MH algorithm is used in each level to generate N_s (a fixed number for each level) samples according to the bridge PDF. The task now is to construct the proposal PDF of the MH algorithm in each level. In the first level, samples are not generated yet, so the information of the posterior PDF for constructing the proposal PDF is not available. A non-informative uniform PDF is used as the proposal PDF. Its range is set to be relatively large to cover a wide region of the parameter space.

For a general level k, the proposal PDF is constructed by kernel density estimation (Lam et al. 2015; Ang et al. 1992; Au & Beck 1999) using the samples from the previous level $k - 1$. The kernel density constructed this way approximates the bridge PDF and thus makes use of the information of the posterior PDF. The procedures of constructing the proposal PDF based on kernel density estimation are summarized as follows.

(1) Given the samples from level $k-1$, $\left\{\boldsymbol{\theta}_{k-1}^{(h)} : h = 1,\cdots,N_s\right\}$, where the superscript h denotes the sample index, calculate the sample covariance matrix of $\left\{\boldsymbol{\theta}_{k-1}^{(h)}\right\}$, $\mathbf{C}_{k-1}$.
(2) The kernel density is built by the following equation

$$q(\boldsymbol{\theta}) = \frac{1}{N_s}\sum_{h=1}^{N_s}\frac{1}{\left(w_o\eta_{k-1}^{(h)}\right)^{N_p}}\mathcal{N}\left(\frac{\boldsymbol{\theta}-\boldsymbol{\theta}_{k-1}^{(h)}}{w_o\eta_{k-1}^{(h)}}\right) \tag{7.75}$$

where w_o is the optimal window width; $\eta_{k-1}^{(h)}$ is the optimal local bandwidth at the h-th sample in level $k-1$; $\mathcal{N}(\cdot)$ denotes the Gaussian distribution with mean 0 and covariance matrix $\mathbf{C}_{k-1}$; and N_p is the dimension of $\boldsymbol{\theta}$. It can be seen that the kernel density in the above equation is the weighted sum of Gaussian PDFs, each of which centers at one sample. The spread of each Gaussian PDF and its weight are determined by the factor $w_o\eta_{k-1}^{(h)}$. The proposal PDF constructed this way can then approximate the bridge PDF to capture the characteristics of parameter space.
(3) The optimal window width is obtained by minimizing the following objective function

$$U(w) = \frac{1}{N_s}\sum_{h=1}^{N_s}\frac{p_{k-1}\left(\boldsymbol{\theta}_{k-1}^{(h)}\right)}{q_{-h}\left(\boldsymbol{\theta}_{k-1}^{(h)}\right)} \tag{7.76}$$

where the kernel density q in the denominator is constructed using Equation (7.75), and the subscript $-h$ means that the h-th sample is excluded when constructing the kernel density. The above equation shows that the objective function is related to the ratio between the bridge PDF and the kernel density, so minimizing it results in approximating the bridge PDF, and keeping the samples in the important regions of the bridge PDF.

Before moving to the next step, we take a detour to provide some background about the objective function used to find the optimal window width. Ang et al. (1992) originally proposed to minimize the following variance of the failure probability to obtain the optimal window width:

$$V(P_F) = \frac{1}{N_s}\left(\int_{\boldsymbol{\Theta}}\left(\frac{1_F(\boldsymbol{\theta})p(\boldsymbol{\theta})}{q(\boldsymbol{\theta})}\right)^2 q(\boldsymbol{\theta})d\boldsymbol{\theta} - P_F^2\right) \tag{7.77}$$

where $1_F(\boldsymbol{\theta}) = 1$ if $\boldsymbol{\theta}\in F$ and $1_F(\boldsymbol{\theta}) = 0$ otherwise denotes the indicator function. However, the above function is difficult to minimize either analytically or numerically. The approximation of Equation (7.77) was done based on cross-validation (Bowman 1984; Rudemo 1982) to give the following function:

$$V(w) = \frac{1}{N_s}P_F\left(U(w) - P_F\right) \tag{7.78}$$

Because the failure probability is a constant, minimizing Equation (7.78) is equivalent to minimizing $U(w)$, i.e., Equation (7.76). To minimize $U(w)$, the following initial window width is proposed (Au & Beck 1999)

$$w_n = \left(\frac{4}{(N_p + 2) M_d} \right)^{\frac{1}{N_p+4}} \tag{7.79}$$

where M_d is the number of distinct samples in each level (note that sampling by the MH algorithm will produce repeated samples due to rejection).

(4) Calculate the optimal local bandwidth $\eta_{k-1}^{(h)}$ for each sample in level $k - 1$ using the following formula

$$\eta_{k-1}^{(h)} = \left(\frac{\left(\prod_{g=1}^{N_s} p_{k-1}\left(\theta_{k-1}^{(g)}\right) \right)^{\frac{1}{N_s}}}{p_{k-1}\left(\theta_{k-1}^{(h)}\right)} \right)^{\alpha} \tag{7.80}$$

where the effect of each sample among all the samples is calculated in terms of the bridge PDF, thus this factor is called optimal local bandwidth; the factor α is taken to be 0.5 according to practical experience (Abramson 1982).

Having obtained the optimal window width w_o by minimizing Equation (7.76) and the optimal local bandwidth $\eta_{k-1}^{(h)}$ for each sample, the kernel density that approximates the bridge PDF is constructed using Equation (7.75). This kernel density is then used as the proposal PDF in the MH algorithm to generate samples in the current level k. The samples generated in level k are again used to construct the kernel density in the next level $k + 1$ for generating samples in the next level. This process is repeated until the samples in the final level are generated.

A stopping criterion is proposed to determine how many sampling levels are needed (Lam et al. 2015). This criterion is based on estimating the variance parameter γ_L^2 in the final level. It is done by first doing a one-time numerical minimization for the function in Equation (7.72) with the initial value of the uncertain parameter $\boldsymbol{\theta}$. Suppose the value of Equation (7.72) at the optimal is J_o. The variance in the final sampling level is then estimated by

$$\gamma_L^2 = \frac{J_o}{N_m} \tag{7.81}$$

The argument of using this estimation is that γ^2 is the variance of the fractional error, and according to Equation (7.72) J_o is the sum of the squared fractional errors of N_m modes, so the average of the squared fractional error over N_m modes is used as the estimation of γ_L^2. Note that this estimation is reasonable even for locally identifiable or unidentifiable problems. For locally identifiable problems, J_o at any local optimum can be used to estimate γ_L^2 because the values of J_o at different local optima should be close to each other. For unidentifiable problems, the minimization may stop at any point on the extended manifold (Katafygiotis & Lam 2002). J_o at any point of this manifold can be used for the estimation because, for unidentifiable problems, the points on the manifold have similar values for J_o.

Using Equation (7.74) the relation of the variance in the final level and the one in level 1 can be obtained

$$\gamma_L^2 = \left(\frac{1}{a}\right)^{L-1} \gamma_1^2 \tag{7.82}$$

so the number of the sampling levels is

$$L = 1 - \frac{\ln\left(\frac{\gamma_L^2}{\gamma_1^2}\right)}{\ln(a)} \tag{7.83}$$

The complete procedures of the proposed multi-level MCMC method are summarized in Algorithm 7.4.

Algorithm 7.4: The multi-level MCMC method

(1) Set the initial uniform PDF of the uncertain parameters that is used as the proposal PDF in the MH algorithm; set the value of a (e.g., $a = 1.2$); do the minimization to get J_o and calculate γ_L^2 by Equation (7.81); determine the number of sampling levels by Equation (7.83); set the number of samples in each level N_s (e.g., $N_s = 1000 \sim 5000$ for complex structures).

(2) Level 1. Generate N_s samples using the MH algorithm (Algorithm 7.2) with the initial uniform PDF used as the proposal PDF and the bridge PDF in level 1 used the target PDF. The form bridge PDF is given in Equation (7.62) and its variance γ_1^2 is set to be relatively large, e.g., $\gamma_1^2 = 3$.

(3) A general Level k. Construct the kernel density in Level k, $q_k(\theta)$, using the samples $\theta_{k-1}^{(h)}$ in Level k – 1 according to Equation (7.75). This kernel density is used as the proposal PDF in the MH algorithm. The procedures of constructing $q_k(\theta)$ and sampling from $q_k(\theta)$ are as follows.

   a. Calculate the sample covariance matrix $\mathbf{C}_{k-1}$ using the samples $\left\{\theta_{k-1}^{(h)}\right\}$ to be used in the Gaussian PDFs in Equation (7.75).
   b. Calculate the optimal local bandwidth by Equation (7.80).
   c. Calculate the optimal window width by minimizing Equation (7.76) with Equation (7.79) as the initial trial.
   d. Use the kernel density to generate candidate samples of the MH algorithm for Level k. To sample from $q_k(\theta)$, a sample is first drawn from the samples in the previous level $\left\{\theta_{k-1}^{(h)}\right\}$, whose probabilities (the weights of the samples) are calculated as follows:

$$W_{k-1}^{(h)} = \frac{\dfrac{1}{\left(w_o \eta_{k-1}^{(h)}\right)^{N_p}}}{\sum_{h=1}^{N_s} \dfrac{1}{\left(w_o \eta_{k-1}^{(h)}\right)^{N_p}}} \tag{7.84}$$

Suppose the sample $\theta_{k-1}^{(l)}$ is drawn from $\left\{\theta_{k-1}^{(h)}\right\}$. The Gaussian PDF in Equation (7.75) that centers at $\theta_{k-1}^{(l)}$ and has covariance matrix $\mathbf{C}_{k-1}$ is employed to generate a candidate sample in Level k. The MH algorithm with the target bridge PDF p_k determines whether this candidate sample is accepted.

e. Repeat Step d until N_s samples $\left\{\theta_k^{(h)}\right\}$ are generated for the current level.

(4) Repeat Step (3) until the samples in the final Level L are generated.

(5) The posterior marginal PDFs of each structural parameter are approximated by the samples in the final Level L. Because Gaussian kernels are used for constructing the kernel densities, the approximated posterior marginal PDF of the i-th component of $\boldsymbol{\theta}$ can be obtained by analytically integrating Equation (7.75) over other components:

$$q(\theta(i)) = \sum_{h=1}^{N_s} W_L^{(h)} \mathcal{N}\left(\frac{\theta(i) - \theta_L^{(h)}(i)}{\sqrt{C_L(i,i)}}\right) \tag{7.85}$$

where $\boldsymbol{\theta}(i)$ denotes the i-th component of $\boldsymbol{\theta}$; here $\mathcal{N}(\cdot)$ denotes the one-dimensional standard Gaussian PDF; $C_L(i,i)$ is the i-th diagonal component of the sample covariance matrix $\mathbf{C}_L$ calculated using the samples in the final level.

(6) The MPVs of the model-predicted modal parameters are calculated by the weighted sum of the modal parameters based on the samples in the final sampling level:

$$\tilde{\omega}_m = \sum_{h=1}^{N_s} W_L^{(h)} \omega_m\left(\theta_L^{(h)}\right) \tag{7.86}$$

$$\tilde{\phi}_m = \sum_{h=1}^{N_s} W_L^{(h)} \phi_m\left(\theta_L^{(h)}\right) \tag{7.87}$$

(7) The posterior standard deviations of the modal parameters are given by

$$s_m \approx \sqrt{\sum_{i=1}^{N_i} W_L^{(h)} \left(\omega_m\left(\theta^{(i)}\right) - \hat{\omega}_m(\theta)\right)^2} \tag{7.88}$$

$$\mathbf{s}_m \approx \sqrt{\sum_{i=1}^{N_i} W_L^{(h)} \left(\mathbf{L}\phi_m\left(\theta^{(i)}\right) - \hat{\phi}_m(\theta)\right)^2} \tag{7.89}$$

7.7.3 Bayesian model updating of a transmission tower

The scaled transmission tower (Figure 7.1) was again used to demonstrate the proposed multi-level MCMC method. In this case, practical conditions in field tests were simulated. The ambient excitation in practice was simulated by a large fan (Figure 7.12). Obtaining detailed mode shapes is important to update complex civil engineering structures, so multiple setups were conducted to identify detailed mode shapes, instead of focusing on only one column as in Section 7.2.1. At the time of the measurement, 20 uniaxial sensors were available. The measurement plan was to measure the vibration along both x- and

Figure 7.12 The scaled transmission tower under the wind excitation (see also Yang & Lam 2018).

y-directions at all nodes except the four nodes of the cross-arms, nodes 37 to 40, and the four nodes at the ground, nodes 1, 10, 19, and 28 (refer to Figure 7.3 for the node numbering). Thirty-two nodes were to be measured, resulting in 64 DOFs to be measured. The measurement was divided into five setups to cover all the desired DOFs. Table 7.3 summarizes the measurement plans for each setup (Yang & Lam 2018). The number-letter combinations in this table indicate how the sensors were installed. The number denotes the sensor number, and the letter x or y indicates that the sensor was installed along the x- or y-direction. For example, "11y" means that in Setup 2, Senor 7 was installed at Node 11 along the y-direction. It can be seen in this table that Sensors 1, 3, 8, 9, 11, and 16 measured the same DOFs in all five setups, i.e., in all five setups these sensors were fixed at Node 2 (the bottom part of the tower), Node 4 (the middle part) and Node 9 (the top) to measure both the x- and y-directions. They were used as the reference sensors such that the partial mode shapes from different setups could be assembled into the global mode shapes.

According to Table 7.3, the column with Nodes 1 to 9 was measured in Setup 1, followed by the column with Nodes 10 to 18, the column with Nodes 19 to 27, and finally the columns with Nodes 28 to 36. Moreover, the x-direction was measured before the y-direction.

The data of the ambient vibration test were measured with the sampling frequency 2048 Hz, and down-sampled to 1024 Hz for analysis. Ten minutes of data were measured for each setup. The measured accelerations were then used for identifying the experimental

Table 7.3 Measurement Plans of the Five Setups

	Sensor number																			
	1	*2*	*3*	*4*	*5*	*6*	*7*	*8*	*9*	*10*	*11*	*12*	*13*	*14*	*15*	*16*	*17*	*18*	*19*	*20*
Setup 1	2x	3x	4x	5x	6x	7x	8x	9x	2y	3y	4y	5y	6y	7y	8y	9y	11x	12x	13x	14x
Setup 2	2x	15x	4x	16x	17x	18x	11y	9x	2y	12y	4y	13y	14y	15y	16y	9y	17y	18y	20x	21x
Setup 3	2x	22x	4x	23x	24x	25x	26x	9x	2y	27x	4y	20y	21y	22y	23y	9y	24y	25y	26y	27y
Setup 4	2x	29x	4x	30x	31x	32x	33x	9x	2y	34x	4y	35x	36x	29y	30y	9y	31y	32y	33y	34y
Setup 5	2x	29x	4x	30x	31x	32x	33x	9x	2y	34x	4y	35x	36x	29y	30y	9y	31y	32y	35y	36y

Note: See also Yang and Lam 2018.

modal parameters for model updating. In total, seven modes were identified. The experimental mode shapes are shown in Figure 7.15 later together with the model-predicted ones for comparison. The seven experimental modes are the first translational modes along the y- and x-directions, respectively (Figure 7.15(a) and (b)); the first torsional mode (Figure 7.15(c)); the second translational modes along the y- and x-directions, respectively (Figure 7.15(d) and (e)); the second torsional mode (Figure 7.15(f)); and the third translational mode along the x-direction (Figure 7.15(g)).

Six uncertain structural parameters were assigned for the model class in model updating. $\boldsymbol{\theta}(1)$ to $\boldsymbol{\theta}(5)$ were used to scale the braces' Young's modulus (i.e., their stiffness) of Levels 1 to 5, respectively, and $\boldsymbol{\theta}(6)$ was used to scale the Young's modulus of the four columns. Uncertain parameters were not assigned for Levels 6 to 8, as it is believed that the members at the upper part of the tower will not affect the dynamic properties much. Note that sampling was only conducted for the structural parameters $\boldsymbol{\theta}$, while the uncertain natural frequencies and system mode shapes (the eigenvalues and eigenvectors of the system matrices of the tower) were computed based on the samples of $\boldsymbol{\theta}$ (refer to Algorithm 7.1 for procedures of computing the natural frequencies and system mode shapes given a particular sample of $\boldsymbol{\theta}$). Using the proposed multi-level MCMC method (Algorithm 7.4), the samples of the uncertain parameters were generated in multiple levels. Figure 7.13(a) shows the samples in the first sampling level. In the first sampling level, a uniform proposal PDF with the range [0, 1.5] was used. The variance of the bridge PDF was set to be relatively large, in this case $\gamma_1 = 1$, so the acceptance rate of the MH algorithm was large and the samples distributed evenly in this part of the parameter space. In the following levels, the variance of the bridge PDF was systematically reduced to let samples explore the parameter space to find the important regions of the posterior PDF. Figure 7.13(b) shows the samples in the final sampling level. It can be seen that the samples distribute differently compared to the samples in the first sampling level. $\boldsymbol{\theta}(1)$ to $\boldsymbol{\theta}(4)$ spread out in the parameter space, while $\boldsymbol{\theta}(5)$ and $\boldsymbol{\theta}(6)$ concentrate in a narrow region.

The posterior marginal PDFs of the structural parameters were constructed by kernel density estimation using the samples in the final sampling level (see Figure 7.14). The posterior marginal PDFs of $\boldsymbol{\theta}(1)$ to $\boldsymbol{\theta}(4)$ are flat and cover a wide range of the parameter space. It indicates that given the data (i.e., the experimental modal parameters and the chosen modeling assumptions, which include the finite element model of the tower), the number of the uncertain parameters and the way these parameters are assigned to the finite element model and how the posterior PDF is formulated, the remaining uncertainties of $\boldsymbol{\theta}(1)$ to $\boldsymbol{\theta}(4)$ are large. The widely spreading posterior PDFs also indicate that multiple models (the finite element model calculated with different model parameter values) in some regions have similar importance. We should consider all these important models in downstream problems, such as assessment, response prediction, and structural health monitoring, rather than picking only one "best" model and discarding others. This example shows that it is important to use Bayesian methods to consider all models in the important regions instead of pinpointing one single "optimal" model. It is fine to use deterministic methods to pinpoint one single "optimal" model when uncertainties of all parameters are small, because in this situation the posterior PDF is sharply peaked in the parameter space. However, it can never be known whether the uncertainties of one problem are small or large unless the Bayesian model updating is conducted. The posterior PDFs of $\boldsymbol{\theta}(5)$ and $\boldsymbol{\theta}(6)$ are sharply peaked, indicating that the remaining uncertainties of these two parameters are small given the data and modeling assumptions.

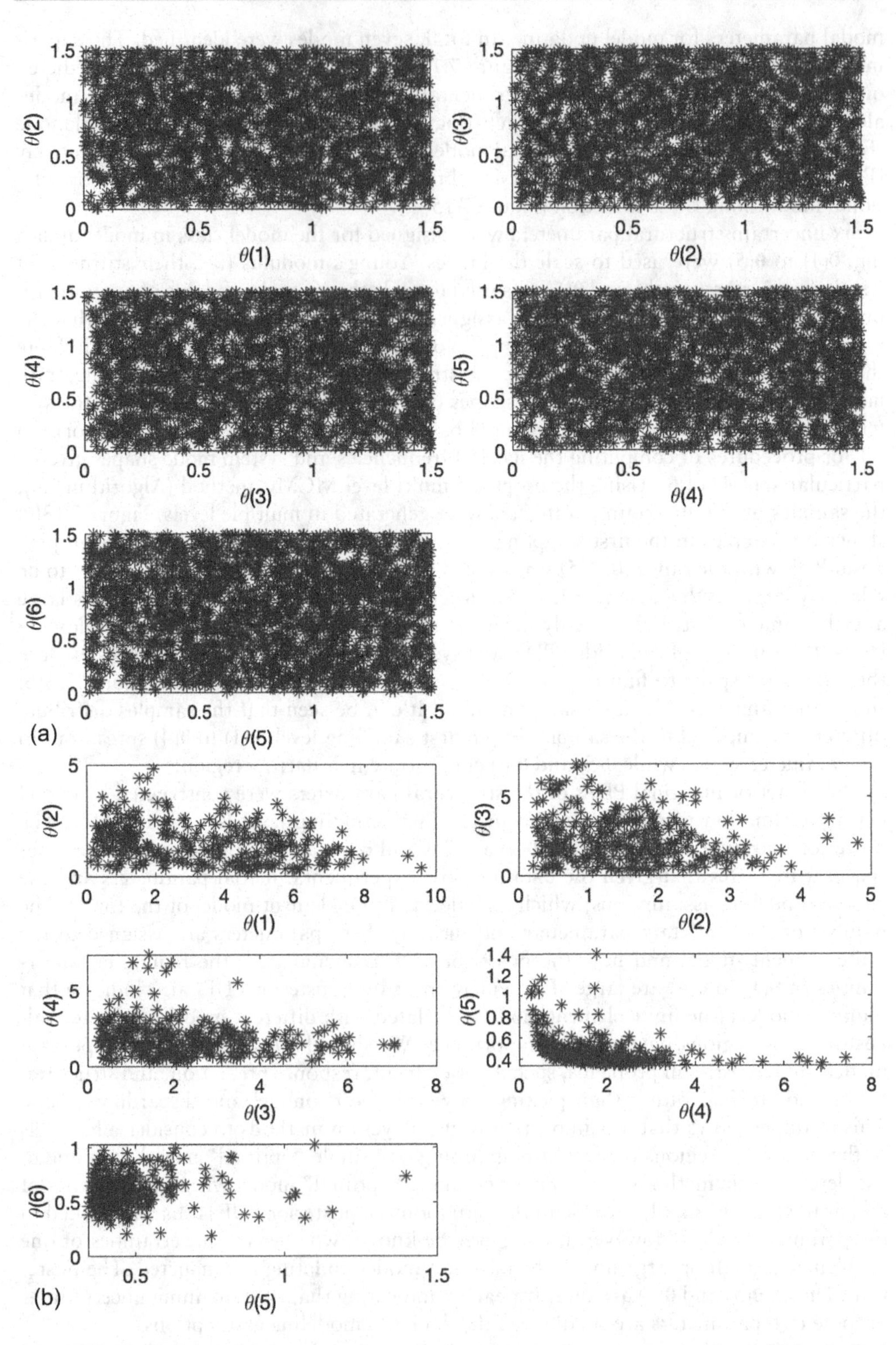

Figure 7.13 The samples of the structural parameters (a) in the first sampling level and (b) in the final sampling level.

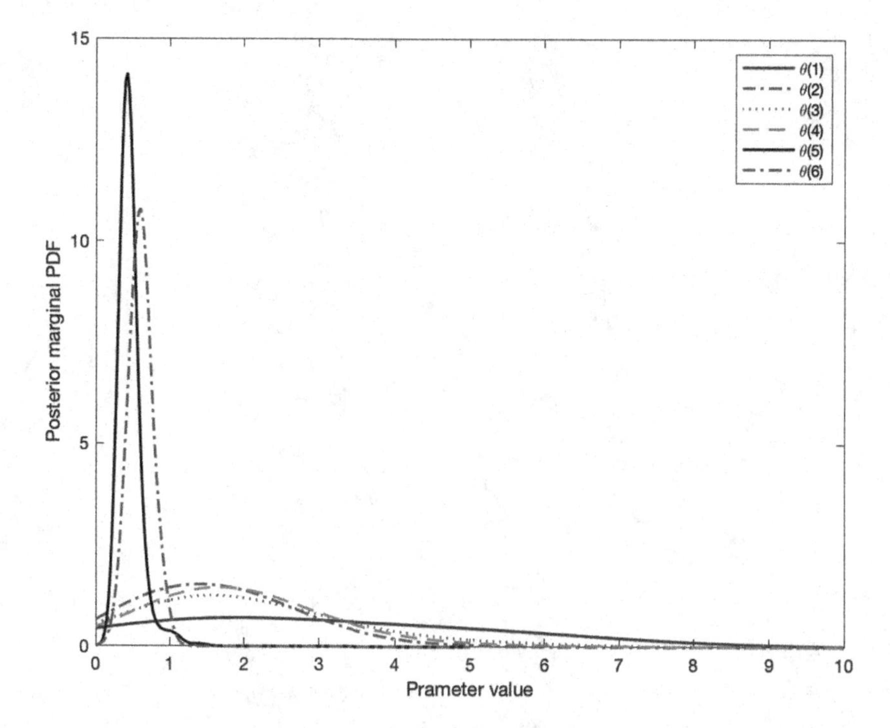

Figure 7.14 The posterior marginal PDFs of the structural parameters.

The MPVs of the modal parameters were compared to the experimental ones to check the performance of the proposed method. The MPVs of the modal parameters were calculated using the weighted sum of the modal parameters calculated based on the samples in the final level (Equations (7.86) and (7.87)). The MPVs calculated this way rigorously consider the posterior uncertainties because these samples were generated according to the posterior PDF. The percentage differences between the MPVs of the model-predicted natural frequencies and the experimental ones are 4.64%, 2.76%, -2.99%, 9.49%, –15.80%, -7.71%, and 3.11% for the seven modes, respectively. The MPVs of the model-predicted mode shapes and the experimental ones at the measured DOFs are compared in Figure 7.15. It can be seen that the model-predicted modal parameters can reasonably fit the experimental ones. The proposed Bayesian method not only can calculate the MPVs of the modal parameters, but also can quantify their posterior uncertainties. The posterior standard deviations of the model-predicted natural frequencies and mode shapes were calculated by Equations (7.88) and (7.89). To compare the uncertainties fairly, the posterior COVs of the modal parameters (the standard deviation divided by the MPV) were calculated and summarized in Table 7.4. Note that for the convenient comparison, the posterior COVs of the mode shapes at different DOFs are averaged over the DOFs to produce one scalar, so this uncertainty reflects the averaged uncertainty among different DOFs. It can be seen from the table that the posterior uncertainties of the mode shapes are much larger than those of the natural frequencies.

7.7.4 The posterior PDF based on the fractional errors of modal parameters

The posterior PDF derived for the eigenvalue problem uses only the information of the finite element model related to the identified experimental modes. Solving the eigenvalue

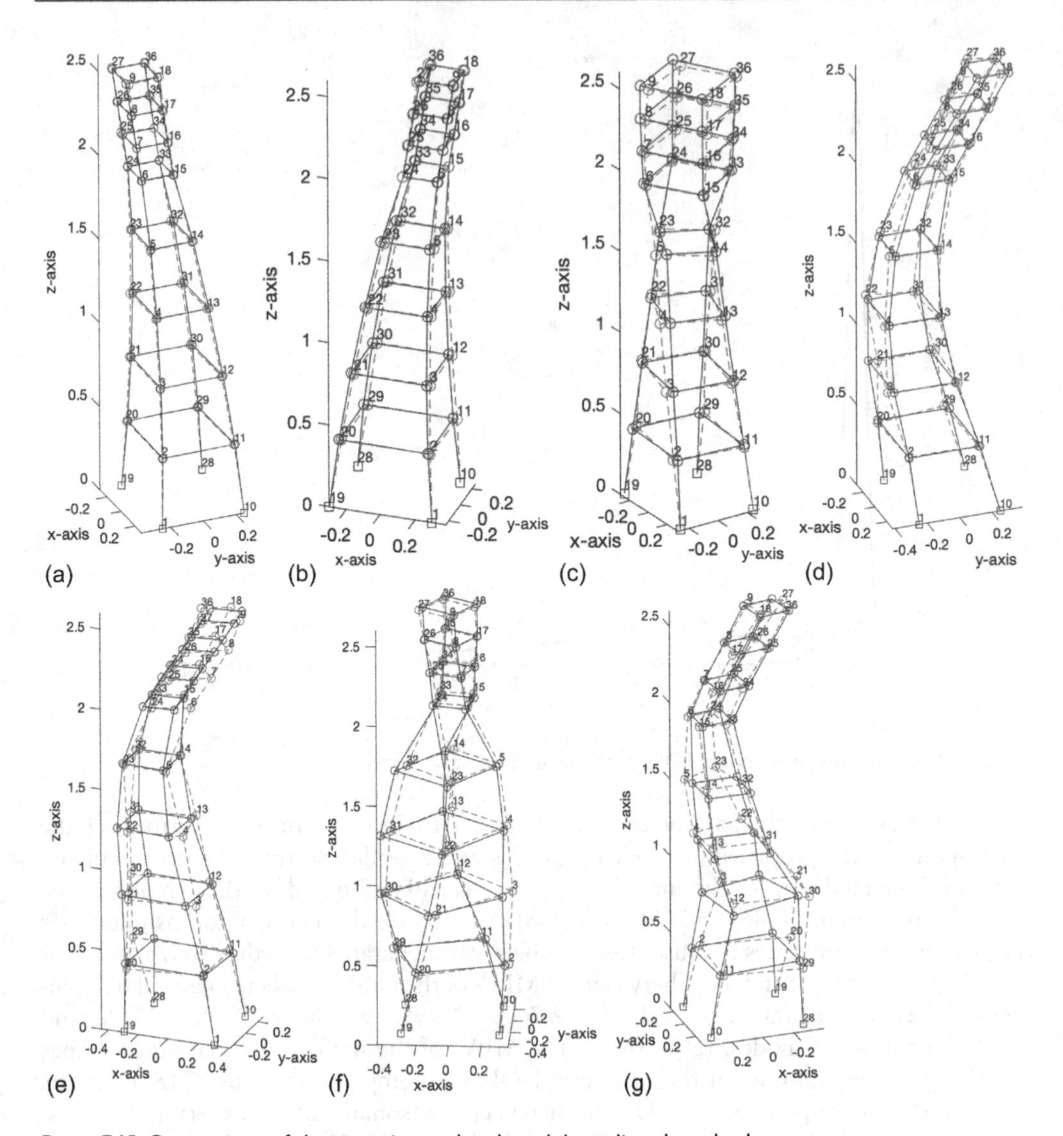

Figure 7.15 Comparison of the experimental and model-predicted mode shapes.

Table 7.4 Posterior COVs of the Model-Predicted Natural Frequencies and Mode Shapes

Mode	*1*	*2*	*3*	*4*	*5*	*6*	*7*
ω	0.0907	0.0870	0.0098	0.0653	0.0549	0.0586	0.0644
$\boldsymbol{\phi}$	0.1570	0.2472	0.3443	13.2502	14.7741	9.2438	24.8645

problem for other modes and matching experimental modes among calculated modes are not required. However, the inverse of the full matrix of a finite element model still needs to be computed. Unless special treatment is done (e.g., efficient computation for matrix inverse or model reduction that reduces the matrix size), this process will be computationally demanded for full-scale structures when the finite element model is large. Directly using experimental data, this section proposes a posterior PDF in terms of the fractional errors of natural frequencies and mode shapes. This formulation effectively considers both measurement noise and modeling errors, and it is suitable for model updating of full-scale structures.

The posterior PDF proposed here is similar to the one in Equation (7.73), except that the eigen equation error is not considered now. Without modeling the eigen equation error, we need to solve the eigenvalue problem of the full finite element model. The stochastic embedding is only done for the fractional errors of natural frequencies and mode shapes (confined at the measured DOFs) in Equations (7.66) and (7.67), i.e., choose the same Gaussian distribution for them (see Equations (7.69) and (7.70)). Be reminded again that we choose the same Gaussian distribution for both natural frequencies and mode shapes because their errors are normalized in the fractional form and have the same scale. The likelihood function can then be obtained assuming that the fractional errors are i.i.d. random variables:

$$p\left(\hat{\omega},\hat{\varphi}\middle|\theta,\mathcal{M}_P\right)=\prod_{m=1}^{N_m}p\left(\hat{\omega}_m\middle|\theta,\mathcal{M}_P\right)p\left(\hat{\varphi}_m\middle|\theta,\mathcal{M}_P\right)=\left(2\pi\gamma^2\right)^{-N_m}\exp\left(-\frac{1}{2\gamma^2}J(\theta)\right) \tag{7.90}$$

where

$$J(\theta)=\sum_{m=1}^{N_m}\left(\left(\frac{\hat{\omega}_m-\omega_m(\theta)}{\hat{\omega}_m}\right)^2+\left(1-\mathrm{MAC}\left(\mathbf{L}\phi_m(\theta),\hat{\varphi}_m\right)\right)\right) \tag{7.91}$$

The actual natural frequencies and mode shapes are no longer included as uncertain parameters, and only the structural parameters $\boldsymbol{\theta}$ are considered as uncertain parameters to be identified. For the current case, the eigenvalue problem of the finite element model must be solved to get $\omega_m(\theta)$ and $\phi_m(\theta)$. Mode matching is also needed for each mode for calculating the second term in Equation (7.91), i.e., for each mode, MAC is calculated between the measured mode shape $\hat{\varphi}_m$ and all the model-predicted mode shapes; and the model-predicted mode shape $\phi_m(\theta)$ that produces the largest MAC value is chosen to calculate the second term in Equation (7.91). The natural frequency $\omega_m(\theta)$ corresponding to $\phi_m(\theta)$ is then used to calculate the frequency fractional error. The posterior PDF based on the fractional errors of modal parameters is obtained:

$$p\left(\theta\middle|\hat{\omega},\hat{\varphi},\mathcal{M}_P\right)=c\exp\left(-\frac{1}{2\gamma^2}J(\theta)\right) \tag{7.92}$$

This posterior PDF has the same form as Equation (7.73), so Algorithm 7.4 can also be used for the posterior PDF in Equation (7.92).

7.7.5 Bayesian model updating of a coupled structural system

In this section, a full-scaled coupled structural system is used to illustrate the multi-level MCMC method using the posterior PDF based on the fractional errors of modal parameters (Equation (7.92)). Modal analysis of this structure was discussed in Chapter 3. The identified modal parameters of this coupled system are used as experimental data to update its finite element model. Figure 7.16 shows the finite element model of this system,

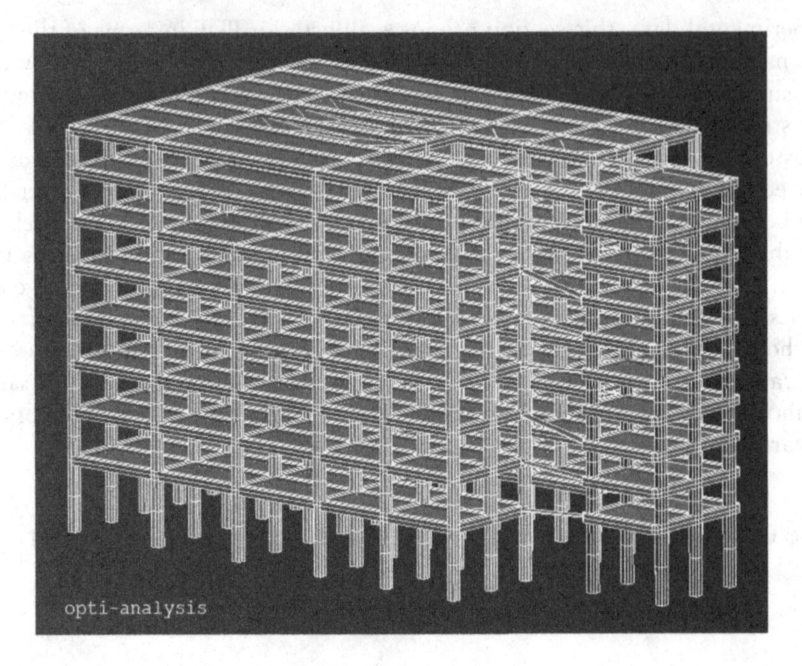

Figure 7.16 The finite element of the coupled structural system.

which has a main building and a complementary building. This model consists of beam elements, for the beam and column members, and shell elements for the floor slabs. It was assumed that the material for all the elements is the C30 concrete. For model updating, two uncertain parameters were used, i.e., $\boldsymbol{\theta}(1)$ is used to scale the Young's modulus of the columns of the main building and $\boldsymbol{\theta}(2)$ is for the Young's modulus of the columns of the complementary building. During model updating, samples of $\boldsymbol{\theta}$ were generated so that the stiffness of this system is adjusted for matching the model-predicted modal parameters to the experimental ones.

By using Algorithm 7.4, the samples of the uncertain parameters were generated. For each level, 1000 samples were generated. In level 1, the variance parameter was set to a relatively large value, i.e., $\gamma_1 = 3$, to let the support of the bridge PDF cover a wide region of the parameter space. Moreover, because no samples were generated in level 1, a uniform PDF with the range [0,3] was used for the proposal PDF in the MH algorithm. Note that the range of this initial uniform PDF will not affect the exploration of the parameter space. The samples will gradually approach the important regions of the posterior PDF because the bridge PDFs gradually approach the posterior PDF by construction. Figure 7.17 shows the samples generated by the proposed multi-level MCMC method. Figure 7.17(a) shows the samples generated at the beginning with the proposal PDF being the uniform PDF. Because the variance parameter of the target PDF (the bridge PDF p_1) is relatively large, most candidate samples were accepted, and approximately the samples distribute evenly. Figure 7.17(b) shows the samples in the middle of the sampling process. It can be seen that the samples shrink to a narrower region than that at the beginning, meaning that the samples move toward the region of high probability of the posterior PDF. The samples in the final sampling level are shown in Figure 7.17(c). They converge in a certain region. The posterior marginal PDFs of the uncertain parameters were constructed based

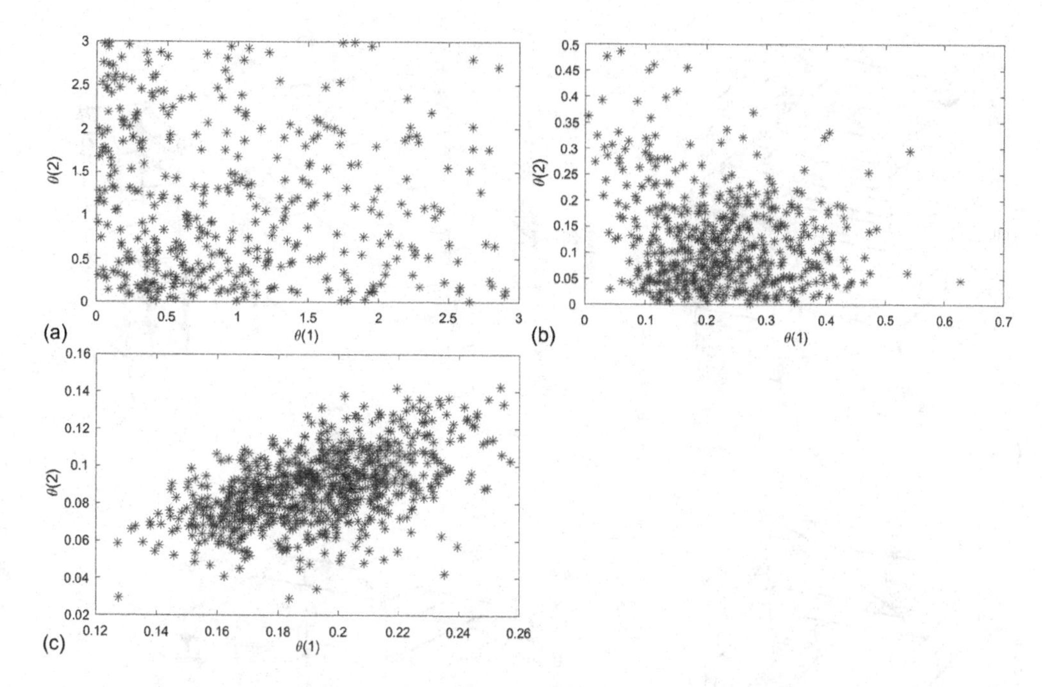

Figure 7.17 Samples of the uncertain parameters: (a) beginning; (b) middle; (c) end.

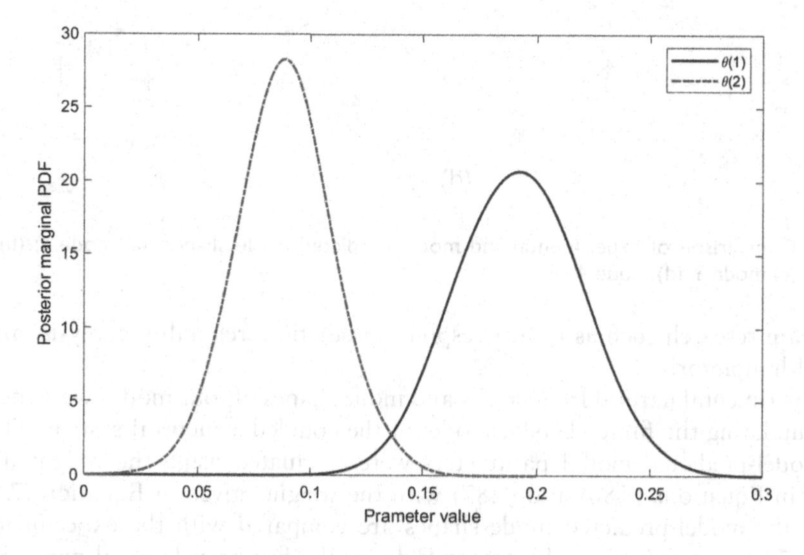

Figure 7.18 Posterior marginal PDFs of the uncertain parameters.

on kernel density estimation using the samples in the final level (see Figure 7.18). The posterior PDFs of both uncertain parameters distribute in a narrow region, meaning that given the experimental data, the uncertain parameters can be identified with small uncertainties. This is reasonable because there are only two uncertain parameters in this case and we have enough information to make inference on these parameters. The posterior PDF identified by the proposed MCMC-based Bayesian updating method is useful for

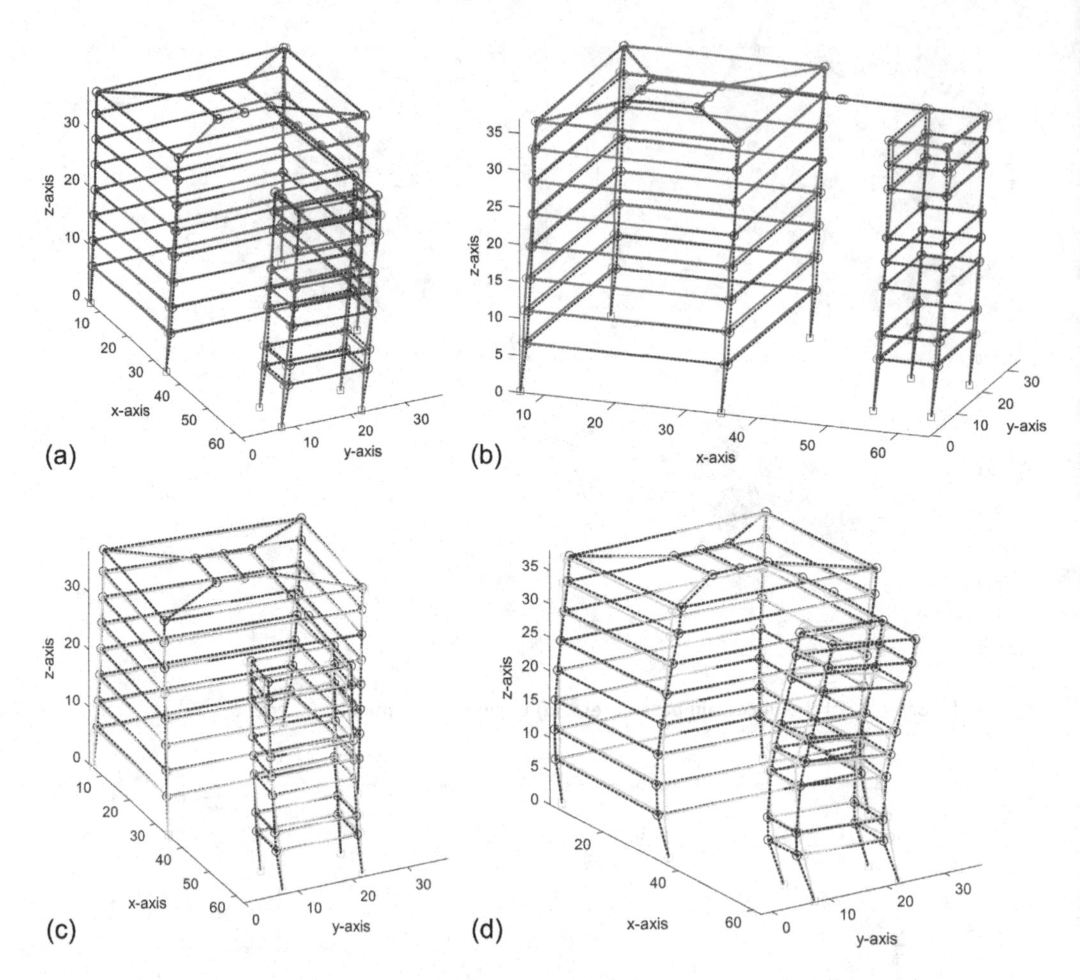

Figure 7.19 Comparison of experimental and model-predicted mode shapes: (a) mode 1; (b) mode 2; (c) mode 3; (d) mode 4.

downstream research such as robust response prediction, reliability analysis, and structural health monitoring.

The experimental natural frequencies and mode shapes of four modes were included as data for updating the finite element model of the coupled structural system. The MPVs of the model-predicted modal parameters were calculated using the weighted average formulas in Equations (7.86) and (7.87) with the weights given in Equation (7.84). The MPVs of the model-predicted mode shapes are compared with the experimental ones in Figure 7.19. Figure 7.19(a), (b), (c), and (d) are the first translational mode along the y-direction, the first translational mode along the x-direction, the torsional mode, and the second translational mode along the y-direction, respectively. The solid lines denote the experimental mode shapes and the dash lines denote the model-predicted ones. It can be seen that the model-predicted mode shapes can well match the experimental ones, indicating the good performance of the proposed method. The MPVs of the model-predicted natural frequencies were also calculated and compared with the experimental ones. The frequency errors of the four modes are 0.125 Hz, 0.044 Hz, 0.231 Hz, and 0.529 Hz, respectively.

REFERENCES

Abramson, I.S., 1982. On bandwidth variation in kernel estimates-a square root law. *Annals of Statistics*, 10(4), 1217–1223.

Ang, G.L., Ang, A.H.S. and Tang, W.H., 1992. Optimal importance-sampling density estimator. *Journal of Engineering Mechanics*, 118(6), 1146–1163.

Au, S.K. and Beck, J.L., 1999. A new adaptive importance sampling scheme for reliability calculations. *Structural Safety*, 21(2), 135–158.

Beck, J.L. and Au, S.K., 2002. Bayesian updating of structural models and reliability using Markov chain Monte Carlo simulation. *Journal of Engineering Mechanics*, 128(4), 380–391.

Bowman, A.W., 1984. An alternative method of cross-validation for the smoothing of density estimates. *Biometrika*, 71, 353–360.

Hastings, W.K., 1970. Monte Carlo sampling methods using Markov chains and their applications. *Biometrika*, 57(1), 97–109.

Katafygiotis, L.S. and Lam, H.F., 2002. Tangential-projection algorithm for manifold representation in unidentifiable model updating problems. *Earthquake Engineering and Structural Dynamics*, 31(4), 791–812.

Lam, H.F. and Yang, J., 2015. Bayesian structural damage detection of steel towers using measured modal parameters. *Earthquakes and Structures*, 8(4), 935–956.

Lam, H.F., Yang, J. and Au, S.K., 2015. Bayesian model updating of a coupled-slab system using field test data utilizing an enhanced Markov chain Monte Carlo simulation algorithm. *Engineering Structures*, 102, 144–155.

Metropolis, N., Rosenbluth, A.W., Rosenbluth, M.N., Teller, A.H. and Teller, E., 1953. Equation of state calculations by fast computing machines. *The Journal of Chemical Physics*, 21(6), 1087–1092.

Meyn, S.P. and Tweedie, R.L., 2012. *Markov Chains and stochastic stability.* Springer Science & Business Media.

Owen, A.B., 2013. Monte Carlo theory, methods and examples. Retrieved from https://statweb.stanford.edu/~owen/mc/.

Robert, C. and Casella, G., 2013. *Monte Carlo statistical methods.* Springer Science & Business Media.

Rudemo, M., 1982. Empirical choice of histograms and kernel density estimators. *Scandinavian Journal of Statistics*, 9, 65–78.

Yang, J.H., 2015. Development of Bayesian structural damage detection methodologies utilizing advanced Monte Carlo simulation (PhD thesis). City University of Hong Kong.

Yang, J.H. and Lam, H.F., 2018. An efficient adaptive sequential Monte Carlo method for Bayesian model updating and damage detection. *Structural Control and Health Monitoring*, 25(12), e2260.

Index

O

P

R

S

T

U

V

W

Taylor & Francis eBooks

www.taylorfrancis.com

A single destination for eBooks from Taylor & Francis with increased functionality and an improved user experience to meet the needs of our customers.

90,000+ eBooks of award-winning academic content in Humanities, Social Science, Science, Technology, Engineering, and Medical written by a global network of editors and authors.

TAYLOR & FRANCIS EBOOKS OFFERS:

A streamlined experience for our library customers

A single point of discovery for all of our eBook content

Improved search and discovery of content at both book and chapter level

REQUEST A FREE TRIAL

support@taylorfrancis.com

Printed in the United States
by Baker & Taylor Publisher Services

Printed in the United States
by Baker & Taylor Publisher Services